Advanced Structured Materials

Volume 26

More information about this series at http://www.springer.com/series/8611

Dan Rosu · Visakh P. M.
Editors

Photochemical Behavior of Multicomponent Polymeric-based Materials

Springer

Editors
Dan Rosu
Advanced Research Center for Bionanoconjugates and Biopolymers
Institute of Macromolecular Chemistry "Petru Poni"
Iassy
Romania

Visakh P. M.
Department of Ecology and Basic Safety
Tomsk Polytechnic University
Tomsk
Russia

ISSN 1869-8433 ISSN 1869-8441 (electronic)
Advanced Structured Materials
ISBN 978-3-319-25194-3 ISBN 978-3-319-25196-7 (eBook)
DOI 10.1007/978-3-319-25196-7

Library of Congress Control Number: 2016952509

Printed on acid-free paper

This Springer imprint is published by Springer Nature
The registered company is Springer International Publishing AG
The registered company address is: Gewerbestrasse 11, 6330 Cham, Switzerland

Contents

About the Editors

Dr. Dan Rosu is a senior scientist at the "Petru Poni" Institute of Macromolecular Chemistry in Iasi, Romania, since 1980, currently working in the department Centre of Advanced Research in Nanoconjugates and Biopolymers on the thermal and photochemical behavior of polymers and polymer-based materials (blends, composites, nanocomposites, and interpenetrated and semi-interpenetrated polymer networks) and polymeric materials derived from bioresources (wood, vegetable oils). He has published over 80 scientific papers and 5 chapters in prestigious international journals and publishers (Springer and Romanian Publishing Houses) and obtained 4 patents. He participated with over 70 oral communications at prestigious national and international conferences and symposia. He was also manager in 4 fundamental research projects and member in other 4. He has 1003 citations, and his h-index is 17. He is a member of the Romanian Chemical Society since 2000 and has won the "Costin D. Nenitescu" award in 2002.

Dr. Visakh P. M. is a prolific editor with more than 20 books already published. Now, he is working as a research associate (under the guidance of Prof. Olga B Nazarenko) in the Department of Ecology and Basic Safety, Tomsk Polytechnic University (TPU), Tomsk, Russia. He obtained his Ph.D., M.Phil., and M.Sc. degrees from the School of Chemical Sciences, Mahatma Gandhi University, Kottayam, Kerala, India. He edited 20 books from Scrivener (Wiley), Springer, Royal Society of Chemistry, Elsevier, and more than 12 books in press (from Wiley, Springer, Royal

Society of Chemistry, and Elsevier). He has been invited as a visiting researcher in Russia (2014 to present), Portugal (2013, 2014), the Czech Republic (2012, 2013), Italy (2009, 2012), Argentina (2010), Sweden (2010, 2011, 2012), Switzerland (2010), Spain (2011, 2012), Slovenia (2011), France (2011), Belgium (2012), and Austria (2012) for his research work. He visited 12 countries; he visited 15 universities in Europe. He published 13 publications, 3 reviews, and more than 18 book chapters. He has attended and presented more than 28 conferences, he has 320 citations, and his h-index is 9. He acts as a guest editor for 4 international journals.

Theoretical Aspects Regarding Polymer Photochemistry

Liliana Rosu, Cristian-Dragos Varganici and Dan Rosu

Abstract The chapter encompasses literature studies related to the basic photochemical decomposition processes occurring in polymers. Ultraviolet (UV) radiations from light, oxygen from air, temperature and humidity are the main factors producing chemical and/or physical irreversible damages in polymers. UVA radiations (λ = 315–400 nm) possess sufficiently high energy for generating most macromolecular chain bonds cleavage leading to initiation of important chemical modifications through radical intermediates such as oxides and hydroperoxides, chromophore carbonyl groups or conjugated double bonds. The presence of oxygen leads to occurrence of photo-oxidative processes through chained mechanisms.

1 Introduction

It is already known that electromagnetic radiations belonging to the visible and/or ultraviolet spectra consist of streams of very small discrete moving particles, characterized by wave-particle properties, named photons [1]. There is an inverse proportionality relationship between photon energy and wavalength of radiation, as described by Eq. (1), where h denotes Planck's constant (6.6249×10^{-27} erg s), λ is the wavelength of electromagnetic radiation, c is light velocity in vacuum (2.9979×10^{10} cm s^{-1}) and ν is the radiation frequency.

L. Rosu (✉) · C.-D. Varganici · D. Rosu
Centre of Advanced Research in Bionanoconjugates and Biopolymers,
"Petru Poni" Institute of Macromolecular Chemistry, 41A Grigore Ghica-Voda Alley,
700487 Iasi, Romania
e-mail: lrosu@icmpp.ro

C.-D. Varganici
e-mail: varganici.cristian@icmpp.ro

D. Rosu
e-mail: drosu@icmpp.ro

D. Rosu and Visakh P. M. (eds.), *Photochemical Behavior of Multicomponent Polymeric-based Materials*, Advanced Structured Materials 26,
DOI 10.1007/978-3-319-25196-7_1

Table 1 Bond dissociation energy values of the main covalent bonds in commercially available organic polymers and the corresponding wavelength [2]

Chemical bond	Bond energy (kJ mol^{-1})	λ (nm)
C–Cl	327	366
C–O (ether)	331	361
C–C (aliphatic)	335	357
N–H	352	340
C–H (allyl)	356	336
S–H and C–O	364	328
Si–O	373	320
C–H (tertiary carbon)	381	314
C–H (secondary carbon)	395	302
C–H (primary carbon)	410	291

$$E = h\frac{c}{\lambda} = hv \quad (1)$$

The ultraviolet (UV) region was formally devided acording to wavelength and energetic content in three different domains: UVA, characterized by wavelengths in the range 315–400 nm, UVB, with wavelengths between 315 and 280 nm and UVC with wavelengths raging from 280 to 100 nm.

Based on Eq. (1) there was calculated the UV radiation energy contained within solar light. The radiation energy corresponding to wavelengths between 280 and 400 nm has values raging between 412 and 300 kJ mol^{-1}, these values being high enough to cleave the majority of covalent bonds specific to organic polymers, as one may observe from Table 1 [2].

Based on values from Table 1 one may conclude that UVA radiations are capable of generating macromolecular chain scissions in a former stage, which may lead to initiation of important chemical modifications via radical intermediates formation [3, 4].

2 Theoretical Aspects Regarding Polymer Photochemistry

Rabek suggested that during the initial stage of polymeric materials photodegradation there occurs the formation of an excited electronic state with singlet structure, resulting after photon absorbtion by the molecules present in the irradiated system [5]. Such a process is schematically given in relation (2):

$$S_0 \xrightarrow{hv} S^* \quad (2)$$

It is known that after absorbtion of luminuous energy any molecule passes from its fundamental electronic state (S_0), characterized by the positioning with spin electrons coupled in bonding orbitals (σ or π), into excited electronic states (S^*), in which the chemical bonds electrons are promoted into different orbitals, such as the

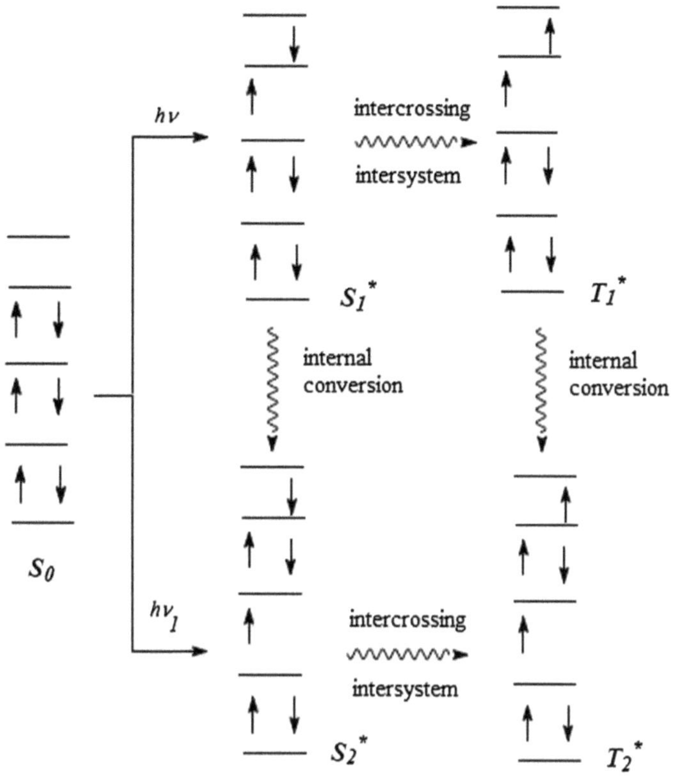

Fig. 1 Electronic excitation and dezexcitation processes, where S_0—fundamental state, S_1^* and S_2^*—singlet excited states, T_1^* and T_2^*—triplet excited states

nonbonding (n) or antibonding (π^* or σ^*) ones, which possess superior energetic content to that of fundamental state. In singlet states chemical bonds electrons decouple remaining antiparallel oriented, whilst in triplet states electronic spins are parallel positioned, as it may be observed from Fig. 1. Possible electronic excitation and dezexcitation processes for a irradiated molecule are also shown in Fig. 1.

Irradiated polymers may suffer the following photophysical processes:

- photon absorbtion and formation of singlet activated entities;
- nonradiative transitions of intercrossing systems with generation of excited triplet states;
- formation of free radicals from the two excited states (singlet and triplet);
- energetic transfer processes between excited molecules (donors) and other molecules present in the system (acceptors).

Basically, polymer photodegradation is an effect of the energy dissipated by photochemically excited molecules and represents one of the first stages of the *primary photochemical process*. This stage may lead to *scission of the excited macromolecule* and appearance in the system of new macromolecular fragments ($P_1, P_2, \ldots, P_i$). Such fragments are characterized by lower molecular weights, as compared to the initial macromolecules, which increase the system polydispersity. Apparition of macroradicals allow the establishing of chemical intercatenary bonding with neighbouring macromolecules leading to obtaining of crosslinked structures, having as consequences increasing of molecular weight and decreasing of solubility.

After initiation through this path, polymer degradation usually continues via *secondary photochemical processes* at which participate both initial macromolecules and products resulted from the *primary photochemical process*.

Returning of the excited macromolecule to its fundamental energetic state is made through *physical radiative photoluminiscence phenomena*, *heat release*, *nonradiative transitions* or even by *energy transfer towards another acceptor molecule* existing in the system (Fig. 2).

There were established three laws which govern photochemical processes [1]. The first one is *the photochemical equivalent law*, also known as the *Einstein–Stark law*. According to it, each molecule participating to a chemical reaction involved in the primary photochemical process absorbs the photon as a necessary condition in initiating photochemical transformations. The photochemical equivalent law is completed by the *Grothus–Draper afirmation* which expresses the necessity of light absorbtion by at least one of the reactants participating in a photophysical process or a photochemical reaction. In conformity with this law, only the light absorbed by the molecule which endures photochemical transformation may determine chemical bond dissociation or generation of structural modifications, such as transpositions.

Although most polymer structures are mainly comprised of C–X type covalent bonds, where X may be C, O, N, P or halogen, and these bonds do not absorb light with wavelengths higher than 200 nm, the observation that polymer photodegradation is generated even at exposure to irradiation wavelengths higher than 300 nm is not considered to disobey the two previous mentioned laws. It was observed that the simple presence of some chromophore carbonyl type groups or conjugated

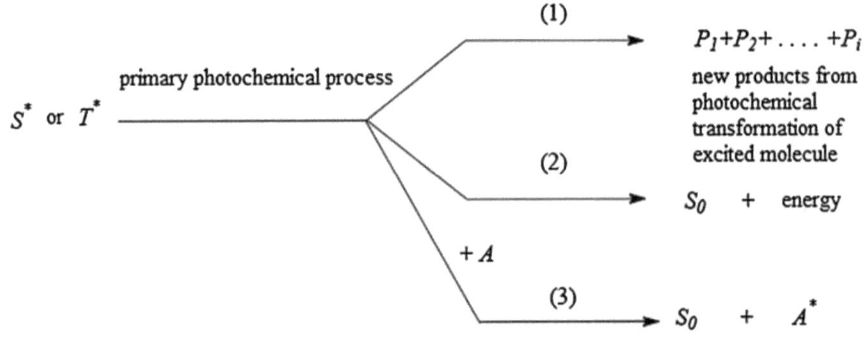

Fig. 2 Dissipation of excited molecule energy during primary photochemical process

double bonds, either in the molecular structure or in the form of impurities in the polymer molecular weight, assure absorbtion of light with wavelengths between values of 300 and 360 nm.

The third law, also known in the literature as the *reciprocity law* was initially elaborated by Bunsen and Roscoe.

From what was previously presented one may conclude that photodegradation of polymeric materials is the result of the following two types of reactions [5]:

- *Primary photochemical reactions* produced under direct absorbed light action, having as result free radicals or transposition products formation;
- *Secondary reactions* initiated by free radicals generated in primary photochemical processes. These reactions are independent of absorbed light. Such an example consists of photooxidation reactions.

The quantitative evaluation of photochemical reactions is expressed through the quantum yield (Φ) which is defined as the relationship between the number of cleaved polymer molecules and number of absorbed photons in the time unity [1, 5]:

$$\Phi = \frac{\textit{number of cleaved macromolecules}}{\textit{number of absorbed photons by the polymer}} \quad (3)$$

The *quantum crosslinking yield* (Φ_c) is defined following a similar manner in Eq. (4):

$$\Phi_c = \frac{\textit{number of crosslinked macromolecules}}{\textit{number of absorbed photons by the polymer}} \quad (4)$$

Quantum yields specific to common photochemical processes are subunitary. An exception consists of the photodegradation reactions occurring through chain mechanism, characterized by supraunitary quantum yields values.

Quantum cleavage yields values of the main commercially available polymers are given in Table 2 [2, 5].

The *quantum cleavage yield* (Φ_{cl}) may be evaluated from experimental data by using the Eq. (5), where I_a is the number of absorbed photons by the polymer and $(M_n)_0$ and $(M_n)_t$ are the averaged numeric molecular weights of the polymer measured before and after irradiation [6]:

$$\Phi_{cl} = \frac{1}{I_a}\left[\left(\frac{1}{\overline{M_n}}\right)_t - \left(\frac{1}{\overline{M_n}}\right)_0\right] \quad (5)$$

The polymer solution viscosity measured before and after exposure to light may also be used in evaluating the quantum cleavage yield of the macromolecular chain, through Eq. (6) in which $(M_v)_0$ and $(M_v)_t$ are viscosity-averaged molecular weights of the initial and irradiated polymer:

Table 2 Quantum cleavage yields of the main commercially available polymers [2, 5]

Name	Wavelength (nm)	Quantum yield
Poly(ethylene terephthalate)	254	1.6×10^{-3}
Poly(methyl methacrylate)	254	1.7×10^{-3}
Poly(acrylonitrile)	254	$2.0–7.7 \times 10^{-4}$
Poly(vinyl acetate) (in air)	254	5×10^{-3}
Poly(α-methylstyrene)	254	7×10^{-3}
Polystyrene	254	5.5×10^{-4}
Poly(methyl vinyl ketone)	254	2.5×10^{-2}
Poly(phenyl isopropyl ketone)	254	1.7×10^{-2}
Polyurethanes	313	5×10^{-3}
	365	2×10^{-3}
Epoxy resins	365	4.5×10^{-4}
Polysulfones	254	8.4×10^{-4}
Poly(ether sulfones)	254	7.7×10^{-4}
Poly(methyl phenyl) silane	313	1.7×10^{-2}

$$\Phi = \frac{1}{I_a}\left[\left(\frac{1}{\overline{M_v}}\right)_t - \left(\frac{1}{\overline{M_v}}\right)_0\right] = \frac{1}{I_a}\left[\frac{(\overline{M_v})_0}{(\overline{M_v})_t} - 1\right] \tag{6}$$

By plotting the linear dependence of $\frac{(\overline{M_v})_0}{(\overline{M_v})_t} - 1$ as function of absorbed photons number, the value of Φ may be determined.

The number of photons absorbed by the polymer is evaluated with the help of light sensitive substances known as chemical actinometers. Such a compound, noted A, transforms into another compound, noted B, after exposure to luminous radiation with known wavelength λ and quantum yield Φ_B. The number of absorbed photons by compound A is determined with relation (7) in which I_a is the number of absorbed photons, I_0 is the number of photons characterizing the incident radiation, ε_A is the molar absorptivity of compound A, c is concentration of compound A and l is the length of the luminous radiation.

$$1 - \left(\frac{I_a}{I_0}\right) = 1 - 10^{\varepsilon} \cdot A^c \cdot A^l \tag{7}$$

The fraction absorbed by the polymer from the incident light may be evaluated either experimentally, by photometric methods, or by the known parameters of Eq. (7).

2.1 *General Photodegradation Mechanisms of Polymers*

2.1.1 Photo-Oxidative Mechanism

As previously stated, primary photochemical reactions occur as a result of macromolecules activation through direct light absorbtion. In inert atmosphere of argon or nitrogen there exists the possibility of specific degradative processes occurence, such as macromolecules scissions and crosslinking, whilst in air oxygen also participates to the photodegradation processes. The oxygen in air may initiate photo-oxidative processes. Photo-oxidative degradation of polymeric materials occurs through radical intermediates, mostly as a chained mechanism [6], as further shown.

2.1.2 Initiation

Initiation is the photodegradation stage in which free radical species are generated and consists of light absorbtion, either by chromophore groups present in the macromolecule structure or by ones present in impurities in the polymer. Thus activated structures with variable stability are generated, capable of reacting with oxygen [reactions (8), (9)]. In the two reactions R and P represent new alkyl radicals resulted from R–H and P–H macromolecules decomposition and $HO_2\cdot$ is hydroperoxide radical.

$$R{-}H + O_2 \xrightarrow{h_v} R\cdot + HO_2\cdot \tag{8}$$

$$P{-}H + O_2 \xrightarrow{h_v} P\cdot + HO_2\cdot \tag{9}$$

There exists the possibility of the newly formed alkyl radicals to react with unactivated macromolecules generating new reactive radicals resulted by hydrogen extraction from the polymer structure, as shown in reaction (10):

$$R\cdot + P{-}H \rightarrow R{-}H + P\cdot \tag{10}$$

Hydrogen atoms are extracted in increasing order of bond energies, the most susceptible being the entities containing tertiary carbon atoms, followed by the ones containing secondary carbon atoms and primary ones.

Also, there exists the possibility of hydroperoxide radicals to react between them leading to H_2O_2 formation and, under H_2O_2 action, even more stable macromolecules may generate new alkyl radicals:

$$\begin{aligned} &HO_2\cdot + HO_2\cdot \rightarrow H_2O_2 + O_2 \\ &H_2O_2 \xrightarrow{h_\nu} 2HO\cdot \\ &R{-}H + HO\cdot \rightarrow R\cdot + H_2O \end{aligned} \tag{11}$$

Free radical generation during initiation is also sustained by other authors. Feller also considers that by exposing the polymeric materials to UV irradiation hydroperoxides and peroxides are generated under oxygen influence through a similar mechanism [7], however the author also stating that free radicals are formed exclusively as result of a C–H bond scission from the macromolecular structure. Such an initiation path is also sustained by data in Table 2 which demonstrate that even UV radiations possess sufficient energy in order to assure a C–H bond scission.

Hydroperoxide formation is essential for explaining photodegradative processes of polymers with aliphatic structure. The presence of hydroperoxides was evidenced during polyethylene, polypropylene and poly(vinyl chloride) photodegradation [8].

2.1.3 Propagation

In the propagation stage the resulted radicals react with oxygen from air to yield new reactive oxi- or peroxi- structures or other macroradicals which finally transform in hydroperoxide and peroxide structures as shown in Eq. (12):

$$\begin{aligned} &P\cdot + O_2 \rightarrow POO\cdot \\ &POO\cdot + RH \rightarrow POOH + R\cdot \\ &POOH \xrightarrow{h_\nu \text{ or } \Delta T} PO\cdot + HO\cdot \\ &PO\cdot + PO\cdot \rightarrow PO{-}OP \end{aligned} \tag{12}$$

Apparition of peroxi radicals is a rapid stage controlled by oxygen diffusion in the polymer bulk. Photodegradation efficiency during propagation is related to the rate of hydroperoxide decomposition and to their capacity of generating new free oxi- and peroxi- radicals, as given in relation (13):

$$PO{-}OH \xrightarrow{h\nu} PO\cdot + \cdot OH \tag{13}$$

The newly generated very reactive radicals stabilize themselves rapidly through extraction of hydrogen atoms from neighbouring macromolecules and continue the degradative process:

$$\sim\!\sim R{-}H + HO\cdot \longrightarrow \sim\!\sim R\cdot + H_2O$$
$$\sim\!\sim R{-}O\cdot + H{-}P \longrightarrow \sim\!\sim R{-}OH + P\cdot \tag{14}$$

Radicals formed through such paths possess the capacity to continue macromolecules cleavage or crosslinking reactions. Another known reaction, produced with the participation of oxi radicals is the one of hydroperoxide decomposition (Eq. 15).

In the case of vicinal peroxide groups, these may stabilize through cyclization when epoxide or peroxide entities are formed (Eq. 16) [5].

In the situation in which the formation of cyclic structures is not sterically favored radical stabilization is achieved via intramolecular interactions (Eq. 17) [5]. In this manner new carbonyl groups are generated in the system.

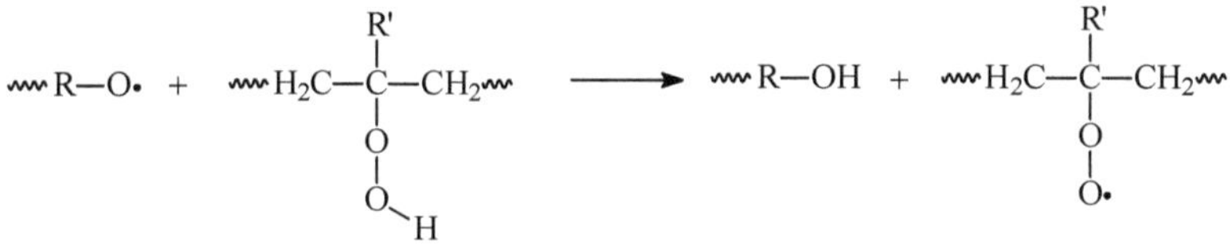

(15)

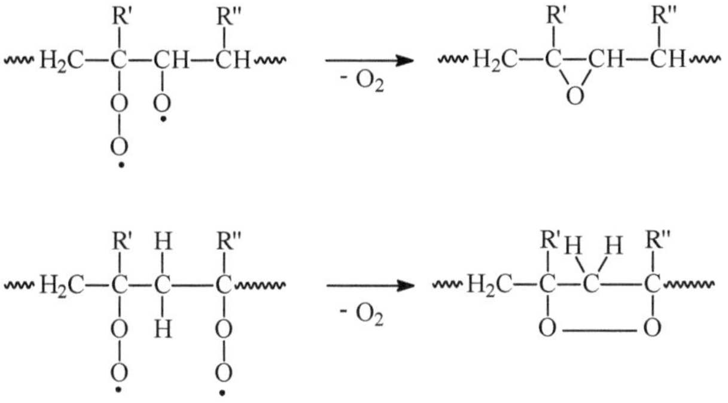

(16)

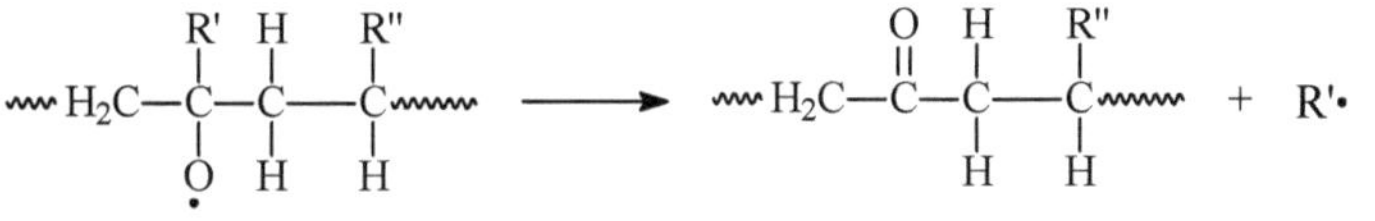

(17)

Another way to stabilize oxi radicals is represented by the β-scission reactions [5]. These transformations result in macromolecular scission and formation of new carbonyl end groups:

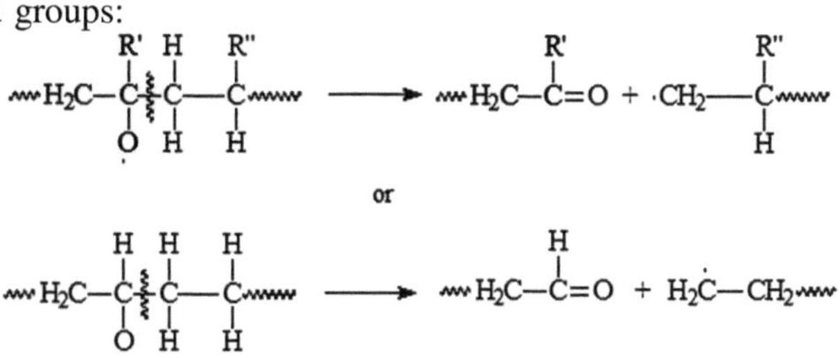

(18)

The photolysis of ketone or ester products, formed via β-scission reactions, occurs through Norrish reactions which may be of type I, when free radicals are generated through decarbonylation, or of type II, occurring via chain scission [9]:

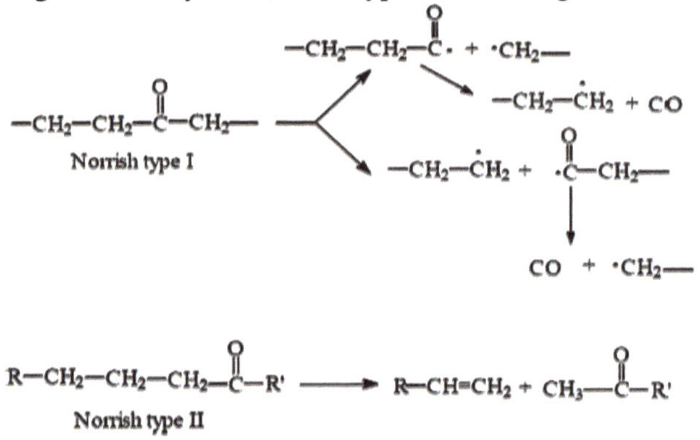

(19)

In Fig. 3 there is given the general photo-oxidation cycle of polymers [9].

2.1.4 Other Aspects Concerning Hydroperoxides Photodecomposition

Solar light may generate sufficient energy to polymeric materials in order to cleave RO–OH and R–OOH bonds in hydroperoxides. So the dissociation energy value of RO–OH bond in hydroperoxides is 176 kJ mol^{-1} and that of R–OOH is 377 kJ mol^{-1} [7]. That is why it is considered that during hydroperoxide irradiation the formation of RO· and HO· radicals is predominant.

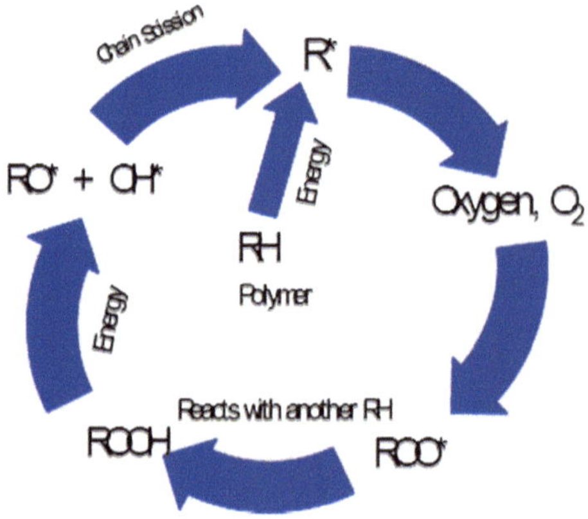

Fig. 3 The general photo-oxidation cycle of polymers [9]

Radiation effect of wavelengths of 254 and 313 nm over photo-oxidation reactions of some elastomers with low aromatic entities content was monitored [8–10]. Authors reported the formation of more or less reactive hydroperoxide groups, dependent on the occupied position in the polymer structure. It was thusly observed that active peroxide groups situated in the vicinity of aromatic entities possess the capacity of easily generating radical species as a result of photolysis reactions and these generated radicals during such dissociations (RO· and HO·) are capable of initiating oxidation reactions in the polymer. The presence of inactive hydroperoxide structures was evidenced by their constant concentration increase with exposure time. Authors also noticed a more increased tendency of active hydroperoxide formation at lower wavelength values. Authors were thus able to demonstrate that the total fraction of active hydroperoxides was higher (0.93) when irradiation was undertaken with radiations at 254 nm in comparison to 0.60, which was the total hydroperoxide fraction at 313 nm exposure of material. Gardette and Lemaire have also debated on different hydroperoxide group types and their different sensitivities to photochemical reactions [11].

It is currently considered that hydroperoxide decomposition represents the main stage involved in the photo-oxidative degradation of polymers.

2.1.5 The Energy Transfer Between Carbonyl Groups and Hydroperoxide Structures

As already mentioned, it was observed that during photo-oxidative degradation of polymers there occur accumulations of new carbonyl groups with ketonic, carboxylic or ester structure [5]. It was noticed that, whilst in most cases carbonyl structure concentrations from polymer bulk increase with exposure time, concentration of hydroperoxide and peroxide structures rapidly reaches a stationary state at relatively low concentrations. An exemplification of such behavior is given in Table 3, in which the variation of hydroperoxide concentration with exposure time (luminous radiation with $\lambda \geq 300$ nm) is represented in the case of photo-oxidative degradation of a vinyl-ester based polymeric structure, crosslinked by end double bonds, where BTAC is benzyltributylammonium chloride (Schemes 1, 2—reproduced with kind permission from Elsevier—license no. 3842460646409) [12].

The increase in hydroperoxide concentration led to a gloss retention decrease and roughness increase of sample surface, as it may be seen from Fig. 4 [12] (reproduced with kind permission from Elsevier—license no. 3842460646409).

Through FTIR and UV–Vis spectroscopy methods there were identified diverse products resulted during photo-oxidative degradation, such as carbonyls, hydroxyl entities and peroxides and hydroperoxides, and photo-oxidative degradation mechanisms were given in Schemes 3 and 4 [12] (reproduced with kind permission from Elsevier—license no. 3842460646409).

Table 3 Hydroperoxide concentration variation during UV exposure [12] (reproduced with kind permission from Elsevier—License no. 3842460646409)

Irradiation time (min)	Hydroperoxydes content mol (–OOH)/g polymer × 10^{-5}
10	7.0
50	9.4
100	12.6
150	14.3
200	15.2
250	15.8
300	16.3
350	16.9
400	17.5
450	17.9
500	17.7
550	17.5
600	17.5

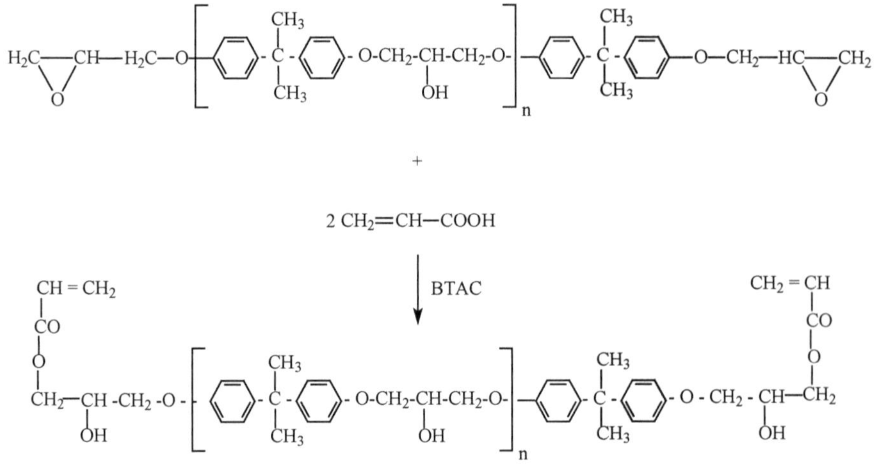

Scheme 1 Synthesis of the vinyl-ester resin [12] (reproduced with kind permission from Elsevier—license no. 3842460646409)

2.1.6 The Photo-Fries Rearrangement

The photo-Fries rearrangement was described as being a rearrangement of phenyl ester entities with their transformation in ortho hydroxy ketones and para hydroxy ketones. The reaction occurs in three steps: formation of two radicals, recombination and hydrogen extraction [13–15]. The general implied reactions are depicted in Scheme 5 [2].

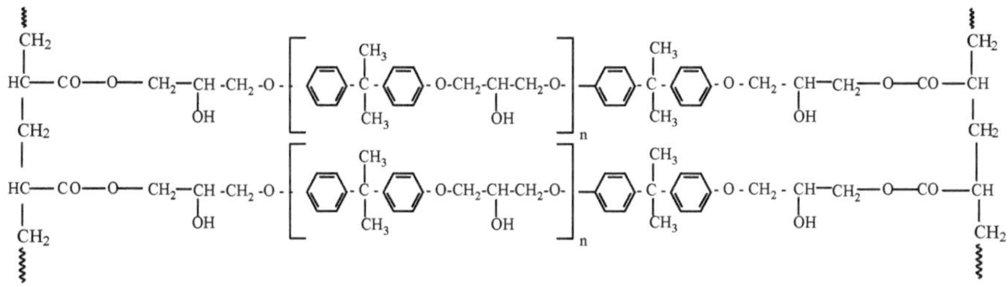

Scheme 2 Structure of the crosslinked vinyl-ester resin [12] (reproduced with kind permission from Elsevier—license no. 3842460646409)

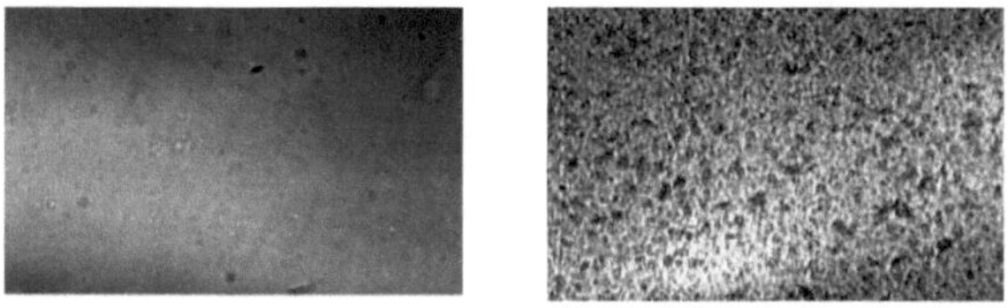

Fig. 4 Optical micrographs of crosslinked vinyl-ester resin film [12] (reproduced with kind permission from Elsevier—license no. 3842460646409)

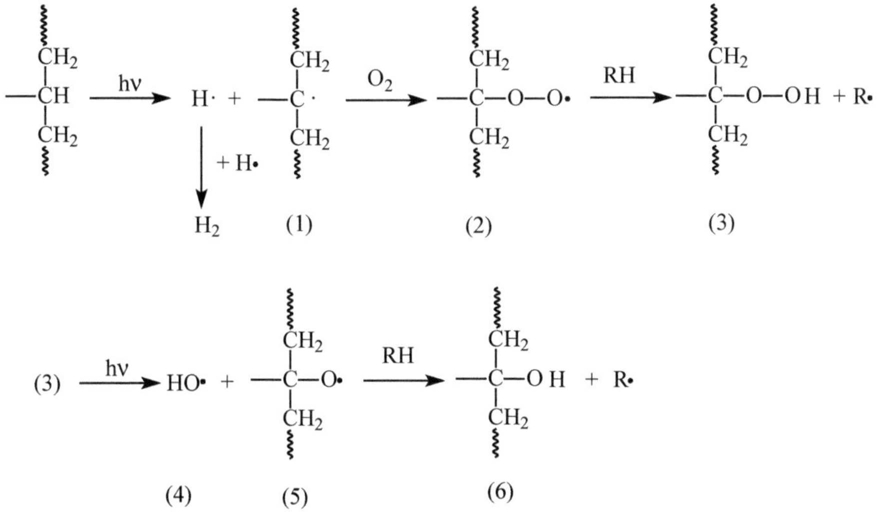

Scheme 3 New hydroxyls, peroxides and hydroperoxides resulted during UV irradiation [12] (reproduced with kind permission from Elsevier—license no. 3842460646409)

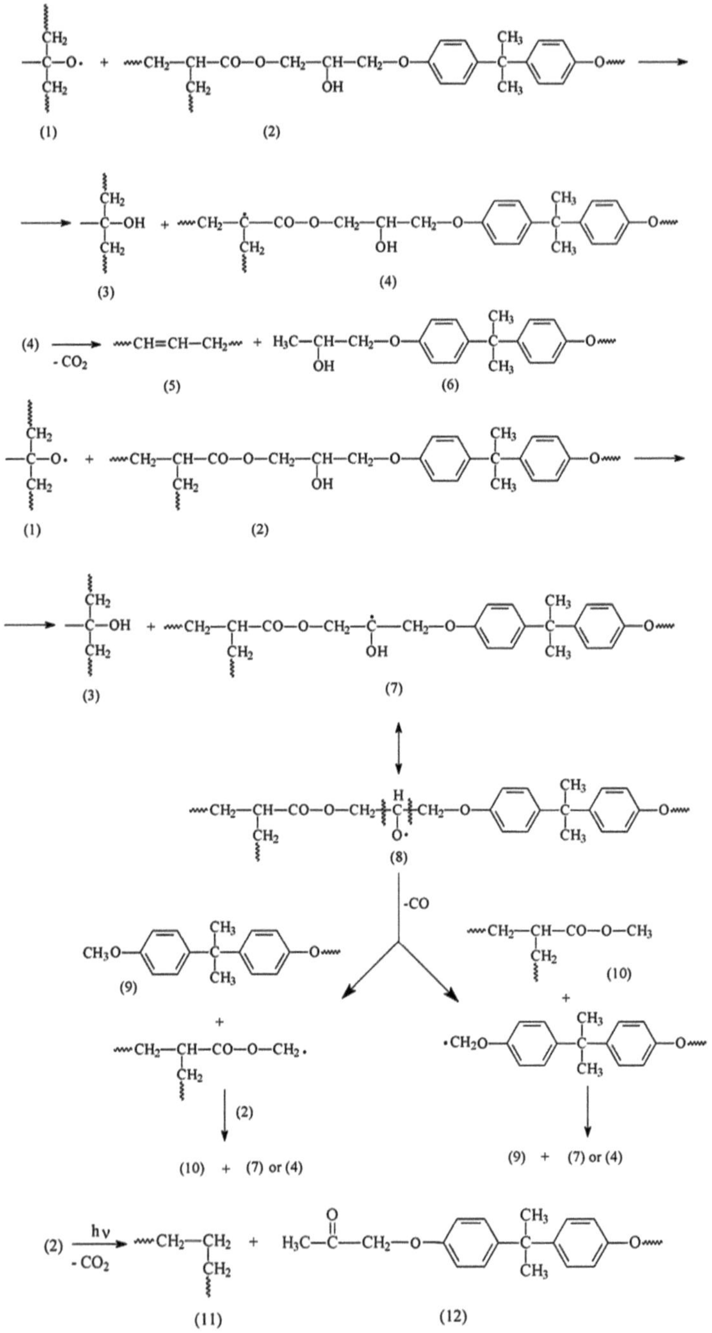

Scheme 4 Photo-oxidative degradation mechanism [12] (reproduced with kind permission from Elsevier—license no. 3842460646409)

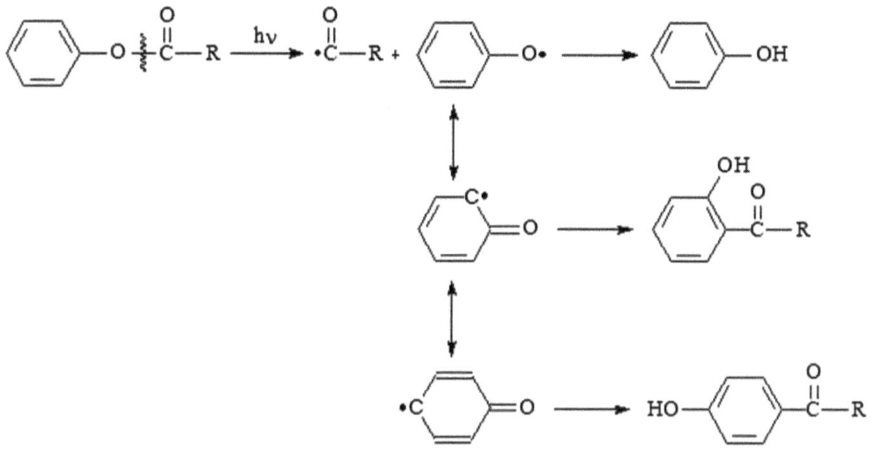

Scheme 5 The photo-Fries rearrangement [2]

According to the literature there exist two types of products which the phenoxy radical may generate during the photodegradation process. In a first instance there is the possibility of the phenoxy radical to generate phenol via extraction of a hydrogen atom from vicinal entities. In a second instance the phenoxy radical may lead to the formation of an acyl phenol through recombination with an acyl radical and hydrogen extraction, after previously passing in the form of one of the two cyclohexadienonic intermediates. Two important factors in radicals rearrangements are temperature and the nature of the substitutes, since high temperatures influence formation of the ortho isomer [13–15].

Norrish type reactions and photo-Fries rearrangements were together observed by Rosu et al. in accelerated ageing studies on semi-interpenetrating polymer networks based on linear semi-aromatic polyurethane and crosslinked epoxy resin, whilst only Norrish type reactions were reported for poly(vinyl alcohol) and cellulose based cryogels by Varganici et al. Further general aspects on the outcome of the two recently reported studies were given in this book.

2.1.7 Termination

Chain termination stage of photo-oxidative processes occurs through recombination of different radical types resulting in linear, branched or crosslinked products. It is already known that during photo-oxidative degradation weak reactive hydroperoxide radicals may accumulate in sufficiently high concentrations. Through their recombination there results a tetraoxyde intermediate which gradually transforms into macroradicals and oxygen [16], as shown in Eq. (20):

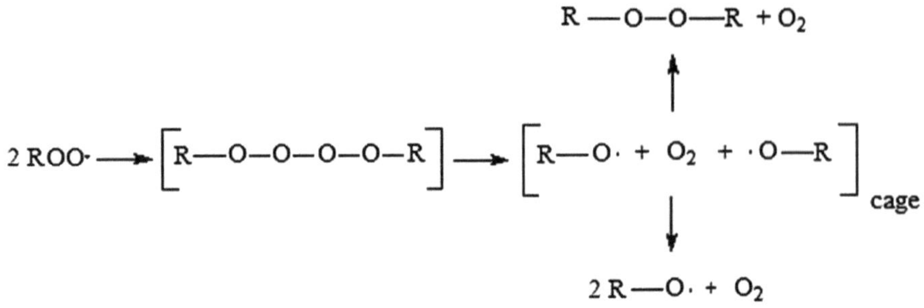

(20)

Small macroradicals reactivity is dependent on electronic density of the atom containing the unpaired electron. If the macroradical reactivity is low, the probability for the macroradical to be encaged before extraction increases with reactivity decrease [17]. The reactivity of radicals formed by polymer degradation decreases in the order: R· > R–O· > R–O–O·.

Light induced degradative effects on the polymeric materials are visible on their surfaces and the depth at which their properties are affected represents an interesting subject [7]. It is why the most affected materials are the transparent or translucid ones. Systems which totally absorb luminous radiation, without occurring of diffusion phenomena, respect the *Lambert–Beer law*, where I_0 is the incident light intensity, I is light intensity at depth x and α is light absorptivity:

$$I = I_0 e^{-\alpha x} \quad \text{or} \quad ln\frac{I_0}{I} = \alpha x \tag{21}$$

The validity of the *Lambert–Beer law* in polymers represented the subject of several photodegradation studies. Jellinek associated this relation with photochemical scissions in macromolecules [18, 19]. Heller used the Lambert–Beer relation for testing protective action efficiency of some UV radiation absorbant materials [20]. The energy absorbed by the vast majority of polymeric materials increases disproportionally with wavelength of the absorbed incident luminous radiation. An example was offered by Müller even from back in 1983, which demonstrated the increase of carbonyl groups content as a consequence of photo-oxidative degradation of polyethylene foils with exposure time increase in a Xenotest device [21]. The author noticed that photodegradation effects significantly decrease after exceeding 200 μm in depth. The increase of nondegraded material content in polymer bulk was also demonstrated through calculus [22, 23]. Polymer surface layers changed their color due to chromophore accumulation. Immediately after their formation, chromophores supplement the absorbed light and favor polymer photodegradation.

Photodegradative processes and the depth to which they affect polymeric materials are dependent on the wavelength of exposure. R.L. Feller demonstrated that the presence of UV radiations with low wavelengths in the light spectrum of a carbon arc lamp determines a in depth crosslinking of butyl acrylate and isobutyl acrylate films, as opposed to high wavelengths radiations, present in the solar light, which only generate crosslinking of thinner films [24].

General photodegradation mechanisms of the most common classes of polymers were vastly detailed by Rabek [5]. These mechanisms include: polyketones, polyolefins, polycarboxylic acids and analogues, poly (vinyl alcohol), polyacrylates and polymethacrylates, polyvinylpyrrolidone, polyacrylonitrile, polystirenes, polydienes, and polyacetylenes.

General oxidation and photooxidation mechanisms of polymers were overviewed in Schemes 6 and 7 [9].

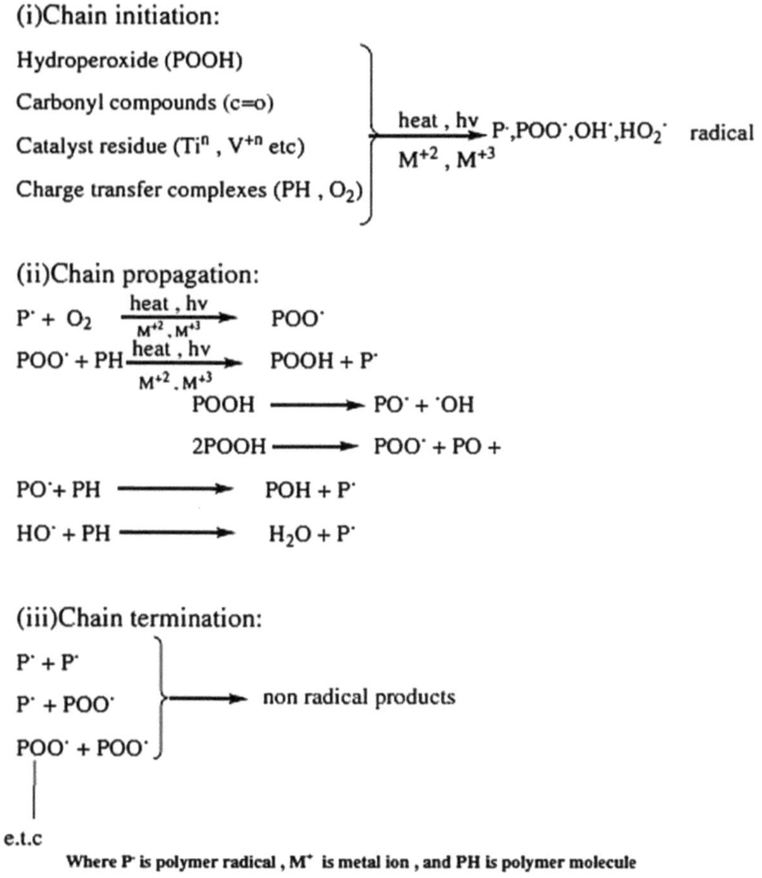

Scheme 6 General oxidation and photooxidation mechanisms of polymers [9]

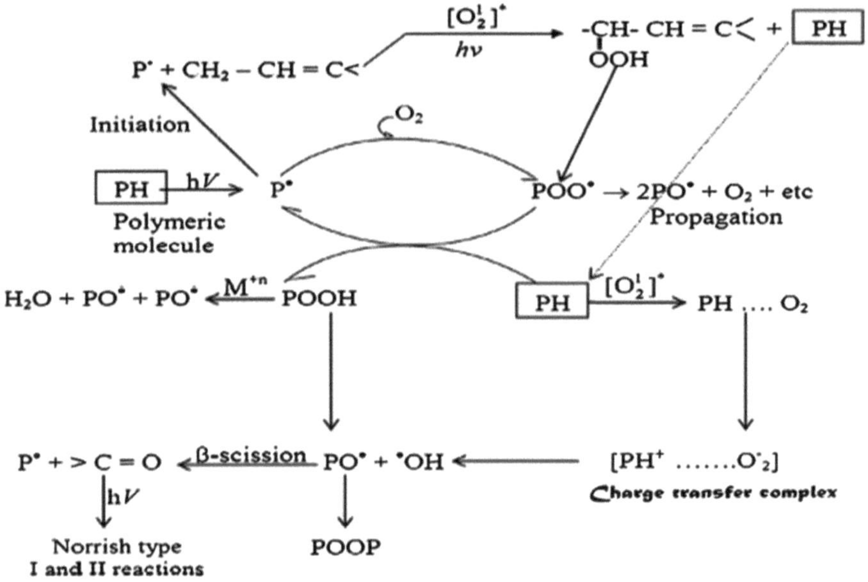

Scheme 7 Photooxidation of polymers [9]

3 Conclusions

Electromagnetic radiations from visible and/or UV spectra are comprised of streams of very small moving particles with wave-particle properties, named photons.

The UV region is devided based on wavelength and corresponding energy in three domains: UVA (315–400 nm), UVB (315–280 nm) and UVC (280–100 nm). The radiation energy of wavelengths in the range 280–400 nm presents values which cleave the majority of covalent bonds in polymers (412–300 kJ mol^{-1}).

Light energy absorbtion leads to the passing of any molecule from its electronic fundamental state into an excited energetically superior electronic state.

During irradiation polymers may undergo different photophysical processes, such as: photon absorbtion and activated singlet entities formation, generation of free radicals from singlet and triplet excited states, energy transfer between donor and acceptor molecules present in the system or nonradiative transitions of inter-crossing systems with excited triplet states formation.

Polymer photodegradation occurs through primary and secondary photochemical processes. A first stage of any primary photochemical process consists of energy dissipation by excited molecules leading to excited macromolecules scissions with formation of new macromolecular fragments of lower molecular weights, increasing the system polydispersity. Higher molecular weight crosslinked fragments may also be generated through chemical intercatenary bonding, leading to decrease in solubility.

Polymer photodegradation continues through secondary photochemical processes with participation of both macromolecules and products resulted during primary photochemical processes. Afterwards, the excited macromolecule returns to its fundamental energetic state via different processes such as heat release, physical radiative photoluminiscence phenomena, nonradiative transitions or energy transfer towards another acceptor molecule in the system.

Carbon–halogen covalent bonds or covalent bonds of carbon with C, O, N and P atoms do not absorb light at higher than 200 nm wavelengths. It is only through the presence of chromophore carbonyl groups or conjugated double bonds that light absorbtion is assured with wavelengths in the rage 300–360 nm.

In inert atmosphere there may occur specific degradative processes (macromolecules scissions and crosslinking). In air atmosphere oxygen also influences photodegradation through photo-oxidative processes occurring via radical intermediates as chained mechanisms.

References

1. Jipa, S., Setnescu, S.R., Zaharescu, T., Gheorghiu, T.L.M.: The Chemical Effect of Ultraviolet and nuclear Radiations. Electra, Bucharest (2004)
2. Varganici, C.-D.: Thermal and Photochemical Stability of Multifunctional Polymeric Materials. Ph.D. thesis, Romanian Academy, Iasi (2014)
3. Dunn, J.R.: Unsolved problems in polymer degradation. Rubber Chem. Technol. **51**(4), 686–703 (1978)
4. Schaaf, R.: Recycling of Termoplastic Wastes. Verein Deutscher Inginieure, Düsseldorf (1979)
5. Rabek, J.F.: Polymer Photodegradation, Mechanisms and Experimental Methods. Chapman & Hall, UK (1995)
6. Rabek, J.F.: Experimental Methods in Polymer Chemistry: Physical Principles and Application. Wiley, Chichester (1980)
7. Feller, R.L.: Accelerated aging, photochemical and thermal aspects. The Paul J Getty Trust, SUA, Ann Arbor, Michigan (1994)
8. Wiles, D.M.: Photostabilization of macromolecules by excited state quenching. Pure Appl. Chem. **50**(4), 291–297 (1978)
9. Yousif, E., Haddad, R.: Photodegradation and photostabilization of polymers, especially polystyrene: review. SpringerPlus **2**(1), 398 (2013)
10. Bousquet, J.A., Fouassier, J.P.: Hydroperoxides as intermediates responsible for wavelength effects in photooxidation reactions. J. Polym. Sci. Polym. Chem. Ed. **22**(12), 3865–3876 (1984)
11. Gardette, J.-L., Lemaire, J.: Oxydation photothermique d'élastomères de polyuréthannes thermoplastiques, 2. Influence de la longueur d'onde et des conditions d'irradiation sur le jaunissement des polyuréthannes à structure aromatique. Die Makromol. Chem. **183**(10), 2415–2425 (1982)
12. Rosu, D., Rosu, L., Cascaval, C.N.: Effect of ultraviolet radiation on vinyl ester network based on bisphenol A. J. Photochem. Photobiol. A Chem. **194**(2–3), 275–282 (2008)
13. Wypych, G.: Handbook of Material Weathering, IV edn. Chemtec, Toronto (2008)
14. Anderson, J.C., Reese, C.B.: Photo-induced Fries rearrangement. In: Proceedings of the Chemical Society, vol. 6, p. 217, London (1960)

15. Diepens, M.: Photodegradation and Stability of Bisphenol A Polycarbonate in Weathering Conditions. Ph.D. thesis, Eindhoven Techincal University, Eindhoven (2009)
16. Nguyen, T.Q.: Polymer degradation and stabilization. In: Meyer, T., Keurentjes, J. (eds.) Handbook of Polymer Reaction Engineering, pp. 757–805. Wiley-VCH Verlag GmbH & Co. KCaA, Weinheim (2005)
17. Male, J.L., Yoon, M., Glenn, A.G., Weakley, T.J.R., Tyler, D.R.: Radical cage effects in the photochemical degradation of polymers: in-cage trapping of photochemically generated radical cage pairs in polymer model compounds. Macromolecules **32**(12), 3898–3906 (1999)
18. Jellinek, H.H.G.: Surfaces (interfaces) and polymer stability. In: Mittal, K.L. (ed.) Physicochemical Aspects of Polymer Surfaces, vol. I, pp. 255–283. Plenum Press, New York (1981)
19. Jellinek, H.H.G.: Fundamental degradation processes relevant to outdoor exposure of polymers. Appl. Polym. Symp. **4**, 41–59 (1967)
20. Heller, H.J.: Protection of polymers against light irradiation. Eur. Polym. J. **5**, 105–132 (1969)
21. Müller, K.: Untersuchung der "werkstoffwirksamkeit" unterschiedlicher strahlungssysteme und bewitterungsverfahren. Angew. Makromol. Chem. **114**(1), 69–82 (1983)
22. Thompson, G.: Penetration of radiation into old paint films. Natl. Gallery Tech. Bull. **3**, 25–33 (1979)
23. Thompson, G.: Topics in the conservation chemistry of surfaces, Application of Science in Examination of Works of Art. Proceedings of the Seminar, Museum of Fine Arts, pp. 78–85. Boston, Massachusetts (1965)
24. Feller, R.L.: New solvent-type varnishes. In: Thompson, G. (ed.) Recent Advances in Conservation, pp. 171–185. Butterworths, London (1963)

Factors Influencing the Photochemical Behavior of Multicomponent Polymeric Materials

Marioara Nechifor

Abstract The study of photo-degradation and stabilization of polymers has an extremely importance from academic, economic and environmental point of view and a better understanding of the mechanism of degradation and correlation between weathering external factors, such as wavelength and irradiation dose, temperature and humidity, and chemical, physical and mechanical effects that they induce are the primary requisites to control the stability and ensure long life of the final product. Although the study on the chemistry of interactions between various constituents (including the polymer itself and the presence of impurities, fillers, antioxidants, photo-stabilizers, dyes, pigments, nucleating agents, flame retardants, and other additives such as catalyst residues) in polymer-based composites exposed to sunlight radiation has attracted a large scientific effort during the last few decades, the changes that occur at molecular level and the degradation kinetics have not been thoroughly elucidated. To predict the photo-damage to a certain formulation under solar UV radiation is still challenging because composite materials are typically exposed to multiple environmental variable during service. While individual degradation variables can lead to formation of chromophores and initiate the photo-degradation mechanism, their combinations may have unknown synergistic or antagonistic effects. This article review sets out to discuss the influence of the primary environmental parameters, i.e., light, heat and moisture, and the polymer structure and chemical make-up of the polymeric composites on fundamental photo-degradation processes. Each of the variables is briefly examined and examples of polymeric material degradation reported by scientific studies during the last two decades are given.

Keywords Photo-chemical degradation · Polymer chemical structure · Impurities · Wavelength and irradiation dose · Temperature · Humidity

---

M. Nechifor (✉)
Department of Photochemistry, "Petru Poni" Institute of Macromolecular Chemistry, 41A Grigore Ghica Voda Alley, Iasi, Romania
e-mail: nechifor@icmpp.ro

D. Rosu and Visakh P. M. (eds.), *Photochemical Behavior of Multicomponent Polymeric-based Materials*, Advanced Structured Materials 26,
DOI 10.1007/978-3-319-25196-7_2

Abbreviations

Φ_R	Reaction quantum yield
AFM	Atomic force microscopy
ATR-FTIR	Attenuated total reflectance-FTIR
CB	Carbon black
CNT	Carbon nanotube
CI	Carbonyl index
CIRA	Quartz with an IR reflecting coating
CL	Chemiluminescence
CT	Charge transfer
DSA	Dielectric sorption analysis
DDS	4,4′-Diaminodiphenyl sulfone
DSC	Differential scanning calorimetry
EPDM	Ethylene-propylene-diene monomer
EVA	Poly(ethylene-*co*-vinyl acetate)
FTIR	Fourier transform infrared spectroscopy
GC	Gas chromatography
HALS	Hindered amine light stabilizers
HDPE	High density polyethylene
LDH	Layered double hydroxide
LDPE	Low density polyethylene
LLDPE	Linear low density polyethylene
m-PE	Metallocene polyethylene
MMT	Montmorillonite
MWCNT	Multi-walled carbon nanotube
OMS	Organically modified clay
OMMT	Organically modified montmorillonite
PA	Polyamide
PA FTIR	Photo acoustic Fourier transformed infrared
PAM	Polyacrylamide
PC	Polycarbonate
PE	Polyethylene
PET	Poly(ethylene terephthalate)
PGC-MS	Pyrolysis gas chromatography-mass spectrometry
PLA	Polylactide
POSS	Polyhedral oligomeric silsesquioxanes
PP	Polypropylene
PS	Polystyrene
PU	Polyurethane
PVA	Poly(vinyl acetate)
PVC	Poly(vinyl chloride)
RH	Relative humidity
SEM	Scanning electron microscopy
SWCNT	Single-walled carbon nanotube

TEM	Tunnelling electron microscopy
TGDDM	Tetraglycidyl-4,4′-diaminodiphenylmethane
TPE	Thermoplastic elastomers
UV	Ultraviolet
UVAb	UV absorber
WF	Wood-floor
WOM	Weather-o-meter
WPC	Wood-polymer composites
XRD	X-ray diffraction

1 Introduction

The commercial importance of polymer-based materials, such as thermoplastics, thermosets and composites, in various fields, such as aerospace, automotive, marine, infrastructure, military, building materials, furniture industry, aeronautics and agricultural, where long-term service in hostile environments is required has been increased. Being light and strong, polymers have been the materials of choice for numerous outdoor uses including gas pipelines and irrigation pipes, water and chemical storage tanks, marine vessel hulls, outdoor carpet and furniture, roofing membrane, geomembranes, underground cabling, agricultural greenhouse applications, automobiles and modern aircraft, playground equipment and artificial turf, rigid and flexible solar photovoltaic and panels, solar water heaters, glazing materials, coatings and mouldings used in vehicles and in traffic signals, cable covering, toys, coated fabric for tents, stadium seating, personnel safety equipment and medical implants, etc. [1].

The durability, performance and rate of deterioration are key features of any composite material that decide the real fate of products designed for outdoor applications. This is especially relevant for synthetic polymer-based materials, which, during the past decades, have increasingly been used in various new fields of applications as replacements for the traditional metal, timber, natural fibers, ceramics, glass, cement-based building materials, and so on. Polymer-based composites undergo degradation when they are used in outdoor applications, and the environment negatively influences their service life. That is because most polymers are inherently unstable to sunlight. The complex nature of the physical environment and factors such as humidity, rain, wind and temperature further complicate the process of natural weathering of polymeric materials. The lifetime of a polymer-based material is determined by chemical degradation upon thermal and light exposure. The stability of polymers is essential, especially for outside use, and many researchers have devoted their efforts to elucidate degradation reactions and mechanisms, to understand the degradation kinetics and the changes that occur at molecular level. However, the natural weathering of thermoplastics, thermosets and their composites in outdoor applications as a result of variations in temperatures, moisture/rainfall and UV radiation, is not fully understood [2].

The degradation of polymeric materials depends on the polymer itself and service environment. The most important polymer-related parameters for deterioration are as follows: the type of polymer (thermoplastics, like polyolefins, elastomers, engineering plastics like polyamides, polyimides or polycarbonates), molecular structure and its defects, polymer morphology (degree of tacticity and crystallinity, the amount of branching), melt flow index, the addition of additives (e.g., pigments, dyes or anti-oxidants), and the presence of chromophoric groups, such as: internal in-chain impurities (e.g., hydroperoxides or carbonyls formed during storage, processing or weathering), external impurities (polymerization catalyst residues), pollutants from the atmosphere or metal traces from processing equipment, parts of the molecular structure of the polymer (i.e., polyaromatics), charge transfer complexes between oxygen and the polymer chain [3].

There are several degradation agents that induce irreversible property changes in polymeric composites and can act alone or concomitantly: temperature—static heat ageing, sub-zero exposure or thermal cycling (thermal degradation), water—exposure to water/humidity (hydrolytic degradation), cyclic conditions: freeze/thaw and dry/wet, weathering (including rain and sand erosion), combined load (i.e., stress) and environmental exposures (mechanical degradation), chemicals (including water, fuel, acids, alkalis, solvents and oxygen) (chemical degradation), ultraviolet (photo-degradation) and high-energy radiation, micro-organisms (fungi, bacteria, yeasts, algae, and their enzymes) (biodegradation) [4].

Weathering is a natural phenomenon that all materials will undergo some degree of ageing during their lifetime, with economic and environmental implications due to extensive deterioration of mechanical properties, such as tensile strength, impact strength and elongation, which lead to useless materials, after an unpredictable time. The more serious effects of photo-degradation include microcracking and embrittlement of polymeric substrate depending on the intensity of the radiation, discoloration, tarnishing, yellowing or darkening of the polymer surface, which affects the aesthetic appeal of the material, and chalking, which appears as a synergistic combination of water and UV radiation, and results in erosion and fading of the surface of certain type of polymers [5].

In light of these observations it should be noted that the degradation process is influenced by numerous variables that are often difficult to elucidate and the factors responsible for degradation mechanism at a given stage of this process are poorly understood. It is necessary to provide controllable experimental conditions and to monitor reactions leading to degradation in experiments simulating real-life degradation processes. The comparison of natural weathering *versus* accelerated weathering testing is very complex because of the inherent variability and complexity of outdoor exposure situations. Accelerated degradation studies are useful if molecular level information can be related to macroscopic observations on well-defined polymeric systems [6].

2 Influence of Chemical Structure

Polymeric composites are almost never produced from pure polymer. Different additives including antioxidants, inorganic reinforcing fillers, thermal and UV stabilizers, flame retardants, colorants, lubricants, processing aids and even biocides are added to the pure polymer to obtain a formulation used to fabricate a product. The chemical formulation of the polymeric material is dictated by the engineering requirements of the product and the environment in which it will be exposed. The database on the effects of UV radiation on various polymeric materials and their formulations is voluminous because each of the additives can significantly alter the photochemical behaviour of base polymer. Outdoor applications of most common polymeric materials are possible because the efficient light stabilizers that can be incorporated into their formulations are various and commercially available. Actually, UV stabilizers, such as triazole, benzophenone, and triazine derivatives, represent the fastest growing segment of the plastic additives market. The most used are the hindered amine light stabilizers (HALS), a class of remarkably efficient radical scavengers that suppress light-induced degradation in polymers. Another important group consists of UV absorbers that are photo-susceptible to the damaging UV-B radiation reaching the surface of material and convert it into thermal energy. Usually, mixtures of thermal and UV stabilizers are added at a level of 1 % by weight or less [7].

Reinforcing fillers are commonly used in polymeric formulations to improve their stiffness and other mechanical properties. Interfacial layer generated between the surface of filler particles and the polymer matrix is responsible for the improvement of mechanical, thermal and physicochemical properties of the composites. As a consequence, a decrease in the average particle size of fillers (high specific surface area) is generally desirable in designing composites. Commercial availability of bulk nanoscale fillers with average particle size from hundreds of nanometers (nanoclays) to tens of nanometers (fumed silica, titanium oxides and carbon nanofillers) are generally used in the chemical make-up of nanocomposites. At the same time, it is well known that filler interactions reduce the stability of the polymeric material which becomes more critical if nano-sized additives are incorporated [8, 9].

Polymer-nanoparticles/nanocomposites have been attracted great interest in the exponentially growing field of research in the development of materials in the last decades, although nanocomposites have shown higher degradability than control samples. Finely dispersing particles in polymer matrices allow an improvement of mechanical properties of the composite materials. Other properties, such as fire resistance, gas barrier properties, thermal properties and electrical conductivity can also be enhanced depending on the nature of the filler. Nanocomposites are characterized by reinforcing agents having nanodimensions added in small amounts (less than 3 vol%) compared with the traditional composites (more than 20 vol%). Incorporating inorganic particles, such as symmetrical or spherical particles (silica, calcium carbonate, etc.), CNTs, halloysite clay nanotubes, chitin nanofibres (whiskers), and lamellar fillers (clay minerals) into the polymer matrices both rigidity and toughness as well as thermal and barrier properties and crystallinity of different thermoplastics are increased [10]. Nanoparticles have a prior role in properties enhancement due to their nanoscale structure and their synergism with the polymer. Generally, the properties of nanocomposites depend significantly on the chemistry of polymer matrices, nature of nanofillers, and their preparation techniques. The uniform and fine dispersion of nanofillers in the polymer matrices is a necessary condition for achieving desired mechanical and physical characteristics. At the same time, it is now generally accepted that nanoparticles, such as OMMT nanofillers, can have negative impact on the durability of the nanocomposite materials under UV-light exposure. The effect of some other nanofillers, such as LDH or CNTs, can be tuned. Meantime, stabilization of polymer nanocomposites is now a bottleneck for industrial development of durable nanocomposites [6, 11, 12].

Flame retardants act in the polymer composites as typical inorganic [$Al(OH)_3$, $AlO(OH)$, $Mg(OH)_2$, silicates, etc.] or organic (melamine cyanurate, melamine polyphosphate etc.) fillers and have to be used in high loadings to achieve the requested flame retardancy. Therefore, these molecules "disturb" the polymer stability and influence the rheological properties during processing, the mechanical properties of the polymeric materials (e.g. improving stiffness but losing elasticity and/or acting as nucleating agent) and the aesthetic quality of the final product. Concomitantly, many flame retardants may cause accelerated degradation of the polymer chain and reduce the long-term performance and/or the flame retardancy over time. The chemical nature of the filler is often less important than the particle size and shape, the surface morphology and dispersion within the polymer matrix [13–15].

Durability problems of composites are specific to these materials used in hostile circumstances, especially when long-term service is required. The extensive outdoor applications require a high resistance of composite structures to degradation agents often within a short period. In some conditions, only a few hours of exposure may result in catastrophic failure or serious damage of the structural integrity. Gradual destruction of materials takes place usually as a result of change in chemical structure induced by simultaneous and interdependent radical photo- and thermo-oxidation processes. The deleterious effects of weathering on polymers have been generally ascribed to a complex set of processes induced by the action of sunlight (particularly UV radiation) combined with the humidity and temperature. Degradation produced by one agent can reduce resistance to other agents. Weathering is of considerable complexity due to natural fluctuations in ultraviolet radiation, temperature, humidity and other environmental factors (especially aggressive and corrosive pollutants), and to the interaction of these factors, and is therefore impossible to simulate [16].

Among various inorganic nanofillers, CB, CNT, fullerenes, layered phyllosilicates (e.g., MMT, OMMT, hectorite, halloysite, saponite, LDH), metal oxides (e.g., TiO_2, SiO_2, Fe_2O_3, Al_2O_3), nanoparticle of metals (e.g., Au, Ag), semiconductors, POSS, etc., are considered to be very important in the formulation of polyolefin-based nanocomposites for industrial applications. The UV durability of the polyolefin-based nanocomposites containing these nanofillers depends on the chemical composition of the nanofiller, its UV and thermal stability, its morphology, and the possible interactions and/or reactions that may occur between the matrix and nanofiller [17].

Fillers added to the formulations can significantly affect the UV-induced degradation of a polymeric material. Inorganic pigments, such as TiO_2 and zinc oxide ZnO, have received great attention in recent years due to their absorption ability [18]. Titania (TiO_2) has attracted a great deal of interest as a photo-catalyst and as a UV screening agent to modify the UV-resistance properties of polymers. Among its polymorphs, anatase TiO_2, usually considered to be more active than rutile TiO_2, has been extensively studied as a photo-catalytic material adapted to photo-catalytic applications, while rutile TiO_2 as a UV absorber due to its high refractive index and hiding power, as well as good chemical stability and UV light screening effects. TiO_2 is considered the most promising photo-catalyst due to its excellent photo-catalytic activity, low cost, nontoxicity, and high stability [19].

Photo-degradation of PLA/TiO_2 nanocomposite films showed that the nanocomposites could be efficiently photo-degraded by UV irradiation compared to pristine PLA. Almost invariable, the obtained results indicated that anatase TiO_2 acts as a photo-catalyst [20–22] while rutile TiO_2 particles have a screening effect [23]. The results reported by Nakayama and Hayashi [20] concluded the rate of PLA degradation can be controlled by the loading of TiO_2 nanofiller, whereas an efficient photo-degradability is finally allowing the eco-friendly disposal of polymer waste. Polymer-TiO_2 composites could be used as photodegradable products depending on the size and uniform dispersion of photo-catalyst in polymer matrices. According to Wang et al. [23], anatase TiO_2 embedded in bulk of the films produces the same effect as rutile TiO_2 as revealed by DSC analysis.

The photocatalytic degradation process of PS-TiO_2 composite sample was much faster than the simple photolysis of a pure PS sample. It was implied that the degradation initially occurred over TiO_2 particles, followed by the diffusion reaction with the aid of reactive oxygen species generated on TiO_2 particle surface [24]. If the particles are well dispersed in the polymer matrix, the efficiency of the photo-degradation of PS-TiO_2 composite can be significantly increased [25, 26].

Studies on photo-catalytic degradation of PVC-TiO_2 composite films evidenced that TiO_2-embedded PVC showed highly enhanced photo-degradation. Since the degradation initiates indirectly through oxidative radicals generated on TiO_2, the presence of chromophoric impurities was not necessary [27]. Outdoor lifetime of the plasticized PVC-TiO_2 system exposed to solar radiation was significantly reduced because the titania protected the PVC from degradation but promoted the photo-degradation of the phthalate plasticizer in the formulation [28]. The influence of TiO_2 on the photo-degradation of PVC was investigated by Gesenhues [29] who found that O_2 was transferred from the TiO_2 surface to the polymer. A novel idea related to preparing the photo-catalytically degradable nanohybrid through TiO_2 nanoparticle-integrated hyperbranched poly (ε-caprolactone) (HPCL–TiO_2) was introduced and the remarkable photo-catalytic degradation was observed in the

PVC/HPCL–TiO_2 nanocomposites. The structural change accompanied by the degradation of the irradiated sample was clearly explained [30].

The TiO_2 particles from PC/TiO_2 composite films play a significant role in the degradation. The weight loss rate of the PC/TiO_2 composite film (33 % weight loss after 300 h UV exposure) with 4 wt% TiO_2 was twice as high as the pure PC films (14 % weight loss under UV light irradiation for 300 h). This indicates that UV irradiated PC/TiO_2 composite film can be applied as an effective photo-catalytic degradation material to control polymer waste effectively [31].

Oxidative degradation can be caused by prodegradants, additives that accelerate the polyolefin deterioration process and promote biodegradation [32]. TiO_2 nanoparticles and their modified forms have been used as prodegradants of PE in outdoor weathering [33]. PE and LDPE with TiO_2 nanoparticles (0.1–1 wt%) composite films prepared by the casting method were exposed to sunlight and UV irradiation to study the solid phase photo-catalytic degradation of the nanocomposites in comparison with neat polymers [33–35]. The results showed that TiO_2 nanoparticles in PE films induced the photocatalytic degradation of the composites compared with the neat polymers and the degradation rate was controlled by the TiO_2 content. The degradation was evidenced by FTIR spectroscopy, which confirmed the appearance of carbonyl groups, by DSC analysis and SEM images, which indicated the formation of cavities.

Generally, nanocomposites of linear PE and LDPE with TiO_2 nanoparticles are prepared by the casting method, which favors inorganic nanoparticles to agglomerate in polyethylene, decreasing the photo-catalytic degradation efficiency of TiO_2 [36]. An improved dispersion and photo-catalytic degradation of the TiO_2 in the LDPE matrix can be achieved by grafting PS (PS-*g*-TiO_2) or PAM (PAM-*g*-TiO_2) and embedding the modified TiO_2 nanoparticles into LDPE by a melting process. The experimental results showed that the efficiency of the photo-catalytic degradation of PS-*g*-TiO_2 and PAM-*g*-TiO_2 films was significantly higher than that of untreated TiO_2 film, and this was related with a good dispersion of the grafted nanoparticles in the LDPE [36, 37].

Photo-degradation experiments of HDPE films embedded with TiO_2 nanospheres, nanoribbons and microspheres were carried out under simulated sunlight and FTIR, TEM, SEM, DSC and GC analysis evidenced that shape, size and concentration of TiO_2 nanoparticles had influence on photo-degradation rate in the crystalline structure of HDPE films. The formation of radicals $HO^{\cdot}$, $O_2^{\cdot-}$, $HOO^{\cdot}$ led to the oxidative degradation of the chains, which occurred preferentially in the amorphous phase of PE, and to increasing the crystalline region of the polymer due to the photocatalytic nature of TiO_2 [38]. The photo-degradation of PE containing TiO_2 takes place when light with energy higher than the energy gap of TiO_2 (band gap ca. 3.2 eV, $\lambda < 387$ nm) promotes an electron from the valence band to the conduction band of the semiconductor, and a simultaneous oxidation of the valence band and reduction of the conduction band of TiO_2 are induced. Water molecules on the TiO_2 surface donate electrons to the oxidized valence band generating hydroxyl radicals $HO^{\cdot}$, and oxygen molecules gain electrons from the reduced conduction band to form active species of oxygen $O_2^{\cdot-}$, which react with hydroperoxyl radicals HOO˙ [39, 40]. These highly reactive species attack nearby polymer chains and initiate the degradation process [20]. The degradation mechanism and degradation rate of neat PE depend on the amount of oxygen that permeates the polymer. Alkyl radicals react with oxygen to form peroxyl radicals, which react with the polymer chains to form hydroperoxides (POOH). The photolysis of the POOH group leads to the generation of CO_2 [41]. In the case of PE/TiO_2 films, CO_2 is generated as a result of photo-catalytic degradation rather than photolysis. GC analysis evidenced that the amount of CO_2 evolved during photo-degradation of all the composites was larger than that of pristine HDPE [38]. The photocatalytic degradation mechanism of the films of the LDPE-TiO_2 was discussed [42]. Study on photo-degradation of PE/transition metal-doped TiO_2 and PVC/transition metal-doped TiO_2 revealed that the nano mixed crystals TiO_2 (rutile/anatase: 20/80) were more photo-active than the most aggressive metal-doped pigments [43, 44]. The effect of nano TiO_2 on the crystalline structure of LDPE has been investigated by Ma et al. [45]. UV light irradiation of PP thin film containing extrafine particles of TiO_2 led to the degradation of the film, as detected by change in color, whitening, and marked decrease in the elongation at break [46]. The particle size and crystal form of TiO_2 played a major role in controlling the photo-activity of TiO_2 in EVA/LDPE composites [47]. The micro rutile TiO_2, anatase TiO_2 and their mixed crystals had a photo-stabilizing effect, and their photo-activity decreased in the following order: micro rutile TiO_2 > micro mixed crystals TiO_2 (rutile/anatase: 13/87) > micro anatase TiO_2. At the same time, the nano mixed crystals TiO_2 (rutile/anatase: 20/80) exhibited a high photocatalytic activity. Studies on the properties of PVA/TiO_2 hybrid nano-fibres were performed by Wu et al. [48]. There are several interesting reports devoted to study of the photo-degradation of nanocomposites containing TiO_2 nanoparticles embedded in PE [49–51], PP [52–54], methyl methacrylate copolymers [55], methyl methacrylate-alkyl acrylate copolymers [56], PS [57–59], PU [60], PVC [61, 62], and poly(ethylene oxide) [63], and a report demonstrated that the polymer-TiO_2 compositing was a practical and promising way to degrade the plastic waste under solar light without any pollution [64].

Similar to TiO_2, ZnO has a photocatalytic activity, as demonstrated by the formation of radicals that are able to degrade the polymer matrix [65]. The addition of ZnO and TiO_2 to PE caused it a faster photo-degradation than the neat sample, but the rate of CO_2 generation was greater with ZnO. It was concluded that ZnO and TiO_2 not only accelerated the rate of photo-degradation but also changed the balance between the two degradation mechanisms. Measurements of molecular mass showed that the reduction in molecular mass of ZnO/LDPE was less than that of TiO_2/LDPE despite the greater CO_2 generated in the ZnO/LDPE. The authors discussed the proposed mechanisms [65].

ZnO polymer nanocomposites are characterized by a large array of properties including intensive ultraviolet absorption. The stabilization effect of ZnO nanoparticles on the photooxidation of LLDPE was reported to be positive and better than that of Al_2O_3 or SiO_2 nanoparticles [66]. The degradation rates of the PLA-ZnO nanocomposites are strongly dependent on the ZnO content, which plays a key role both in the photo-catalytic effect and in the UV-screen effect and, consequently, in the distribution of the photoproducts into the polymer bulk. The prodegradant role arises from the generation of active species of ZnO that can initiate the radical oxidation of PLA [67]. Silica coated ZnO nanoparticles blended into PP matrix introduced an inert silica shell between the ZnO and polymer chains and protected the polymer against photo-degradation. The results indicated that ZnO nanoparticles had a strong UV-light screening effect and reduced polymer photo-degradation and improved the UV stability of the PP nanocomposites [68, 69]. Coated and dispersed nanoparticulate ZnO was shown to improve UV stability of PP and HDPE. The nanoparticles have been found to have superior resistance to UV degradation compared to organic HALS at appropriate loading levels [70]. Incorporation of ZnO into PP matrix significantly reduced photo-degradation and surface cracks [71]. The performance of wood coated with ZnO nanoparticles dispersed in maleic anhydride modified polypropylene (MAPP) can be enhanced by incorporation of nanoparticles in the coatings. Dispersion of ZnO nanoparticles in MAPP restricted the colour changes and photo-degradation of wood polymers. Incorporation of ZnO nanoparticles in PU exterior clear coatings also enhanced its photo-stability [72]. Three different pigments (Fe_2O_3, TiO_2 and ZnO) were incorporated into the shell layer of the PP-based composites. SEM study evidenced that the density of surface cracking was dramatically reduced in the composites containing inorganic pigments in comparison with the reference composite after 500 h of weathering. The composites containing TiO_2 exhibited better colour stability than the composites containing the other pigments. The FTIR analysis revealed that ZnO had a stabilizing effect on polypropylene photo-degradation [73].

Studies on the photo-ageing of CNT-modified polyolefins have mentioned that the durability of CNT-based nanocomposites depends on the nanotubes dimensions and their morphology, and the possible presence of oxidized groups and/or impurities on the polymer matrix surface that may negatively affect its resistance to photo-oxidation [74, 75]. The higher UV resistance of polymer/CNT composites has been attributed to the CNTs ability to disperse and filter radiation energy as well as the strong interaction between free radicals (generated during irradiation) and CNTs [76]. Absorption of UV-light by CNTs can induce two opposite effects. The first is an inner filter effect which probably produces a decrease of the photo-chemically induced oxidation. The second is due to an increase of the local temperature which can produce a thermo-oxidation of the polymer matrix or increase in the rate of the photo-thermal oxidation. Since the CNTs are converting light into heat, the composites have a higher temperature during the degradation experiment and exhibit a higher thermo-oxidation. For the PE composites, the positive effect of the CNTs seems to be balanced by the negative effects of the higher temperature. The antagonism between the two effects produces either a stabilizing or a prodegradant effect [6, 77, 78]. Whatever the effect, it is clear that the presence of CNTs in the polymers retards their oxidation [79].

Thin films of different HDPE nanocomposites containing 2.5 wt% of MWCNTs, pristine and OMMT, and SiO_2 nanoparticles were exposed to accelerating weathering. The FTIR spectroscopy confirmed the MWCNTs stabilization effect on

HDPE but indicated that OMMT and especially SiO_2 acted as catalysts and had an accelerated effect on the photo-oxidation of HDPE during UV irradiation. Micro-Raman analysis confirmed that untreated montmorillonite had a small influence on UV damaging, while neat HDPE and its nanocomposites containing MWCNTs had the highest UV stability and the strongest accelerating effect came from SiO_2 nanoparticles. This contradiction was explained by thermal properties analysis, which showed an increase in crystallinity. The lowest increase in crystallinity was observed in the HDPE/MWCNTs nanocomposite and was attributed to the stabilization effect of MWCNT nanoparticles. Although MWCNTs nanoparticles contained inorganic impurities such as Al or Fe, resulted from the catalysts used for their preparation, they conferred good protection to the HDPE polymer matrix towards UV radiation, neat MMT nanoparticles offered low protection to HDPE, slowing down the degradation process, while nanocomposites containing OMMT and SiO_2 accelerated the oxidation [80]. The stabilization effect of MWCNTs was also proved in other polymers, such as PE [74], EVA [75], PP [81], PE and PP [77], epoxy [82, 83], PU [84], PA6 [85], PP, EVA, and PA6 [86], PLA [87], PS [88]. Epoxy containing MWCNTs exposed to UV radiation degraded at a much slower rate than the unfilled epoxy or the epoxy/nanosilica composite [89].

Results reported on the photo-oxidation of EVA/CNT nanocomposites [74] showed that the global effect of CNTs resulted mainly in stabilization of the nanocomposites when the CNT amount was higher than 1 % w/w. SWCNTs, MWCNTs, and hydroxylated MWCNTs were used as fillers and reinforcements into HDPE to investigate the influence of wall number and surface functionalization of CNTs on their antioxidant behaviour [90]. Based on measurements of the oxidation induction temperature and oxidation induction time of CNT/HDPE composites, the antioxidant ability of the three kinds of CNTs decreased in the following order: MWCNTs-OH > MWCNTs > SWCNTs. A free radical scavenging mechanism was proposed to explain the antioxidant behavior of CNTs. CNTs with a great number of walls and surface hydroxyl groups had more structural defects and exhibited higher antioxidant ability. These results indicated the possibility that CNTs would improve antioxidant properties and mechanical properties of polymer matrix. Grafting HALS stabilizer onto MWCNTs was reported to lead to UV-durable polymer nanocomposites [78].

Carbon black is one of the most effective stabilizers for most polymers because it exerts a light-filtering action and is able to protect the polymer matrix against photo-induced degradation. Its effectiveness depends on its purity, the particle size and the degree of dispersion within the polymer matrix [91]. The most important drawback of carbon black is that this stabilizer at high concentration affects negatively the mechanical properties of the resultant composites [92]. The effects of TiO_2 and CB at variable loadings were also studied during the photo-oxidation process of a blend of styrene-ethylene-butylenes-styrene (SEBS)/EVA and poly (phenylene ether)-PS [92] and poly(propylene-*co*-ethylene) [93]. In TiO_2 and CB containing blends, the level of degradation and decreasing mechanical properties were found to be dependent on the quantity of pigment and UV absorber.

Blending with phthalates and terephthalate was found to decrease the stability of the irradiated poly(fluorostyrene) isomers following a mechanism of photo-degradation similar to that of PS [94]. The same polystyrene isomers were blended with dioctyl phthalate and dioctyl terephthalate plasticizers and the photo-degradation processes in polymeric chains increased [95].

Inorganic nanofillers as POSS with different inorganic framework and pendant organic groups have been used in the PS nanocomposite preparation and the formulated films were subjected to accelerated weathering. Compared to pristine PS, the POSS-containing PS showed a significantly improved resistance to photo-degradation. The investigated POSS compounds played a protective role and could extend the in-use lifetime of the polymeric matrix [96].

The photo-oxidation of HDPE microcomposites filled with $CaCO_3$, considered photo-chemically neutral in term of chemical reactivity, under accelerated ageing showed that the presence of this conventional filler changed the HDPE photo-degradation mechanism and contradictory explanations were given with an acceleration of chain scission reactions and a protective barrier effect for UV light penetration in HDPE [97]. A study on the natural photo-oxidation of HDPE composites showed that some inorganic fillers, e.g., $CaCO_3$ and wollastonite, produced lower degrees of HDPE degradation than other inorganic fillers, e.g., kaolin, diatomite, mica and talc, which accelerated the photo-oxidation degradation. The FTIR, UV, SEM and PGC-MS analyses showed that $CaCO_3$ and wollastonite could reflect nearly all the ultraviolet light and protected HDPE from photo-degradation because the possible oxidation reactions were centered on the surface, while for HDPE filled with diatomite the surface damage was much slighter than the former two [98]. The role of two different OMS and polymer matrix type (LDPE and EVA), in photo-degradation behavior of polyolefin/clay nanocomposite films was investigated [99]. The polyolefin films with high concentration of organo-modifier and hydrophobicity showed accelerated photo-oxidation rate, i.e., a faster loss of mechanical performance and higher carbonyl formation with respect to the films with low concentration of organo-modifier and hydrophobicity. The LDPE based nanocomposite underwent photo-oxidation more rapidly than the EVA based one. The drop of the elongation at break and increase of the CI of the EVA based nanocomposite films occurred at photo-oxidation times much longer than those based on LDPE. This behaviour was based on matrix polarity and its ability to facilitate the solubilizing of alkyl radicals coming from the degradation of the organo-modifier.

Hindered amine light stabilizers have attracted enormous scientific and commercial interest due to their efficiency as photo-antioxidants. The stabilization process in the presence of HALS involves the interference of secondary >NH or tertiary (>NCH_3) groups with hydroperoxides POOH generated by oxidation chain mechanism of polymer alkyl radicals $P^{\cdot}$. HALS-derived nitroxides >$NO^{\cdot}$, formed through scavenging peroxyls $POO^{\cdot}$ by hydroxylamine species >NOH, are the key intermediates in the HALS stabilization mechanism because they are able to quench alkyl radicals arising in the initiation step of oxidation [100, 101]. The involvement of HALS in the stabilization process can be attributed to their ability to remain an active participant in the regeneration cyclic activity mechanism. Photo-stability to simulated solar radiation in an accelerated weathering chamber of a series of HDPE monofilaments of varying thickness photostabilized with different concentrations of UVAbs, HALS, CB and their combinations in various ratios was evaluated by testing their tensile strength and elongation. The UV protective ability at a concentration of 0.2 % of various HALS on the HDPE films was assessed by measuring ultraviolet protection factor (UPF). The results showed a sharp increase of UPF values with film thickness [102]. Influence of various HALS on the photo-oxidation process of HDPE and PP [103], PP, PE, PS and poly(ethylene-*co*-norbornene) [101], PP [104], PP/EPR copolymer (elastomeric propylene-ethylene nodules embedded in isotactic PP) [105], PE-PP copolymers [106], acrylic coatings [107], PS [108], PE [109] was studied.

Besides HALS, another class of photostabilizers used commercially in plastics, elastomer-modified plastics and coatings include preventive stabilizers acting by a preferential absorption of the radiation and transforming the harmful radiation into harmless thermal energy. The role of UVAbs is to inhibit the photo-oxidation of polymeric materials, therefore, they are very important additives used to prevent plastics from photo-degradation by UV. The UVAbs include derivatives of benzophenone, triazole, triazine, *N*-(substituted phenyl) itaconimides, etc. [110–113]. Stabilizers embedded into the polymer matrix are the most photo-reactive components of the system [114–120].

3 Presence of Impurities

Photo-degradation (photolysis and photo-oxidation) of most commercial organic polymers is controlled by their chemical structure and is dependent on radiation wavelength. UV light with shorter wavelengths and higher photon energies is more strongly absorbed in most polymeric materials, and has a greater potential to break chemical bonds in polymer structure. As usually, the initiation of photolysis is attributed to impurities and chromophores formed during polymerization, processing and storage. Moreover, impurities imbedded into materials complicate the scene of the photochemical reactions, all the more so as the kinds of impurities and their influence on whole system can't be certain.

In the case of polyolefins and PVC, which do not absorb the UV radiation present in sunlight, photo-initiation can occur via the activation of chromophores present as impurities in the polymer formulation, which are able to supply radicals in a photochemical way. The photo-oxidation of polymers is a chain oxidation reaction involving hydrogen abstraction from the polymeric backbone and formation of oxidized groups, chain scissions, and cross-links. Most of the aromatic polymers have intrinsic chromophoric groups located on the macromolecular chain, which contain functional groups with π electrons, like carbonyls and carbon-carbon double bonds that can absorb the near-UV photons of the sunlight. The oxidation of the polymer is initiated by radicals that are formed by direct absorption of light by the chromophoric units. On the contrary, in the case of nonabsorbing polymers (polyolefins and nonaromatic polymers), the photo-degradation is initiated by radical species resulting from the absorption of incident light by chromophoric impurities, formed during the processing of the polymers or resulting from the thermal history of the polymeric materials, that then react with the polymer. The direct absorption of light by the intrinsic chromophores or by chromophoric impurities is responsible for photo-physical processes and photochemical reactions, which generally result in a loss of the initial physical and mechanical properties of the polymeric material.

According to Schnabel [121], impurities responsible for causing photo-degradation of polymeric materials can be originated from various sources and can be divided as follows: (1) internal impurities introduced into macromolecule structure during polymerization processing and storage, such as: hydroperoxides, carbonyl groups, C=C double bonds, catalyst residues, CT complexes formed between the polymer and oxygen, traces of solvents, traces of metals and metal oxides from processing equipment and containers, such as Fe, Ni or Cr; (2) external impurities (environmental contaminants) which are compounds from a polluted urban atmosphere and smog; (3) additives (pigments, dyes, thermal stabilizers, photo-stabilizers, fillers, etc.).

PEs are quite stable during natural photo-oxidation, and the main reaction is the backbone scission to form a series of *n*-alkanes, with few oxides in the sample after half a year of exposure. HDPE consists of only C–C and C–H bonds which absorb light in the wavelength region shorter than 190 nm. However, polyolefins can be used in their applications only after the addition of certain amounts of UV stabilizers. The UV decomposition of HDPE, without stabilizers, is so fast that the life time of an article could be shorter than one year. That is because polyolefins contain different impurities acquired during their synthesis, like aldehyde, peroxide, hydroxyl, ester, ketone, hydroxyperoxide, carbonyl, acid and unsaturated groups, etc., which absorb the light at higher wavelength and attack the polymer main chain to form free radicals. These impurities act as initiators for photo-degradation of HDPE and this process can be accelerated by metal ions which can be found in catalyst residues (Zr, Cr, Ti, etc.) or in various additives like nanoparticles [122].

PEs are usually produced in the presence of Ziegler-Natta, chromium and metallocene and post-metallocene catalysts impregnated on a suitable support, mostly silica because of its desired particulate morphology and low cost. The catalyst sites, monomer feed composition, polymerization process type, and reactor configuration and operating conditions have a significant influence on the molecular structure of the resulted polymers. The average molecular weight and molecular weight distribution, copolymer composition, methylene sequence length, side-chain branching, and chain unsaturation have a direct influence on thermal, rheological, mechanical, and chemical properties, the processing characteristics and the applications of the end-products [123].

Polymerization reaction is initiated by the active sites located inside the catalyst pores where ethylene and the alkyl-aluminum cocatalyst penetrate. As the polyolefin chains grow, stress develops inside the catalyst pores and small particles are produced by its fracture. These particles reduce their dimensions to a critical size around which polymer forms, and replicate the shape of the solid catalyst. This residual catalyst having its characteristic surface chemistry and solid-state electronic environment is an integral part of the resulting polymer. The support type, catalyst activity, and polymerization process and conditions influence the concentration and distribution level of the residual catalyst. Consequently, the polymer and end-products eventually contain residual catalysts. The effects of two catalysts used for the synthesis of HDPE on the UV-induced oxidative degradation of the resulting polymer films were studied by exposing them in an accelerated weathering tester equipped with fluorescent UV lamps (315–400 nm) at 60 °C for 0–360 h. The metallocene catalyst was found to be more susceptible to UV-induced degradation than the Ziegler-Natta catalyst, despite the residual Zr level was 1/15th of the Ti content. Therefore, from a mechanistic view point, the zirconocene residual catalyst proved more efficient in (1) photo-initiation reaction (by transferring the energy of the excited Zr to a polymer chain), and (2) decreasing the activation energy required for the decomposition of the resulting hydroperoxide. The explanation resides in the differences between (1) electronic configuration and atomic radius, and (2) surface chemistry and solid-state electronic environment of both catalysts. Both polymer films produced the same products, such as carbonyl, ester, carboxyl, lactone, vinyl, and hydroperoxide, as a result of photo-oxidation. In both polymer films, peak melting point and peak crystallization point did not significantly vary with the exposure time. The change of % crystallinity was found to be irregular and disturbed and was attributed (in the literature) to the simultaneous occurrence of cross-linking and chain scission [124].

The same research group prepared an ethylene homopolymer and an ethylene/1-hexene copolymer using a vanadium post-metallocene catalyst in order to investigate the effect of backbone inhomogeneity on the UV oxidative degradation of both PEs films. Evaluation of chemical/structural changes due to oxidative degradation of the polymer films was monitored by FTIR spectroscopy where the intensities of the peaks for various carbonyls (ketone, ester, lactone, and carboxyl), hydroperoxide, and vinyl unsaturation were measured and calculated by dividing their corresponding absorbance peaks with that of the reference peak $\nu = 2019\ cm^{-1}$, which corresponded to alkane CH stretching vibrations of the $-CH_2-$ groups. The homopolymer had linear structure while the copolymer was inhomogeneous. The inhomogeneity was evidenced in the distributions of short chain branch, 1-hexene composition, and methylene sequence length. Both PEs showed very low unsaturation in the backbone. The major functional group present in the branched copolymer film was carbonyl followed by carboxyl, then vinyl/ester, while in the linear homopolymer film the carboxyl group was dominant. The distributions of the tertiary ≡C–H sites and methylene sequence length in the branched copolymer film favored abstraction of H, generation and decomposition of hydroperoxide group, and formation of carbonyl compounds in comparison with the linear homopolymer film, which established the role that the backbone inhomogeneity played in the photo-degradation process. The accumulation of the above oxygenated functionalities and its effect on % crystallinity was explained in terms of polyethylene UV autooxidation mechanism, and Norrish I and Norrish II chain scissions [125].

The photo-oxidative behaviour of three PEs (linear low, metallocene and high density) has been investigated by FTIR spectroscopy and the data related to the CL and thermal analysis of the polymers. Polymer films were irradiated in the wavelength region 300–800 nm. CI analysis showed that metallocene polymer was the most unstable of the three PEs and hydroperoxide analysis evidenced a higher light stability for LLDPE compared to HDPE and m-PE. CL analysis of the irradiated polymer films showed a higher light stability of LLDPE, which did not exhibit any CL emission at lower temperatures. CL emission was also observed for m-PE and HDPE below the melting point, with a shorter inhibition period for m-PE, which was in agreement with the measurement of carbonyl group values obtained by FTIR. The activation energies calculated at different irradiation times using TGA were found to decrease following their instability, i.e. m-PE > HDPE > LLDPE which confirm the data obtained by CL analysis, where higher stability for LLDPE under UV irradiation was evidenced [126]. Studies by FTIR, DSC, and DMA analysis and hydroperoxide concentration on various PEs established the following order light stability LLDPE > HDPE > m-PE [127]. Unlike PE, the outdoor aging of some ethylene-octene copolymers having different degrees of crystallinity and thermal characteristics gave rise to a lot of peroxide derivatives after about 3000 h of exposure [128].

The most studied polymer-clay nanocomposites have concerned the nanocomposites with OMMT, as a layered silicate filler, and polyolefin matrices. Regarding their photo-degradation, most of scientific papers have concluded that nanocomposites degrade faster than the neat polymers and the presence of OMMT decreased the efficiency of conventional stabilizers such as phenolic and redox antioxidants. One of the causes responsible for the higher level of photo-oxidation of nanocomposites was attributed to the catalytic effect of the metal ion impurities, such as iron, which was present in naturally occurring clays [129–134]. In PP/MMT nanocomposites, iron could catalyze the decomposition of the hydroperoxides formed by photo-oxidation of PP, which would accelerate the additive consumption and decrease the period until the permanent regime of oxidation is reached, leading to an unexpected decrease of the durability of the material. Some authors have ruled out that the degradation may be mainly attributed to photo-degradation of alkylammonium surfactant (the organic part of the organo-MMT) [129, 135] and others [132, 136, 137] suggest that the decomposition of the ammonium-based organo-modifier can lead to the formation of unsaturated groups and catalytic acidic sites on the silicate layers.

Similar findings were reported for EPDM/MMT clay nanocomposites [116, 138] and for other clay-like minerals such as talc (hydrated magnesium silicate, the impurity Fe_2O_3), which also catalyzed the oxidative degradation in PP [73, 139]. Study on the degradation of PLA/Cloisite 30B nanocomposites under natural weathering, investigated as a function of clay loadings (1, 3 and 5 wt%), revealed that introducing the organophilic clay into PLA matrix promoted the degradation rates of the nanocomposite materials; this effect was much more pronounced for the samples containing 3 and 5 wt% clay [140].

On the contrary, the organoclay improved the photo-oxidation resistance of the OMMT-filled PE/PA blends [141] and PVC-laponite nanocomposites [142].

The prodegradant effect of clays was surmounted by adding anti-UV and a metal deactivator [143, 144]. The chemical modifications of LLDPE/nanoblend nanocomposites exposed to accelerated ageing and natural weathering were studied by IR spectroscopy, which showed that the presence of an organo-clay led to the decrease of the oxidation induction time of the polymer matrix and, as a consequence, to lower durability of the nanocomposites. Protection against photo-oxidation was tested with different UV stabilizers and with a metal deactivator (Irganox MD-1024). The metal deactivator proved to be very efficient in stabilizing the nanocomposites because the prodegradant effect of the organoclay was canceled. The catalytic effect of iron impurities of nanoclays on hydroperoxide decomposition was supposed to be the main effect on the polymer oxidation in the LLDPE/nanoblend nanocomposites, both in accelerated and natural conditions In addition, the metal deactivator confirmed the role played by iron impurities in natural clays and offered a new insight in the stabilization strategy of polymer/clay nanocomposites.

Photo-oxidation of some acrylic-urethane thermoset networks was induced by chromophoric impurities that absorb UV light and produce radicals, initiating a radical oxidation of the polymer [145]. The authors introduced a quantitative kinetic model based on the identified mechanisms and a multi-scale approach from the molecular to the macroscopic level.

Mechanisms of photo-degradation initiated by chromophoric impurities that absorb UV light and produced radicals, initiating a radical oxidation of the polymer, were proposed and introduced in order to explain the photo-oxidation of some acrylic-urethane thermoset networks [145], and poly(lactic acid) nanocomposites [22, 146–149].

Photo-degradation of WPC largely depends on the ability of chromophores present in lignin to absorb UV light [150, 151] because lignin is prone to photo-degradation and can be degraded into water soluble products, which eventually lead to the formation of chromophoric functional groups such as carboxylic acids, quinones, and hydroperoxy radicals. This process is claimed to be responsible for discoloration of WPC [152]. Some reports claim that wood fibers or flours are effective photo-sensitizers (i.e., chromophore materials), and their incorporation into the polymer matrix promotes and/or accelerates photo-degradation of the polymer [153, 154]. The CI was found to be higher for composites of WF and HDPE than that of pristine HDPE, i.e., WF had more chromophores than HDPE because carbonyl groups were present in lignin, a component of WF. As a consequence, the addition of WF to an HDPE matrix was detrimental to weathering because increases in carbonyl groups in WF/HDPE composites in comparison with neat HDPE made it more susceptible to attack by UV light [155].

TiO_2 nanoparticles and their modified forms have been used as prodegradants of polyethylene in outdoor weathering [33, 35]. Linear (PE) and low density polyethylene (LDPE) with TiO_2 nanoparticles (0.1–1 wt%) composite films prepared by the casting method were exposed to sunlight and UV irradiation to study the solid phase photo-catalytic degradation of the nanocomposites in comparison with neat polymers. The results showed that TiO_2 nanoparticles in polyethylene films induced the photocatalytic degradation of the composites compared with the neat polymers and the degradation rate was controlled by the TiO_2 content. The degradation was evidenced by FTIR spectroscopy, which confirmed the appearance of carbonyl groups, by DSC analysis and SEM images, which indicated the formation of cavities. Photo-oxidation of PLA-based nanocomposites prepared by adding MMT, OMMT and SiO_2, monitored by FTIR analysis, confirmed an increase in anhydride formation rate based on hydroperoxide decomposition. The main responsible for this acceleration was identified as transition metals contained in the nanofillers as impurities and involved in the catalytic hydroperoxide decomposition [147].

4 Influence of Wavelength and Irradiation Dose

The photon energy in solar radiation is the most damaging component of the outdoor environment and can initiate various chemical changes in polymeric materials. Sun emits radiation over a wide range of wavelengths ranging from below 100 nm to over 3000 nm, but radiation of wavelength less than 290 nm is filtered by the stratosphere and does not reach the Earth's surface. The UV-B (290–315 nm) component of sunlight is the main agent responsible for weathering of polymeric materials but some degradation is attributed to the solar UV-A (315–400 nm) radiation component. The energy of ultraviolet photons is comparable with the dissociation energy of covalent bonds, which lies in the range of 290–412 kJ/mol. This high energy is capable of initiating bond breaking within the polymer backbone (photolysis) or chain reactions through the radical species in the presence of oxygen (photo-oxidation).

In order to be effective and to induce a chemical change, light must be absorbed by the polymer substrate. A material totally transparent in the UV range will not exhibit photo-initiation reaction. Saturated hydrocarbon materials, such as the polyolefins, do not absorb light above 250 nm and photolysis cannot play an important role, but these materials undergo degradation through photo-oxidation mechanism. The light-absorbing entities in a polymer can be either the monomeric units comprising the main chain which are themselves able to absorb photo-chemically significant wavelengths to initiate photo-deterioration (for example PET and PS) or chromophoric (light-absorbing, color-producing) groups located at the chain ends or attached along the chain, often present naturally or introduced accidentally into polymers (during storage and processing). In the latter situation, it is difficult to identify the photosensitive group or impurity responsible for initiating photochemical process. The photo-degradation of the polymer involves exposure under energy with wavelengths necessary to photo-activate one or more components of the material which are capable of directly degrading that component and/or initiating a reaction which may or may not damage the absorbing component but does damage other components. The first process is called *photolysis* and involves the *homolytic* breaking (scission) of a bond and takes place in the absence of oxygen. The second is a photochemical process characterized by four stages: (1) *initiation*, when the UV energy generates a free alkyl radical R˙. The excited singlet or triplet states generated may release a certain fraction of absorbed energy and become precursors of all photochemical reactions that produce deterioration: chain scission, cross-linking and oxidation. Due to photolytic reactions of these UV-absorbing species (carbonyl group, atmospheric impurities, aromatic hydrocarbons, singlet oxygen), radicals are formed and they initiate the photo-oxidation reaction; (2) *propagation*, when R˙ reacts very quickly with oxygen to generate a peroxyl radical ROO˙. This abstracts an H atom from the polymer to form an alkyl radical and a hydroperoxide ROOH, which decomposes homolytically into alkoxy RO˙ and hydroxyl HO˙ radicals, which can initiate another propagation cycle. The fundamental process in the mechanism of photo-oxidative degradation is the formation of hydroperoxides during propagation. These radicals are highly unstable and provide the necessary mechanistic route for chain scission and cross-linking reactions; (3) *chain branching*, where very reactive radicals RO˙ and HO˙ further abstract hydrogen atoms from the polymer to yield new alkyl radicals R˙. Carbonylic products (such as ketonics) lead to in-chain scission reactions reducing the polymer molecular weight and sometimes induce cross-linking reactions with property losses. Many polymers undergo both chain scission and cross-linking and this mitigates the changes in molecular weight because the two types of molecular modifications provide increments of opposite sense and partly cancel out one another; (4) *termination*, when reactions between radicals occur and inert products are formed [156, 157].

Unlike polyolefins, the most engineering plastics (aromatic PEs, PAs, PUs, PCs, polyketones, etc.) exhibit UV absorptions at wavelengths present in terrestrial sunlight, so that for these polymers photolysis plays an important role and leads to significant changes in the molecular structure. Concomitant, photo-oxidation initiated by photolysis reaction of the polymer itself or by external impurities can take place. The mechanism of photolysis is characterized by the Norrish I reaction, that evidences the chain cleavage, radical formation and photo-oxidation, Norrish II reaction, that is a non-radical intramolecular process ending in chain scission, and photo-Fries reaction specific to aromatic polyesters (PET) when chain rupture is accompanied by production of more photo-reactive ketones. Unlike thermo-oxidative degradation which occurs in the whole bulk of polymer, photo-degradation takes place mostly on its surface and subsurface layer, because of the limited possibility of UV radiation to penetrate the deeper layers of material [158].

The most of the polymers used as matrices in polymer composites (mainly PE, PP, PVC, PC) are inherently photolabile materials, slowly losing their desirable aesthetic, physical, chemical, optical, electrical and mechanical properties reflected in crazing, cracking, erosion, discoloration and phase separation on routine exposure to solar UV radiation.

An appropriate combination of testing conditions and methods for assessing weather-induced changes provides a better understanding of mechanism of degradation and correlation between chemical, physical and mechanical effects induced. The necessity to evaluate the activity of the above parameters more rapidly is the reason behind the choice of artificial weathering. The main goal of accelerated ageing is to improve lifetime prediction under service conditions. Because the most appropriate conditions are those that match the service environment exactly, they are most easily obtained by naturally exposing specimens at outdoor exposure sites. The exposure time required will be at least as long as the life-expectance of the product, and that is a serious drawback. Natural weathering is therefore normally accelerated by exposure to climatic conditions that are more severe than those that are expected in service [159].

The main goal of laboratory-accelerated weathering experiments is often to predict outdoor weathering performance. Many variables have to be well understood and properly applied in order for the simulated tests to be successful. The most critical variable is the light and the effect of the spectral distribution of light both outdoors and in artificial weathering, although this remains an imperfectly solved problem.

Artificial weathering involves exposing test specimens to an artificial light source in a laboratory in which the temperature, humidity and water spray are controlled. During artificial weathering is easier to manipulate the various weathering agents to an extent where they could be comparable to the natural effects and to accelerate the testing because all of the exposure conditions are under control. The other benefits of this approach are the reproducibility and repeatability of the experiments. The negative aspect is to evaluate how these different parameters interact in the weathering process. Therefore, detailed knowledge of ageing mechanisms is important to predict the lifetime of the material and performance at the standard outdoor site during product development, to short the design-development-production cycle and also to provide guidance on service conditions to be avoided and on improvements that should be considered when introducing replacement materials [160]. Correlations between outdoor and accelerated weathering testing are an interesting but contradictory subject. There are authors who state that it is possible to correlate the results obtained in different weathering conditions. Some others have an opposite opinion, or indeed impossible at all, for some polymers.

The deteriorating effect of UV radiation depends on its wavelength. It is generally accepted that the degradation effect increases exponentially with lower wavelengths, hence higher energies, but it depends on the material which energies are the most harmful. The wavelength dependence of the damage susceptibility may be described by a spectral response specific to the material and the type of damage. Two basic forms for the spectral response exist: an activation spectrum and an action spectrum, the latter also known as a spectral sensitivity or wavelength sensitivity. The distinction between the two forms consists in the principle followed in derivation of the response. The activation spectrum has been settled for a specific light source, and is therefore valid under exposure of that particular source. The action spectrum accounts for the radiant exposure used in the experiment, and it is source independent and useful in any exposure conditions [161, 162].

The relative efficiency of different wavelengths of light in causing a specific type of degradation in material is quantified in an action spectrum of the material. The action spectra are plots of sensitivity (or moles of photons) *versus* the wavelength of radiation, and in the case of polymers, generally show an exponential increase in efficiency of degradation with decreasing wavelength. These differ from a plot of wavelength-dependent quantum efficiency in that it is the incident rather than the absorbed radiation that is used in generating action spectra. The extent of degradation in materials depends on the dose of solar radiation absorbed and therefore on that incident on the material. The action spectra are different for different modes of damage and the presence of additives such as UV-stabilizers can dramatically change them. The published action spectra strictly reflect the wavelength sensitivity of specific formulations of the polymers for a given type of damage. They serve only as a guideline for the behaviour of the class of polymers in question exposed to solar radiation [2].

Current devices employed in studies on wavelength dependent sensitivities of materials fall into two main categories [163]. The first category includes filters with known radiation transmission characteristics. Filters are used to spectrally limit the radiation exposing the material samples. Two different operating principles apply to these filters. Cut-on (cut-off) filters pass through only the wavelengths longer than the cut-on (cut-off) wavelength, whereas bandpass filters transmit only certain wavelengths on a narrow band. The second category of methods used for spectral sensitivity studies are based on spectrographs, where the radiation is diffracted to a spectrum by gratings or prisms [164–166].

Activation and action spectra give valuable insight into the degradation behaviour of a material. However, the applicability of the source-dependent responses is more limited than those independent from the source, i.e., in the prediction of the ageing behaviour of a material in its service environment or in a weathering test. Exposure conditions similar to those given by a laboratory source do not prevail in natural outdoor weathering. Moreover, no two perfectly identical artificial sources exist. Hence, in order to make predictions on the ageing of the material, response that is not tied with a particular source of radiation is required. It has been recognized that this type of action spectra do not deal with the long-term development of the property change caused by radiation. They cannot be therefore used for predicting the long-term ageing behaviour of the materials.

For the method used in deriving an action spectrum of a damage that may be considered an indication of ageing, two major requirements hereby unfold. The follow-up of the temporal development of the damage should cover the region beyond the initiation phase. In addition, the method should be able to account for the radiant exposure, i.e., both the irradiance of the exposing radiation and the duration of exposure [167].

Investigations of the effects of different radiation wavelengths and radiation intensity on the degradation mechanism and quantum efficiency of various polymeric materials have been performed over the last decades. Normally, the relation between degradation rate and intensity obeys Schwarzschild's law, which is given by Eq. (1):

$$k = AI^p \tag{1}$$

where:

k the reaction rate;
A a proportionality constant;
I intensity;
p the Schwarzschild coefficient.

The p-coefficient depends on the material and in some cases even radiation wavelength. When p = 1, the Schwarzschild's law becomes the reciprocity law. Based on the reciprocity law, the photoresponse of a material is dependent only on the total energy to which the specimen is exposed, and is independent of the exposure time and the intensity of the radiation taken separately.

A critical consideration is whether or not a material obeys reciprocity, that is, whether a doubling of the light intensity results in a doubling of the degradation rate. Little information is available on the photo-degradation behavior of polymeric materials with respect to the effects of light intensity or dose rate on polymer weathering. Martin et al. [168] have published a comprehensive review of the literature and found very few examples of actual polymer degradation studies in which effects of light intensity were studied. Some materials obeyed reciprocity and others did not. For most (unstabilized, stabilized or pigmented) polymeric systems studied, the p-coefficient ranges between 0.5 and 1.0.

The reciprocity law is the relationship between irradiance I (photons/cm^2), exposure duration t (h) and the resulting photo-damage obtained in an exposure experiment. In instances where the law holds, the quantity (It) is directly proportional to extent of damage, implying that a short exposure at high intensity and a longer exposure at a correspondingly low intensity would yield the same amount of material damage. Even where the reciprocity law generally applies, deviations are expected at both very high and very low irradiance. When deviation is observed at moderate irradiance, a modified form of the reciprocity rule, where the extent of photo-damage is proportional to (I^pt) applies. A compilation of available data on materials photo-degradation reported the value of p ~0.9–1.0 for a majority of polymers for which data were available, confirming that reciprocity is a reasonable assumption for most materials [168].

A p-coefficient of 1 was reported for a non-pigmented, unstabilized acrylic–melamine coating exposed to six different UV radiation intensities ranging from 36 to 322 W/m^2, and in the spectral region between 290 nm and 400 nm. The reciprocity law was verified for different photo-degradation parameters, such as chain scission, oxidation, and mass loss [169].

A *p*-coefficient of 0.5 was found for the rate of carbon dioxide generation in unpigmented, anatase-pigmented and rutile-pigmented acrylic film. As the rate of carbon dioxide generation was a measure of paint film degradation, the rate of film degradation was proportional to the square root of UV intensity. This dependence is a result of the concentration of radicals formed that lead to carbon dioxide, which would be the degradation rate determining step [170].

An activation spectrum is the most common representation of the effect of UV wavelength on polymer degradation. It is a qualitative or semi-quantitative measure of the response of a material to radiation from a particular light source, usually xenon arc or sunlight, and can be determined by exposing samples behind a series of sharp cut-off filters [171]. The different degradation rate (or amount) of samples located behind certain filters is due to the difference in the light passed by those filters. Because the activation spectrum is dependent on the light source it can be very different when determined under different conditions. In this way, the effect of various virtual wavelength bands can be calculated. Various methods for determining spectral sensitivities of polymers as well as their limitations have been published [163].

Generally, quantifying the deterioration induced by the exposure of polymeric materials to solar radiation requires knowledge of: (a) the wavelength dependence of sensitivity of the material to the particular type of damage of interest; (b) a dose-response relationship that is applicable to the exposure and deviations from the reciprocity law applicable to that photo-degradation processes; and (c) the validity of assumptions of additivity of damage at different wavelengths for the system of interest. Data for wavelength dependence of sensitivity in the form of action spectra [172] or as activation spectra [171] for several polymeric materials have been published. However, only very limited information is available on dose-response relationships of relevant materials for modes of photo-damage of interest.

The effect of irradiation conditions on the degradation of 48 samples across a range of aromatic engineering thermoplastics was studied based on their activation spectra. Using a Xe arc weathering device filtered with CIRA/soda lime filter combinations to irradiate various pigmented polycarbonates, different irradiation intensities were obtained by using different quartz neutral density filter combinations [160].

Information on the degradation mechanism and quantum efficiency of some acrylic polyol-based PU coatings is based on Fourier infrared spectroscopy analysis in the transmission mode, which measures chemical changes on the entire film thickness. The quantum efficiency at different wavelengths is estimated by the Eq. (2):

$$\Phi(\lambda) = D_{dam}(\lambda, t)/D_{dos}(\lambda, t) \tag{2}$$

where:

$\Phi(\lambda)$ apparent quantum efficiency within the exposed radiation wavelengths, in A/mJ (A is the infrared absorbance);
D_{dam} damage, in infrared absorbance units;
D_{dos} the dose, i.e., the amount of energy absorbed by the chromophoric species during a duration t, in J/m^2.

The total absorbed dosage is the total number of quanta absorbed by a material and is given by the Eq. (3):

$$D_{dos}(t) = \int_{0}^{t} \int_{\lambda min}^{\lambda max} E_0(\lambda)(1 - 10^{-A(\lambda,t)})d\lambda dt \qquad (3)$$

where:

λ_{min} and λ_{max} minimum and maximum photolytically effective UV-visible wavelengths, in nm;
$A(\lambda, t)$ absorbance of the sample at specified UV wavelength and at time t (dimensionless);
$E_0(\lambda, t)$ incident spectral UV-visible radiation dose on sample surface at time t, in J/cm^2;
t exposure time, in s.

The total absorbed dosage is obtained by integrating the product of the spectral irradiance, $E_0(\lambda, t)$ and the spectral absorption of the polymer, $(1 - 10^{-A(\lambda,t)})$, over the wavelengths of irradiation impinging on the sample for the exposure time at a particular humidity and temperature. Both quantities were measured by UV-visible spectroscopy, while the damage was measured by FTIR. The quantum efficiency is considered as an 'apparent" value and is expressed as the change in FTIR intensity per unit thickness per amount of radiation absorbed by the polymer. FTIR intensity of the degradation is expressed in absorbance units (A), thickness in meters (m), absorbed radiation dosage in Joules (J), and apparent quantum efficiency in A/mJ. Since both the absorbed dosage and material damage were measured on exposed samples, the quantum efficiency accounts for the effects of relative humidity, temperature and radiant flux on the degradation [173, 174].

The Φ_R, a fundamental parameter when it comes to quantifying photo-stability and understanding the reaction mechanisms, is less frequently reported probably due to the complexity of the experimental design. In polymer systems, the Φ_R may be defined as the ratio between the number of molecules undergoing chain scission, cross-linking, or any other relevant photo-degradation process *per* photon absorbed. The Φ_R represents a universal and straightforward parameter to assess and compare polymer photo-stability.

The Φ_R for chain scission was determined to be 7.40×10^{-8} at 313 nm for pure PVA irradiated at $\lambda \geq 300$ nm, which indicated a different mechanism from that studied with irradiation involving 254 nm. No molecular evidence was obtained concerning the formation of other carbonyl functions, the disappearance of the PVA carbonyl, or the formation of hydroperoxides; the only observation is the slight decrease in intensity of the entire spectrum [175].

The wavelength dependence of the photo-degradation of poly(2,6-butylene-naphthalate-*co*-tetramethyleneglycol) (PBN-PTMG), was investigated using monochromatic radiation, and the results were as follows: (1) main chain scission occurred when PBN-PTMG was exposed to $\lambda \leq 380$ nm, the formate, propyl, aldehyde and the aliphatic ester groups were all formed and gel fraction was greatly enhanced; (2) when the PBN-PTMG was irradiated at $\lambda \geq 390$ nm, gel formation was low and carboxyl groups appeared; (3) the threshold wavelength for the photo-degradation of PBN-PTMG was about 380 nm [176].

The comparison of degradation behavior of PP and PP/talc composites was carried out with one outdoor weathering test and three accelerated weathering (xenon, metal halide and carbon arc lamps) tests, respectively. The outdoor exposure provoked drastically degradation with the lowest amount of UV exposure energy. The degradation rate was affected by the visible light intensity in the light sources. In the case of PP/talc composites, the degradation was synergistically accelerated by the exposures to the sunshine, the xenon and the metal halide lamps having higher visible light intensity [139].

5 Influence of Temperature

The deleterious effects of weathering on polymers have been generally ascribed to a complex set of processes induced by the combined action of sunlight (particularly UV radiation), humidity and temperature. Degradation produced by one agent can reduce resistance to other agents. The activity of these factors depends on geographical location, day/night temperature, varying seasons, pollution, etc. and need to be taken into account when designing these materials. Weathering is of considerable complexity due to natural fluctuations in UV radiation, temperature, humidity and other environmental factors (especially aggressive and corrosive pollutants), and to the interaction of these factors, and is therefore impossible to simulate.

The time reduction of lab tests is usually possible due to the intensification of UV radiation, usually emitted by xenon lamps equipped with suitable set of filters, imitating sun light spectrum. Lab tests are performed in various climate chambers, that allow programming and constant control of radiation intensity, temperature and humidity inside test chamber, temperature of sample surface, time of water sprinkling (imitating rain) as well as airflow (imitating wind). Generally, simulation is achieved with devices equipped with filtered xenon arcs or fluorescent tubes of the type UVA-340 and having an automatic control of humidity and temperature. Sources having high concentration of UV-B radiation provide fast but unrealistic results due to the promotion of unnatural photo-triggered processes [177, 178].

Among weathering variables, temperature seems to be a less complex test parameter than light, but its evaluation is quite difficult, because temperature and light do not act as independent variables. We can record the total light or UV dose in the accelerated tests and compare them with annual dosages at outdoor sites, but temperature and light intensity are constantly changing outdoors. An "effective temperature" (T_{eff}) for exposure is defined as an average temperature measured for light dose over a certain period. The outdoor temperatures usually are reported only as the daily highs and lows, so information about the temperatures at certain times with particular light intensities is lost. It has been suggested that average daily high temperatures would be good substitutes for the effective temperatures. Seasonal average daily high air temperatures and average seasonal UV doses to calculate correlation factors for automotive coatings have been reported [179]. Offsets from this temperature to estimate the temperatures of coated have been suggested [180]. The daily high temperatures would seem to be a good way to correlate temperature with light intensity; the temperatures are higher on sunny days and cooler on cloudy days. The average daily high temperature is important if temperature affects primarily the photochemical reactions. The daily average temperature is more important if temperature affects both photochemical and dark reactions.

The damage caused by the combination of light and heat over a period of time can be described by Eq. (4) [181, 182]:

$$D = \sum_t I_t A \exp\left(-\frac{E_a}{RT_t}\right) \Delta t \tag{4}$$

where:

- D the total damage;
- I the light intensity at time t;
- A an Arrhenius pre-exponential factor, which combines a proportionality constant from pure irradiation;
- E_a the activation energy;
- R the gas constant;
- T_t the sample temperature at time t;
- Δt the time interval for the observations.

Activation energies for weathering under accelerated conditions are known for a few materials. For aromatic engineering thermoplastics, E_a of 10–30 kJ/mol (3–7 kcal/mol) for yellowing and 0–20 kJ/mol (0–5 kcal/mol) for gloss loss across several resin types and colors [183], 16–20 kJ/mol (4–5/kcal mol) for polycarbonates [180], 27 kJ/mol (6.5 kcal/mol) for automotive coatings [179], and approx. 60 kJ/mol for PE [184] have been reported. Fischer and Ketola [185] reported that average thermal response during weathering for 50 reference materials was a rate increase of 1.41 × per 10 °C, which corresponded to an activation energy of approx. 30 kJ/mol.

Although the temperature varies outdoors, there should be a constant effective temperature (T_{eff}), which would result in the same amount of damage after exposure to the same amount of sunlight (Eq. 5). The T_{eff} represents a constant temperature that creates the same amount of photo-degradation as the naturally varying temperature and provides a benchmark temperature for making lifetime predictions from accelerated laboratory exposures. Sample temperatures in xenon arc exposures usually are higher than the outdoor T_{eff}, so Arrhenius temperature corrections need to be carried out to relate accelerated to outdoor exposures. Temperatures in xenon arc exposure tests often correspond more closely to maximum temperatures that samples might encounter for only a few hours per year.

$$D = \sum_t I_t A \exp\left(-\frac{E_a}{RT_{eff}}\right)\Delta t \tag{5}$$

Setting Eqs. (4) and (5) equal to each other and rearranging gives Eq. (6), where the left term is the annual radiant exposure. This is a way to estimate T_{eff}. It is necessary a data set extending for at least one year with light intensity (irradiance) and temperature data taken at small time intervals, Δt. T_{eff} defined in this way depends on the temperature sensitivity of the material, i.e., the activation energy. However, the results were relatively insensitive to values between 10 and 40 kJ/mol (3 and 10 kcal/mol), which cover many materials. The annual effective ambient air temperature was approx. 30 °C and the effective black panel temperature approx. 42 °C outdoor test site for an activation energy of 21 kJ/mol (5 kcal/mol). T_{eff} during outdoor exposures were lower than the temperature samples usually exposed to xenon arc weathering, making Arrhenius rate correction an essential part of lifetime prediction [186].

$$\sum_t I_t \Delta t = \sum_t I_t \exp\left[\frac{E_a}{R}\left(\frac{1}{T_{eff}} - \frac{1}{T_t}\right)\right]\Delta t \tag{6}$$

Reaction rate is expected to increase with increasing temperature according to the Eyring equation or the simpler Arrhenius equation (Eq. 7):

$$K = A \exp\left(-\frac{E_a}{RT}\right) \tag{7}$$

where:

K the reaction rate;
A a pre-exponential factor;
E_a the activation energy (cal/mol or J/mol);
R the gas constant;
T the absolute temperature (K).

Equation (8) shows the relative reaction rate resulting from increased temperature. The relationship between the rate of reaction and the absolute temperature is exponential; a small change in temperature can therefore result in a large increase in the rate of degradation reactions, for reactions of low activation energy.

$$\frac{K_1}{K_2} = \exp\left[\left(\frac{E_a}{1.987}\right)\left(\frac{1}{T_1} - \frac{1}{T_{2)}}\right)\right] \tag{8}$$

Unlike purely photochemical reactions, which should have very low activation energies, photo-degradation processes are rarely simple, one-step reactions. Measuring E_a is a complex process of many steps, some of them not involving light. As a result, the net E_a is strictly phenomenological and has little mechanistic significance. Degradation due to multiple reactions can result in non-linear Arrhenius plots [187], but this is not a problem over the relatively narrow temperature range of outdoor or artificial weathering experiments.

E_a is determined by plotting rate data according to Eq. (9). The natural logarithm of the relative rates is plotted against the inverse of the absolute temperature. The slope is multiplied by the negative of the gas constant, ($-R = -1.987$), to give E_a.

$$lnK = -\left(\frac{E_a}{1.987}\right)\frac{1}{T} + A \tag{9}$$

E_a of yellowing and gloss loss for a large number of TPE and blends under accelerated weathering conditions were calculated [188]. Often, the E_a depend on the property measured and exposure conditions, although they were ≤ 5 kcal/mol for gloss loss and yellowing for the most polymeric materials tested under the CIRA/sodalime-filtered xenon arc conditions. A reaction with an E_a of 5 kcal/mol will increase its rate by about 33 % for each 10 °C increase in temperature near room temperature. For PCs, temperature is not an overwhelming factor that controls the rate of yellowing [189].

The measurement of ambient temperature can be only used as a climatic descriptor because ambient temperature is not coincidental with material surface temperature. Weathering is mainly a surface effect and material surface temperature has a greater influence on the weathering properties of the material than the ambient temperature [190]. The temperature that a material reaches is dependent on several factors, such as color, absorptivity in the infrared, light intensity, orientation, wind speed, thermal conductivity, and backing. The temperature of the irradiated polymer depends on the spectral distribution of solar radiation and on the absorbance of the polymer, as well as on the long wave infrared radiation of the environment. Infrared radiation is a source of differences measured between environmental temperature and surface temperature of the irradiated polymer. These influencing factors have to be measured when simulating environmental factors in laboratory devices. The radiation source has a severe heating effect on the surface temperature of a polymer. Each different polymer is unique in its response to weathering. The following equation (Eq. 10) was proposed as a first approximation to defining the absorbed radiation which contributed to the heating of the material [191]:

$$E_{abs} = \int E_{\lambda}\epsilon(\lambda)d\lambda \qquad (10)$$

where:

E_λ the thermal conductivity;
$\varepsilon(\lambda)$ the absorbance.

The influence of the weathering stress factors UV light, RH and temperature on optical properties of new silvered polymeric reflector materials was investigated in terms of their loss of performance with exposure time. A stress time performance model was built based on the cumulative dosage of ultraviolet radiation between 290–320 nm (UV-B), together with the synergistic effects of temperature and relative humidity. This model developed for accelerated laboratory data was verified for data obtained from outdoor sites, and a high level of agreement was found [192].

6 Influence of Humidity

Most polymer composites absorb small, but potentially damaging amounts of moisture from the surrounding environments. The absorbed water may negatively affect the material by dimensional changes (swelling), decreasing the glass transition temperature of the polymer, and diminution of mechanical and physical properties (i.e., stiffness, strength and hardness). In the case of fibre reinforced polymers, capillary action along the fibres can account for a significant amount of initial moisture uptake, even though a chemically resistant matrix may encapsulate the fibres. Shrinkage of the resin away from the fibres during curing is an additional factor to the capillary effect. The effect of moisture determines hydrolytic breakdown of the fibre-matrix interface, which diminishes the efficiency of load transfer between the matrix and the fibre.

Often, water reacts with the polymer matrix and causes irreversible chemical changes and diminishes performance. The process of moisture absorption and desorption on the surface layers takes place almost immediately on contact with the environment, but moisture diffusion into the bulk material is usually a slow process. It takes weeks to months for a substantial amount of moisture to be absorbed by the composite and long time (i.e., 1–2 years) before the material reaches saturation. The rate of moisture uptake by a composite laminate depends on the relative humidity, temperature, exposure time and mechanical load. Elevated temperatures accelerate the rate of moisture uptake and material degradation. Similarly, tensile loads accelerates moisture uptake by opening existing internal cavities or voids, and by micro-crack formation. The effect of moisture on polymer composites is potentially greater under natural and accelerated weathering conditions (i.e., combination of moisture and UV light). Absorbed water is generally less damaging to polymeric materials than sunlight, although most polymers absorb some moisture through diffusion in high humidity environments. The extent of moisture absorption depends on polymer hydrophilicity. Since absorption of moisture into the material does not result in main chain bond scission, this rarely causes degradation leading to failure. Rainfall, however, may act in concert with photo-initiated oxidation to produce erosion of the polymeric material, serving to wash away the embrittled surface layer so as to expose new material to direct sunlight [4].

At the same time, for those materials that require the presence of water, it is an extremely important environmental parameter, because water provokes damage to materials by both physical and chemical processes. Absorption of water into a material, from humid air or surface deposition induces a volume expansion resulting in mechanical stress within the material. A wet period followed by a dry period leads to a volume contraction of the surface layers setting up more stress within the material. Several cyclic dry/wet conditions can eventually result in fatigue, and favor further chemical and mechanical change or degradation. Generally, the composite matrix is the component most vulnerable to environmental attack, with UV light and moisture being two of the primary environmental factors contributing to material degradation. Increased humidity is well-known to accelerate the photo-thermal breakdown of polymers. Although plastics are hydrophobic materials, they absorb moisture and are stressed due to the swell/shrink or freeze/thaw cycles. The evaporation of the absorbed water initially dries the surface layer resulting in stress at the interface with underlying hydrated layers. In addition to the physical impacts such as matrix swelling, erosion, and stabilizer extraction, humidity can also play a chemical role in chalking of materials such as TiO_2 pigmented coating and synthetic materials. These materials are decomposed under influence of the weather and the TiO_2 particles are set free on the surface where they create a dull layer that can be easily wiped off. It has been observed that the chalking is strongest where more water is available on the surface. No chalking can be observed in dry atmospheres. An electron hole pair, which is created in the TiO_2 lattice through radiation by short wavelength light, reacts with the hydroxide groups present on the surface and the titanium ions, and with oxygen and water resulting in hydroxide and hydroperoxide radicals, whereby the TiO_2 surface again assumes the hydrated form, generating the so-called chalking cycle. In natural ageing, water may be delivered to the material surface as relative humidity, as dew or condensation, or rainwater. In accelerated weathering, outdoor moisture attack is simulated via a pure water spray through nozzles located in the top of the chamber. Spray can be programmed to operate during either the dark or light cycle. The duration of the cycles and the temperature can be selected. Spray cycles may simulate the rainfall, and condensation may also be possible depending on the desired simulation. Some accelerated test results are considered variable because relative humidity is not controlled during the exposure, but many accelerated ageing devices are available with precise control of relative humidity [190].

The influence of the combined action of water, temperature and radiation in the oxidative ageing of various TiO_2 pigmented PVC formulations designed for outdoor applications was studied using infrared spectroscopy to evaluate the photo-oxidation mechanism, and scanning spectroscopy and colourimetry measurements to evaluate the aesthetic properties. The presence of water during artificial ageing created large changes in the stoichiometry and concentration of the oxidation products, evidenced in FTIR spectra by an initial decrease in the magnitude of the maximum absorption bands in the carbonyl range, resulting from water spray, and no significant changes during the subsequent weathering. The colourimetric changes that were observed during artificial ageing, without intervention of water, consisted of brief initial whitening, followed by a recovery of the initial appearance and subsequent yellowing. The introduction of cyclic water spray (3550 h with cycles of water spray for 18 min every 2 h), which occurred after 2660 h of xenon irradiation, led to chalking and bleaching in all PVC formulations. Generally, the photo-activity of TiO_2 pigments should result in the formation of oxidation products, such as alcohols, ketones, aldehyde and other functional groups, which are responsible for the yellowing in PVC and for the loss of adhesion to the polymeric matrix surface by water action. This effect and the fact that no new band in the infrared spectrum was detected in FTIR spectra confirmed that the decrease in the concentration of the oxidation products formed when titanium dioxide was present in the PVC formulations, after initiation of the cyclic water spray, resulted from an erosion phenomenon [193].

The effects of relative humidity on polymer photo-degradation have been studied on PVC and PE, both containing photoactive TiO_2 pigment, using a closed loop photo-reactor which allowed measurements of evolved CO_2. With both polymers, there was an initial CO_2 evolution rate which was less dependent on humidity. After a short incubation time, the CO_2 evolution changed to a higher secondary rate, which was more sensitive to the relative humidity. For PE, increasing humidity led to a general increase in CO_2 evolution rate and a faster transition to a higher secondary degradation rate, for humidity above 60 %. The dominant effect of humidity in the presence of photoactive TiO_2 was to increase the production of hydroxide radicals due to the photo-catalytic breakdown of adsorbed water on the TiO_2 particles. Concomitantly, faster breakdown of esters to CO_2 in the presence of water was probably to occur. For PVC, photo-degradation was more complex. At very low humidity, the CO_2 evolution rate was low, but increased significantly at around 35 % RH. At higher humidity levels, the CO_2 evolution rate decreased. This effect was attributed to the catalytic effect of hydrochloric acid. In this case, the dominant effect was an autocatalytic degradation involving hydrochloric acid, although the production of hydroxide radicals on the TiO_2 surface could not be neglected. In dry conditions, there was no water to form hydrochloric acid and, consequently, CO_2 evolution rates were low and there was no transition to a faster evolution rate. As humidity increased, more hydrochloric acid was formed in the PVC film and this led to a higher CO_2 evolution rate. At humidity above 35 %, the additional water effectively dilutes the hydrochloric acid and the degradation rate slowed down. The addition of hydrotalcite, an inert easily handled mineral pigment, reduced efficiently this acid catalyzed degradation, and addition of 10 % completely removed hydrochloric acid catalysis even using a highly photoactive TiO_2. The hydrotalcite acted via an ion exchange mechanism, whereby the chloride ions were removed from the film and replaced with carbonate ions, resulting in a much weaker acid which was unable to catalyze the degradation [194].

Photo-degradation of some clear and pigmented coatings, investigated by DSA, was proved to affect the water sorption into these coatings, due to increased porosity and hydrophilicity. Sorption curves for various organic coatings and filters measured by DSA showed clear trends with increasing photo-degradation time. At the same time, desorption was found to occur after longer degradation times (40 days UV irradiation) and was caused by swelling of the coating and squeezing out the excess of water. Differences in dynamics of absorption, due to increased hydrophilicity, crosslinking and porosity, were found between UV degraded and unexposed epoxy coatings. Due to crack formation during DSA measurements, clear coatings showed inconstant data, as a result of release of tension of the UV degraded clear coating by humidified nitrogen. Pigmented coatings did not show this behaviour, resulting in an increasing water absorption trend with increasing degradation times [195].

Ageing of polyester-urethane coatings (poly(neopentylisophthalate) non-cross-linked or cross-linked with hexamethylenediisocyanurate) that were degraded under different accelerated laboratory weathering conditions, such as aerobic or anaerobic conditions as well as wet or dry conditions, was monitored using a combination of FTIR and UV-Vis spectroscopy. The experimental results showed that both evaporation and water-caused removal of degraded material caused a particularly pronounced decay in the thickness of the coating. In the WOM, the degraded material was removed by water during wet cycles as a result of the water spray. In addition, this material removal became more severe as weathering proceeded; more urethane bonds brook and cross-link density reduced. The FTIR analysis of the extracted material showed that the material washed out was essentially the same as the degraded material. In the absence of water spray, the material loss took place in the same period as urethane groups decomposed and stoped afterwards, even though the ester bond scission proceeded with higher rates. This supports the hypothesis of photo-oxidative pathways for the urethane group decomposition and photolytic mechanisms for ester bond scission [196].

The photochemical ageing of expanded polytetrafluoroethylene laminated to a Nomex fabric has been investigated by FTIR, XRD, DSC, SEM, AFM and permeability measurements, and the results showed that UV light irradiation had a significant effect on permeability, morphology, structure and mechanical properties of the membrane. The dramatic decrease of vapor permeability after photo-ageing was due to the gradual decrease of size and number of pores of the membrane which led to the decrease of diffusion capacity of water vapor through the membrane and thus decreasing water vapor permeability. The photo-oxidation of the membrane led to a significant reduction in water vapor permeability and it was found that the higher light intensity is, the greater the drop in permeability [197].

The hydrolytic and photochemical ageing behavior of a Kevlar-PBI blend fabric was investigated by ATR-FTIR and tensile tests. Irradiation with UV light or exposure to high humidity resulted in a continuous decrease of the yarn breaking force with exposure time. ATR-FTIR analyses of photo-chemically aged samples evidenced a photo-oxidative reaction initiated by the cleavage of the amide bond of Kevlar polymer. At constant temperature the breaking force was unaffected by the variation of the relative humidity, which suggested that the absorption of water was not the rate-controlling step in the degradation kinetics. ATR-FTIR analyses revealed the presence of a new absorption band ascribed to carboxylic acid end groups produced during the hydrolysis of the amide linkage that occurred after humidity ageing. The relative intensity of the –COOH band tended to a constant value as exposure times increased, suggesting that in addition to the hydrolysis, a competing recombination reaction took place during degradation. The authors introduced a kinetic model for the hydrolytic degradation process [198].

Degradation due to exposure to UV radiation and moisture of a bisphenol A based epoxy polymer, monitored by PA FTIR spectroscopy, evidenced the presence of oxirane ring and polyamine cross-linker, which were reaction sites responsible for crosslinking reactions of epoxy polymers. This indicated that the ring opening reactions of oxirane groups of bisphenol A epoxy polymer occurred further away from the surface, thus resulting in the increase in hydroxyl group content. PA FTIR spectroscopy performed on the surface of specimens indicated that UV exposure in the presence of water condensation resulted in the formation of hydroxyl groups on the surface, which further promoted cross-linking reactions on the surface resulting in degradation. At the same time, spectra recorded by FTIR spectroscopy detected higher amine content, indicating that the formation of amides, via chain scission, had a greater contribution to epoxy degradation [199].

Similar exposure studies on carbon/epoxy laminates subjected to UV radiation exposure suggested increased crosslink density on the surface of the epoxy matrix when analyzed with FTIR. The C–N stretching vibrations due to amide formation indicated the presence of chain scission reactions. Both of these studies indicated that cross-linking and chain scission mechanisms operated concomitantly during the degradation process. Increased cross-linking dominated in the early stages of degradation, after which carbonyl amide formation by chain scission took over. Both of these mechanisms explained increased micro-cracking and surface deterioration, ultimately reducing the mechanical strength of the composite structure. Specimens of carbon fiber-reinforced epoxy composite exposed to 500 h of only UV radiation exhibited a minor 0.27 % decrease in weight, which was attributed to the expulsion of volatiles and residual moisture. The same specimens that were subjected to only water vapor condensation gained 0.89 % by weight and approached complete saturation after 500 h of exposure to condensation. When specimens were exposed sequentially to UV radiation followed by condensation, they initially lost weight during the UV radiation cycle and subsequently gained weight during the condensation cycle. When the samples were cyclically exposed to both UV radiation and condensation, they exhibited continuous weight loss at a steady rate throughout the exposure duration, which indicated that material was being removed from the composite specimens. Based on micrographic observations and FTIR analysis it was concluded that UV radiation and condensation operated in a synergistic manner that led to extensive matrix erosion, matrix micro-cracking, fiber debonding, fiber loss and void formation [200].

TGDDM epoxy resin cured with aromatic hardener DDS was exposed to UV radiation and humidity and several different degradation mechanisms were proposed based on FTIR analysis. One of them involved hydrogen abstraction on the methylene group, scission of the carbon-nitrogen bond and formation of an aldehyde (carbonyl group). Another one involved hydrogen abstraction of the CH–OH bond and a similar chain scission reaction at the carbon-nitrogen bond, resulting in the formation of a ketone (carbonyl group). The third involved chain scission at the carbon-carbon bond and generation of a carboxylic acid (carbonyl group) and an amide linkage. The main pathway for amide formation occurred by chain scission of the carbon-carbon bond, rather than the carbon-nitrogen bond, producing amide molecules which may propagate the photo-oxidative sequence. The authors concluded that photo-oxidative degradation of TGDDM/DDS could potentially involve several different mechanisms, which ultimately bring about chain-scission, leading to the formation of amide and carbonyl groups [201].

Two types of thermosets that consist of a mixture of acrylate-polyol copolymers, cross-linked with oxylated melamine and with isocyanate trimers, respectively, were subjected to accelerated weathering and action of water after the UV light exposure. By decoupling photo-degradation and the action of water, the authors proved that UV-light without water had important consequences on polymer degradation and water played the key role in the cracking process for the polymer network coatings. Although chain scissions led to the formation of several photo-products, a significant increase in the mechanical properties was observed. This increase was attributed to an overall densification of the network induced by cross-linking. This phenomenon provoked a drastic increase of the internal strain in the polymer. For a determined critical value of the mechanical properties, water produced latent damage by leading to the formation of cracks. Therefore, when the strain reached a maximum (i.e., micro-hardness was high) in the polymer, the evaporation of the water, which had a good affinity with the material, disorganized the network. This occurred because the system could not accept the stress induced by the water inflation/deflation process and led to the formation of cracks [202].

A study regarding the influence of water on the chemical modifications of a phenoxy resin (bisphenol-A with epichlorohydrin) during accelerating ageing and presence of water (alternating or simultaneously: UV light irradiation under dry conditions, alternating irradiation/immersion cycles in water and irradiation in water) showed that water had a double effect. The first one was a partial hydrolysis of the main photoproducts (phenyl formats) generated during UV light exposure, which led to macromolecular phenols and formic acid that was easily extracted from the polymer matrix by water. The second one was an increase in the polymer photo-oxidation rate as a consequence of the formation of phenols, which acted as initiators of polymer photo-oxidation [203].

High temperature and high humidity increase hydrolytic degradation of the polymers. Temperature of the material and the presence of moisture show a considerable synergistic effect on the photo-degradation of the polymeric materials [204]. In the case of stabilized polymers, water induces photo-degradation because high humidity favours soluble photo-stabilizers to leach out of the plastic matrix, decreasing the efficiency of the light stabilizer. High humidity, especially at the higher temperatures, provoked increased photo-damage in polymers such as thermoplastic polyester elastomers [205].

7 Conclusions

External environmental factors can induce dramatic changes in physicochemical properties of polymer-based materials, which may limit their applicability. Chemical stability of polymeric materials exposed to UV light remains a key element in determining their useful life. Gradual deterioration of these materials takes place usually as a result of change in polymer chemical structure induced by radical photo- and/or thermo-oxidation processes. The polymer composites of most useful polymers have exhibited less stability than pristine polymers. Although composites may get highly improved material properties by filling the polymer matrix with various fillers, their durability in outdoor applications is still a challenge.

The present chapter has provided a brief review concerning atmospheric components involved in the weathering of stabilized polymeric materials and their effects on polymer degradation. Basic degradation processes and examples of degradation under various environmental conditions have been presented. The text has outlined the relationship between exposure conditions (radiation, temperature, and humidity), chemical composition of formulation and presence of impurities, photochemistry of pigments, fillers or stabilizers, and chemical changes occurring as a result of degradation.

References

1. Andrady, A.L., Hamid, H.S., Torikai, A.: Effects of stratospheric ozone depletion and climate change on materials damage. Photochem. Photobiol. Sci. **6**, 311–318 (2007)
2. Andrady, A.L., Hamid, H., Torikai, A.: Effects of solar UV and climate change on materials. Photochem. Photobiol. Sci. **10**, 292–300 (2011)
3. Pandey, J.K., Reddy, K.R., Kumar, A.P., Singh, R.P.: An overview on the degradability of polymer nanocomposites. Polym. Degrad. Stab. **88**, 234–250 (2005)
4. Maxwell, A.S., Broughton, W.R., Dean, G., Sims. G.D.: Review of Accelerated Ageing Methods and Lifetime Prediction Techniques for Polymeric Materials. NPL report DEPC MPR 016 (2005)

5. Roylance, M., Roylance, D.: Forms of polymer degradation: overview. In: Hihara, L.H., Adler, R.P.I., Latanision, R.M. (eds.) Environmental Degradation of Advanced and Traditional Engineering Materials, p. 337. CRC Press, Taylor and Francis Group, Boca Raton (2014)
6. Bussière, P.-O., Peyroux, J., Chadeyron, G., Therias, S.: Influence of functional nanoparticles on the photostability of polymer materials: Recent progress and further applications. Polym. Degrad. Stab. **98**, 2411–2418 (2013)
7. Gugumus, F.: Light stabilizers. In: Zweifel, H., Maier, R.D., Schiller, M. (eds.) Plastic Additives Handbook, 6th edn, p. 139. Carl Hanser Verlag, Munchen (2009)
8. Hohenberger, W.: Fillers and reinforcements/coupling agents. In: Zweifel, H., Maier, R.D., Schiller, M. (eds.) Plastic Additives Handbook, 6th edn, p. 919. Carl Hanser Verlag, Munchen (2009)
9. Pfandner, R.: Stabilisation of polymer nanocomposites. In: Beyer, G. (ed.) Industry Guide to Polymer Nanocomposites, p. 117. Plastics Information Direct, Bristol (2009)
10. Tjong, S.C.: Structural and mechanical properties of polymer nanocomposites. Mater. Sci. Eng. R Rep. **53**, 73–197 (2006)
11. Paul, D.R., Robeson, L.M.: Polymer nanotechnology: nanocomposites. Polymer **49**, 3187–3204 (2008)
12. Pfandner, R.: Nanocomposites: industrial opportunity or challenge? Polym. Degrad. Stab. **95**, 369–373 (2010)
13. Braun, U., Wachtendorf, V., Geburtig, A., Bahr, H., Schartel, B.: Weathering resistance of halogen-free flame retardance in thermoplastics. Polym. Degrad. Stab. **95**, 2421–2429 (2010)
14. Chantegraille, D., Morlat-Therias, S., Gardette, J.-L.: Photochemical behaviour of fire-retarded polymers. Polym. Degrad. Stab. **95**, 274–277 (2010)
15. Pfandner, R.: (Photo)oxidative degradation and stabilization of flame retarded polymers. Polym. Degrad. Stab. **98**, 2430–2435 (2013)
16. Sobkow, D., Barton, J., Czaja, K., Sudol, M., Mazon, B.: Studies of material resistance against natural environmental factors. CHEMIK **68**, 347–354 (2014)
17. Kumar, A.P., Depan, D., Tomer, N.S., Singh, R.P.: Nanoscale particles for polymer degradation and stabilization—trends and future perspectives. Prog. Polym. Sci. **34**, 479–515 (2009)
18. Seentrakoon, B., Junhasavasdikul, B., Chavasiri, W.: Enhanced UV-protection and antibacterial properties of natural rubber/rutile TiO_2 nanocomposites. Polym. Degrad. Stab. **98**, 566–578 (2013)
19. Egerton, T.A.: UV-absorption—the primary process in photocatalysis and some practical consequences. Molecules **19**, 18192–18214 (2014)
20. Nakayama, N., Hayashi, T.: Preparation and characterization of poly(L-lactic acid)/TiO_2 nanoparticle nanocomposite films with high transparency and efficient photo-degradability. Polym. Degrad. Stab. **92**, 1255–1264 (2007)
21. Fujishima, A., Zhang, X., Tryk, D.A.: TiO_2 photocatalysis and related surface phenomena. Surf. Sci. Rep. **63**, 515–582 (2008)
22. Man, C., Zhang, C., Liu, Y., Wang, W., Ren, W., Jiang, L., Reisdorffer, F., Nguyen, T.P., Dan, Y.: Poly(lactic acid)/titanium dioxide composites: preparation and performance under ultraviolet irradiation. Polym. Degrad. Stab. **97**, 856–862 (2012)
23. Wang, W.-W., Man, C.-Z., Zhang, C.-M., Jiang, L., Dan, Y., Nguyen, T.-P.: Stability of poly(L-lactide)/TiO_2 nanocomposite thin films under UV irradiation at 254 nm. Polym. Degrad. Stab. **98**, 885–893 (2013)
24. Shang, J., Chai, M., Zhu, Y.: Solid-phase photocatalytic degradation of polystyrene plastic with TiO_2 as photocatalyst. J. Solid State Chem. **174**, 104–110 (2003)
25. Zan, L., Tian, L., Liu, Z., Peng, Z.: A new polystyrene-TiO_2 nanocomposite film and its photocatalytic degradation. Appl. Catal. A **264**, 237–242 (2004)

26. Zan, L., Wang, S., Fa, W., Hu, Y., Tian, L., Deng, K.: Solid-phase photocatalytic degradation of polystyrene with modified nano-TiO_2 catalyst. Polymer **47**, 8155–8162 (2006)
27. Cho, S., Choi, W.: Solid-phase photocatalytic degradation of PVC-TiO_2 polymer composites. J. Photochem. Photobiol., A **143**, 221–228 (2001)
28. Searle, J., Worsely, D.: Titanium dioxide photocatalysed oxidation of plasticizers in thin poly(vinyl chloride) films. Plast. Rubber. Compos. Macromol. Eng. **31**, 329–335 (2002)
29. Gesenhues, U.: Influence of titanium dioxide pigments on the photodegradation of poly (vinyl chloride). Polym. Degrad. Stab. **68**, 185–196 (2000)
30. Kim, S.H., Kwak, S.-Y., Suzuki, T.: Photocatalytic degradation of flexible PVC/TiO_2 nanohybrid as an eco-friendly alternative to the current waste landfill and dioxin-emitting incineration of post-use PVC. Polymer **47**, 3005–3016 (2006)
31. Hwang, D.K., Shul, Y.G., Chu, Y.H.: Photodegradation behavior of the polycarbonate/TiO_2 composite films under the UV irradiation in ambient air condition. Polym. Compos. **36**, 1462–1468 (2015)
32. Ammala, A., Bateman, S., Dean, K., Petinakis, E., Sangwan, P., Wong, S., Yuan, Q., Yu, L., Patrick, C., Leonh, K.H.: An overview of degradable and biodegradable polyolefins. Prog. Polym. Sci. **36**, 1015–1049 (2011)
33. Zhao, X., Li, Z., Chen, Y., Shi, L., Zhu, Y.: Enhanced of photocatalytic degradation of polyethylene with CuPc modified TiO_2 photocatalyst under solar light irradiations. Appl. Surf. Sci. **254**, 1825–1829 (2008)
34. Thomas, R.T., Sandhyarani, N.: Enhancement in the photocatalytic degradation of low density polyethylene-TiO_2 nanocomposite films under solar irradiation. RSC Adv. **3**, 14080–14088 (2013)
35. Thomas, T.R., Nair, V., Sandyarani, N.: TiO_2 nanoparticles assisted solid phase photocatalytic degradation of polyethylene film: a mechanistic investigation. Colloids Surf. A **422**, 1–9 (2013)
36. Liang, W., Luo, Y., Song, S., Dong, X., Yu, X.: High photocatalytic degradation activity of polyethylene containing polyacrylamide grafted TiO_2. Polym. Degrad. Stab. **98**, 1754–1761 (2013)
37. Luo, Y., Dong, X.M., Liu, Y.J., You, J.E.: Solid-Phase photocatalytic degradation of PELDPE film filled with grafted nano-TiO_2. China Plast. **23**, 58–60 (2009)
38. da Silva, K.I.M., Santos, M.J.L., Gil, M.P.: Dependence of the photodegradation rate on the crystalline portion of PE films obtained through in situ polymerization in the presence of TiO_2 nanospheres, nanoribbons and microspheres. Polym. Degrad. Stab. **112**, 78–85 (2015)
39. Allen, N.S., Edge, M., Corrales, T., Childs, A., Liauw, C.M., Catalina, F., Peinado, C., Minihan, A., Aldcroft, D.: Ageing and stabilisation of filled polymers: an overview. Polym. Degrad. Stab. **61**, 183–199 (1998)
40. Allen, N.S., Edge, M., Ortega, A., Sandoval, G., Liauw, C.M., Verran, J., Stratton, J., McIntyre, R.B.: Degradation and stabilisation of polymers and coatings: nano versus pigmentary titania particles. Polym. Degrad. Stab. **85**, 927–946 (2004)
41. Gijsman, P., Meijers, G., Vitarelli, G.: Comparison of the UV-degradation chemistry of polypropylene, polyethylene, polyamide 6 and polybutylene terephthalate. Polym. Degrad. Stab. **65**, 433–441 (1999)
42. Zan, L., Fa, W., Wang, S.: Novel photodegradable low-density polyethylene-TiO_2 nanocomposite film. Environ. Sci. Technol. **40**, 1681–1685 (2006)
43. Kemp, T.J., McIntyre, R.A.: Transition metal-doped titanium (IV) dioxide: characterization and influence on photodegradation of poly(vinyl chloride). Polym. Degrad. Stab. **91**, 165–194 (2006)
44. Kemp, T.J., McIntyre, R.A.: Influence of transition metal-doped titanium(IV) dioxide on the photodegradation of polyethylene. Polym. Degrad. Stab. **91**, 3020–3025 (2006)
45. Ma, D., Akpalu, Y.A., Li, Y., Siegel, R.W., Schadler, L.S.: Effect of titania nanoparticles on the morphology of low density polyethylene. J. Polym. Sci., Part B: Polym. Phys. **43**, 488–497 (2005)

46. Ohtani, B., Azuma, S., Nishimoto, S., Kagiya, T.: Photocatalytic degradation of polypropylene film by dispersed titanium dioxide particles. Polym. Degrad. Stab. **23**, 271–278 (1989)
47. Liu, Z., Jin, J., Chen, S., Zhang, J.: Effect of crystal form and particle size of titanium dioxide on the photodegradation behaviour of ethylene-vinyl acetate copolymer/low density polyethylene composite. Polym. Degrad. Stab. **96**, 43–50 (2011)
48. Wu, N., Shao, D.F., Wei, Q.F., Cai, Y.B., Gao, W.D.: Characterization of PVAc/TiO_2 hybrid nanofibers: from fibrous morphologies to molecular structures. J. Appl. Polym. Sci. **112**, 1481–1485 (2009)
49. Jin, C., Christensen, P.A., Egerton, T.A., Lawson, E.J., White, J.R.: Rapid measurement of polymer photo-degradation by FTIR spectrometry of evolved carbon dioxide. Polym. Degrad. Stab. **91**, 1086–1096 (2006)
50. Pandey, J.K., Kim, M.H., Chun, D.M., Lee, C.S., Ahn, S.H.: Surface modification of polyethylene (PE) by the deposition of titanium dioxide (TiO_2) nanoparticles to enhance the photocatalytic activities. Surf. Rev. Lett. **16**, 259–263 (2009)
51. Zapata, P.A., Rabagliati, F.M., Lieberwirth, I., Catalina, F., Corrales, T.: Study of the photodegradation of nanocomposites containing TiO_2 nanoparticles dispersed in polyethylene and in poly(ethylene-*co*-octadecene). Polym. Degrad. Stab. **109**, 106–114 (2014)
52. White, J.R., Shyichuk, A.V., Turton, T.J., Syrotynska, I.D.: Effect of stabilizer and pigment on photodegradation of polypropylene as revealed by macromolecule scission and crosslinking measurements. Polym. Degrad. Stab. **91**, 1755–1760 (2006)
53. Mina, M.F., Seema, S., Matin, R., Rahaman, MdJ, Sarker, R.B., Gafur, MdA, Bhuiyan, M. A.H.: Improved performance of isotactic polypropylene/titanium dioxide composites: effect of processing conditions and filler content. Polym. Degrad. Stab. **94**, 183–188 (2009)
54. Miyazaki, K., Arai, T., Nakatani, H.: Polypropylene plasticization and photodegradation with a TiO_2/poly(ethylene oxide)/methyl linoleate paint photocatalyst system. J. Appl. Polym. Sci. (2014). doi:10.1002/APP.39909
55. Vinu, R., Madras, G.: Photocatalytic degradation of methyl methacrylate copolymers. Polym. Degrad. Stab. **93**, 1440–1449 (2008)
56. Konaganti, V.K., Madras, G.: Photooxidative and pyrolytic degradation of methyl methacrylate-alkyl acrylate copolymers. Polym. Degrad. Stabil. **94**, 1325–1335 (2009)
57. Fa, W., Guo, L., Wang, J., Guo, R., Zheng, Z., Yang, F.: Solid-phase photocatalytic degradation of polystyrene with TiO_2/$Fe(St)_3$ as catalyst. J. Appl. Polym. Sci. **128**, 2618–2622 (2013)
58. Nakatani, H., Miyazaki, K.: Polystyrene photodegradation with a novel titanium dioxide/poly(ethylene oxide)/methyl linoleate paint photocatalyst system. J. Appl. Polym. Sci. **129**, 3490–3496 (2013)
59. Lei, Y., Lei, H., Huo, J.: Innovative controllable photocatalytic degradation of polystyrene with hindered amine modified aromatic polyamide dendrimer/polystyrene-grafted-TiO_2 photocatalyst under solar light irradiation. Polym. Degrad. Stab. **118**, 1–9 (2015)
60. Chen, X.D., Wang, Z., Liao, Z.F., Mai, Y.L., Zhang, M.Q.: Roles of anatase and rutile TiO_2 nanoparticles in photooxidation of polyurethane. Polym. Testing **26**, 206–208 (2007)
61. Jin, C.Q., Christensen, P.A., Egerton, T.A., White, J.R.: Rapid measurement of photocatalytic oxidation of poly(vinyl chloride) by in situ FTIR spectrometry of evolved CO_2. Mater. Sci. Technol. **22**, 908–914 (2006)
62. Yang, T.-C., Noguchi, T., Isshiki, M., Wu, J.-H.: Effect of titanium dioxide on chemical and molecular changes in PVC sidings during QUV accelerated weathering. Polym. Degrad. Stab. **104**, 33–39 (2014)
63. Miyazaki, K., Nakatani, H.: Preparation of degradable polypropylene by an addition of poly (ethylene oxide) microcapsule containing TiO_2. Polym. Degrad. Stab. **94**, 2114–2120 (2009)
64. Zhao, X., Li, Z., Chen, Y., Shi, L., Zhu, Y.: Solid-phase photocatalytic degradation of polyethylene plastic under UV and solar light irradiation. J. Mol. Catal. A: Chem. **268**, 101–106 (2007)

65. Yang, R., Christensen, P.A., Egerton, T.A., White, J.R.: Degradation products formed during UV exposure of polyethylene-ZnO nanocomposites. Polym. Degrad. Stab. **95**, 1533–1541 (2010)
66. Yang, R., Li, Y., Yu, J.: Photostabilization of linear low density polyethylene by inorganic nanoparticles. Polym. Degrad. Stab. **88**, 168–174 (2005)
67. Therias, S., Larché, J.-F., Bussière, P.-O., Gardette, J.-L., Murariu, M., Dubois, P.: Photochemical behavior of polylactide/ZnO nanocomposite films. Biomacromolecules **13**, 3283–3291 (2012)
68. Wen, B., Wang, F., Xu, X., Ding, Y., Zhang, S., Yang, M.: The effect of encapsulation of nano zinc oxide with silica on the UV resistance of polypropylene. Polym. Plast. Technol. Eng. **50**, 1375–1382 (2011)
69. Jiang, J., Li, G., Ding, Q., Mai, K.: Ultraviolet resistance and antimicrobial properties of ZnO-supported zeolite filled isotactic polypropylene composites. Polym. Degrad. Stab. **97**, 833–838 (2012)
70. Ammala, A., Hill, A.J., Meakin, P., Pas, S.J., Turney, T.W.: Degradation studies of polyolefins incorporating transparent nanoparticulate zinc oxide UV stabilizers. J. Nanopart. Res. **4**, 167–174 (2002)
71. Zhao, H., Robert, K., Li, Y.: A study on the photo-degradation of zinc oxide (ZnO) filled polypropylene nanocomposites. Polymer **47**, 3207–3217 (2006)
72. Salla, J., Pandey, K.K., Srinivas, K.: Improvement of UV resistance of wood surfaces by using ZnO nanoparticles. Polym. Degrad. Stab. **97**, 592–596 (2012)
73. Butylina, S., Martikka, O., Karki, T.: Weathering properties of coextruded polypropylene-based composites containing inorganic pigments. Polym. Degrad. Stab. **120**, 10–16 (2015)
74. Dintcheva, N.T., La Mantia, F.P., Malatesta, V.: Photo-oxidation behaviour of polyethylene/multi-wall carbon nanotube composite films. Polym. Degrad. Stab. **94**, 162–170 (2009)
75. Morlat-Therias, S., Fanton, E., Gardette, J.L., Peeterbroeck, S., Alexandre, M., Dubois, P.: Polymer/carbon nanotube nanocomposites: influence of carbon nanotubes on EVA photodegradation. Polym. Degrad. Stab. **92**, 1873–1882 (2007)
76. Najafi, E., Shin, K.: Radiation resistant polymer-carbon nanotube nanocomposite thin films. Colloids Surf. A **257**, 333–337 (2005)
77. Bocchini, S., Di Blasio, A., Frache, A.: Influence of MWNT on polypropylene and polyethylene photooxidation. Macromol. Symp. **301**, 16–22 (2011)
78. Lonkar, S.P., Kushwaha, O.S., Leuteritz, A., Heinrich, G., Singh, R.P.: Self photostabilizing UV-durable MWCNT/polymer nanocomposites. RSC Adv. **2**, 12255–12262 (2012)
79. Watts, P.C.P., Fearn, P.K., Hsu, W.K., Billingham, N.C., Kroto, H.W., Walton, D.R.M.: Carbon nanotubes as polymer antioxidants. J. Mater. Chem. **13**, 491–495 (2003)
80. Grigoriadou, I., Paraskevopoulos, K.M., Chrissafis, K., Pavlidou, E., Stamkopoulos, T.-G., Bikiaris, D.: Effect of different nanoparticles on HDPE UV stability. Polym. Degrad. Stab. **96**, 151–163 (2011)
81. Guadagno, L., Naddeo, C., Raimondo, M., Gorrasi, G., Vittoria, V.: Effect of carbon nanotubes on the photo-oxidative durability of syndiotactic polypropylene. Polym. Degrad. Stab. **95**, 1614–1626 (2010)
82. Asmatulu, R., Mahmud, G.A., Hille, C., Misak, H.E.: Effects of UV degradation on surface hydrophobicity, crack, and thickness of MWCNT-based nanocomposite coatings. Prog. Org. Coat. **72**, 553–561 (2011)
83. Petersen, E.J., Lam, T., Gorham, J.M., Scott, K.C., Long, C.J., Stanley, D., Sharma, R., Alexander Liddle, J., Pellegrin, B., Nguyen, T.: Methods to assess the impact of UV irradiation on the surface chemistry and structure of multiwall carbon nanotube epoxy nanocomposites. Carbon **69**, 194–205 (2014)
84. Wohlleben, W., Meier, M.W., Vogel, S., Landsiedel, R., Cox, G., Hirth, S., Tomovic, Z.: Elastic CNT-polyurethane nanocomposite: synthesis, performance and assessment of fragments released during use. Nanoscale **5**, 369–380 (2013)

85. Vilar, G., Fernandez-Rosas, E., Puntes, V., Jamier, V., Aubouy, L., Vazquez-Campos, S.: Monitoring migration and transformation of nanomaterials in polymeric composites during accelerated aging. J. Phys: Conf. Ser. **429**, 012044 (2013)
86. Busquets-Fite, M., Fernandez, E., Janer, G., Vilar, G., Vazquez-Campos, S., Zanasca, R., Citterio, C., Mercante, L., Puntes, V.: Exploring release and recovery of nanomaterials from commercial polymeric nanocomposites. J. Phys: Conf. Ser. **429**, 012048 (2013)
87. Gorrasi, G., Sorrentino, A.: Photo-oxidative stabilization of carbon nanotubes on polylactic acid. Polym. Degrad. Stab. **98**, 963–971 (2013)
88. Dintcheva, N.T., Arrigo, R., Catalanotto, F., Morici, E.: Improvement of the photo-stability of polystyrene-block-polybutadiene-block-polystyrene through carbon nanotubes. Polym. Degrad. Stab. **118**, 24–32 (2015)
89. Nguyen, T., Pellegrin, B., Bernard, C., Gu, X., Gorham, J.M., Stutzman, P., Stanley, D., Shapiro, A., Byrd, E., Hettenhouser, R., Chin, J.: Fate of nanoparticles during life cycle of polymer nanocomposites. J. Phys: Conf. Ser. **304**, 012060 (2011)
90. Shi, X., Jiang, B., Wang, J., Yang, Y.: Influence of wall number and surface functionalisation of carbon nanotubes on their antioxidant behavior in high density polyethylene. Carbon **50**, 1005–1013 (2012)
91. Pena, J.M., Allen, N.S., Edge, M., Liauw, C.M., Roberts, I., Valange, B.: Triplet quenching and antioxidant effect of several carbon black grades in the photodegradation of LDPE doped with benzophenone as a photosensitizer. Polym. Degrad. Stab. **70**, 437–454 (2000)
92. Gupta, S., Chandra, T., Sikder, A., Menon, A., Bhowmick, A.K.: Accelerated weathering behavior of poly(phenylene ether)-based TPE. J. Mater. Sci. **43**, 3338–3350 (2008)
93. Maia, D.R.J., Balbinot, L., Popi, R.J., De Paoli, M.A.: Effect of conducting carbon black on the photostabilization of injection molded poly(propylene-*co*-ethylene) containing TiO_2. Polym. Degrad. Stab. **82**, 89–98 (2003)
94. Al Ani, K.E., Anabtawi, M.: UV-irradiation effect on the photodegradation of plasticized poly (*ortho, meta, and para*-fluorostyrene) isomers in solid films. Int. J. Chem. **4**, 62–84 (2012)
95. Al Ani, K.E., Ramadhan, A.E., Anabtawi, M. Irradiation effect on stability of plasticized poly(fluorostyrene) isomers in solution. Int. J. Photochem. (2014). Article ID 536068
96. Dintcheva, N.T., Morici, E., Arrigo, R., La Mantia, F.P., Malatesta, V., Schwab, J.J.: UV-stabilisation of polystyrene-based nanocomposites provided by polyhedral oligomeric silsesquioxanes (POSS). Polym. Degrad. Stab. **97**, 2313–2322 (2012)
97. Valadez-Gonzalez, A., Cervantes-Uc, J.M., Veleva, L.: Mineral influence on the photooxidation of high density polyethylene. I: accelerated UV chamber exposure test. Polym. Degrad. Stab. **63**, 253–260 (1999)
98. Yang, R., Yu, J., Liu, Y., Wang, K.: Effects of inorganic fillers on the natural photo-oxidation of high-density polyethylene. Polym. Degrad. Stab. **88**, 333–340 (2005)
99. Botta, L., Dintcheva, N.T., La Mantia, F.P.: The role of organoclay and matrix type in photo-oxidation of polyolefin/clay nanocomposite films. Polym. Degrad. Stab. **94**, 712–718 (2009)
100. Pospísil, J.: Aromatic and heterocyclic amines in polymer stabilization. Adv. Polym. Sci. **124**, 87–189 (1995)
101. Pilar, J., Michálková, D., Sedenková, I., Pfleger, J., Pospísil, J.: NOR and nitroxide-based HAS in accelerated photooxidation of carbon-chain polymers; Comparison with secondary HAS: an ESRI and ATR FTIR study. Polym. Degrad. Stab. **96**, 847–862 (2011)
102. Butola, B.S., Joshi, M.: Photostability of HDPE filaments stabilized with UV absorbers (UVA) and light stabilizers (HALS). J. Eng. Fibers Fabr. **8**, 61–68 (2013)
103. Pilar, J., Michalkova, D., Slouf, M., Vackova, T.: Long-term accelerated weathering of HAS stabilized PE and PP plaques: compliance of ESRI, IR, and microhardness data characterizing heterogeneity of photooxidation. Polym. Degrad. Stab. **120**, 114–121 (2015)
104. Christensen, P.A., Egerton, T.A., Fernando, S.S., White, J.R.: Photogeneration of carbon dioxide from polypropylene film stabilized by hindered amines. Polym. Degrad. Stabil. **94**, 1999–2003 (2009)

105. Haillant, O.: Spectroscopic characterization of the stabilising activity of migrating HALS in a pigmented PP/EPR blend. Polym. Degrad. Stab. **93**, 1793–1798 (2008)
106. Haillant, O., Lemaire, J.: Natural and artificial photo-aging of non-stabilized and pigmented, hindered-amine stabilized propylene-ethylene copolymers. Polym. Degrad. Stab. **91**, 2748–2760 (2006)
107. Forsthuber, B., Grüll, G.: The effects of HALS in the prevention of photo-degradation of acrylic clear topcoats and wooden surfaces. Polym. Degrad. Stab. **95**, 746–755 (2010)
108. Marek, A., Kapralkova, L., Schmidt, P., Pfleger, J., Humlicek, J., Pospisil, J., Pilar, J.: Spatial resolution of degradation in stabilized polystyrene and polypropylene plaques exposed to accelerated photodegradation or heat aging. Polym. Degrad. Stab. **91**, 444–458 (2006)
109. Horrocks, A.R., Liu, M.G.: UV stabilizing synergies between carbon black and hindered light stabilizer in linear low density polyethylene. Macromol. Symp. **202**, 199–219 (2003)
110. Dobashi, Y., Kondou, J., Ohkatsu, Y.: Photo-antioxidant abilities of 2-hydroxybenzoyl compounds. Polym. Degrad. Stab. **89**, 140–144 (2005)
111. Pospísil, J., Pilar, J., Billingham, N.C., Marek, A., Horak, Z., Nespurek, S.: Factors affecting accelerated testing of polymer photostability. Polym. Degrad. Stab. **91**, 417–422 (2006)
112. Fahmi, M.M., Mohamed, N.A.: *N*-(Substituted phenyl) itaconimides as organic stabilizers for plasticized poly(vinyl chloride) against photo-degradation. Polym. Degrad. Stab. **92**, 733–740 (2007)
113. Dobashi, Y., Ohkatsu, Y.: Dependence of ultraviolet absorbers' performance on ultraviolet wavelength. Polym. Degrad. Stab. **93**, 436–447 (2008)
114. Fechine, G.J.M., Rabello, M.S., Souto Maior, R.M., Catalani, L.H.: Surface characterization of photodegraded poly(ethylene terephthalate). The effect of ultraviolet absorbers. Polymer **45**, 2303–2308 (2004)
115. Pickett, J.E.: UV absorber performance and coating lifetimes. J. Test. Eval. **32**, 240–245 (2004)
116. Lonkar, S.P., Kumar, A.P., Singh, R.P.: Photo-stabilization of EPDM-clay nanocomposites: effect of antioxidant on the preparation and durability. Polym. Adv. Technol. **18**, 891–900 (2007)
117. Shenoy, M.A., Marathe, Y.D.: Studies on synthetic effect of UV absorbers and hindered amine light stabilizers. Pigm. Resin Technol. **36**, 83–89 (2007)
118. Takenaka, H., Mizokawa, S., Ohkatsu, Y.: Interaction of hindered amine light stabilizers and UV absorbers. J. Jpn. Petrol. Inst. **50**, 8–15 (2007)
119. Santos, R., Botelho, G., Machado, A.V.: Role of ultraviolet absorbers (UVA) and hindered amine light stabilizers (HALS) in ABS stabilization. Mater. Sci. Forum **636–637**, 772–778 (2010)
120. Ge, F., Ding, Y., Yang, L., Huang, Y., Jiang, L., Dan, Y.: Effect of the content and distribution of ultraviolet absorbing groups on the UV protection and degradation of polylactide films. RSC Adv. **5**, 70473–70481 (2015)
121. Schnabel, W.: Polymer Degradation: Principles and Practical Applications. Hanser International, München (1981)
122. Fernando, S.S., Christensen, P.A., Egerton, T.A., White, J.R.: Carbon dioxide evolution and carbonyl group development during photodegradation of polyethylene and polypropylene. Polym. Degrad. Stab. **92**, 2163–2172 (2007)
123. Hussein, I.A.: Rheological investigation of the influence of molecular structure on natural and accelerated UV degradation of linear low density polyethylene. Polym. Degrad. Stab. **92**, 2026–2032 (2007)
124. Hussain, I., Atiqullah, M., Fazal, A., Alam, K., Hossaen, A.: Effect of selected residual Ziegler-Natta and metallocene catalysts on the UV-induced degradation of unstabilized ethylene homopolymer films. Polym. Degrad. Stab. **95**, 2289–2299 (2010)
125. Atiqullah, M., Winston, M.S., Bercaw, J.E., Hussain, I., Fazal, A., Al-Harthi, M.A., Emwas, A.-H.M., Khan, M.J., Hossaen, A.: Effects of a vanadium post-metallocene catalyst-induced

polymer backbone inhomogeneity on UV oxidative degradation of the resulting polyethylene film. Polym. Degrad. Stab. **97**, 1164–1177 (2012)
126. Corrales, T., Catalina, F., Peinado, C., Allen, N.S., Fontan, E.: Photooxidative and thermal degradation of polyethylenes: interrelationship by chemiluminescence, thermal gravimetric analysis and FTIR data. J. Photochem. Photobiol., A **147**, 213–224 (2002)
127. Allen, N.S., Edge, M., Holdsworth, D., Rahman, A., Catalina, F., Fontan, E., Escalona, A. M., Sibon, F.F.: Ageing and spectroscopic properties of polyethylenes: comparison with metallocene polymer. Polym. Degrad. Stab. **67**, 57–67 (2000)
128. Severini, F., Gallo, R., Ipsale, S., Nisoli, E., Pardi, M.: Natural weathering of some new ethylene copolymers. Polym. Degrad. Stab. **65**, 329–335 (1999)
129. Morlat-Therias, S., Mailhot, B., Gonzalez, D., Gardette, J.-L.: Photooxidation of polypropylene/montmorillonite nanocomposites. 2. Interactions with antioxidants. Chem. Mater. **17**, 1072–1078 (2005)
130. Chmela, S., Kleinova, A., Fiedlerova, A., Borsig, E., Kaempfer, D., Thomann, R., Mulhaupt, R.: Photo-oxidation of sPP/organoclay nanocomposites. J. Macromol. Sci. Part A Pure Appl. Chem. **42**, 821–829 (2005)
131. Qin, H., Zhang, S., Liu, H., Xie, S., Yang, M., Shen, D.: Photo-oxidative degradation of polypropylene/montmorillonite nanocomposites. Polymer **46**, 3149–3156 (2005)
132. Dintcheva, N.T., Al-Malaika, S., La Mantia, F.P.: Effect of extrusion and photo-oxidation on polyethylene/clay nanocomposites. Polym. Degrad. Stab. **94**, 1571–1588 (2009)
133. Bottino, F.A., Di Pasquale, G., Fabbri, E., Orestano, A., Pollicino, A.: Influence of montmorillonite nano-dispersion on polystyrene photo-oxidation. Polym. Degrad. Stab. **94**, 369–374 (2009)
134. Gaume, J., Rivaton, A., Thérias, S., Gardette, J.-L.: Influence of nanoclays on the photochemical behaviour of poly(vinyl alcohol). Polym. Degrad. Stab. **97**, 488–495 (2012)
135. Morlat, S., Mailhot, B., Gonzalez, D., Gardette, J.L.: Photooxidation of polypropylene/montmorillonite nanocomposites: part I: influence of nanoclay and compatibilising agent. Chem. Mater. **16**, 377–383 (2004)
136. Qin, H., Zhao, C., Zhang, Z., Chen, G., Yang, M.: Photo-oxidative degradation of polyethylene/montmorillonite nanocomposite. Polym. Degrad. Stab. **81**, 497–500 (2003)
137. Qin, H., Zhang, Z., Feng, M., Gong, F., Zhang, S., Yang, M.: The influence of interlayer cations on the photo-oxidative degradation of polyethylene/montmorillonite composites. J. Polym. Sci., Part B: Polym. Phys. **42**, 3006–3012 (2004)
138. Morlat-Therias, S., Mailhot, B., Gardette, J.-L., Da Silva, C., Haidar, B., Vidal, A.: Photooxidation of ethylene-propylene-diene/montmorillonite nanocomposites. Polym. Degrad. Stab. **90**, 78–85 (2005)
139. Azuma, Y., Takeda, H., Watanabe, S., Nakatani, H.: Outdoor and accelerated weathering tests for polypropylene and polypropylene/talc composites: a comparative study of their weathering behavior. Polym. Degrad. Stab. **94**, 2267–2274 (2009)
140. Zaidi, L., Kaci, M., Bruzaud, S., Bourmaud, A., Grohens, Y.: Effect of natural weather on the structure and properties of polylactide/Cloisite 30B nanocomposites. Polym. Degrad. Stab. **95**, 1751–1758 (2010)
141. Dintcheva, N.T., Filippone, G., La Mantia, F.P., Acierno, D.: Photo-oxidation behaviour of polyethylene/polyamide 6 blends filled with organomodified clay: improvement of the photo-resistance through morphology modification. Polym. Degrad. Stab. **95**, 527–535 (2010)
142. Essawy, H.A., Abd El-Wahab, N.A., Abd El-Ghaffar, M.A.: PVC-laponite nanocomposites: enhanced resistance to UV radiation. Polym. Degrad. Stab. **9**, 1472–1478 (2008)
143. La Mantia, F.P., Dintcheva, N.T., Malatesta, V., Pagani, F.: Improvement of photostability of LLDPE-based nanocomposites. Polym. Degrad. Stab. **91**, 3208–3213 (2006)
144. Morlat-Therias, S., Fanton, E., Gardette, J.-L., Dintcheva, N.T., La Mantia, F.P., Malatesta, V.: Photochemical stabilization of linear low-density polyethylene/clay nanocomposites: towards durable nanocomposites. Polym. Degrad. Stab. **93**, 1776–1780 (2008)

145. Larché, J.-F., Bussière, P.-O., Gardette, J.-L.: Photo-oxidation of acrylic-urethane thermoset networks. Relating materials properties to changes of chemical structure. Polym. Degrad. Stab. **96**, 1438–1444 (2011)
146. Janorkar, A.V., Metters, A.T., Hirt, D.E.: Degradation of poly(L-lactide) films under ultraviolet-induced photografting and sterilization conditions. J. Appl. Polym. Sci. **106**, 1042–1047 (2007)
147. Bocchini, S., Fukushima, K., Di Blasio, A., Fina, A., Frache, A., Geobaldo, F.: Polylactic acid and polylactic acid-based nanocomposite photooxidation. Biomacromolecules **11**, 2919–2926 (2010)
148. Gardette, M., Thérias, S., Gardette, J.-L., Murariu, M., Dubois, P.: Photooxidation of polylactide/calcium sulphate composites. Polym. Degrad. Stab. **96**, 616–623 (2011)
149. Araújo, A., Botelho, G.L., Silva, M., Machado, A.V.: UV stability of poly(lactic acid) nanocomposites. J. Mater. Sci. Eng. B **3**, 75–83 (2013)
150. Ndiaye, D., Fanton, E., Morlat-Therias, S., Vidal, L., Tidjani, A., Gardette, J.-L.: Durability of wood polymer composites: Part 1. Influence of wood on the photochemical properties. Compos. Sci. Technol. **68**, 2779–2784 (2008)
151. Matuana, L.M., Jin, S., Stark, N.M.: Ultraviolet weathering of HDPE/wood-flour composites coextruded with a clear HDPE cap layer. Polym. Degrad. Stab. **96**, 97–106 (2011)
152. Peng, Y., Wang, W., Cao, J., Guo, X.: Effects of a layered double hydroxide (LDH) on the photostability of wood flour/polypropylene composites during UV weathering. RSC Adv. **5**, 41230–41237 (2015)
153. Matuana, L.M., Kamdem, D.P.: Accelerated ultraviolet weathering of PVC/wood flour composites. Polym. Eng. Sci. **42**, 1657–1666 (2002)
154. Stark, N.M., Matuana, D.P.: Surface chemistry changes of weathered HDPE/wood-flour composites studied by XPS and FTIR spectroscopy. Polym. Degrad. Stab. **86**, 1–9 (2004)
155. Stark, N.M., Matuana, L.M.: Surface chemistry and mechanical property changes of wood-flour/high-density-polyethylene composites after accelerated weathering. J. Appl. Polym. Sci. **94**, 2263–2273 (2004)
156. Yousif, E., Haddad, R.: Photodegradation and photostabilization of polymers, especially polystyrene: review. Springerplus **2**, 398 (2013)
157. Gulmine, J.V., Janissek, P.R., Heise, H.M., Akcelrud, L.: Degradation profile of polyethylene after artificial accelerated weathering. Polym. Degrad. Stab. **79**, 385–397 (2003)
158. Feldman, D.: Polymer weathering: photo-oxidation. J. Polym. Environ. **10**, 163–173 (2002)
159. Pickett, J.E., Gardner, M.M., Gibson, D.A., Rice, S.T.: Global weathering of aromatic engineering thermoplastics. Polym. Degrad. Stab. **90**, 405–417 (2005)
160. Pickett, J.E., Gibson, D.A., Gardner, M.M.: Effects of irradiation conditions on the weathering of engineering thermoplastics. Polym. Degrad. Stab. **93**, 1597–1606 (2008)
161. Andrady, A.L.: Wavelength sensitivity in polymer photodegradation. Adv. Polym. Sci. **128**, 47–94 (1997)
162. Searle, N.D.: Activation spectra: techniques and applications to stabilization and stability testing of materials. In: Ketola, W.D., Evans, J.D. (eds.) Durability 2000: Accelerated and Outdoor Weathering Testing, p. 41. ASTM, West Conshohocken, PA (2000)
163. Trubiroha, P., Geburtig, A., Wachtendorf, V.: Determination of the spectral response of polymers. In: Martin, J.W., Ryntz, R.A., Dickie, R.A. (eds.) Service Life Prediction: Challenging the Status Quo, p. 241. Federation of Societies for Coatings Technology, Bluebell, PA (2005)
164. Chin, J., Byrd, E., Embree, N., Garver, J., Dickens, B., Finn, T., Martin, J.: Accelerated UV weathering device based on integrating sphere technology. Rev. Sci. Instrum. **75**, 4951–4959 (2004)
165. Kärhä, P., Heikkilä, A., Ruokolainen, K., Kaunismaa, M.: A novel facility for ageing materials with narrow-band ultraviolet radiation exposure. Rev. Sci. Instrum. **82**, 023107 (2011)

166. Geburtig, A., Wachtendorf, V.: Determination of the spectral sensitivity and temperature dependence of polypropylene crack formation caused by UVirradiation. Polym. Degrad. Stab. **95**, 2118–2123 (2010)
167. Heikkilä, A., Kärhä, P.: Photoyellowing revisited: determination of an action spectrum of newspaper. Polym. Degrad. Stab. **99**, 190–195 (2014)
168. Martin, J.W., Chin, J.W., Nguyen, T.: Reciprocity law experiments in polymeric photodegradation: a critical review. Prog. Org. Coat. **47**, 292–311 (2003)
169. Chin, J., Nguyen, T., Byrd, E., Martin, J.: Validation of the reciprocity law for coating photodegradation. J. Coat. Technol. Res. **2**, 499–508 (2005)
170. Christensen, P.A., Dilks, A., Egerton, T.A., Temperley, J.: Infrared spectroscopic evaluation of the photodegradation of paint Part II: the effect of UV intensity & wavelength on the degradation of acrylic films pigmented with titanium dioxide. J. Mater. Sci. **35**, 5353–5358 (2000)
171. Searle, N.D.: Activation spectra of polymers and their application to stabilization and stability testing. In: Hamid, S.H. (ed.) Handbook of Polymer Degradation, 2nd edn, p. 605. Marcel Dekker Inc, New York (2000)
172. Torikai, A.: Wavelength sensitivity of polymers. In: Hamid, S.H. (ed.) Handbook of Polymer Degradation, 2nd edn, p. 773. Marcel Dekker Inc, New York (2000)
173. Nguyen, T., Martin, J. W., Byrd, E., Embree, E. Effects of spectral UV on degradation of acrylic-urethane coatings. In: Proceedings of the 80th Annual Meeting of the Program of FSCT (2002)
174. François-Heude, A., Richaud, E., Desnoux, E., Colin, X.: Influence of temperature, UV-light wavelength and intensity on polypropylene photothermal oxidation. Polym. Degrad. Stab. **100**, 10–20 (2014)
175. Ferreira, J.L., Melo, M.J., Ramos, A.M.: Poly(vinyl acetate) paints in works of art: a photochemical approach. Part 1. Polym. Degrad. Stab. **95**, 453–461 (2010)
176. Nagai, Y., Nakamura, D., Ueno, H., Matsumoto, N., Ohishi, F.: Photodegradation mechanisms in poly(2,6-butylenenaphthalate-*co*-tetramethyleneglycol) (PBN-PTMG). II: wavelength sensitivity of the photodegradation. Polym. Degrad. Stab. **88**, 256–260 (2005)
177. Kockott, D.: Weathering. In: Brown, R. (ed.) Handbook of Polymer Testing. Physical Methods, p. 697. Rapra Technology, Shawbury (1999)
178. Pospisil, J., Nespurek, S.: Photostabilization of coatings. Mechanisms and performance. Prog. Polym. Sci. **25**, 1261–1335 (2000)
179. Bauer, D.R.: Interpreting weathering acceleration factors for automotive coatings using exposure models. Polym. Degrad. Stab. **69**, 305–316 (2000)
180. Pickett, J.E., Sargent, J.R.: Life predictions for hardcoated polycarbonate. In: Martin, J.W., Ryntz, R.A., Chin, J., Dickie, R.A. (eds.) Service Life Prediction of Polymeric Materials: Global Perspectives, p. 153. Springer, Berlin (2008)
181. Meeker, W.Q., Escobar, L.A., Chan, V.: Using accelerated tests to predict service life in highly variable environments. In: Martin, J.W., Bauer, D.R. (eds.) Service Life Prediction Methodology and Metrologies. ACS Symposium Series 805, p. 396 (2002)
182. Bauer, D.R.: Combining failure models, environmental load distributions, and customer perception of failure to generate real world durability requirements. In: Martin, J.W., Ryntz, R.A., Dickie, R.A. (eds.) Service Life Prediction: Challenging the Status Quo, p. 39. Federation of Societies for Coatings Technology, Bluebell, PA (2005)
183. Pickett, J.E.: Highly predictive accelerated weathering of engineering thermoplastics. In: Martin, J.W., Ryntz, R.A., Dickie, R.A. (eds.) Service Life Prediction: Challenging the Status Quo, p. 93. Federation of Societies for Coatings Technology, Bluebell, PA (2005)
184. Cruz-Pinto, J.J.C., Carvalho, M.E.S., Ferreira, J.F.A.: The kinetics and mechanism of polyethylene photo-oxidation. Angew. Makromol. Chem. **216**, 113–133 (1994)
185. Fischer, R.M., Ketola, W.D.: Error analysis and associated risks for accelerated weathering results. In: Martin, J.W., Ryntz, R.A., Dickie, R.A. (eds.) Service Life Prediction: Challenging the Status Quo, p. 79. Federation of Societies for Coatings Technology, Bluebell, PA (2005)

186. Pickett, J.E., Sargent, J.R.: Sample temperatures during outdoor and laboratory weathering exposures. Polym. Degrad. Stab. **94**, 189–195 (2009)
187. Gillen, K.T., Bernstein, R., Celina, M.: Non-arrhenius behavior for oxidative degradation of chlorosulfonated polyethylene materials. Polym. Degrad. Stab. **87**, 335–346 (2005)
188. Pickett, J.E., Gibson, D.A., Rice, S.T., Gardner, M.M.: Effects of temperature on the weathering of engineering thermoplastics. Polym. Degrad. Stab. **93**, 684–691 (2008)
189. Pickett, J.E., Sargent, J.R., Blaydes, H.A., Babbie, N.: The effect of temperature on the weathering lifetime of coated polycarbonate. Polym. Degrad. Stab. **94**, 1085–1091 (2009)
190. Jacques, L.F.E.: Accelerated and outdoor/natural exposure testing of coatings. Prog. Polym. Sci. **25**, 1337–1362 (2000)
191. Kockott, D.: Factors influencing the reliability of results in accelerated weathering tests. In: Durability Testing on Nonmetallic Materials, pp. 24–39. ASTM STP 1294 (1996)
192. Kim, H.M., Jorgensen, G.J., King, D.E., Czanderna, A.W.: Development of a methodology for service lifetime prediction of renewable energy devices. In: Durability testing of nonmetallic materials. ASTM STP 1294, pp. 173–189 (1996)
193. Real, L.P., Gardette, J.-L., Rocha, A.P.: Artificial simulated and natural weathering of poly (vinyl chloride) for outdoor applications: the influence of water in the changes of properties. Polym. Degrad. Stab. **88**, 357–362 (2005)
194. James, S.L., Robinson, A.J., Arnold, J.C., Worsley, D.A.: The effects of humidity on photodegradation of poly(vinyl chloride) and polyethylene as measured by the CO_2 evolution rate. Polym. Degrad. Stab. **98**, 508–513 (2013)
195. Penon, M.G., Picken, S.J., Wubbenhorst, M., van Turnhout, J.: Water sorption in UV degraded clear and pigmented epoxy coatings assessed by dielectric sorption analysis. Polym. Degrad. Stab. **92**, 1247–1254 (2007)
196. Makki, H., Adema, K.N.S., Peters, E.A.J.F., Laven, J., van der Ven, L.G.J., van Benthem, R. A.T.M., de With, G.: Quantitative spectroscopic analysis of weathering of polyester-urethane coatings. Polym. Degrad. Stab. **121**, 280–291 (2015)
197. Aidani, R.E., Nguyen-Tri, P., Malajati, Y., Lara, J., Vu-Khanh, T.: Photochemical aging of an e-PTFE/NOMEX® membrane used in firefighter protective clothing. Polym. Degrad. Stab. **98**, 1300–1310 (2013)
198. Arrieta, C., David, E., Dolez, P., Vu-Khanh, T.: Hydrolytic and photochemical aging studies of a Kevlar®-PBI blend. Polym. Degrad. Stab. **96**, 1411–1419 (2011)
199. Kim, H.M., Urban, M.W.: Molecular level chain scission mechanisms of epoxy and urethane polymeric films exposed to UV/H_2O. Multidimensional spectroscopic studies. Langmuir **16**, 5382–5390 (2000)
200. Kumar, B.G., Singh, R.P., Nakamura, T.: Degradation of carbon fiber-reinforced epoxy composites by ultraviolet radiation and condensation. J. Compos. Mater. **36**, 2713–2733 (2002)
201. Musto, P., Ragosta, G., Abbate, A., Scarinzi, G.: Photo-oxidation of high performance epoxy networks: correlation between the molecular mechanisms of degradation and the viscoelastic and mechanical response. Macromolecules **41**, 5729–5743 (2008)
202. Larché, J.-F., Bussière, P.-O., Gardette, J.-L.: How to reveal latent degradation of coatings provoked by UV-light. Polym. Degrad. Stab. **95**, 1810–1817 (2010)
203. Malajati, Y., Therias, S., Gardette, J.-L.: Influence of water on the photooxidation of KHJ® phenoxy resins. 1. Mechanisms. Polym. Degrad. Stab. **96**, 144–150 (2011)
204. Andrady, A.L., Hamid, H.S., Torikai, A.: Effects of climate change and UV-B on materials. J. Photochem. Photobiol. Sci. **2**, 68–72 (2003)
205. Nagai, Y., Ogawa, T., Nishimoto, Y., Ohishi, F.: Analysis of weathering of a thermoplastic polyester elastomer II. Factors affecting weathering of a polyether-polyester elastomer. Polym. Degrad. Stab. **65**, 217–224 (1999)

Multicomponent Polymer Materials: Photodegradation Mechanism

Dan Rosu, Cristian-Dragos Varganici and Liliana Rosu

Abstract The chapter follows two pathways into polymer photodegradation. The first one consists of an introduction into the obtaining of photodegradable polymers by inserting several photolabile entities, such as o-nitrobenzyl, coumarin-4-ylmethyl, p-hydroxyphenacyl or acetal, with emphasis on the o-nitrobenzyl group. The second and main pathway of the chapter encompasses literature studies related to photochemical decomposition mechanisms occurring in different types of multicomponent polymer systems. UV radiations, oxygen, temperature and humidity represent the main driving forces in polymer irreversible photochemical deterioration, occurring through radical intermediates (hydroperoxides, chromophore carbonyl groups or conjugated double bonds) as chained mechanisms.

1 Introduction

The designing of multicomponent polymeric systems is an interesting research field because of the multiple possibilities for obtaining new materials without resorting to complicated synthesis methods of new polymers. Polymer mixtures, interpenetrated and semi-interpenetrated polymeric networks are some of the most popular multicomponent polymeric systems. Also, the polymeric materials usually contain many other micromolecular substances such as: plasticizers, colorants, antioxidants, fillers and various types of additives that may affect the photochemical stability [1]. Although the interest in photochemical stability of polymeric materials is still limited to some photodegradation topics, such as degradative processes, the linkage

D. Rosu (✉) · C.-D. Varganici · L. Rosu
Centre of Advanced Research in Bionanoconjugates and Biopolymers, "Petru Poni" Institute of Macromolecular Chemistry, 41A Grigore Ghica-Voda Alley, 700487 Iasi, Romania
e-mail: drosu@icmpp.ro

C.-D. Varganici
e-mail: varganici.cristian@icmpp.ro

L. Rosu
e-mail: lrosu@icmpp.ro

D. Rosu and Visakh P. M. (eds.), *Photochemical Behavior of Multicomponent Polymeric-based Materials*, Advanced Structured Materials 26,
DOI 10.1007/978-3-319-25196-7_3

between the degradative processes and chemoluminescence effects, modifications induced by light on chemical bonds, the sensitivity of polymers to different wavelengths and environmental issues, these still represent state of the art challenges [2]. The interest in photodegradable and/or photosensible polymeric materials significantly increased in recent years due to advantages of ultraviolet or visible light as trigger agent that can be easily controlled both spatially and temporarily [3–5]. Photodegradable polymeric materials are usually designed to check their degradability during exposure to sunlight.

Design of new polymers who are able of splitting into toxic fragments after overcoming their lifetime is, along with recycling, the main technique used to avoid the accumulation of plastic waste in the environment [6]. In the range of degradable plastics photodegradable polymers occupy a special place due to their multiple practical applications in agriculture as ground cover with plastic sheets (plasticulture) [7], in packaging industry and in household where the waste materials left after unpacking and the kitchenware plastics or the trash bags should quickly degrade without negative impact on the environment, in smart materials manufacture for microelectronics devices and in biomedical applications [8]. Hydrogels with photo-degradable bonds have high potential of applications in tissue engineering due to their possibilities of release on demand of bio-active principles or of cells included in the tridimensional polymer structure [9]. Some papers shows that small particles encapsulated in light sensible polymers with different structures have been designed with the purpose to decompose through photo-activation for releasing the active principles [10–12]. A careful handling of several parameters such as wavelength, light intensity, irradiation dose, temperature, exposure time and mechanical stress, the stability of the polymeric material, the degradation rate and its shape may be controlled [13]. By including in the polymer main chain of some photolabile chemical structures, such as o-nitrobenzyl derivatives (I), compounds based on coumarin-4-ylmethyl (II), p-hydroxyphenacyl (III) or acetal (ketal) moieties IV), polymeric materials with controlled photo-degradability have been obtained.

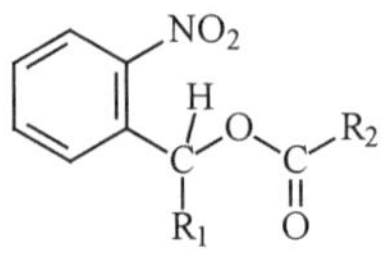

I

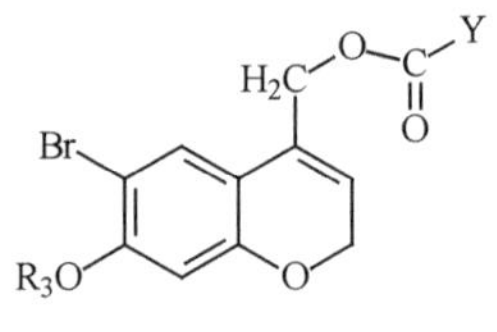

II

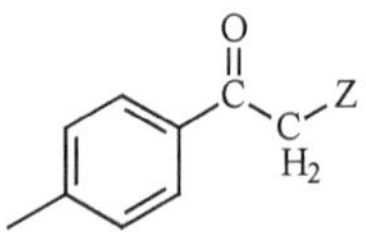

III

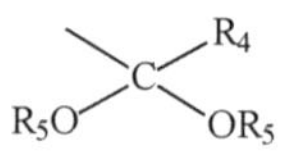

IV

Photo reconfigurable polymeric materials contain some chemical structures such as cinnamate derivatives (V) or functionalized coumarin (VI), which are enabled via reversible photochemical dimerization [13].

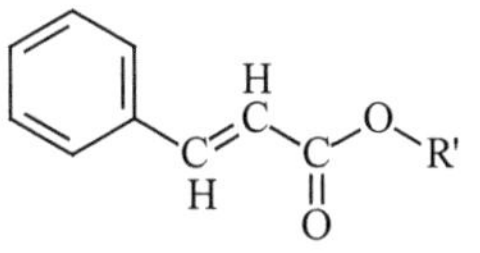

V

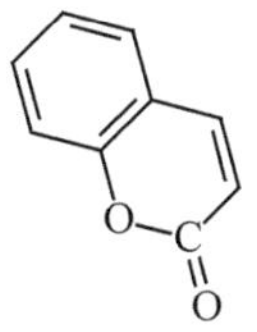

VI

Photoizomerization is another type of reversibile structural modification supported by the macromolecular chains during exposure to light [8, 13].

Some examples on photodegradative processes of different photochemical decomposition mechanisms of multicomponent based polymeric materials, either made photodegradable via incorporated specific entities or not, are further depicted in this chapter.

2 Photodegradation Mechanisms of Multifunctional Polymeric Materials

It is first of all worth mentioning that polymeric materials during photodecomposition are not fully degraded to lower molecular mass polymer chains by different photochemical reactions, rather photochemical reactivity introduces different functions, such as aldehyde, carboxylic acid, alcohol or ketone end-groups [14, 15]. Generation of these oxygen based end-groups may also account for further cellular β-oxidation processes occurrence [16, 17].

There are two basic pathways for obtaining photodegradable polymers, either by chemically modifying the polymer main chains with the insertion of a light responsive photodegradable chromophore entity, such as polyolefins or carbonyls [18], or by blending them with specific additives able of initiating photochemical decomposition processes (typically radical autoxidation reactions) within the polymer [6, 19, 20].

As previously discussed in the introductory chapter, the Norrish type reactions account for photodecomposition of polymers with carbonyl moieties, such as polyacrylates, polyesters and polyketones, and through abstraction of the hydrogen atom [21, 22]. After forming of free radicals through Norrish I reactions, photodegradation further proceeds via autoxidation.

Other chromophore entities than carbonyl ones may be used for obtaining photodegradable polymers, such as metal–metal bonds in the main chains, which undergo homolytic scission by absorbtion in the visible light domain [23–26].

Another pathway implied in the obtaining of photodegradable polymers consists in the blending of light responsive radical initiators. A very often used compound in this sense is TiO_2 which generates free radicals capable of reacting with oxygen to yield hydroperoxides, as shown below.

$$TiO_2 \xrightarrow{hv} TiO_2^* \xrightarrow{O_2,RH} TiO_2 + R^{\cdot} + HOO^{\cdot}$$

Cleavage of O–O bonds from hydroperoxides further assure photodecomposition together with alkoxy radicals β-scissions afterwards. Another route to polymer photodegradation resides in the further reactions of the hydroperoxides to generate other chromophores and/or new functional entities, such as carbonyls, carboxylic acids, alcohols and olefins, which further contribute to main chain scissions [6] (Scheme 1).

As previously discussed, the generated carbonyl entities further cleave main chains through Norrish reactions, while β-scissions are the main driving force in most polymer backbones photo-oxidative degradation [27] (Scheme 1). Specific UV absorbers may also be blended into the polymer for lowering or retarding the photodecomposition rate by reducing light intensity to carbonyl entities.

Another approach for obtaining photodegradable polymers consists in using the sol–gel technique to covalently couple different inorganic photosensitizers to specific polymer pendant moieties [28, 29], such initiators possessing solid state selectivity regarding inorganic aggregates [6].

Lei et al. [3]. reported the first photodegradable polymers multifunctionalization for micropatterns by firstly obtaining three ortho-nitrobenzaldehyde monomers, containing allyl, propargyl and epoxy entitites, through reacting 5-hydroxy-2-nitrobenzaldehyde with bromides of allyl, propargyl and with epichlorohydrin (Scheme 2) (Reproduced with permission. Copyright 2014, American Chemical Society [3]).

The obtained monomers were further polymerized through the Passerini multicomponent polymerization pathway by further reactions with 1,6-hexanedioic acid

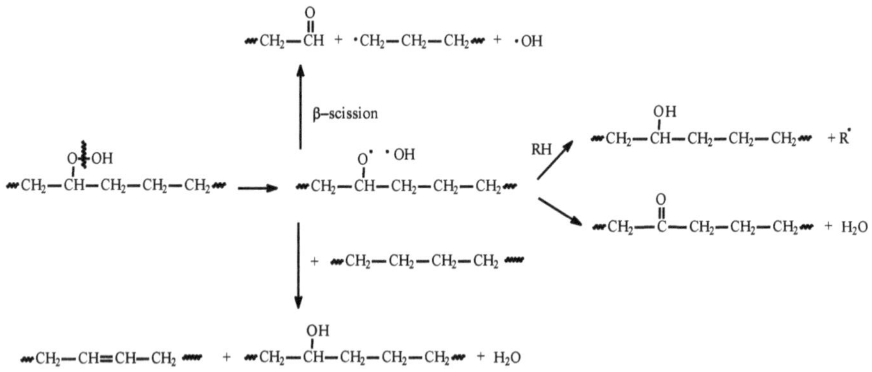

Scheme 1 Different reactions of hydroperoxides

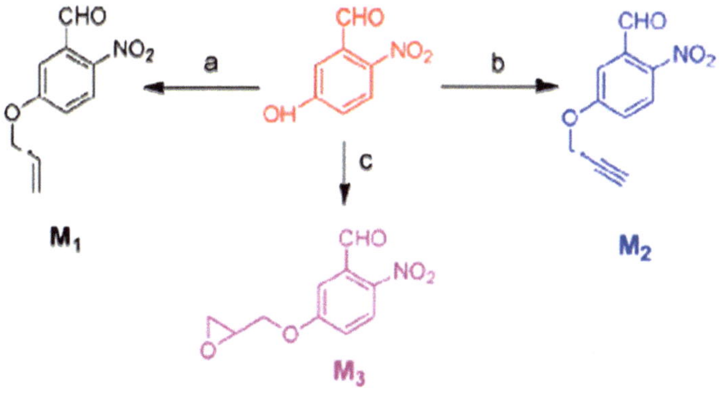

Scheme 2 Obtaining of functional monomers from 5-hydroxy-2-nitrobenzaldehyde. Reactions conditions: (*a*) allyl bromide, K_2CO_3, DMF, room temperature, overnight; (*b*) Propargyl bromide, K_2CO_3, DMF, room temperature, overnight; (*c*) epychlorohydrin, K_2CO_3, DMF, 80 °C, 24 h. Reproduced with permission. Copyright 2014, American Chemical Society [3]

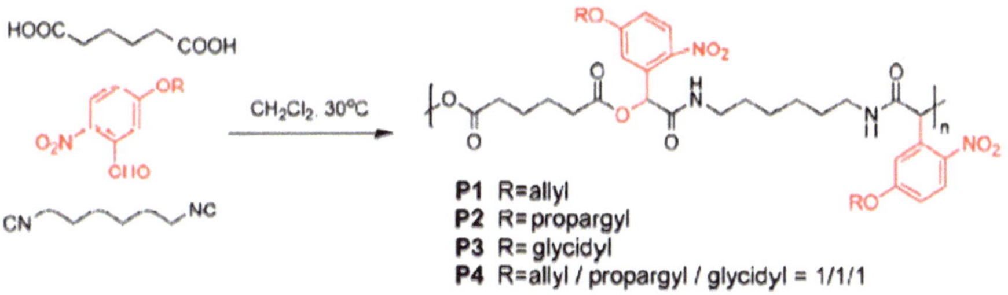

Scheme 3 Photocleavable polymers via the Passerini synthesis pathway. Reproduced with permission. Copyright 2014, American Chemical Society [3]

and 1,6-diisocyanohexane to yield three poly(ester-amides). The obtained polymers are thusly photodegradable through the ester bonds containing ortho-nitrobenzyl entities (Scheme 3) (Reproduced with permission. Copyright 2014, American Chemical Society [3]). Since the obtaining of photocleavable polymers consists of multistage time consuming processes, through the Passerini approach such disadvantages are greatly reduced by directly obtaining the desired polymers from o-nitrobenzaldehyde and further expanding their applications range [30].

The functional groups from the starting monomers may be further modified either by click chemistry, such as thiol-ene, epoxy-amine reaction and copper catalyzed azide–alkyne cycloaddition. Authors also applied the Passerini synthesis pathway to a mixture of the three initial monomers to obtain a combined photodegradable polymer with triple functionality (P4) (Scheme 4). Reproduced with permission. Copyright 2014, American Chemical Society [3]. The photocleavable character may be determined by UV–Vis and gel permeation chromatography (GPC) techniques, when the polymers cleave into the corresponding ortho-nitrosobenzaldehydes and carboxylic acids.

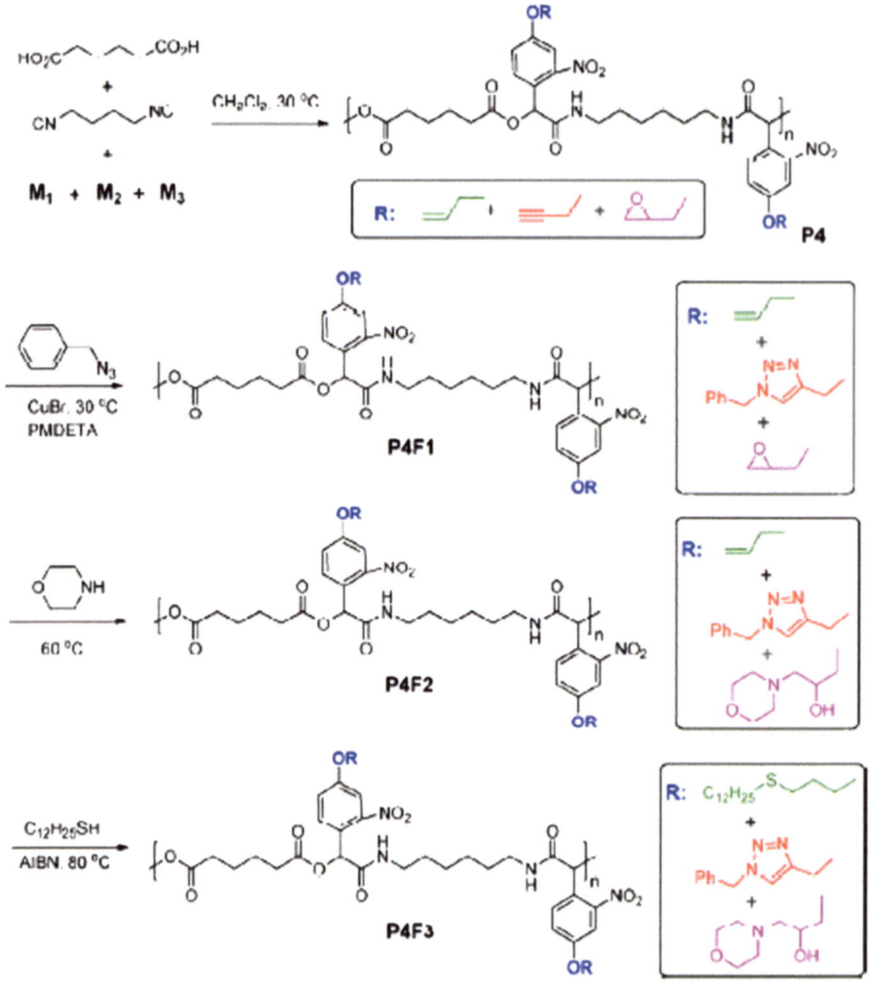

Scheme 4 Obtaining of the photocleavable triple function polymer and the sequential functionalizations. Reproduced with permission. Copyright 2014, American Chemical Society [3]

The modification of different structural properties upon irradiation could lead to polymer disruption or release of different loadings thus making such materials good candidates for a wide palette of applications which range from controlled drug delivery to surface modifications [31–36]. Obtaining of various other such polymers was reported in the literature [37, 38].

The photodegradable characteristic of such polymers may be exploited as photoresists. In this sense, the triply functionalized polymer was tested as photoresist for obtaining reactive patterns, since the multiple groups may be used in surface modification. Photopatterning was undertaken by a UV lamp and a TEM grid photomask. This may be achievable, as previously mentioned, by:

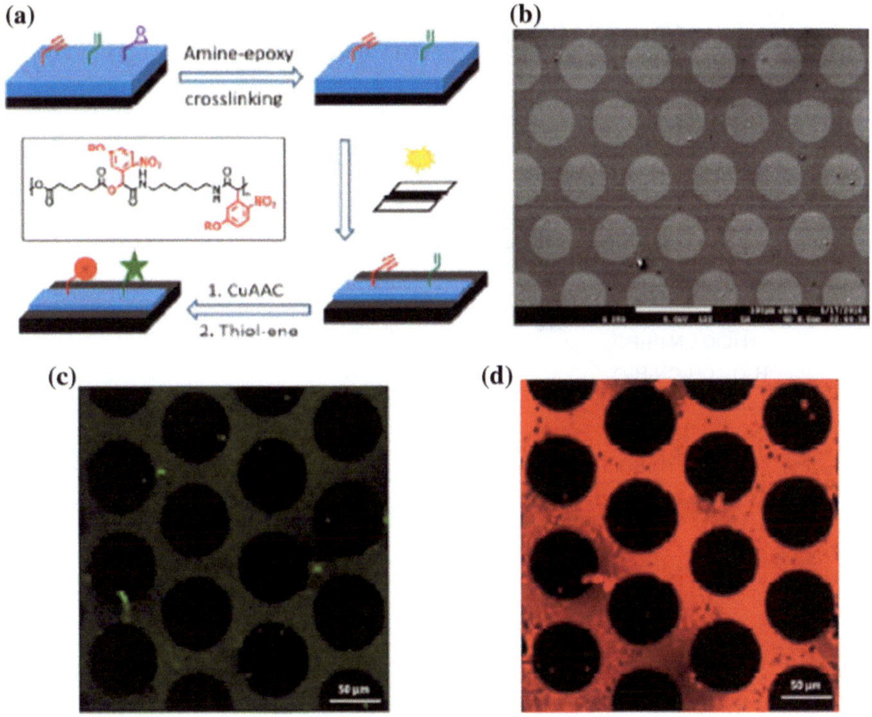

Scheme 5 **a** Multifunctional patterning on a silicon wafer using the photoresist polymer P4. **b** SEM micrograph of the patterned film; **c** Confocal fluorescence of the patterned silicon wafer film functionalized with Azido-FITC and **d** rhodamine dye. Reproduced with permission. Copyright 2014, American Chemical Society [3]

(1) crosslinking the film with amine–epoxy process (2) directly patterning of the surface due to photocleavable property; (3) copper catalyzed azide–alkyne cycloaddition (CuAAC) and (4) thoil-ene reaction. The surface used was a silicon wafer (Scheme 5) Reproduced with permission. Copyright 2014, American Chemical Society [3].

Another challenging and emerging field is that of obtaining phototriggable polymers from renewable bioresources. This is to address the issues surrounding synthetic commercially available polymers (e.g. poor degradability and sustainability) by obtaining biobased polymers with photocleavable units which may be preprogrammed for light decomposition. Two key compounds in doing so are 5-hydroxymethylfurfural (HMF) and one of its oxidation products, 2,5-furandicarboxylic acid (FDCA). For example, terephthalic acid may be replaced by FDCA in PET, with similar final properties. Rajendran et al. obtained a polymer based on FDCA with nitrobenzyl group for triggering photodecomposition upon light irradiation. Firstly, authors obtained model compounds to test the phototriggering efficiency of the widely mentioned 2-nitrobenzyl group. Authors converted

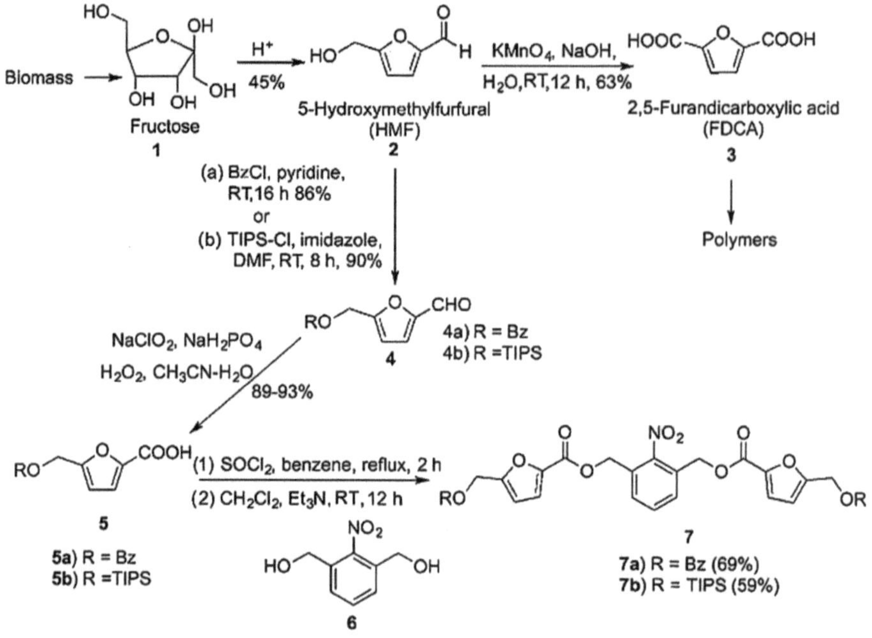

Scheme 6 Synthesis of biobased model compounds for phototriggered degradation, where Bz is benzyl, TIPS is triisopropylsilyl and DMF is N,N-dimethylformamide [39] (reproduced with kind permission from John Wiley and Sons—License no. 3879230249556)

fructose into HMF and FDCA and further functionalized them with 2-nitro-1,3-benzenedimethanol as phototrigger [39]. Photoirradiation of the obtained polymer was undertaken at 350 nm, monitored by ^{1}H–NMR and UV–Vis methods and was reported as being very efficient. There was yielded nitrosoaldehyde which degraded to furan carboxylic acid. (Scheme 6—reproduced with kind permission from John Wiley and Sons—License no. 3879230249556) [39]. Its unsymmetrically substituted ester photodecomposed with the same efficiency (Scheme 7(b)—reproduced with kind permission from John Wiley and Sons—License no. 3879230249556) [39].

With the success obtained after testing the model compounds, authors further synthesized polymer/oligomers from FDCA, not only to assess photocleavage efficiency, but also to attempt monomer recovery. The reaction is given in Scheme 8 [39] (reproduced with kind permission from John Wiley and Sons—License no. 3879230249556). The product was washed with acetone and CH_2Cl_2 for removing the low molecular weight oligomers. Its photodegradability was monitored in both suspension and solid state, in the latter occurring much slower. The photodegradation of the polymer was complete with the yielding of the starting monomer FDCA. Authors further expanded with success their approach to copolymers with glycols as phototriggers [39].

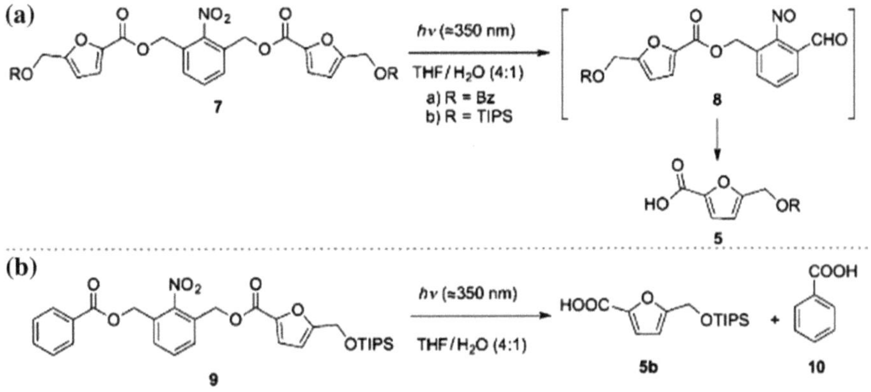

Scheme 7 Photodegradation of biobased model compounds: **a** symmetrical ester; **b** unsymmetrically ester [39] (reproduced with kind permission from John Wiley and Sons—License no. 3879230249556)

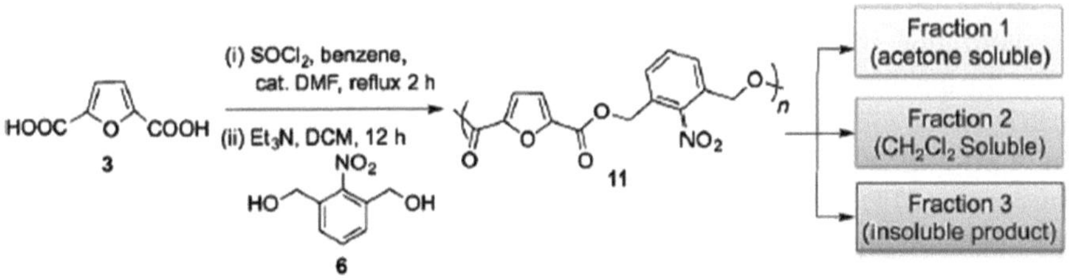

Scheme 8 Synthesis of the biobased polymer/oligomer [39] (reproduced with kind permission from John Wiley and Sons—License no. 3879230249556)

Christensen et al. [40] studied the UV photodecomposition behavior of poly (vinyl chloride) (PVC) and poly(caprolactone) (PCL) polymers in a 1:1 w/w blend and the corresponding homopolymers by monitoring the CO_2 emission during irradiation. The authors observed that the blend yielded lower CO_2 emission, thus the existing of specific beneficial morphological interactions between blend components which reduce photodegradation by a control over photo-oxidation. In individual PCL, chains entanglements reduce the possibility of crystalline regions formation, as opposed to PCL in the 1:1 ratio blend with PVC, in which PVC perturbs carbonyl entities vibration of PCL due to their interactions with chlorine atoms, thus acting as plasticizer. If PCL would have exhibited a higher crystallinity degree as PVC in the blend, then it would reduce oxygen diffusion. If PVC would have exhibited a higher crystallinity degree as PCL in the blend and phase-separated into partially or full crystalline domains, as suspected by the authors, this would greatly reduce the overall oxygen diffusion rate, effect which would be greater enhanced if both polymers would yield phase separation phenomena. Crystallite shapes are very important, since they cause light to scatter into the blend, thus generating the opposite by favoring photo-oxidation.

Li et al. [41] studied the natural photo–aging for polypropylene (PP) and PP nanocomposites with $CaCO_3$ and SiO_2 exposed for 88 days and characterized them by means of FTIR, high temperature gel permeation chromatography (HGPC) and pyrolysis gas chromatography–mass spectrometry (PGC–MS). It was found that the PP nanocomposites were more prone to photodecomposition than the initial polymers and with higher oxidation rates, through many chain scissions with formation of much lower molecular mass alcohol, ester and ketone entities and unsaturated double bonds which recrystallized and enhanced the nanocomposites crystallinity. Authors gave three possible reasons for the faster photodegradation rate of the PP nanocomposites compared to the unfilled polymers, with further research needing to be conducted on this aspect in the future for confirmation. The first reason refers to the functional groups covering the nanoparticles which may catalyze photo-oxidation. Hydroxyl groups are absorbed on the nano filler surfaces, more pregnant for SiO_2, thus activating photodegradation reactions. The second reason may reside in the agglomeration of filler nanoparticles leading to their weak adhesion to the PP matrix and interfacial incompatibility under stress due to crystallization contraction, more enhanced in the lesser compatible SiO_2 nanoparticles to the PP matrix. The weaker SiO_2 nanoparticles compatibility to the PP matrix is due to a higher number of hydroxyl groups on the surface. Authors based the third reason on their finding of the nanoparticle acting as a nucleation agent. This reduces PP spherulites sizes, enhancing interface crystallinity, which, together with the amorphous domains, is another photo-oxidation site.

Rosu et al. [42]. studied the photochemical stability of a cotton cellulose substrate painted with a series of reactive azo-triazine dyes (Table 1—reproduced with permission of Springer—License no. 3887660087491) [42]. The materials were exposed to UV irradiations with $\lambda > 300$ nm. The purpose of the study consisted in establishing of a photodecomposition mechanism of the dyed textiles, since other studies reported that aryl amines and free radicals, resulted during photodegradation of such dyes, are toxic to human skin after their absorbtion during perspiration. Details of the attachment mechanisms of the dyes to the cellulose substrate, dyeing process and toxicological studies of the dyes were reported in a Ph.D. thesis [43]. The same thesis reports colorimetric analyses of the dyed textiles during the UV irradiation, followed by dye extractions in aqueous media with acid and alkaline pH for human perspiration simulation and spectrophotometric identification of the dyes in solutions.

Colorimetric studies indicated a darkening tendency of surfaces painted with RY-143 and RO-13 dyes (light color dyes) and insignificant color modifications of surfaces painted with RR-2 and RR-183 dyes (dark color dyes), due to accumulation of blue and green chromophores on cellulose surface during 100 h irradiation time due to cellulose surface photo–oxidation reactions. Surface coloring in complementary colors (blue and green) during photoirradiation led to simultaneous hypochrome and bathochrome displacements of the absorbtion maxima. There could also be observed a widening in the absorbtion spectrum towards higher wavelengths, up to

Table 1 Characteristics of the four reactive azo–triazine dyes used in the study [42] (reproduced with permission of Springer—License no. 3887660087491)

Name/manufacturer	Chemical formula	λ_{max}	M_w
Reactive Yellow 143 (RY-143)/Hutsman textile effects	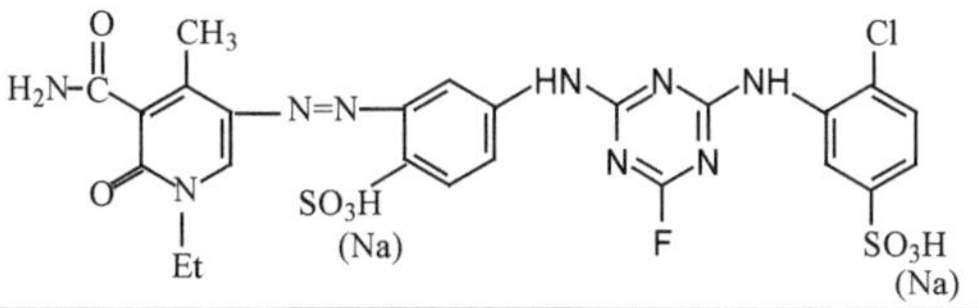	217, 265, 422	742.02
Reactive Orange 13 (RO-13)/Hutsman textile effects	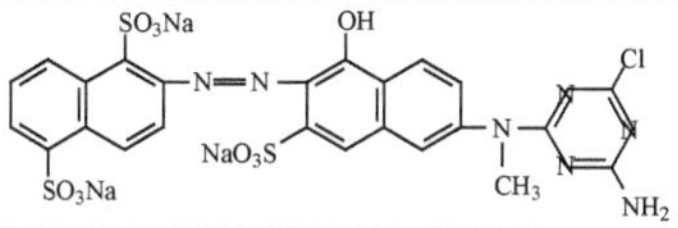	226, 284, 488	762.03
Reactive Red 183 (RR-183)/Hutsman textile effects	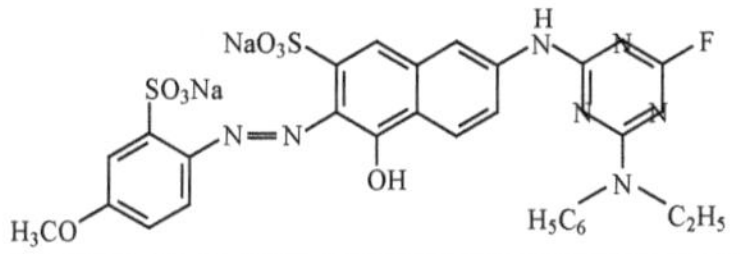	227, 280, 503	713.62
Reactive Red 2 (RR-2)/PRO chemical and dye	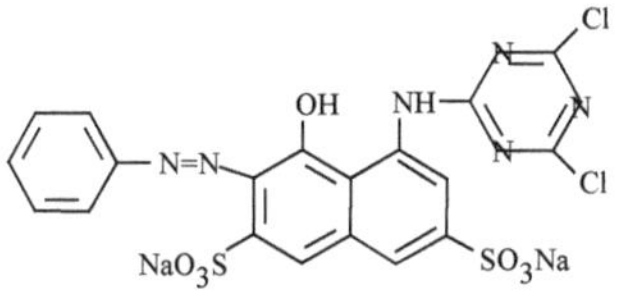	217, 289, 544	615.33

800 nm, which is a supplementary indication of new chromophores formation on the irradiated surfaces. The darkening of samples surfaces was attributed to C–N bond scissions from amine entities with the resulting of new chromophore structures [42] (Scheme 9—reproduced with permission of Springer—License no. 3887660087491) and was associated with the bathochrome displacement. Also, this darkening was also correlated with the substitution degree and number of amine entities. Scheme 10 [42] (reproduced with permission of Springer—License no. 3887660087491) shows the formation of intermediate blue chromophores from radical IX in Scheme 9 during photoirradiation of RY-143 dye, the other dyes exhibiting similar photochemical behavior [42].

Initiation of the photodegradation process may be undertaken through hydroxyl radical intermediates resulted after hydroperoxides decomposition. The general photodecomposition mechanism of RY-143 dye from the textile substrate is given in Scheme 11 [42] (reproduced with permission of Springer—License no. 3887660087491). The dye's chromophore entities, consisting of azo groups and aromatic rings, are destroyed after the attack of intermediate free radicals with the resulting of nitrogen, process associated with the hypochrome effect. The FTIR spectrum recorded after photoirradiation indicated significant modifications in the absorbtion ranges specific to glucose ring and carbonyl groups for the cellulose substrate painted with RY-143 dye. Results have demonstrated the apparition and intensification of photo-oxidative processes by the appearance of new absorbtion bands specific to carbonyl entities and an increase in absorbance signals specific to

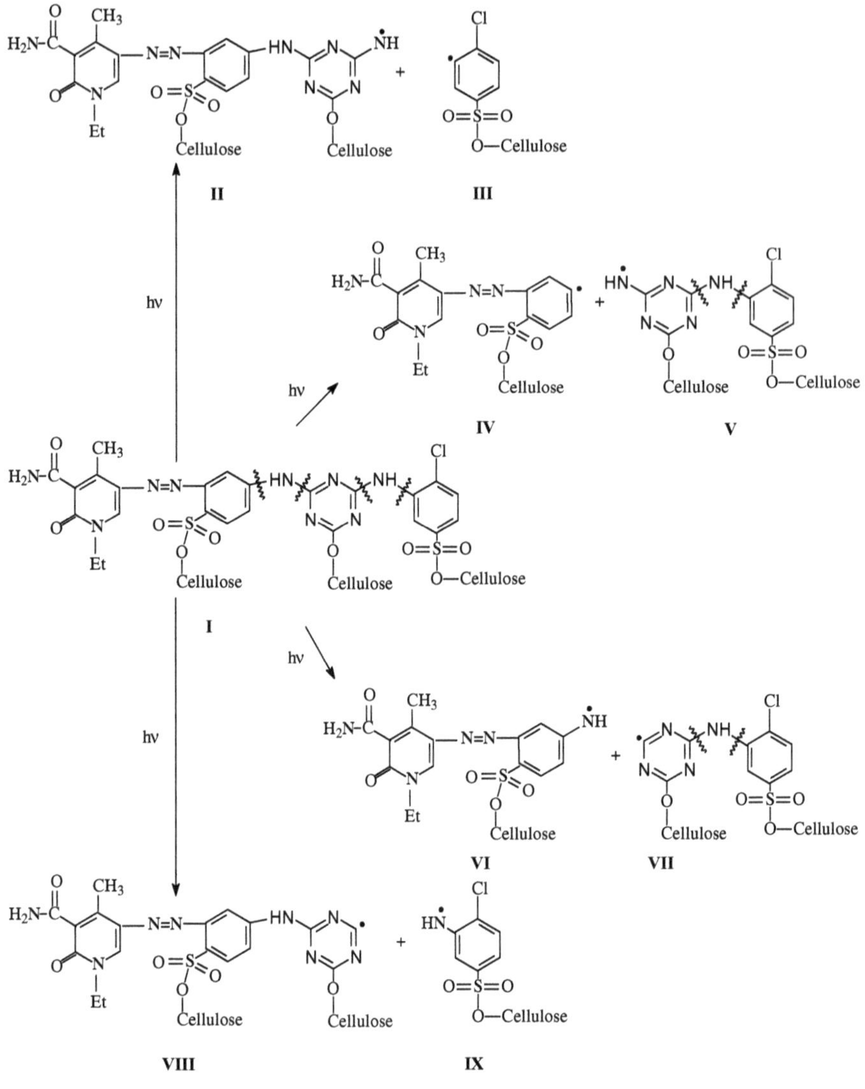

Scheme 9 Scission of C–N bonds during photodecomposition of RY-143 painted cellulose [42] (reproduced with permission of Springer—License no. 3887660087491)

glycosidic rings, suggesting cellulose chain-end initiated depolymerization processes and lowering of molecular mass. The dye molecules from the cellulosic substrate may be dissolved by human perspiration thus provoking hazardous health issues. Authors have observed an increase in the extracted dye quantity with irradiation dose and pH increase [42, 43].

Collagen photochemical behavior in multicomponent polymeric materials is another extensively debated topic. One must first comprehend basic photochemical

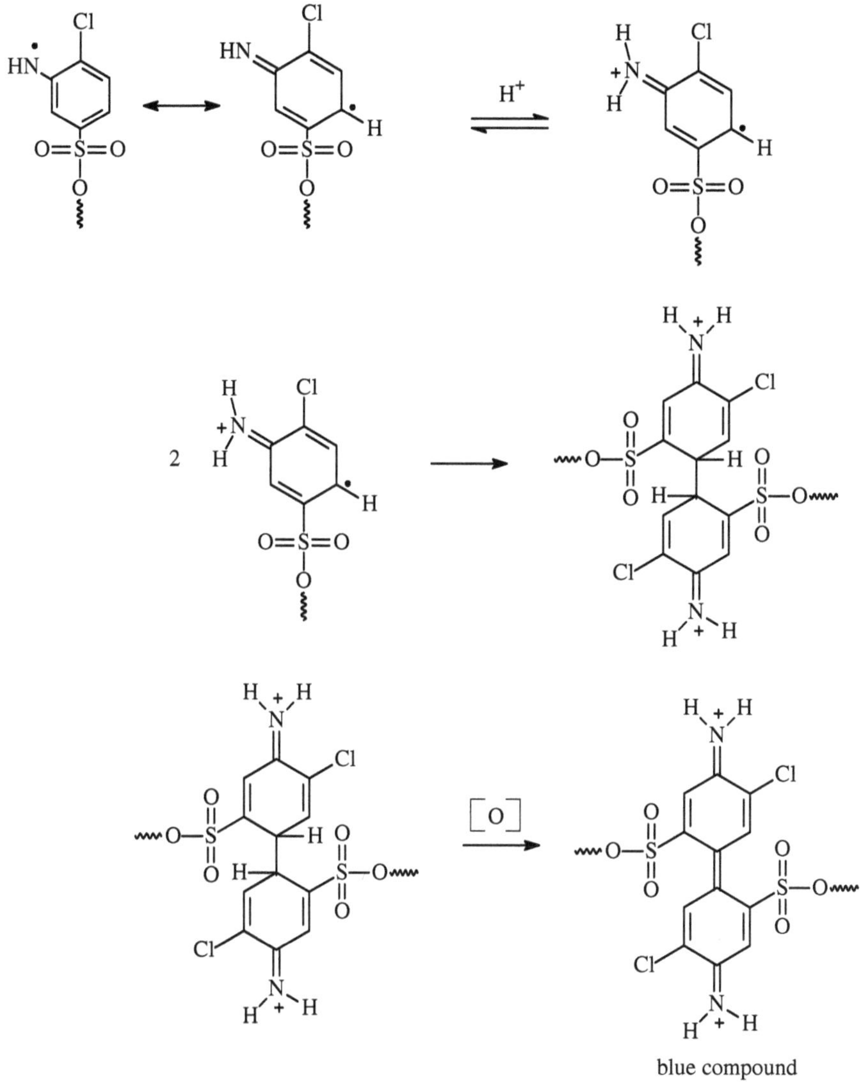

Scheme 10 The generating of blue chromophores during RY-143 dye photoirradiation [42] (reproduced with permission of Springer—License no. 3887660087491)

processes occurring in individual pure collagen. Rabotyagova et al. [44] studied the photochemical behavior of pure collagen irradiated at 254 nm and found changes at every structural level. In early photodegradation stages collagen shifts from the triple helix conformation, protector of single collagen macromolecules against peptide cleavage, to random coil, associated with water elimination (Fig. 1) [44] (Reproduced with permission of Springer—License no. 3890680622983).

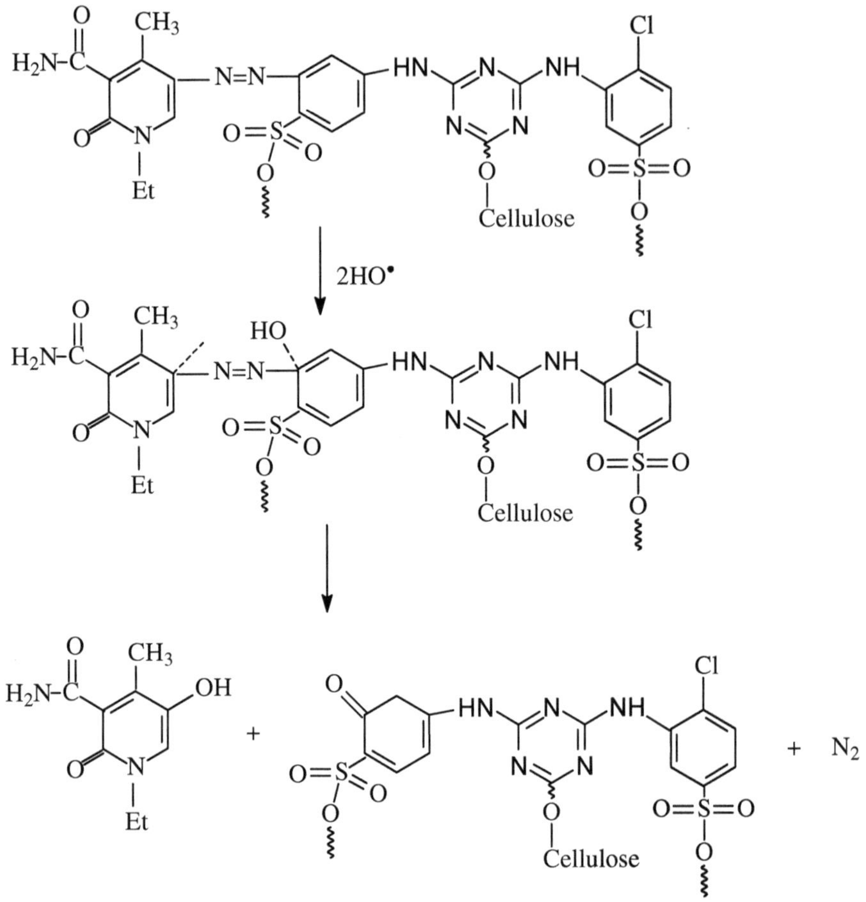

Scheme 11 Photodecomposition mechanism of RY-143 dye [42] (reproduced with permission of Springer—License no. 3887660087491)

With the collagen triple helix once degraded, free radicals generate massive peptide bond scission (Fig. 2) [44] (Reproduced with permission of Springer—License no. 3890680622983. Hydroxyl radicals resulted from the water round the collagen backbone make interactions with its macromolecules to yield new radicals.

The chemically destructive UV irradiation effect on the collagen hierarchical structure was confirmed by FTIR, AFM and sodium dodecyl sulfate-polyacrylamide gel electrophoresis (SDS-PAGE) studies. These negative effects of electromagnetic radiation on collagen matrices are an important study in tissue remodeling.

Sionkowska [45] studied the photochemical behavior of collagen and PEG based blends by FTIR, UV–Vis, viscometry and microscopy before and after UV irradiation (λ = 254 nm) and observed that the two components were immiscible in the blend, thus exhibiting lower photochemical stability and poor mechanical properties compared to individual components. The photochemical stability of collagen was

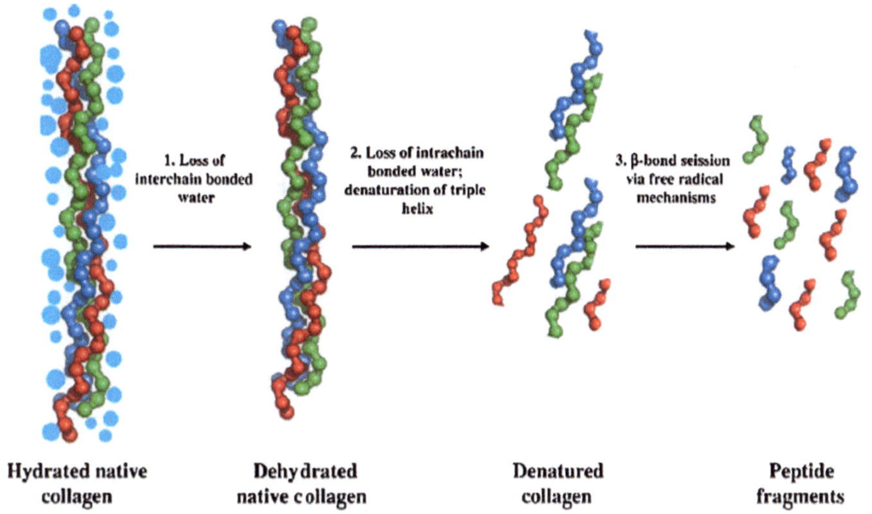

Fig. 1 Possible UV decomposition mechanism of collagen at 254 nm [44] (reproduced with permission of Springer—License no. 3890680622983)

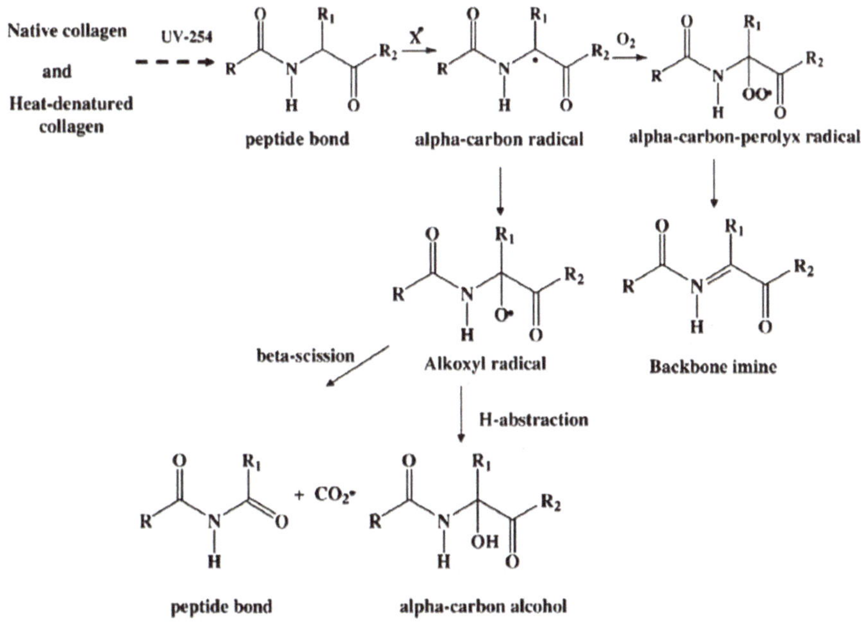

Fig. 2 Collagen backbone scission mechanisms [44] (reproduced with permission of Springer—License no. 3890680622983)

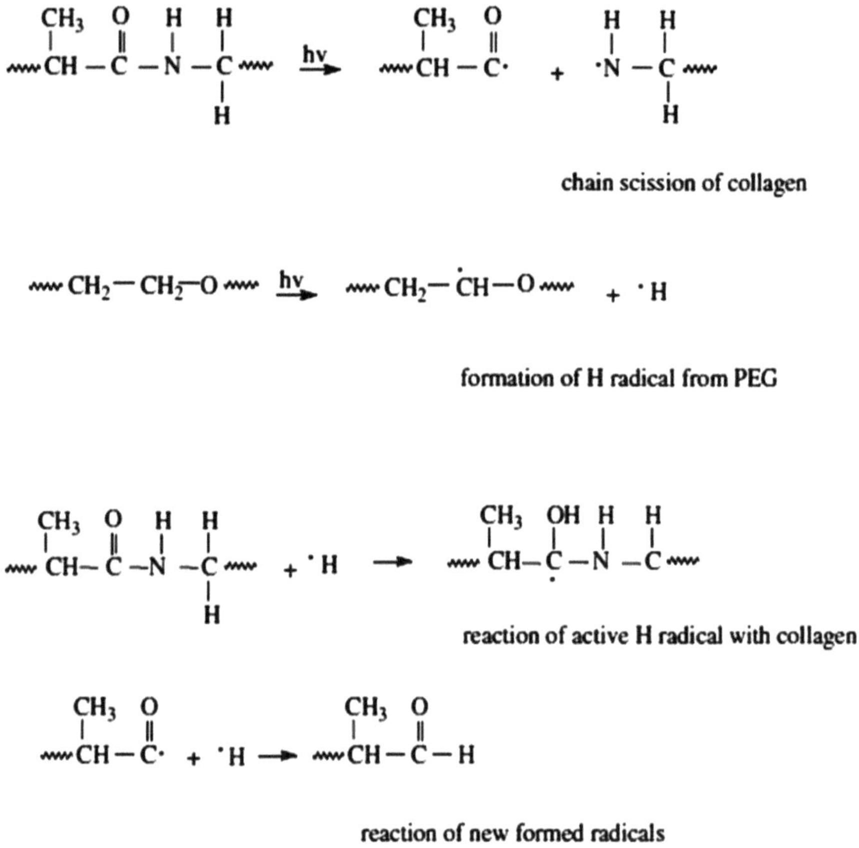

Fig. 3 Possible UV degradation mechanism in PEG/collagen films [45] (reproduced with kind permission from Elsevier—License no. 3890091014498)

dependent on PEG concentration in the blend. Photodegradation in collagen is initiated from the side groups through the amino acids tyrosine and phenylalanine chromophores which absorb light under 300 nm, while in PEG photodegradation occurs through main chain scission initiated by weak chemical linkages. Active primary macroradicals are produced during photo–oxidation of one of the blend components and initiate photodegradation of the second component. Free active radicals from both components react with each other leading to blend properties alteration (Fig. 3) [45] (reproduced with kind permission from Elsevier—License no. 3890091014498). FTIR spectra indicated that the films were not so photochemically degraded compared to pure collagen and that changes in the properties of blends in solution occur at a lower irradiation time compared to the ones in solid state. Optical micrographs indicated that surface characteristics of collagen and the blends as films are not strongly altered after UV exposure. (Fig. 4) [45] (reproduced with kind permission from Elsevier—License no. 3890091014498).

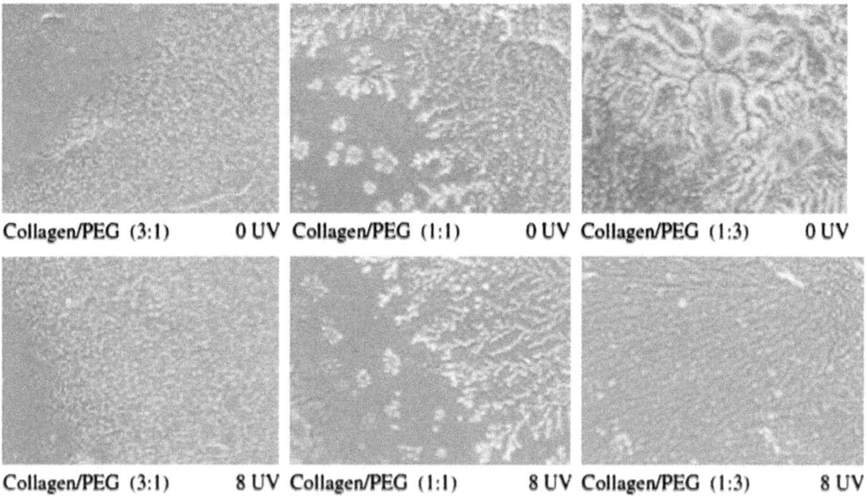

Fig. 4 Optical micrographs of the PEG/collagen films before and after UV irradiation [45] (reproduced with kind permission from Elsevier—License no. 3890091014498)

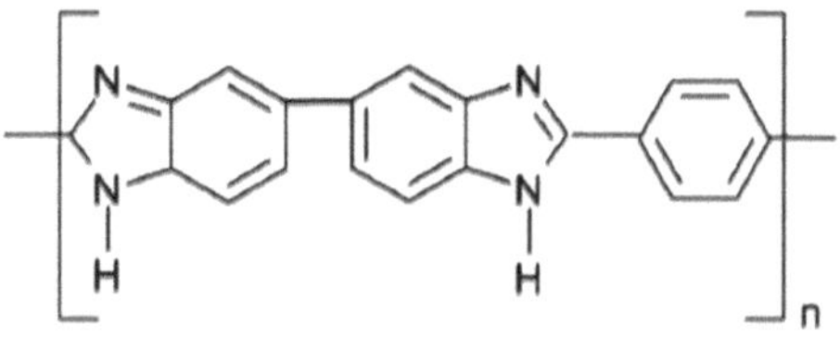

Fig. 5 Chemical structure of PBI [47] (reproduced with kind permission from Elsevier—License no. 3890161358622)

Sionkowska et al. [46] further studied the photochemical stability of poly(vinyl pyrrolidone) in the presence of 1, 3 and 5 % collagen irradiated at $\lambda = 254$ nm in air for 24 h and characterized the samples by by UV–Vis and FTIR spectroscopy and thermal analysis (TGA, DTG). It was observed that PVP was thermally and photochemically less stable in the presence of collagen. Collagen led to crosslinking of PVP and increased its polarity during irradiation, indicating surface photo-oxidation.

Arrieta et al. [47] conducted hydrolytic and photochemical aging studies for a Kevlar–Polybenzimidazole (PBI) blend in yarn form. UV irradiation was conducted at $\lambda = 340$ nm and four different temperatures (50, 60, 70 and 80 °C) between 4 and 31 days. The PBI structure is given in Fig. 5 [47] (reproduced with kind permission from Elsevier—License no. 3890161358622).

The UV exposure led to a decrease in breaking force. New FTIR absorption bands appeared, corresponding to C=O bond vibration in end groups with carboxylic acid. A photodecomposition mechanism was proposed (Fig. 6) [47] (Reproduced with kind permission from Elsevier—License no. 3890161358622), occurred through the Photo-Fries rearrangement, which yielded 2-aminobenzo phenones and carboxylic acid end groups (Fig. 7) [47] (reproduced with kind permission from Elsevier—License no. 3890161358622).

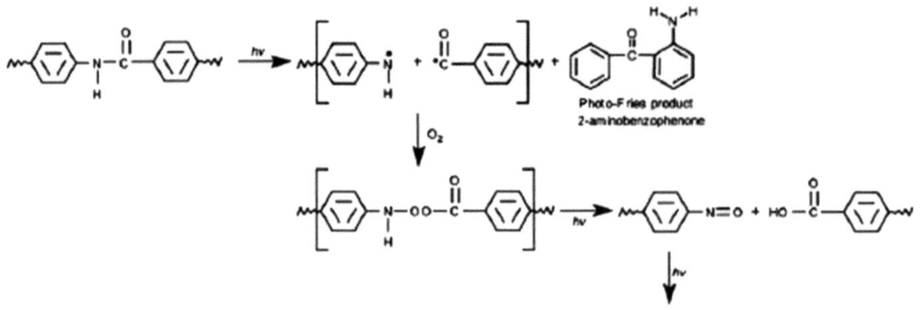

Fig. 6 Photochemical aging mechanism of Kevlar [47] (Reproduced with kind permission from Elsevier—License no. 3890161358622)

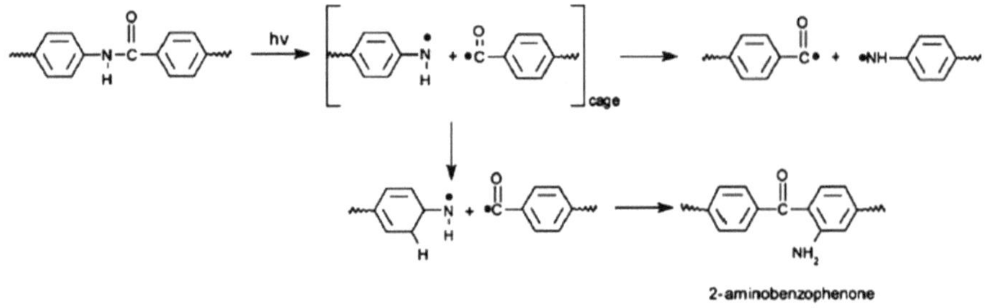

Fig. 7 Photo-Fries rearrangement mechanism in Kevlar [47] (reproduced with kind permission from Elsevier—License no. 3890161358622)

The accumulation of high absorbent products from the Photo–Fries rearrangement led to the occurrence of a "screening" effect at the yarns' surface which protected the Kevlar underlying layers by preventing amide bonds cleavage. The material was afterwards exposed to relative humidity levels of 60 and 80 % and temperatures of 50, 60 and 80 °C) during which the breaking force decreased continuously. FTIR spectra indicated new carboxylic acid groups resulted from amide bond hydrolysis under acid catalyst. Results indicated that hydrolysis competed with a recombination reaction by consuming –COOH containing moieties. Authors elaborated a differential kinetic equation of the aging process based on the amount of resulted –COOH which they solved by analytic methods.

Cui et al. [48] studied UV radiation effect on renewable biobased polymer sheets by SEM and photoacoustic FTIR (PAS-FTIR) (Fig. 8) [48] (reproduced with kind permission from Elsevier—License no. 3891281355385).

UV irradation was undertaken at 300 nm on a 12 h cycle: of 8 h UV at 60 °C, followed by 4 h condensation at 50 °C, at exposure times of 250, 500, 750, 1000 and 2000 h. UV irradiation and cycled weathering led to crosslinking and surface cracking. The main driving forces of the photodegradation process were hydroperoxide formation and decomposition, the latter occurring through formation of oxidation products, such as ketones, alcohols and aldehydes (Fig. 9) [48]

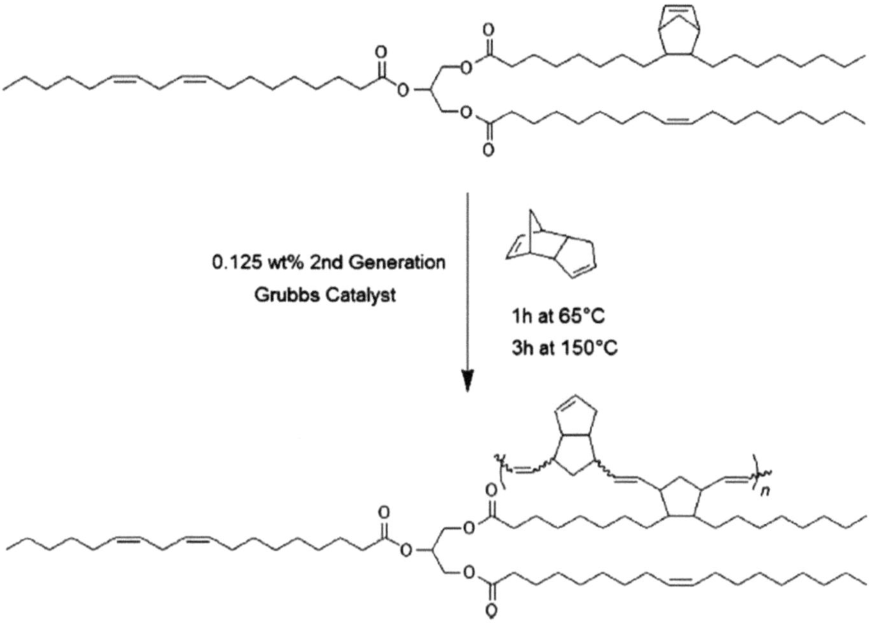

Fig. 8 Ring-opening metathesis polymerization of Dilulin and dicyclopentadiene [48] (reproduced with kind permission from Elsevier—License no. 3891281355385)

(reproduced with kind permission from Elsevier—License no. 3891281355385), and decreasing with penetration depth increase, due to lowering of oxygen concentration.

Through oxygen consumption, the formed hydrocarbon radicals convert into peroxy radicals which further lead to the formation of hydrocarbon radicals and hydroperoxides via hydrogen atoms abstraction from the polymer chain. The new carbonyl entities generated during photo-oxidation also act as UV absorbers, decomposing afterwards through a Norrish reaction and yielding esters, acids, alkenes and enols, as general products resulted during photodecomposition. SEM and depth profile studies results indicated decomposition occurrence in thin layer surface due to limited oxygen diffusion through the interior layers. (Fig. 10) [48] (reproduced with kind permission from Elsevier—License no. 3891281355385). The formation of hydroperoxides and new carbonyl entities from main chain decomposition was demonstrated by PAS–FTIR.

Further various recent examples on the photochemical behavior of multicomponent polymer based materials were given in *Chap.* 1. *Theoretical aspects regarding polymer photochemistry* and *Chap.* 8. *Lifetime prediction of multicomponent polymeric materials.*

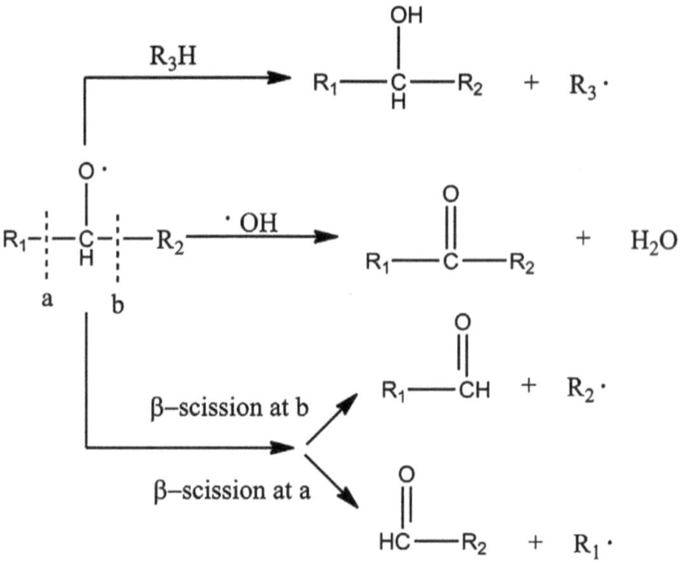

Fig. 9 Second oxidation reactions in alkoxy radical [48] (reproduced with kind permission from Elsevier—License no. 3891281355385)

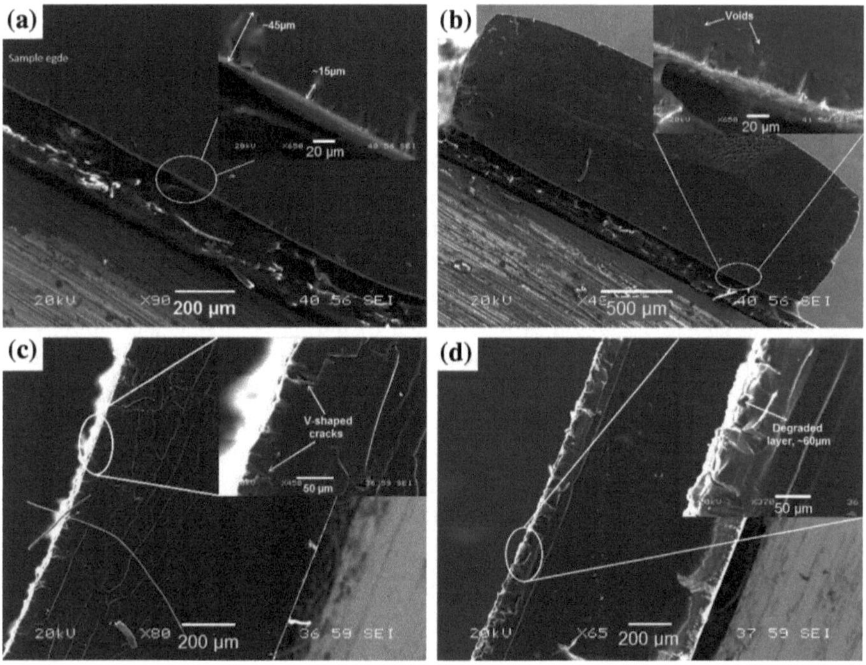

Fig. 10 Cross-section SEM images of UV-degraded structures at **a** 250 h; **b** 500 h; **c** 1000 h; **d** 2000 h [48] (reproduced with kind permission from Elsevier—License no. 3891281355385)

3 Conclusions

Photodegradable polymers and polymer based materials have witnessed a growing interest in the last two decades, since their properties may be controlled both spatially and temporarily by phototriggering groups or agents. A continuously troubling issue is that of the hazardous environmental impact of polymeric materials. That is why much effort is undertaken into the developing and design of polymers able of splitting into comprising corresponding monomers after service lifetime end. Photodegradable polymers possess a wide palette of applications, such as in agriculture as plastic sheets for ground cover (plasticulture), in the packaging industry and household where they must degrade fast and without an environmentally negative impact, in the obtaining of smart materials for microelectronics and in biomedical applications. Hydrogels with phototriggered controlled release of bioactive principles or cells have high applications potential in tissue engineering. By including photolabile chemical structures in the polymer main chain [o-nitrobenzyl, coumarin-4-ylmethyl, p-hydroxyphenacyl or acetal (ketal)] and by monitoring parameters such as wavelength, light intensity, irradiation dose, temperature, exposure time and mechanical stress, one may control the stability, degradation rate and polymer shape.

References

1. Paul, D.R., Sperling, L.H.: Multicomponent Polymer Materials. American Chemical Society, Washington, DC (1986)
2. Allen, N.: Polymer photochemistry. In: Dunkin, I. (ed.) Photochemistry, vol. 35, p. 206. The Royal Society of Chemistry, London (2005)
3. Li, L., Deng, X.-X., Li, Z.-L., Du, F.-S., Li, Z.-C.: Multifunctional photodegradable polymers for reactive micropatterns. Macromolecules **47**, 4660–4667 (2014)
4. Griffin, D.R., Kasko, A.M.: Photo-selective delivery of model therapeutics from hydrogels. ACS Macro Lett. **1**(11), 1330–1334 (2012)
5. Griffin, D.R., Kasko, A.M.: Photodegradable macromers and hydrogels for live cell encapsulation and release. J. Am. Chem. Soc. **134**(31), 13103–13107 (2012)
6. Daglen, B.C., Tyler, D.R.: Photodegradable plastics: end of life design principle. Green Chem. Lett. Rev. **3**(2), 69–82 (2010)
7. Gilead, D.: Photodegradable plastics in agriculture. In: Scott, G., Gilead, D. (eds.) Degradable polymers, p. 186. Springer, Amsterdam (1995)
8. Zhu, C., Ninh, C., Bettinger, C.J.: Photoreconfigurable polymers for biomedical applications: chemistry and macromolecular engineering. Biomacromolecules **15**(10), 3474–3494 (2014)
9. Ki, C.S., Shih, H., Lin, C.-C.: Facile preparation of photodegradable hydrogels by photopolymerization. Polymer **54**, 2113–2122 (2013)
10. Lv, C., Wang, Z., Wang, P., Tang, X.: Photodegradable polyester for triggered release. Int. J. Mol. Sci. **13**(12), 16387–16399 (2012)
11. Fomina, N., McFearin, C., Sermasakdi, M., Edigin, O., Almutairi, A.: UV and near-IR triggered release from polymeric nanoparticles. J. Am. Chem. Soc. **132**(28), 9540–9542 (2010)

12. de Gracia, L.C., McFearin, C.L., Almutairi, A.: Single UV or near IR triggered event leads to polymer degradation into small molecules. ASC Macro Lett. **1**(7), 922–936 (2012)
13. Volgraf, M., Banghart, M., Trauner, D.: Switchable proteins and channels. In: Feringa, B. (ed.) Molecular Switches, vol. 2, 2nd edn, p. 563. Wiley, Weinheim (2011)
14. Scott, G., Gilead, D.: Degradable Polymers: Principles and Applications, 2nd edn. Chapman & Hall, New York (2002)
15. Wiles, D.M., Scott, G.: Polyolefins with controlled environmental degradability. Polym. Degrad. Stab. **91**, 1581–1592 (2006)
16. Albertsson, A.-C., Barenstedt, C., Karlsson, S., Lindberg, T.: Degradation product pattern and morphology changes as means to differentiate abiotically and biotically aged degradable polyethylene. Polymer **36**(16), 3075–3083 (1995)
17. Steber, J., Herold, C.P., Limia, J.M.: Comparative evaluation ofanaerobic biodegradability of hydrocarbons and fatty derivatives currently used as drilling fluids. Chemosphere **31**(4), 3105–3118 (1995)
18. Sommazzi, A., Garbassi, F.: Olefin-carbon monoxide copolymers. Prog. Polym. Sci. **22**(8), 1547–1605 (1997)
19. Grassie, N., Scott, G.: Polymer Degradation and Stabilization. Cambridge University Press, New York (1985)
20. Guillet, J.: Polymer Photophysics and Photochemistry: An Introduction to the Study of Photoprocesses in Macromolecules. Cambridge University Press, New York (1985)
21. Schnabel, W.: Polymers and Light: Fundamentals and Technical Aspects. Wiley, Weinheim (2007)
22. Hartley, G.H., Guillet, J.E.: Photochemistry of ketone polymers. I. Studies of ethylene-carbon monoxide copolymers. Macromolecules **1**(2), 165–170 (1968)
23. Tyler, D.R.: Photochemically degradable polymers containing metal–metal bonds along their backbones. Coord. Chem. Rev. **246**(1–2), 291–303 (2003)
24. Chen, R., Tyler, D.R.: Origin of tensile stress-induced rate increases in the photochemical degradation of polymers. Macromolecules **37**, 5430–5436 (2004)
25. Meyer, T.J., Caspar, J.V.: Photochemistry of metal–metal bonds. Chem. Rev. **85**(3), 187–218 (1985)
26. Tenhaeff, S.C., Tyler, D.R.: Photochemically reactive polymers; synthesis and characterization of polyurethanes containing Cp2Mo2(CO)6 or Cp2Fe2(CO)4 molecules along the polymer backbone. Organometallics **10**(2), 473–482 (1991)
27. Rabek, J.F.: Photodegradation of Polymers: Physical Characteristics and Applications. Springer, New York (1996)
28. Matsukawa, K.J.: Development of photo-functional materials from organic/inorganic nano-hybrids. J. Photopolym. Sci. Technol. **18**(2), 203–210 (2005)
29. Xu, J., Shi, W., Gong, M., Yu, F., Yan, L.: Preparation of poly(methyl methacrylate-co-maleic anhydride)/SiO_2–TiO_2 hybrid materials and their thermo– and photodegradation behaviors. J. Appl. Polym. Sci. **97**(4), 1714–1724 (2005)
30. Li, L., Lv, A., Deng, X.X., Du, F.S., Li, Z.C.: Facile synthesis of photo-cleavable polymers via Passerini reaction. Chem. Commun. **49**(76), 8549–8551 (2013)
31. Koylu, D., Thapa, M., Gumbley, P., Thomas, S.W.I.I.I.: Photochemical disruption of polyelectrolyte multilayers. Adv. Mater. **24**, 1451–1454 (2012)
32. Yamaguchi, S., Yamahira, S., Kikuchi, K., Sumaru, K., Kanamori, T., Nagamune, T.: Photocontrollable dynamic micropatterning of non-adherent mammalian cells using a photocleavable poly(ethylene glycol) lipid. Angew. Chem. Int. Ed. **51**, 128–131(2012)
33. Pauloehrl, T., Delaittre, G., Bruns, M., Meißler, M., Börner, H.G., Bastmeyer, M., Barner-Kowollik, C.: (Bio)Molecular surface patterning by phototriggered oxime ligation. Angew. Chem. Int. Ed. **51**, 9181–9184 (2012)
34. Jiang, J.Q., Tong, X., Zhao, Y.: A new design for light-breakable polymer micelles. J. Am. Chem. Soc. **127**(23), 8290–8291 (2005)

35. Pasparakis, D.G., Manouras, T., Selimis, A., Vamvakaki, M., Argitis, P.: Laser-induced cell detachment and patterning with photodegradable polymer substrates. Angew. Chem. Int. Ed. **50**, 4142–4145 (2011)
36. Cui, J.X., Miguel, V.S., Campo, A.D.: Light-triggered multifunctionality at surfaces mediated by photolabile protecting groups. Macromol. Rapid Commun. **34**, 310–329 (2013)
37. Liu, G., Liu, W., Dong, C.M.: UV- and NIR-responsive polymeric nanomedicines for on-demand drug delivery. Polym. Chem. **4**, 3431–3443 (2013)
38. Gohy, J.F., Zhao, Y.: Photo-responsive block copolymer micelles: design and behavior. Chem. Soc. Rev. **42**, 7117–7129 (2013)
39. Rajendran, S., Raghunathan, R., Hevus, I., Krishnan, R., Ugrinov, A., Sibi, M.P., Webster, D. C., Sivaguru, J.: Programmed photodegradation of polymeric/oligomeric materials derived from renewable bioresources. Angew. Chem. **127**, 1175–1179 (2015)
40. Christensen, P.A., Egerton, T.A., Martins-Franchetti, S.M., Jin, C., White, J.R.: Photodegradation of polycaprolactone/poly(vinyl chloride) blend. Polym. Degrad. Stab. **93**, 305–309 (2008)
41. Li, J., Yang, R., Yu, J., Liu, Y.: Natural photo-aging degradation of polypropylene nanocomposites. Polym. Degrad. Stab. **93**, 84–89 (2008)
42. Rosu, L., Rosu, D., Gavat, C.-C., Varganici, C.-D.: Photochemical stability of cellulose textile surfaces painted with some reactive azo-triazine dyes. J. Mater. Sci. **49**(13), 4469–4480 (2014)
43. Gavat, C.-C.: Contributions for knowledge on some reactive dyes implications in the determination of biological effects, in correlation with their physico-chemical properties. In: Ph.D. Thesis, “Gr. T. Popa” University of Medicine and Pharmacy, Iaşi, Romania (2011)
44. Rabotyagova, O.S., Cebe, P., Kaplan, D.L.: Collagen structural hierarchy and susceptibility to degradation by ultraviolet radiation. Mater. Sci. Eng. C **28**, 1420–1429 (2008)
45. Sionkowska, A.: The influence of UV light on collagen/poly(ethylene glycol) blends. Polym. Degrad. Stab. **91**, 305–312 (2006)
46. Sionkowska, A., Kozlowska, J., Planecka, A., Skopinska-Wisniewska, J.: Photochemical stability of poly(vinyl pyrrolidone) in the presence of collagen. Polym. Degrad. Stab. **93**, 2127–2132 (2008)
47. Arrieta, C., David, É., Dolez, P., Vu-Khanh, T.: Hydrolytic and photochemical aging studies of a Kevlar–PBI blend. Polym. Degrad. Stab. **96**, 1411–1419 (2011)
48. Cui, H., Hanus, R., Kessler, M.R.: Degradation of ROMP-based bio-renewable polymers by UV radiation. Polym. Degrad. Stab. **98**, 2357–2365 (2013)

Photochemical Behavior of Wood Based Materials

Carmen-Alice Teacă and Ruxanda Bodîrlău

Abstract Wood, as a natural composite material used mainly in exterior construction and building applications, requires long-term durability under exposure to environmental factors including solar UV radiation, temperature, humidity and pollutants. The outdoor service life of wood and wood based materials, namely wood–thermoplastic polymer composites, is strongly related to their deterioration under weathering conditions. Wood is extremely sensitive to UV radiation in the range from 300 to 400 nm. Photo-chemical degradation of wood causes significant structural and color changes, lignin component being the most susceptible to de-polymerization reactions by which phenoxy radicals are generated as intermediates and further oxidized into colored chromophores (quinones). Effective treatments, e.g. UV photo-stabilizers, or surface coatings of wood products, e.g. paints, varnishes, stains or water repellents, may provide significant wood protection against weathering in long-term outdoor applications.

1 Introduction

Polymer based materials (wood, plastics) used in the exterior of buildings and in construction have a service life determined by their exposure to UV radiation from sunlight and consequently their further deterioration through weathering process. Both climate changes which determine a substantial temperature increment, as well as increasing of the UV radiation levels, mainly UV-B radiation component, cause a significant diminution of service life for such materials in outdoor applications.

Dedicated to the memory of my dear friend and colleague, Dr. Ruxanda Bodîrlău, who recently passed away.

C.-A. Teacă (✉) · R. Bodîrlău
Advanced Research Center for Bionanoconjugates and Biopolymers, "Petru Poni" Institute of Macromolecular Chemistry, Gr. Ghica Voda Alley, 41A, 700487 Iasi, Romania
e-mail: cateaca14@yahoo.com

D. Rosu and Visakh P. M. (eds.), *Photochemical Behavior of Multicomponent Polymeric-based Materials*, Advanced Structured Materials 26,
DOI 10.1007/978-3-319-25196-7_4

Through appropriate stabilization methodologies, e.g. modification of polymer properties and use of high quality light stabilizers, the mitigation of deleterious effects induced by increased UV radiation levels upon polymer materials can be attained with maintenance of their lifetime under outdoor exposure conditions [1].

Investigation and knowledge of wood aging and its properties changes are of great significance for conservation of cultural heritage objects and historic buildings made from wood and also for the restoration of ancient wood constructions. In this context, a recently thorough literature review presents the different aspects of wood aging [2]. A new methodology using Fourier transform infrared spectroscopy (FTIR) in order to study the age of the carving phase for wood objects can be effectively applied, with analysis of samples from different depths and localizations for an eventual gradient of water absorbed (namely, hydroxyl groups OH) between the surface and the inner part of both an ancient and a modern wood sculptures [3].

The commercial value and the usage of the wood is strongly related to the colour changes caused by UV radiation on the surface, these further affecting wood's esthetical appearance and its entire properties (e.g. physical, mechanical). The colour changes may be correlated and validated by statistical analysis with chemical changes evidenced in wood structure after photo-degradation process through spectral analysis, an effective method to estimate the preservation stage for the wood surfaces [4].

Wood is a natural composite material consisting of different polymers, including polysaccharides, namely cellulose and hemicelluloses, and an amorphous three-dimensional network polymer comprising different linked phenyl-propane units, namely lignin, which act as binder in wood composite structure, and extractives (Fig. 1).

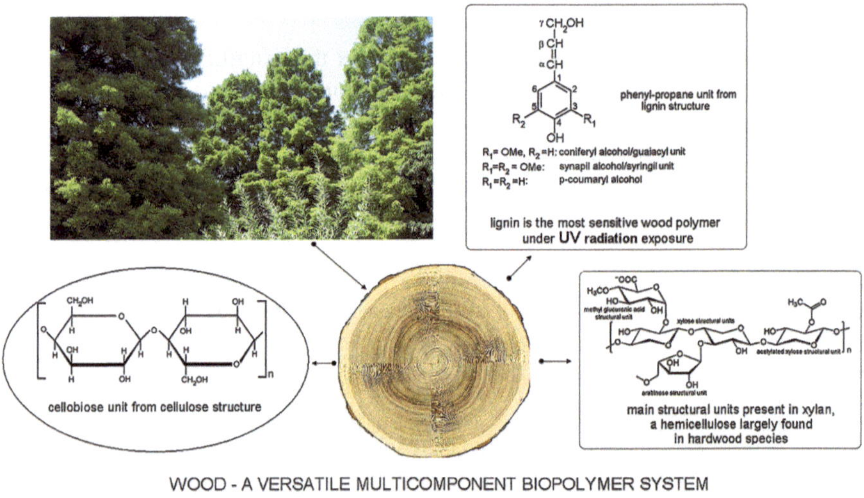

Fig. 1 Schematic representation of wood chemical structure

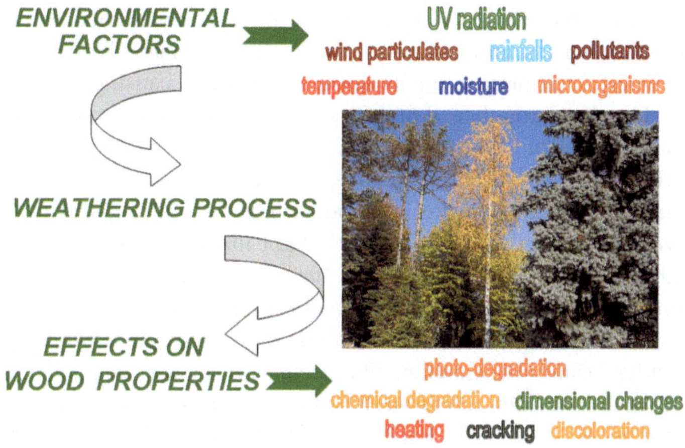

Fig. 2 Schematic representation of wood weathering process under environmental factors action

Wood is the most versatile and widely exploited engineering material for both indoor and outdoor applications, mainly in building constructions. Wood has many advantages including aesthetic appearance, easy processing, low density, low thermal expansion, and good mechanical properties. Wood and wood based materials (e.g. polymer composites) are susceptible to undergo surface degradation processes, mainly under exposure to outdoor environmental conditions which involve a complex combination of chemical, mechanical and light factors. These include: solar radiation—mainly UV part; winds; particulates; humidity occurred through dewing, raining, and snowing; temperature; atmospheric oxygen; pollutant gases such as sulphur dioxide, and nitrogen dioxide. All of these factors significantly contribute to the *weathering process* [5–9]. A schematic representation of this process is presented in Fig. 2.

Wood interacts with the electromagnetic spectrum of sunlight, namely ultraviolet (UV) radiation, being a very good light absorber, and further determines photochemical reactions (photolysis, photo-oxidation, and thermal oxidation) in the weathering or oxidative degradation processes of wood [5, 8, 10–12]. Usually, this very complex process relies upon activation of the chromophore groups which in the case of wood and wood based materials are represented by the phenolic groups commonly present in lignin biopolymer from wood structure. Among constituent wood biopolymers, lignin is the most reactive through its chromophore groups which are the most sensitive under sunlight exposure, especially UV radiation [13–16]. It is also well-known that the presence of extractives, mainly phenolic type ones, strongly influences the photo-degradation process, these kinds of compounds being very sensitive to the UV light action [17, 18]. These chromophores generate free radicals through chains cleavage with occurrence of low molecular weight compounds which are yellow coloured, mainly quinones, and stilbenes [19]. The colour changes are also determined by occurrence of carbonyl groups through photo-degradation process [15]. These substances represent a good substrate for the fungal growth (blue staining fungi) that further initiates biodegradation thus increasing wood deterioration on its surface and even causing entirely structure destruction. Usually, a roughened texture and cracks on the wood surfaces followed the combined action of the UV radiation from sunlight and water from rainfalls [7, 8, 20]. This rapid degradation affects to a large extent the mechanical behaviour of wood under outdoor exposure conditions, thus limiting its service life. That's why it is very important to control the UV light impact upon wood surfaces [10].

Physical degradation features of wood surface include bleaching (usually called discoloration), micro-structure checking, and erosion on the top surface (evidenced through enhanced roughness, loss of gloss and lightness) which are readily observed under visual investigation. The occurrence of chemical changes in wood structure is mainly evidenced through colour changes with the progress of the weathering process [16, 21–24]. Further, significant deleterious effects upon wood's aesthetic value and performance can be observed, wood being subjected to a more enhanced and progressive photo-oxidative degradation process.

2 Photochemical Behaviour of Wood Based Materials

Conversion of matter (organic, inorganic) implies occurrence of ageing and deterioration processes of materials. Weathering is the dominant process of natural ageing caused by abiotic and biotic factors mainly when inorganic materials are involved. This process implies chemical transformation by changing the original state, this meaning not only degradation but also occurrence of new compounds and even their desirable stabilization. In this context, the multivariate data analysis can be an effective tool to indicate the contribution of different environmental factors [25].

Modification of biopolymers (e.g. wood or cellulose component) through different chemical and physical routes is usually applied for increasing to some extent their resistance against biological degradation caused by microbial communities. Wood modification through chemical reactions including esterification (using organic anhydrides, e.g. acetic, succinic, or aromatic vinyl esters), or reaction of wood with furfuryl alcohol, and resins treatment, or physical treatments (e.g. thermal, plasma) can contribute to an increased lifetime of wood based materials and products by improving the resistance of wood to photo-degradation [13, 26–29]. However, these structural diversification routes generate some problems related to the environmental degradation of the resulted modified wood materials.

The degradation processes of plastics, rubber and wood materials are oxidative reactions which are accelerated under exposure to increased levels of UV radiation (especially UV-B component—280–315 nm—that is efficiently absorbed by chromophores present in these materials), mainly at high ambient temperatures (thermal oxidation). As immediate consequences, useful service life-time periods for such materials in outdoor applications are significantly reduced [10, 30, 31].

Chemical additives are usually used for light-stabilization of plastic materials, while UV resistant coatings (e.g. clear polyurethane coatings) are applied for wood products protection, since photo-degradation processes cause increased water sensitivity and further their biodegradation under outdoor exposure conditions [32]. In most cases, it is preferable to use materials with greater resistance to UV radiation in order to prevent the increased damage of these materials under outdoor exposure to solar radiation.

The technical advances in degradation and stabilization techniques which impact the lifetimes of plastics and wood products routinely exposed to solar UV radiation and the protection to humans offered by materials against solar UV radiation were recently assessed [1].

Photo-degradation process in wood and wood-based materials is usually limited to the surface layer, this being the location for many phenomena including physical and chemical interactions between a material and its surrounding environment. Nevertheless, the process is strongly related to the depth of penetration of the UV radiation. Therefore, accelerated testing of wood in laboratory conditions requires use of UV light sources that generate similar depths of penetration as resulted in natural conditions [10].

The exposure of wood to indoor and especially outdoor conditions (namely moisture, sunlight radiation, heat/cold, wind abrasion, biological agents which exert their action as individual or in combination) significantly affects its native durability (dimensional stability, surface integrity), thus occurring natural degradation process [5, 7, 8, 33].

Wood weathering is initiated primarily by solar radiation, mainly the UV portion of the solar spectrum. It can be considered as a photo-oxidation or photo-chemical process that starts immediately after wood exposure to sunlight action, affecting only the surface of the wood. Initially, the colour changes occur and these are visually noticeable, further a slow erosion process being induced and accompanied by increased loss of the surface wood fibers [7, 8]. The main features of the weathering process are strongly influenced and determined by the wood properties including density (one of the most important factors affecting the process), grain characteristics (presence of early wood and latewood), texture (hardwood or softwood species), extractives, resins, and oils content [7].

Investigation related to the photo-degradation process of wood under UV radiation exposure was initiated some decades ago [12, 34]. This type of investigation implies usually two distinct stages, as follows:

1. *irradiation* of wood surface using different UV light sources;
2. *monitoring* and evaluation of the changes (structural, colour, morphological) occurred under UV radiation exposure.

The absorption of UV light by wood occurs mainly due to the absorption by lignin (an aromatic polymer wood component) that is a good absorber of UV radiation comparatively with cellulose (the main polysaccharide from wood structure). It seems that lignin acts as a photo-sensitizer in the degradation of the

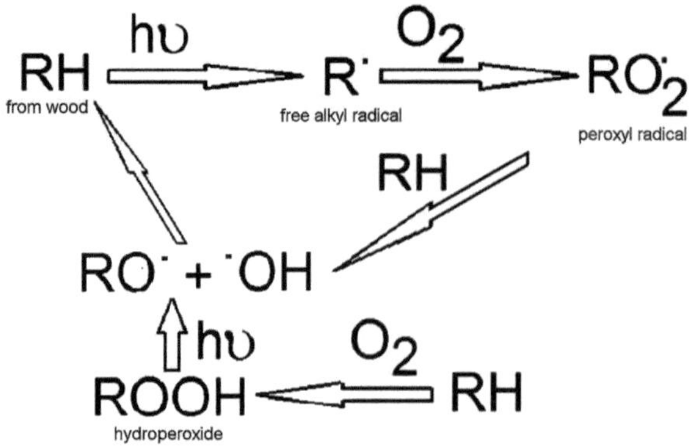

Fig. 3 Schematic representation of photo-oxidative pathway during photo-degradation process

other wood components by transferring light energy [12]. After absorption of a quantum of energy hυ (photon) by wood, the occurrence of the free radicals through main chain scission (dissociation of chemical bonds) is observed, these resulted radicals being easily detected using electron spin resonance (ESR) method [8].

Among the intermediate products formed, there are mainly those containing oxygen with low activation energy, namely peroxides, hydroperoxides, peroxyl, and alkoxyl radicals (schematically presented in Fig. 3), these being involved into further chemical and photo-chemical reactions.

Alkoxyl radicals (RO·) are well-known as intermediate products involved in the pyrolysis and photolysis reactions of organic compounds (peroxides, esters, ethers), as well as in the autoxidation and combustion processes [12, 20]. In the case of wood photo-degradation process, alkoxyl radicals are responsible for producing methanol (from the methoxyl groups present in lignin), formaldehyde and hydrogen (from the photo-dissociation of methanol), carbon monoxide (from carbonyl groups through photo-oxidation), and carbon dioxide (from carboxyl acid groups evidenced in wood by irradiation).

The first research studies were performed on wood and unbleached mechanical pulps and described the role of lignin and the effects of ambient oxygen, temperature and humidity [35–37]. The chemical changes in lignin associated with light-induced yellowing were evidenced, the reduction in methoxyl groups content being strongly related to the brightness decreasing. It is well-known that lignin presents a phenyl propane as basic unit (softwood species have methoxyl-phenyl propane units—namely guaiacyl propane units, while hardwood species contains an additional methoxyl group). Besides these, there are also present various reactive groups including carbonyls, carboxyls, and ethers [38–40].

As evidenced by early studies [41, 42], free radicals are not usually present in wood. A small amount of free radicals can result from wood surface activation in the presence of oxygen from air, this acting as a catalyst or promoter. Usually, these free radicals are phenoxy radicals resulted from oxidation of the phenyl type moieties from lignin component in wood [8, 43].

On the other hand, considering the significant UV absorption ability of lignin, it may also protect the cellulose component from photo-degradation process [43]. The depletion of the relatively hydrophobic lignin from the wood surfaces can occur during wood exposure to UV light action [21]. Given the interaction of wood and UV radiation which is mainly a surface process when UV light penetrates slightly wood [44], the generation of the free radical intermediates originating from lignin is the main promoter for the wood surface deterioration and discoloration [8, 12, 45]. The free radicals rapidly generate hydroperoxide type moieties in the presence of oxygen from surrounding atmosphere these further producing chromophore groups, namely carboxyl and conjugated carbonyl groups.

The occurrence of the photochemical reactions on the wood surfaces implies the absorption of an amount of energy by different chemical moieties, this energy having to be capable to disrupt a chemical bond [7]. The dissociation energies for many of carbon–oxygen bonds usually found in lignin moieties are present in the UV radiation range (295–400 nm). The elevated temperature generates higher level of vibration energy. The interaction of these two effects may result in higher photo-degradation rate at 80 °C than at 30 °C, for example [46].

Some significant features of the photochemical reactions in wood were previously evidenced [9, 39] and briefly summarized [20]. Thus, significant colour changes occurs, these being determined by the chromophore groups formation through lignin degradation process under wood exposure to light having wavelength <350 nm. Discoloration (or photo-bleaching) process is observed when wood is exposed to light with wavelength value >400 nm. As consequence, the methoxyl groups' content of lignin is reduced, while phenoxy radicals are readily formed from phenolic hydroxyl groups. The last mentioned intermediates generate o-quinonoid structures through demethylation reaction promoted by oxygen atmosphere [39]. Phenoxy radicals could be also involved in radical transfer reactions which may further confer stabilization to wood, as suggested by Kamoun et al. [47].

Several test methods have been developed using artificial light sources to provide accelerated weathering testing procedures and for comparison purposes when considering natural environmental weathering. Usually, all these methods imply the monitoring of the main features reflecting an ageing process which include mechanical properties and optical characteristics (roughness increase with evidence of crack formation, changes in colour and gloss). The selection of an artificial weathering method is frequently preferred due its shortened duration by comparison with a natural weathering test which instead provides more accurate and reproducible data. The durability of non-finished and finished wood products that will be used outdoors can be effective evaluated through artificial weathering methods [4, 22, 30, 46, 48–52]. These valuable tests accelerate the effects of natural weathering up to 20 times as function of the established conditions for exposure which can be controlled and reproduced. Generally, the accelerated weathering tests envisage investigation of the effects exerted by UV light, humidity and temperature. Simulation of all factors causing natural weathering degradation including UV light, temperature variations, wetting by water, discoloration by mould and stain fungi is not readily achievable under accelerated weathering testing conditions.

Photo-degradation of wood surfaces under UV radiation action is usually affected in outdoor conditions by other environmental factors including mainly humidity, temperature, visible light, and infrared radiation. The wavelength of the radiation can be known, its energy and intensity at the surface can be also determined by using UV-laser lamps [53, 54] instead of traditional UV lamps (xenon and mercury vapour lamps) as radiation sources during irradiation testing. In this case, the duration of accelerated weathering experiments can be significantly reduced.

An increased lifetime of wood is an essential prerequisite for sustainable outdoor applications (e.g. roofs, doors, windows type frames, bridge type structures built over rivers, roads, or railways). Wood presents a natural durability in terms of biodegradation, this feature being strongly related to the wood species (is higher in hardwoods than in softwoods) and mainly to the extractives content and composition [55]. This durability can be predicted by infrared spectroscopy [56, 57].

The lifetime of wood can be efficiently improved through biological control of wood resistance against fungal decay [58], and different treatments applied to wood [26] including thermal—e.g. heating [59, 60] and chemical modification using anhydrides, acid chlorides, and silanes. The chemical routes include treatments such as acetylation [26, 27, 29, 61–63], succinylation [13], butyrylation [64], furfurylation [65], or silylation [66, 67].

Chemical modification of wood represents an effective manner to improve its weathering resistance and photostability [68, 69]. Effective and easily modification of wood can be achieved by using vinyl esters that can contribute to wood photostability and generates acetaldehyde as a desirable non-acidic by-product [70, 71]. Modification of wood using aromatic vinyl esters can also protect wood from photo-degradation [28], this protection being strongly related to the type of the ester used for wood treatment and the degree of modification (evidenced by weight gain values). It was concluded that wood protection from photo-degradation requires a more complete chemical modification of wood polymer constituents when comparing with the degree of modification demanded by achieving significant prerequisites such as wood dimensional stability and prevention of its biological degradation. In this context, a more feasible and effective way to improve the wood photostability implies use of aromatic compounds having a simple, non-branched structure and a low molecular weight that can modify wood by easily reacting with its polymer constituents when a high degree of substitution of accessible hydroxyl groups could be achieved [28]. Another promising way to improve wood photostability is conferred by use of polymeric polyester-type UV absorbers (UVA), produced by reaction of an epoxy-functionalized UV absorber (a benzophenone type compound with anhydrides—maleic, phthalic, or succinic [72]. It was evidenced that the most effective UV absorber for wood photostability (by preserving the benzophenone group and hindering weight and tensile strength losses during accelerated weathering) was the polyester film resulted from reaction of the epoxy-functionalized UV absorber with phthalic anhydride.

The most sensitive indicator that evidence occurrence of wood degradation process under UV light exposure is the colour change of wood [24, 45, 73–75]. The rapid colour differences evidenced after irradiation as primary consequences of the photo-oxidation phenomena can be useful for establishment a correlation between testing of accelerated and natural durability [76].

Chemical modification of wood (e.g. with succinic anhydride) may confer it more stability in colour under artificial light action comparatively with the non-modified wood, this feature being due to a slight increase in lignin stability under exposure to UV action [13]. Successive treatment stages (with succinic anhydride and epoxy functionalized soybean oil) provided protection of wood from photo-oxidation reactions under UV radiation exposure as related to the colour preservation, expressed by a reduced fading tendency [77]. As expected, lignin underwent significant photo-oxidation reactions, exhibited as increase of the redness tendency of wood surfaces. Nevertheless, the modified wood surface was protected by the coating layer consisting of epoxy functionalized soybean oil through a screening effect.

A promising way to reduce the rate of free radical formation and thereby minimize the occurrence of wood discoloration (photo-yellowing) was recently presented [66] and involves the silylation of the UV radiation sensitive groups at the wood surface (in air and at room temperature) in order to interfere with the photo-degradation process, the resulted photo-induced free radicals being determined using ESR spectroscopy.

Traditional methods applied to protect wood against biodegradation (process that may enhance susceptibility to further photo-degradation) involve chemicals which usually pose toxicological effects upon human health and environment [78–80]. Recently, alternative methods for wood protection against biodegradation were reviewed and presented with focus upon significant opportunities of using natural products as biocides and environmental biocompatible compounds [81]. Nevertheless, these organic biocides may be susceptible to photo-degradation [82] and have to be co-impregnated with protecting agents. A potential route to be efficiently applied for conversion of softwood species into self-antibacterial and sustainable wood materials (e.g. for decking and fencing purposes) was recently presented [83]. This novel process involve chemical attachment of quaternary ammonium compounds via a carbamate/urethane linkage to softwood (in this case, hemlock) using supercritical carbon dioxide. The resulted chemically modified softwood exhibited excellent antibacterial activity and enhanced dimensional stability.

Lignin is a photo-labile aromatic polymer component from wood structure which presents chromophore groups with great affinity to absorb UV radiation that mainly initiate further degradation process of wood and wood-based materials (e.g. wood/wood composites, wood/polymer composites) under solar light and artificial light exposure [5, 14, 15, 84–89]. Wood/polymer composites (usually coded as WPC's) are emerging hybrid materials comprising wood as filler, and thermoplastics as polymer matrix (polyethylene PE, polypropylene PP, polyvinylchloride PVC, polystyrene PS), produced through different processes (e.g. extrusion, injection moulding) and presenting a better dimensional stability by comparison with the usual wood products. Incorporation of wood fibers within a polymer matrix has positive and significant advantages including possibility to use recycled raw materials with large volume and low cost, reduced energy and production costs, improved environmental behaviour (being environmental-friendly materials) and properties (mechanical: stiffness; physical: resistance against moisture, biological decay, UV radiation; thermal stability; easy processing ability; dimensional stability). The properties of WPC's (e.g. appearance, mechanical) are strongly dependent by the interaction between wood fibers and the polymer matrix, this interaction being improved by addition of coupling agents.

Exposure of wood/polymer composites to weathering in outdoor conditions [90] generates significant changes including colour with impact upon aesthetic appearance, and reduction in mechanical performance. These changes are the consequence of the surface chemistry changes including mainly oxidation processes occurred on the composites surfaces which affect crystallinity of the polymer matrix and cause degradation at the interface areas. Both components, wood and thermoplastic polymer undergo photo-degradation processes in the presence of water and UV radiation. These two factors act in a synergistic manner by surface erosion and increasing wettability, thus causing mechanical properties loss and reducing their lifetime. This synergism is more harmful than exposure of composites to UV radiation only [91]. Cellulose component from wood structure is the main responsible for water absorption through its hydroxyl groups and further causes wood fibers' swelling on the composites surface with formation of more pores susceptible to allow UV light penetration. Under UV radiation exposure, the polymer matrix undergoes intense cracking processes.

A measure of the WPC's surface degradation is represented by oxidation that increases following weathering process. Photo-degradation disrupts molecular bonds with formation of the free radicals on the surface. These radicals interact through chain scission, generating shorter molecular chains with high mobility and susceptible to re-crystallization phenomena [92]. Changes in the chemical composition at the surface of WPC's can be investigated by Fourier transform infrared (FTIR) spectroscopy, with a simultaneous evaluation of the changes in the polymer matrix crystallinity.

During a prolonged exposure to UV radiation, significant colour changes occur with increasing carbonyl groups content and wood loss on the WPC's weathered surfaces [14]. Wood species, namely their lignin and wood extractives contents, are responsible for WPC's discoloration through photo-degradation during weathering process [8, 88, 93], expressed by oxidation of structures originated mainly from lignin (α-carbonyl, biphenyl, ring-conjugated double bond structures) with generation of carbonyl and carboxyl groups (as chromophoric groups). The photo-degradation of wood and wood/polymer composites can be attenuated by using different UV stabilizers, including UVA—hydroxybenzophenones and hydroxybenzotriazoles; pigments (used in coatings; for WPC's are used titanium dioxide and carbon black), and hindered amine light stabilizers (HALS—triazine-based and diester-based for WPC's), the last ones mentioned being applied alone or in combination with bridged amines [94]. For a best UV protection, these UV stabilizers are usually combined, because UVA enable formation of the free radicals, while HALS compounds inhibit the photo-degradation reactions [95]. The addition of pigments and mineral fillers can represent an efficient method to improve the weathering properties of WPC's [96, 97], fact evidenced by enhancement of their color stability and moisture resistance especially in outdoor weathering. Usually, pigments, inorganic and organic types, are often added to inhibit the photo-degradation of polymer matrix in WPC's [98]. They are usually applied alone or in combination with other protective agents such as UVA in printing inks, paints, stains, and coatings [99, 100]. Addition of a low percentage of nano-sized UVA in coatings composition (below 5 %) contributes to improvement of wood characteristics including weathering performance, physico-mechanical and thermal properties [100–103]. Their efficiency is determined by a good dispersion, that's why water-based media are preferred in order to avoid coagulation of nano-sized UVA in paints, both in the wet and dry states. Titanium dioxide (TiO_2) and zinc oxide (ZnO) nanoparticles are widely used in paints and coatings to block harmful portions of the solar radiation and to reduce the rate of photo-degradation processes [100, 104].

The application of a maintenance coating is highly recommended when an extended life-time of protected wood materials is envisaged, mainly in outdoor exposure conditions which require a better performance and durability during weathering. Most coating systems have no ability to control the moisture content variations in the wood surfaces which significantly affect the long-term performance of both substrate (wood) and coating [105]. The durability of wood coating systems is expressed by reaching different types of limit states (e.g. micro cracks that are hardly visible) which require restoration [106]. A developed coating material of known and controlled rate of decomposition, based on poly (vinyl acetate) copolymer binder in combination with Fe^{2+} as redox catalyst, can be used as weathering indicator in artificial and natural (outdoor) exposure [107]. This weathering indicator is useful for comparison purposes, e.g. different locations as weathering intensity, artificial weathering devices, and time periods by monitoring significant weathering factors such as UV irradiation, humidity, precipitation. Further, it could indicate the necessity of a sustainable maintenance of coated wood surfaces in outdoor applications (e.g. constructions). The data obtained by using the weathering indicator can be combined in terms of time with the durability of wood coatings in order to reach a limit state [106, 108].

3 Conclusions

Wood is a renewable material that is produced through natural processes in forest ecosystems. Wood chemistry and structure are both strongly related to its degradability that is also influenced by the environmental conditions (humidity, temperature, UV radiation, pollutants). In order to fulfill the requirement of sustainability, a well-established equilibrium between biosynthesis, transformation and degradation should exist. Modification of wood or its main polymer component, cellulose, is an effective way to improve its stability against biodegradation processes caused by different biological agents. This method can be chemical (e.g. acetylation) or physical (thermal treatment), protecting wood to some degree from biological degradation, with a positive effect on the lifetime of wood based materials. UV weathering, a process initiated primarily by the UV portion of the solar spectrum, causes photo-chemical degradation reactions on the wood surfaces. Knowledge of the chemical structure of wood polymer components and their interactions with UV radiation is essential for understanding the chemistry of UV photo-degradation of wood. Photo-chemical degradation of wood causes significant structural and color changes, lignin component being the most susceptible to de-polymerization reactions by which phenoxy radicals are generated as intermediates and further oxidized into colored chromophores (quinones). This pathway is well-known as the repeating phenoxy quinone redox cycle, by which under UV radiation, hydroquinones are oxidized to paraquinones which are further reduced to hydroquinones by the effect of hydroperoxy radicals. Chemical changes can be evidenced by FTIR spectroscopy. Effective treatments, e.g. UV photo-stabilizers, or surface coatings of wood products, e.g. paints, varnishes, stains or water repellents, may provide significant wood protection against weathering in long-term outdoor applications. The application of a maintenance coating is highly recommended when an extended life-time of protected wood materials is envisaged, mainly in outdoor exposure conditions which require a better performance and durability during weathering.

References

1. Andrady, A.L., Torikai, A., Redhwi, H.H., Pandey, K.K., Gies, P.: Consequences of stratospheric ozone depletion and climate change on the use of materials. Photochem. Photobiol. Sci. **14**, 170–184 (2015)
2. Kranitz, K., Sonderegger, W., Bues, C.T., Niemz, P.: Effects of aging on wood: a literature review. Wood Sci. Technol. **50**, 7–22 (2016)
3. Vartanian, E., Barres, O., Roque, C.: FTIR spectroscopy of woods: a new approach to study the weathering of the carving face of a sculpture. Spectrochim. Acta, Part A **136**, 1255–1259 (2015)
4. Bonifazi, G., Calienno, L., Capobianco, G., Lo Monaco, A., Pelosi, C., Picchio, R., Serranti, S.: Modeling color and chemical changes on normal and red heart beech wood by reflectance spectrophotometry, Fourier Transform Infrared spectroscopy and hyperspectral imaging. Polym. Degrad. Stab. **113**, 10–21 (2015)
5. Evans, P.D.: Weathering of wood and wood composites. In: Rowell, R.M. (ed.) Handbook of wood chemistry and wood composites, p. 151. CRC Press, New York (2012)
6. George, B., Suttie, E., Merlin, A., Deglise, X.: Photodegradation and photostabilisation of wood—the state of the art. Polym. Degrad. Stab. **88**(2), 268–274 (2005)
7. Williams, R.S.: Weathering of wood. In: Rowell, R.M. (ed.) Handbook of Wood Chemistry and Wood Composites, p. 142. CRC Press, Boca Raton (2005)

8. Hon, D.N.S.: Weathering and photochemistry of wood. In: Hon, D.N.S., Shiraishi, N. (eds.) Wood and Cellulosic Chemistry, 2nd edn, p. 513. Marcel Dekker, New York (2001)
9. Feist, W.C., Hon D.N.S.: Chemistry of weathering and Protection. In: Rowell, R.M. (eds.) The Chemistry of Solid Wood. ACS Advances in Chemistry Series No. 207, p. 401. American Chemical Society, Washington, DC (1984)
10. Andrady, A.L., Hamid, H., Torikai, A.: Effects of solar UV and climate change on materials. Photochem. Photobiol. Sci. **10**, 292–300 (2011)
11. Andrady, A.L., Hamid, S.H., Hu, X., Torikai, A.: Effects of increased solar ultraviolet radiation on materials. J. Photochem. Photobiol. B **46**, 96–103 (1998)
12. Kalnins, M.A.: Surface Characteristics of Wood As They Affect Durability of Finishes. Part II. Photochemical Degradation of Wood, pp. 23–60. U.S. Forest Service Research Paper no. 57 Forest Products Laboratory, Madison, Wisconsin (1966)
13. Roşu, D., Teacă, C.-A., Bodirlău, R., Roşu, L.: FTIR and color change of the modified wood as a result of artificial light irradiation. J. Photochem. Photobiol. B **99**, 144–149 (2010)
14. Fabiyi, J.S., McDonald, A.G., Wolcott, M.P., Griffths, P.R.: Wood plastic composites weathering: visual appearance and chemical changes. Polym. Degrad. Stab. **93**, 1405–1414 (2008)
15. Pandey, K.K.: Study of the effect of photo-irradiation on the surface chemistry of wood. Polym. Degrad. Stab. **90**(1), 9–20 (2005)
16. Müller, U., Rätzsch, M., Schwanninger, M., Steiner, M., Zöbl, H.: Yellowing and IR changes of spruce wood as result of UV-irradiation. J. Photochem. Photobiol. B **69**, 97–105 (2003)
17. Chang, T.C., Chang, H.T., Wu, C.L., Chang, S.T.: Influences of extractives on the photodegradation of wood. Polym. Degrad. Stab. **95**, 516–521 (2010)
18. Pandey, K.K.: A note on the influence of extractives on the photo-discoloration and photo-degradation of wood. Polym. Degrad. Stab. **87**(2), 375–379 (2005)
19. Jaaskelainen, A.-S., Saariaho, A.-M., Vyorykka, J., Vuorinen, T., Matousek, P., Parker, A. W.: Application of UV–Vis and resonance Raman spectroscopy to study bleaching and photoyellowing of thermomechanical pulps. Holzforschung **60**(3), 231–238 (2006)
20. Feist, W.C.: Outdoor wood weathering and protection. In: Rowell, R.M., Barbour, R.J. (eds.) Archaeological Wood: Properties, Chemistry, and Preservation. ACS Advances in Chemistry Series No. 225, p. 263. American Chemical Society, Washington, DC (1990)
21. Teacă, C.-A., Roşu, D., Bodîrlău, R., Roşu, L.: Structural changes in wood under artificial UV light irradiation determined by FTIR spectroscopy and color measurements—a brief review. BioResources **8**(1), 1478–1507 (2013)
22. Persze, L., Tolvaj, L.: Photodegradation of wood at elevated temperature: colour change. J. Photochem. Photobiol. B **108**, 44–47 (2012)
23. Mitsui, K., Takada, H., Sugiyama, M., Hasegawa, R.: Changes in the properties of light-irradiated wood with heat treatment. Part 1: effect of treatment conditions on the change in color. Holzforschung **55**, 601–605 (2001)
24. Tolvaj, L., Faix, O.: Artificial ageing of wood monitored by DRIFT spectroscopy and CIE L* a* b* color measurements. I. Effect of UV light. Holzforschung **49**, 397–404 (1995)
25. Smidt, E., Schwanninger, M., Tintner, J., Böhm, K.: Ageing and deterioration of materials in the environment—application of multivariate data analysis. In: Leandro Valim de Freitas, Ana Paula Barbosa Rodrigues de Freitas (eds.) Multivariate Analysis in Management, Engineering and the Sciences, p. 133. InTech, Rijeka (2013)
26. Bryne, L.E., Lausmaa, J., Ernstsson, M., Englund, F., Wälinder, M.E.P.: Ageing of modified wood. Part 2: determination of surface composition of acetylated, furfurylated, and thermally modified wood by XPS and ToF-SIMS. Holzforschung **64**, 305–313 (2010)
27. Mohebby, B., Militz, H.: Microbial attack of acetylated wood in field soil trials. Int. Biodeterior. Biodegrad. **64**, 41–50 (2010)
28. Jebrane, M., Sèbe, G., Cullis, I., Evans, P.D.: Photostabilisation of wood using aromatic vinyl esters. Polym. Degrad. Stab. **94**, 151–157 (2009)

29. Hill, C.A.S.: Why does acetylation protect wood from microbiological attack? Wood Mater. Sci. Eng. **4**, 37–45 (2009)
30. Tolvaj, L., Persze, L., Albert, L.: Thermal degradation of wood during photodegradation. J. Photochem. Photobiol. B **105**, 90–93 (2011)
31. Mitsui, K., Tsuchikawa, S.: Low atmospheric temperature dependence on photodegradation of wood. J. Photochem. Photobiol. B **81**, 84–88 (2005)
32. Chou, P., Chang, H., Yeh, T., Chang, S.: Characterizing the conservation effect of clear coatings on photodegradation of wood. Bioresour. Technol. **99**, 1073–1079 (2008)
33. Evans, P.D.: Weathering and photoprotection of wood. In: Schultz, T.P., Militz, H., Freeman, M.H., Goodell, B., Nicholas, D.D. (eds.) Development of Commercial Wood Preservatives. ACS Symposium Series No. 982, p. 69. American Chemical Society, Washington, DC (2008)
34. Hon, D.N.S., Chang, S.T.: Surface degradation of wood by ultraviolet light. J. Polym. Sci. Polym. Chem. **22**, 2227–2241 (1984)
35. Leary, G.J.: The yellowing of wood by light. Tappi **50**, 17–19 (1967)
36. Leary, G.J.: The yellowing of wood by light. Part II. Tappi **51**(6), 257–260 (1968)
37. Forman, L.V.: Action of ultraviolet light on lignin. Pap. Trade J. **111**(21), 34–40 (1940)
38. Sarkanen, K.V., Ludwig, C.H.: Lignins: Occurrence, Formation, Structure and Reactions. Wiley, New York (1971)
39. Hon, D.N.S.: Weathering reactions and protection of wood surfaces. J. Appl. Polym. Sci. **37** (1), 845–864 (1983)
40. Heitner, C.: Light-induced yellowing of wood-containing papers—an evolution of the mechanism. In: Heitner, C., Scaiano, J.C. (eds.) Photochemistry of Lignocellulosic Materials. ACS Symposium Series No. 531, p. 2. American Chemical Society, Washington DC (1993)
41. Ranby, B., Krinstad, K.P., Cowling, E.B., Lin, S.: The free radical content in wood and lignins. Acta Chem. Scand. **23**, 3257–3275 (1969)
42. Hon, D.N.S., Ifju, G., Feist, W.C.: Characteristics of free radicals in wood. Wood Fiber **12** (2), 121–130 (1980)
43. Hon, D.N.S.: Formation of free radicals in photo irradiated cellulose. VI. Effect of lignin. J. Polym. Sci. Polym. Chem. Ed. **13**, 2641–2652 (1975)
44. Hon, D.N.S., Ifju, G.: Measuring penetration of light into wood by detection of photo-induced free radicals. Wood Sci **11**, 118–127 (1978)
45. Oltean, L., Teischinger, A., Hansmann, C.: Wood surface discolouration due to simulated indoor sunlight exposure. Holz als Roh Werkst. **66**, 51–56 (2008)
46. Tolvaj, L., Molnar, Z., Nemeth, R.: Photodegradation of wood at elevated temperature: infrared spectroscopic study. J. Photochem. Photobiol. B **121**, 32–36 (2013)
47. Kamoun, D., Merlin, A., Deglise, X., Urizar, S.H., Fernandez, A.M.: Electron paramagnetic resonance spectroscopy study of photodegradation of lignins extracted and isolated from *Pinus radiata* wood. Ann. For. Sci. **56**(7), 563–578 (1999)
48. Tolvaj, L., Popescu, C.-M., Molnar, Z., Preklet, E.: Effects of air relative humidity and temperature on photodegradation processes in beech and spruce wood. BioResources **11**(1), 296–305 (2016)
49. Timar, M.C., Varodi, A.M., Gurău, L.: Comparative study of photo degradation of six wood species after short-time UV exposure. Wood Sci. Technol. **50**, 135–163 (2016)
50. Calienno, L., Lo Monaco, A., Pelosi, C., Picchio, R.: Colour and chemical changes on photodegraded beech wood with or without red heartwood. Wood Sci. Technol. **48**(6), 1167–1180 (2014)
51. Srinivas, K., Pandey, K.K.: Photodegradation of thermally modified wood. J. Photochem. Photobiol. B **117**, 140–145 (2012)
52. Xiao, Z., Xie, Y., Adamopoulos, S., Mai, C.: Effects of chemical modification with glutaraldehyde on the weathering performance of Scots pine sapwood. Wood Sci. Technol. **46**, 749–767 (2012)

53. Barta, E., Papp, G., Preklet, E., Tolvaj, L., Berkesi, O., Nagy, T., Szatmari, S.: Changes of absorption in infrared spectra of softwood materials irradiated by UV-laser as a function of energy. Acta Silv. Lignaria Hung. **1**(1), 83–91 (2005)
54. Papp, G., Barta, E., Tolvaj, L., Berkesi, O., Nagy, T., Szatmári, S.: Changes in DRIFT spectra of wood irradiated by UV laser as a function of energy. J. Photochem. Photobiol. A **173**(2), 137–142 (2005)
55. Windeisen, E., Wegener, G., Lesnino, G., Schumacher, P.: Investigation of the correlation between extractives content and natural durability in 20 cultivated larch trees. Holz als Roh Werkst. **60**, 373–374 (2002)
56. Sykacek, E., Gierlinger, N., Wimmer, R., Schwanninger, M.: Prediction of natural durability of commercial available European and Siberian larch by near-infrared spectroscopy. Holzforschung **60**, 643–647 (2006)
57. Gierlinger, N., Schwanninger, M., Hinterstoisser, B., Wimmer, R.: Rapid determination of heartwood extractives in *Larix* sp. by means of Fourier transform near infrared spectroscopy. J. Near Infrared Spectrosc. **10**, 203–214 (2002)
58. Susi, P., Aktuganov, G., Himanen, J., Korpela, T.: Biological control of wood decay against fungal infection. J. Environ. Manag. **92**, 1681–1689 (2011)
59. Kocaefe, D., Huang, X., Kocaefe, Y., Boluk, Y.: Quantitative characterization of chemical degradation of heat-treated wood surfaces during artificial weathering using XPS. Surf. Interface Anal. **45**, 639–649 (2013)
60. Esteves, B.M., Domingos, I.J., Pereira, H.M.: Pine wood modification by heat treatment in air. BioResources **3**, 142–154 (2008)
61. Evans, P.D., Wallis, A.F.A., Owen, N.L.: Weathering of chemically modified wood surfaces. Natural weathering of Scots pine acetylated to different weight gains. Wood Sci. Technol. **34**, 151–165 (2000)
62. Hill, C.A.S., Forster, S.C., Farahani, M.R.M., Hale, M.D.C., Ormondroyd, G.A., Williams, G.R.: An investigation of cell wall micropore blocking as a possible mechanism for the decay resistance of anhydride modified wood. Int. Biodeterior. Biodegrad. **55**, 69–76 (2005)
63. Papadopoulos, A.N.: Chemical modification of solid wood and wood raw material for composites production with linear chain carboxylic acid anhydrides: a brief review. BioResources **5**, 499–506 (2010)
64. Chang, S.T., Chang, H.T.: Inhibition of the photo-discoloration of wood by butyrylation. Holzforschung **55**, 255–259 (2001)
65. Esteves, B., Nunes, L., Pereira, H.: Properties of furfurylated wood (*Pinus pinaster*). Eur. J. Wood Wood Prod. **69**, 521–525 (2011)
66. Baur, S.I., Easteal, A.J.: Improved photoprotection of wood by chemical modification with silanes: NMR and ESR studies. Polym. Adv. Technol. **24**, 97–103 (2013)
67. Donath, S., Militz, H., Mai, C.: Weathering of silane treated wood. Holz als Roh Werkst. **65**, 35–42 (2007)
68. Rowell, R.M.: Chemical Modification of Wood. Handbook of Wood Chemistry and Wood Composites, 2nd edn, p. 537. CRC Press, Boca Raton (2012)
69. Hill, C.A.S.: Wood Modification: Chemical, Thermal and Other Processes. Wiley, Chichester (2006)
70. Jebrane, M., Sèbe, G.: A new process for the esterification of wood by reaction with vinyl esters. Carbohydr. Polym. **72**(4), 657–663 (2008)
71. Jebrane, M., Sèbe, G.: A novel simple route to wood acetylation by transesterification with vinyl acetate. Holzforschung **61**(2), 143–147 (2007)
72. Evans, P.D., Chowdhury, M.: Photoprotection of wood using polyester-type UV-absorbers derived from the reaction of 2-hydroxy-4(2,3-epoxypropoxy)-benzophenone with dicarboxylic acid anhydrides. J. Wood Chem. Technol. **30**, 186–204 (2010)
73. Hansmann, C., Deka, M., Wimmer, R., Gindl, W.: Artificial weathering of wood surfaces modified by melamine formaldehyde resins. Holz als Roh Werkst. **64**, 198–203 (2006)

74. Sharratt, V., Hill, C.A.S., Kint, D.P.R.: A study of early colour change due to simulated accelerated sunlight exposure in Scots pine (*Pinus sylvestris*). Polym. Degrad. Stab. **94**, 1589–1594 (2009)
75. Oltean, L., Hansmann, C., Nemeth, R., Teischinger, A.: Wood surface discolouration of three Hungarian hardwood species due to simulated indoor sunlight exposure. Wood Res. Slovak. **55**, 49–58 (2010)
76. Chang, H.T., Chang, S.T.: Correlation between softwood discoloration induced by accelerated lightfastness testing and by indoor exposure. Polym. Degrad. Stab. **72**, 361–365 (2001)
77. Roşu, D., Bodîrlău, R., Teacă, C.-A., Roşu, L., Varganici, C.D.: Epoxy and succinic anhydride functionalized soybean oil for wood protection against UV light action. J. Clean. Prod. **112**, 1175–1183 (2016)
78. de Meijer, M.: Review on the durability of exterior wood coatings with reduced VOC-content. Prog. Org. Coat. **43**, 217–225 (2001)
79. Katz, S.A., Salem, H.: Chemistry and toxicology of building timbers pressure-treated with chromated copper arsenate: a review. J. Appl. Toxicol. **25**, 1–7 (2005)
80. Obanda, D.N., Shupe, T.F., Barnes, H.M.: Reducing leaching of boron-based wood preservatives—a review of research. Bioresour. Technol. **99**, 7312–7322 (2008)
81. Singh, T., Singh, A.P.: A review on natural products as wood protectant. Wood Sci. Technol. **46**, 851–870 (2012)
82. Ruddick, J.N.R.: Biocide depletion: chemical, physical, and photodegradation. In: Schultz, T.P., Militz, H., Freeman, M.H., Goodell, B., Nicholas, D.D. (eds.) Development of Commercial Wood Preservatives: Efficacy, Environmental and Health Issues, p. 285. American Chemical Society, Washington, DC (2008)
83. Xu, W.Z., Yang, L., Charpentier, P.A.: Preparation of antibacterial softwood via chemical attachment of quaternary ammonium compounds using supercritical CO_2. ACS Sustain. Chem. Eng. **4**(3), 1551–1561 (2016)
84. Peng, Y., Liu, R., Cao, J.: Characterization of surface chemistry and crystallization behaviour of polypropylene composites reinforced with wood flour, cellulose, and lignin during accelerated weathering. Appl. Surf. Sci. **332**, 253–259 (2015)
85. Peng, Y., Liu, R., Cao, J., Chen, Y.: Effects of UV weathering on surface properties of polypropylene composites reinforced with wood flour, lignin, and cellulose. Appl. Surf. Sci. **317**, 385–392 (2014)
86. Chaochanchaikul, K., Jayaraman, K., Rosarpitak, V., Sombatsompop, N.: Influence of lignin content on photodegradation in wood/HDPE composites under UV weathering. BioResources **7**, 38–55 (2012)
87. Matuana, L.M., Jin, S., Stark, N.M.: Ultraviolet weathering of HDPE/wood–flour composites coextruded with a clear HDPE cap layer. Polym. Degrad. Stab. **96**, 97–106 (2011)
88. Fabiyi, J.S., McDonald, A.G.: Effect of wood species on property and weathering performance of wood plastic composites. Compos. A Appl. Sci. Manuf. **41**, 1434–1440 (2010)
89. Stark, N.M.: Photodegradation and Photostabilization of Weathered Wood Flour Filled Polyethylene Composites. Dissertation. Michigan Technological University (2003)
90. Stark, N.M., Gardner, D.J.: Outdoor durability of wood–polymer composites. In: Oksman Niska, K., Sain, M. (eds.) Wood Polymer Composites, p. 142. CRC Press, Boca Raton (2008)
91. Stark, N.M.: Effect of weathering cycle and manufacturing method on performance of wood flour and high-density polyethylene composites. J. Appl. Polym. Sci. **100**, 3131–3140 (2006)
92. Stark, N.M., Matuana, L.M.: Characterization of weathered wood-plastic composite surfaces using FTIR spectroscopy, contact angle, and XPS. Polym. Degrad. Stab. **92**, 1883–1890 (2007)

93. Temiz, A., Terziev, N., Eikenes, M., Hafren, J.: Effect of accelerated weathering on surface chemistry of modified wood. Appl. Surf. Sci. **253**, 5355–5362 (2007)
94. Hyvärinen, M., Väntsi, O., Butylina, S., Kärki, T.: Ultraviolet light protection of wood-plastic composites. A review of the current situation. Adv. Sci. Lett. **19**(1), 320–324 (2013)
95. Schiller, C., Rogez, D., Braig, A.: Hindered amine light stabilizers in pigmented coatings. J. Coat. Technol. Res. **6**(1), 81–88 (2009)
96. Butylina, S., Hyvärinen, M., Kärki, T.: Accelerated weathering of wood–polypropylene composites containing minerals. Compos. A Appl. Sci. Manuf. **43**(11), 2087–2094 (2012)
97. Butylina, S., Hyvärinen, M., Kärki, T.: Weathering of wood-polypropylene composites containing pigments. Eur. J. Wood Wood Prod. **70**, 719–726 (2012)
98. Stark, N.M., Matuana, L.M.: Influence of photostabilizers on wood flour–HDPE composites exposed to xenon-arc radiation with and without water spray. Polym. Degrad. Stab. **91**, 3048–3056 (2006)
99. Forsthuber, B., Schaller, C., Grüll, G.: Evaluation of the photo stabilising efficiency of clear coatings comprising organic UV absorbers and mineral UV screeners on wood surfaces. Wood Sci. Technol. **47**, 281–297 (2013)
100. Cristea, M.V., Riedl, B., Blanchet, P.: Enhancing the performance of exterior water-borne coatings for wood by inorganic nanosized UV absorbers. Prog. Org. Coat. **69**(4), 432–441 (2010)
101. Fufa, S.M., Jelle, B.P., Hovde, P.J.: Effects of TiO_2 and clay nanoparticles loading on weathering performance of coated wood. Prog. Org. Coat. **76**, 1425–1429 (2013)
102. Fufa, S.M., Jelle, B.P., Hovde, P.J.: Weathering performance of spruce coated with water based acrylic paint modified with TiO_2 and clay nanoparticles. Prog. Org. Coat. **76**, 1543–1548 (2013)
103. Lowry, M., Hubble, D., Wressell, A., Vratsanos, M., Pepe, F., Hegedus, C.: Assessment of UV-permeability in nano-ZnO filled coatings via high throughput experimentation. J. Coat. Technol. Res. **5**(2), 233–239 (2008)
104. Cristea, M.V., Riedl, B., Blanchet, P.: Effect of addition of nanosized UV absorbers on the physico-mechanical and thermal properties of an exterior waterborne stain for wood. Prog. Org. Coat. **72**, 755–762 (2011)
105. Miller, E.R.: Wood substrate—the foundation. Surf. Coat. Int. Part B Coat. Trans. **88**, 157–161 (2005)
106. Grüll, G., Truskaller, M., Podgorski, L., Bollmus, S., Tscherne, F.: Maintenance procedures and definition of limit states for exterior wood coatings. Eur. J. Wood Wood Prod. **69**, 443–450 (2011)
107. Grüll, G., Forsthuber, B., Tscherne, F., Spitaler, I.: Weathering indicator for artificial and natural weathering of wood coatings. Eur. J. Wood Wood Prod. **72**, 681–684 (2014)
108. Grüll, G., Tscherne, F., Spitaler, I., Forsthuber, B.: Comparison of wood coating durability in natural weathering and artificial weathering using fluorescent UV-lamps and water. Eur. J. Wood Wood Prod. **72**, 367–376 (2014)

Photochemical Behavior of Synthetic Polymeric Multicomponent Materials Composites and Nanocomposites

Madalina Zanoaga and Fulga Tanasa

Abstract Multicomponent materials based on synthetic polymers were designed and used in a wide variety of common and hi-tech applications, including the outdoor applications as well. Therefore, their response to the UV radiation and complex weathering conditions (temperature, seasonal or freeze–thaw cycles, humidity, pH, pollutants, ozone, microorganisms) is a matter of utmost importance in terms of operational reliability and lifetime, protection of the environment and health safety. This chapter offers an overview of this subject and a critical assessment of more particular topics related to this issue. Thus, various types of multicomponent systems based on thermoplastic and thermosetting polymer matrices were subjected to natural and/or simulated UV radiation and/or weathering conditions. Their behavior was evaluated in correlation with their complex formulation and taking into consideration that the overall effect is a sum of the individual responses and interactions between components. The nature and type of the matrix, the nature, type and size distribution of the filler, the formation of the interphase and its characteristics, the interfacial adhesion and specific interfacial interactions, they all were considered as factors that influenced the materials behavior, and, at the same time, were used as classification criteria for this review.

Abbreviations

AELO	Acrylated epoxidized linseed oil
AESO	Acrylated epoxidized soybean oil
AHK	α-Hydoxy ketone
BAPO	Bisacylphosphine oxide
BEMA 1400	Oligomer containing bisphenol A and terminal α,ω methacrylic groups
BPO	Benzoyl peroxide

M. Zanoaga · F. Tanasa (✉)
"Petru Poni" Institute of Macromolecular Chemistry, Iasi, Romania
e-mail: ftanasa@icmpp.ro

M. Zanoaga
e-mail: zanoaga@icmpp.ro

D. Rosu and Visakh P. M. (eds.), *Photochemical Behavior of Multicomponent Polymeric-based Materials*, Advanced Structured Materials 26,
DOI 10.1007/978-3-319-25196-7_5

CB	Carbon black
CNT	Carbon nanotube
DVE	Triethylene glycol divinyl ether
E	Modulus of elasticity
ELO	Epoxidized linseed oil
ESO	Epoxidized soybean oil
EVA	Ethylene vinyl acetate
EVO	Epoxidized vegetal oils
FTIR	Fourier transform infrared spectroscopy
GPTS	Glycidyl-propyl-triethoxysylane
GO	Graphite oxide
HALS	Hindered Amine Light Stabilizer
HBVE	Hydroxybutyl vinyl ether
HDPE	High density polyethylene
iPP	Isotactic polypropylene
IPT	3-Isocyanatopropyltriethoxysilane
IT	Information technology
LDPE	Low-density polyethylene
LLDPE	Linear low-density polyethylene
MA-OH	3-(Acryloyloxy)-2-hidroxy-propyl methacrylate
MAPO	Monoacylphosphine oxide
MAPP	Maleic anhydride-grafted polypropylene (maleated polypropylene)
MEMO	Methacryloyl-oxypropyltrimethoxysilane
MMT	Montmorillonite
MPTS	3-Trimethoxysilyl propyl methacrylate
MWCNT	Multiwalled carbon nanotubes
ODA	Octadecylammonium
PA6	Polyamide 6
PA66	Polyamide 66
PA12	Polyamide 12
PAN	Polyacrylonitrile
PBT	Polybutylene therephtalate
PC	Polycarbonate
PCL	Polycaprolactone
PCT	Polycyclohexylenedimethylene terephtalate
PCTA	Copolyester poly(cyclo-hexane-dimethanol-terephthalate acid); acid modified PCT (polycyclohexane dimethanol terephthalate)
PE	Polyethylene
PEEK	Poly(ether-ether-ketone)
PEGDA 600	Poly(ethylene glycol) 600 α,ω diacrylate
PE-*g*-MA	Polyethylene-grafted maleic anhydride
PEN	Poly(ethylene 2,6-naphthalate)
PEO	Poly(ethylene oxide)
PET	Polyethylene terephthalate

PGC-MS	Pyrolysis gas chromatography-mass spectroscopy
PHB	Polyhydroxybutyrate
phr	Parts per hundred resin
PLA	Polylactides
PMMA	Polymethyl methacrylate
PP	Polypropylene
PP-*g*-MA	Polypropylene-grafted maleic anhydride (maleated polypropylene)
PS	Polystyrene
PVA	Poly(vinyl alcohol)
PVC	Poly(vinyl chloride)
RIDFT	Resin infusion between double flexible tooling
SEM	Scanning electron microscopy
SO	Soybean oil
TEOS	Tetraethoxysilane
TEM	Transmission electron microscopy
T_g	Glass transition temperature
T_m	Melting temperature
TPGDA	Tripropylene glycol diacrylate
2 K PU	Two-pack polyuretanes
UV	Ultraviolet
VARTM	Vacuum assisted resin transfer molding
VBRI	Vacuum bag resin infusion
XPS	X-ray photoelectron spectroscopy
ZnO	Zinc oxide
WPC	Wood–polymer composite
wt	Weight

1 Introduction

Multicomponent materials based on synthetic polymer matrices are increasingly used in our day-to-day life, in common or hi-tech applications ranging from tools and house appliances to electronics, automotive, defense and aerospace industry, information technology (IT) and nanorobotics, medicine and healthcare, etc. Optimized complex composite formulations are required for these materials to make them able to respond to specific demands, according to the final end envisaged since the product design stage.

In most cases, these multicomponent systems consist of several different phases between which interfaces occur and mediate specific interactions. Depending on the scale determined by components characteristics, the type and strength of these interactions at the interface may vary within large limits and strongly influence the bulk properties.

All these materials, whether particulate filled or fiber reinforced, single polymer or polymer blend matrices based on, thin single layered or multilayered or sandwich-like structured, filled with organic, inorganic or renewable reinforcing components, have one major characteristic: all components act in synergy. Hence, the importance of studying the complex phenomena taking place at the interfaces and inside interphases of heterogeneous multicomponent systems [1–3]. It is still difficult to estimate the effects of these interactions at their full length, despite the advances in terms of structural characterization and property investigation techniques, due not only to their number, but their variety as well. Therefore, most studies tend to approach these interactions separately, without extensive correlations between phenomena and processes.

Materials used in outdoor applications must respond to a series of particular requirements, considering the specific conditions of service. One of the most important factors is the UV radiation which is known to be beneficial in small doses for various biochemical processes, but may cause severe damage in all types of living organisms and materials upon prolonged exposure. High levels of UV radiations, associated with other environmental factors, such as temperature, ozone, moisture, etc., promote and accelerate degradation mechanisms in plastics—a complex phenomenon known as weathering, thus drastically limiting their service life. Therefore, plastics are UV-stabilized by the use of specific additives; wood containing materials are using coated or pre-treated wood in order to limit their susceptibility toward water; other formulations may include UV-resistant additives, may use UV-absorbing coatings, etc. All these product-oriented solutions yield in increased production costs, so an optimization which considers effectiveness of the selected approach over the estimated damage is always required.

The subject of photochemical behavior of natural and synthetic polymers is well-known and documented, but studies in recents decades focused mainly on narrow domains, given the wide range of new materials developed and processes considered [3–12]. Still, a few basics must be emphasized. The UV radiation represents only 4.6 % of the solar spectrum and it ranges between 280 and 400 nm; the most aggressive component is the UVB radiation (very short wavelenghts, at 280–315 nm) [13]. UV radiation causes an intense photooxidative stress that entails degradation processes which result in breaking of the polymer chains and molecular weight reduction, along with the formation of free radicals able to undergo various secondary reactions, all yielding in the decay of mechanical properties [14–17]. Initiation reactions may be promoted by hydroperoxide or carbonyl groups, residues of metal oxides acting as catalysts, or charge transfer complexes containing oxygen, as well as other molecules which may contain chromophoric groups: traces of solvents, components of the polluted urban atmosphere (e.g., polynuclear hydrocarbons), pigments, dyes, various stabilizers, etc. The free radicals formed contribute to the propagation of the chain reaction under the favouring effect of other environmental factors that add to the UV exposure. The termination stage evolves with the formation of new oligomers and other molecules [17].

Stabilization of synthetic polymers may be acquired using screeners, UV-absorbers, excited-stated quenchers, peroxide decomposers, free radical scavengers, pigments, metal chelates, opacizers, all aimed at limiting the UV radiation effects [14].

As for the multicomponent systems, their photochemical behavior must be assessed considering that the overall process is the sum of processes taking place in each component, at the interface between components and in the corresponding interphases. It is noteworthy that, in the case of particular materials, the degradation under the UV radiation may take place localized, in the outer layers; the photochemical reactions that occur yield in products that absorb the UV energy and limit the in-depth penetration of the UV radiation, thus protecting the inner layers.

This chapter is aiming to systematically present an overview of some recent advances in this very active field of research. For a better understanding, the information was structured in sections using basic discriminating criteria such as: the nature of matrix and filler, the scale of interactions (micro- or nanoscale), the processing approach, etc.

2 Photochemical Behavior of Thermoplastic Matrix Based Composites

2.1 General Considerations

Thermoplastic polymers based composites are a class of materials ranging from engineering to hi-tech, which can be processed at temperatures either higher than the matrix glass transition temperature (T_g)—amorphous thermoplastics, or higher than its corresponding melting temperature (T_m)—semi-crystalline thermoplastics. Given the dependency of modulus of elasticity (E) on T_g and T_m (Scheme 1), it is obvious that semi-crystalline matrices will allow composites with E highly increased above T_g, hence materials able to work under high stress.

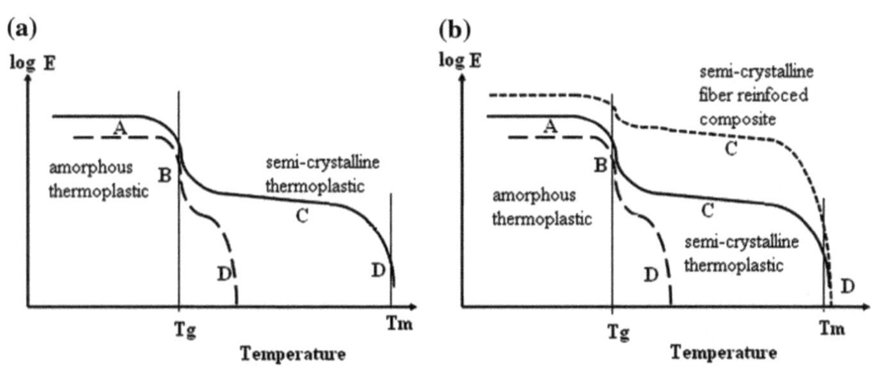

Scheme 1 The elastic modulus (log E) versus temperature in: **a** amorphous and semi-crystalline thermoplastics; **b** their corresponding fiber-reinforced composites (domains: *A*—glass plateau, *B*—glass transition, *C*—semi-crystalline plateau, *D*—flow; http://www.pluscomposites.eu/publications)

Most thermoplastic composites exposed to UV irradiation are used in outdoor applications (sidings, windows and doors, exposed pipes, decking, panels for the automotive industry, furnishing, packaging etc.), so the longer the exposure, the higher the extent of degradation.

The major part of the thermoplastic synthetic polymers (polyolefins and vinylic polymers—PE, PP, PVC, PVA; polyesters—PET, PBT, PC, polybutyrate PEN, PCTA, etc.; polylactides, (PLA), (co)polyamides—PA6, PA6,6, PA12, etc.; thermoplastic polyurethanes, etc.), as well as the biopolymers obtained either from biosynthesis or through chemical modification of the natural ones, are susceptible to photochemical degradation upon absorption of the UV radiation. Their structural features include photolabile moieties and different additives, which, associated with other aggravating weathering factors, such as environmental humidity and salty atmosphere, pH and pollutants, wind and particle abrasion, microorganisms attack, temperature variation and freeze–thaw cycles [5], contributes to their degradation upon prolonged UV exposure. In the case of biopolymers from natural resources, their photochemical degradation depends highly on their origin and processing.

The photochemical degradation of thermoplastic multicomponent systems evolves gradually, starting with spots of discoloration at the exposed surface, spots that expand and fuse; then microcracks occur, yielding in crevices, when the drop in mechanical properties (tensile strength, impact strength, elongation at break) is noticeable. Different solutions were designed to protect such materials, mainly (a) the inclusion in their formulations of different additives and fillers able to grant them some UV resistance and (b) the use of protective coatings. Thus, based on the impressive amount of data regarding the features of photochemical degradation underwent by various polymers and composites, it was possible to select a wide range of formulations that include: thermal stabilizers, antioxidants, flame retardants, colorants, biocides, lubricants, etc. It is also of great interest that in composites the photochemical degradation is a sum of processes that occur not only inside the matrix, but in the filler as well, whether the filler is another polymer or an organic UV-sensitive material (particulate wood—chips and powder; long or short ligno-cellulosic fibers—wood, hemp, kenaf; proteic fibers—silk, wool), or inorganic materials, either particles (clay, graphite, talc, calcium carbonate, etc.) or fibers (glass and carbon fibers).

2.2 *Photochemical Behavior of Thermoplastic Matrices*

Thermoplastics exposed to environmental factors are highly susceptible to undergo degradation processes that yield in significant changes in appearance (discoloration, yellowing, turnings to redish or grayish), chemical and physical properties alteration, failure of mechanical characteristics (embrittlement, decrease of elasticity and impact strength, etc.) [18–22]. Yellowing and discoloration, although surface phenomena, may be indications of a more in-depth degradation. The color changes are caused by the increased absorption of UV radiation, but a higher absorption reduces the overall optical transmission in transparent thermoplastics, such as some aliphatic copolyamides [23], and changes the color temperature of the transmitted light [24]. In opaque materials, the degradation is no progressing too deep, but it favours the formation of micro-cracks that merge in larger cracks where the degradative processes may continue under weathering agents and due to the presence of microorganisms.

Photochemical degradation may cause changes in the monomer unit (by oxidation reactions), the macromolecular chain (through crosslinking or chain scission reactions), and even on the macroscopic scale [7, 25]. Thermal-oxidative degradation, hydrolysis and microbial attack are simultaneously promoted by the presence of the other associated factors (oxygen and ozone, temperature and freeze–thaw cycles, environmental moisture and microorganisms) and contribute to the overall effect [26–29].

Oxygen is present at the surface of the composite and is involved in photo-oxidation reactions so that only a small amount is able to further diffuse toward the inner layers. Thus, the degradation is limited to the composite-environment interface and material outer layers. The UV-induced oxidation also occurs in amorphous regions due to their higher permeability to oxygen [30]. Given the oxidative degradation, oxygen gradients occur and may cause localized material density gradients; they yield in local stress concentrators which, combined with the effects of chain scission reactions, will promote and propagate the cracks formation [25]. Cracks generate multiple facets on the surface, alter the light diffusion and create a whitening effect in the composite appearance, followed in time by the decrease in mechanical properties [31].

Temperature regime strongly influences the photochemical behavior of thermoplastic matrices in multicomponent systems exposed to UV radiation, as higher temperatures highly accelerate the degradation. The chain of reactions is propagated *via* free radical mechanism and proceed through oxidation reactions and scission of the macromolecules, followed by crosslinking and recombination reactions [32].

The environmental moisture catalyzes the UV-induced degradation, especially in the presence of acid or alkali traces found in atmosphere as air pollutants, due to the high polarity of the functional groups susceptible to undergo hydrolysis. The process is further enhanced by the thermoplastics capacity to absorb various amounts of water depending on their polarity (higher amounts as compared to polyolefins which are non-polar) and affinity toward water [33, 34]. If the reaction sites are in the main chain, the macromolecules are cleaved and the degradation has stronger effects than if it occurs in the pendant branches.

Simultaneously to the UV degradation, microbial attack may also occur and add to the overall yellowing effect and embrittlement [29, 33].

Other impurities (catalytic residues, reactive oxygen-containing groups resulted during processing, etc.) that may be present in the composite formulations can contribute to the initiation of the photochemical reactions, even upon reduced UV-exposure [35].

A more detailed presentation of the factors influencing the photochemical behavior of polymers is extensively presented in another chapter of this handbook (see Chap. 2).

2.3 *Thermoplastic Composites with Natural Fillers from Renewable Resources*

2.3.1 Wood Particles (Chips and Flour)

Wood–polymer composites (WPCs) are a typical example of multicomponent systems as most formulations may contain, aside from matrix and filler, various additives with specific functions: compatibilizers, stabilizers, plasticizers, pigments and dyes, flame retardants and fireproofing agents, etc. Numerous studies on the photochemical behavior of WPCs were reported given the extent of their range of outdoor applications (decking, siding, railings, fencing, windows, roof tiles, door frames, etc.) due to the convenient properties:cost ratio. (Scheme 2)

In wood flour filled composites, which are typical particulate composites, interfacial interactions are of utmost importance, given the micrometric size of wood particles. Particle–particle interactions lead to secondary aggregation, but interactions at the matrix–filler interface determine processes at micrometric scale and macroscopic properties [3].

Major components of wood are cellulose, lignin and hemicellulose, and all of them are susceptible of photochemical degradation, behavior manifested by discoloration and loss of gloss and lightness, roughening and loss in mechanical properties [22, 36].

Scheme 2 PVC-based WPCs panels for outdoor decking (www.globalsources.com)

Pure cellulose scarcely absorbs UV light, probably due to the accidental presence of some carbonyl groups formed during processing, but lignin absorbs 80–95 % of the incident UV radiation due to its constitutive chromophoric moieties (phenolic, hydroxylic, carbonylic groups, double bonds). By consequence, the rate and extent of degradation are higher in the lignin-rich lamella and, upon further exposure, in the secondary walls. The mechanism involves the energy transfer from outer layers to the inner ones along with the formation of free radicals which can migrate deeper and further promote the photochemical reactions [36]. At the same time, the UV-induced degradation of lignin yields in low molecular compounds able to support growth of fungi and other microorganisms that contribute to the overall effect [18].

Extensive information on the photochemical reactions of wood is provided elsewhere in this handbook (see Chap. 4).

Recent studies on WPCs made of PE, PP and PVC and submitted to outdoor conditions evidenced that their photodegradation, as an overall degradation of all components, is accentuated by the presence of atmospheric moisture and temperature gradient, and the primary and most noticeable effect is the loss of colour and surface smoothness and gloss [25, 31, 32, 37–41]. It was also demonstrated that crystallinity of PP in pine-PP formulations increased along with the weathering interval as a result of oligomeric chains realignment upon degradation [41]. At the same time, the loss in mechanical properties is significant. For HDPE composites containing 50 % wood flour, the flexural elasticity decreased by 35 % [42].

Other factors that may affect the photochemical response of WPCs may be the nature of wood particles, as well as the source of wood and manufacturing methods. Thus, the lignin content in WPCs is suggested to influence the photochemical stability of these materials, depending on the nature of particles. In example, it was found that wood shavings were able to enhance the photostability of wood chips-PE composites [43]. Other studies indicated that the wood filler rather reduced the photostability [44] of HDPE-based composites and the rate of degradation depended directly on the volume fraction of wood in the composite formulation.

The processing techniques, such as injection molding or extrusion, were also found to significantly influence the WPCs behavior under UV radiation. Samples of HDPE-wood obtained by injection molding showed an increased susceptibility to undergo photochemical degradation [45], evidenced by discoloration and decrease of mechanical properties. A comparison between samples made by injection molding and extrusion indicated that injection-moulded composites have a lower rate of degradation than the extruded ones, probably due to the formation of a hydrophobic HDPE coating at the surface of wood particles [25] and this interphase layer limited the energy transfer and the extent of the degradation reactions chain. By the same criterion, PP-based composites degraded at a higher rate than the corresponding HDPE-based ones [25, 31]. Data reported on coextruded and uncapped WPCs revealed that coextrusion may limit the surface erosion and removal of wood debris by enhancing the adhesion between matrix and filler [25]. At the same time, the cap layer contributed to the protection of the composite as it absorbed a fraction of the incident UV radiation and reduced the amount of oxygen at the surface of the sample.

The effect of the stabilization approaches was also considered in various studies. Lignin was reported to wield different effects depending on the amount and the nature of matrix. Thus, an addition of less than 10 % lignin to PP proved to be beneficial as it stabilized the composite toward UV light, while amounts higher than 10 % lignin to PE produced photochemical degradation at high rates [36, 46], effect enhanced by the presence of moisture. The same stabilizing effect of lignin was noticed in the case of PS, LDPE and LLDPE-based formulations [47], and PP-EVA blends used as matrix in wood composites [20]. The protection against the UV radiation may be achieved by the use of HALS [48, 49]. The effectiveness of these additives for HDPE-hardwood powder composites was estimated to be enhanced by the supplemental presence of benzotriazole UV absorber, due to a synergistic action [50]. Other stabilization methods employed a compatibilizer to improve the adhesion between wood and HDPE, HALS and UV absorber (Tinuvin) to enhance the UV protection [51], or functional silica–lignin hybrid fillers for PP-based composites [52].

A combination of additives, namely hindered amine light stabilizers (HALS), an UV filter and antioxidant, was used for improving the durability of WPCs [53]. The mixture displayed a higher efficiency than HALS alone, as all components contributed to the protective effect by specific mechanisms. WPCs exhibited no cracks or fractures during exposure and, although some discoloration was noticed, the colour modification in stabilized formulations was significantly lower than in neat composites.

Pigments, just like UV absorbers and light stabilizers, can influence the colour changes in WPCs upon their exposure to UV radiation [54]. It was demonstrated that pigments are more effective than UV absorbers in terms of limiting discoloration in wood flour-HDPE composites exposed to accelerated weathering [55]. On the other hand, some pigments, such as red and black ones, proved to be effective in tempering discoloration in both polyethylene and PP-based wood–plastic composites [56], while metal oxide pigments may add to the UV-induced degradation by introducing supplemental chromophores into the considered formulations [19].

The effect of three pigments, namely green, brown and gray, on the photochemical behavior of extruded wood-PP composites submitted to weathering was also investigated. The composite formulations were based on neat and recycled PP, and included also a compatibilizer (maleated polypropylene, MAPP) and a lubricant. The darker pigments contributed to a significant retention of the composites colour during weathering, in comparison to composites with green pigment or without pigment. Composites made of recycled PP displayed an improved colour stability and maintenance of flexural strength after weathering as compared with other composites due, most probably, to their higher density which limited the water uptake and oxygen diffusion, factors that can aggravate the UV-initiated degradation [57].

A special mention has to be made concerning the PVC-wood composites subjected to accelerated weathering: they retained their initial mechanical properties in a higher proportion than the originating polymer [35, 58, 59]. The possible explanation of this behavior may be that the photo-oxidation takes place only at the surface of the composite where wood particles, considered as a chromophore material because of the presence of carbonyl groups, undergo photochemical degradation. Hence, all composite samples exhibited a more intense discoloration than the neat PVC.

2.3.2 Wood Fibers and Other Ligno-cellulosic Fibers

In the case of fiber-reinforced composites, even where the filler consists of wood fibers and other ligno-cellulosic or proteic fibers, the main issue is the interfacial adhesion between matrix and fibers, as a *sine qua non* requirement for the load transfer, as the fibers carry the load, while the matrix distributes and transfers it from fiber to fiber. Hence, the relevance of the interphase that may be formed spontaneously, no matter the fibers are raw or modified by a surface treatment, and which strongly affects the properties of the composite through its thickness, structure and properties [3].

Studies on the photo-oxidation of the fiber reinforced polymer composites reported a variety of results. Some of them are consistent with data registered for composites with particulate wood; others differ. The photochemical degradation of thermoplastics reinforced with wood fibers and other ligno-cellulosic fibers comprises colour changes which are an indication that the outer layers of composites were altered [7, 35]. These changes are due to the formation of chromophoric groups, quinones and hydroperoxydic radicals in lignin upon its exposure to UV radiation.

A parallel reaction that occurs during UV degradation is the reduction of the *p*-quinones to hydroquinones, leading to photo-bleaching. The longer the exposure, the more intense the phenomenon. Thus, it was shown that a higher content in wood fibers may yield in significant colour changes in PP-based composites [57]. The yellowing fades along with the increasing time of UV-exposure, followed by the photo-bleaching and more intense degradation of the exterior layers. When the amount of fibers is over 40 %, their wettability significantly decreases, the matrix loses its continuity around fibers and the rigidity exceeds the elasticity of the composite, limiting the load transfer.

These active oxygen-containing species may promote even the matrix UV-initiated photochemical degradation, acting as catalysts. In the presence of moisture, natural fibers favour water sorption and swelling during weathering [7]. Aside colour and appearance changes, natural fibers-reinforced thermoplastics exhibit a loss in mechanical properties as well, which may be explained by the surface oxidation upon UV exposure, changes in the matrix crystallinity, loss in the interfacial adhesion [60]. The oxidation rate in such composites increases along with the fiber amount in the formulation and correspondingly decreases when a compatibilizer is employed for enhancing interfacial interactions between phases, by improving the fibers dispersion and their wettability by the matrix.

But, depending on the nature of fibers, they may exert different effects on the composite properties. Thus, when kenaf fibers were employed in significant amount (49 % by weight) in PE composites, they acted as a photostabilizer limiting the UV-induced degradation [43]. Similar results were reported for the PP/palm-derived fiber composites submitted to both natural and artificial weathering [61]. On the other hand, when the matrix was unplasticized PVC, the same palm-derived fibers added in various amounts (10–40 % by weight) increased the photochemical degradation as proved by the composite discoloration, while the mechanical properties remained unchanged during accelerated weathering [62]. The same behavior was noted for PP-based composites containing 25 and 50 % w/w, respectively, wood fibers [63]. These contradictory data may be explained by the differences in fibers composition, in strong correlation with the fibers source.

When cellulose fibers were partially photochemically degraded by exposure to UVB radiation prior to their use in composite formulations with PP, they proved to enhance the biodegradation of composites as their surface became more hydrophilic after exposure [64].

Other studies revealed that increasing the amount of fiber load, it was possible to preserve the most of the tensile strength of the neat polymer. Thus, natural weathering tests run for 3 months on neat PP indicated a loss in tensile strength of 92.57 %. The percent decreased significantly to 58, 37 and 23 % when sisal fibers were employed in the formulation in 10, 20 and 30 %, respectively [27].

The presence of atmospheric moisture influences the UV-stability of plastic composites with natural fibers. Films made of hemp-reinforced polyhydroxybutyrate (PHB) were submitted to accelerated weathering. They underwent simultaneous photochemical degradation reactions, initiated by the UV radiation, and ester hydrolysis due to the environmental moisture [7]. Subsequently, composites exhibited cracks caused by the fibers swelling and loss in the matrix elasticity and fibers wettability. At this stage, the matrix stability is less important, as the driving force of these changes is the interfacial interaction at the fiber/matrix interface and the water uptake in fibers. The increased crystallinity and embrittlement caused by photo-oxidation contributed to the higher rate of composite degradation [65].

As in the case of composites with particulate wood, the increase in carbonyl groups (present in both carboxylic and ester moieties) is an indication that ligno-cellulosic fibers were degraded along with thin layers of polymer. Therefore, it is recommended for fibers to be treated prior to their use in composites. The fibers surface treatments refer to the improvement of their UV-resistance/absorbance (depending on the fibers effect on composite degradation) and fiber/matrix interfacial interactions, as well. Thus, bleached fibers have a lower content of lignin and may interact strongly with the matrix [7]. On the other hand, mercerized fibers contain higher amounts of lignin than the bleached ones and, subsequently, are more susceptible to undergo degradation processes upon UV and moisture exposure [66].

Composites with bleached fibers exhibited a significant decrease in tensile properties as compared with untreated fibers, due to the morphology alteration during UV irradiation. Sisal fibers-reinforced polycaprolactone (PCL) composites were produced using bleached sisal fibers and their study revealed that a certain degree of PCL protection against UV radiations was achieved by the interaction of PCL with the cellulosic component in the treated fibers [60]. These carboxyl-hydroxyl groups interactions, associated with enhanced fiber/matrix interfacial interactions, allowed composites to preserve their mechanical properties (elasticity) and crystallinity, as well as morphology, upon UV exposure.

When raw and bleached Kraft wood fibers were employed for fiber reinforced PP composites, a compatibilizer, namely maleated polypropylene (MAPP), was also considered [30]. Submitted to accelerated weathering, composites containing 40 % fibers showed decreased tensile and impact strength, regardless the fibers treatment, which lead to the conclusion that the composites degradation occurred due to both lignin and PP degradation, along with a feeble interfacial interaction between fibers and matrix, despite the presence of the compatibilizer.

Various additives (UV absorbers and stabilizers, pigments, etc.) may also enhance the resistance to UV irradiation of composites with ligno-cellulosic fibers, but they may also interfere with the mechanical strength if added in exceeding amounts. Hence, their addition must be tuned as to grant simultaneously strength and durability to the considered composites [7]. Irgastab and Tinuvin-783 were employed in palm fibers-PP composites, prepared with and without MAPP as compatibilizer. Their behavior during natural and accelerated weathering indicated that compatibilized composites were more susceptible to UV degradation as MAPP is less stable. The UV stabilizers enhanced the composites stability, due to the interfacial adhesion, as well as the presence of lignin as natural anti-oxidant, the dark colour and the interphase at the fibers surface. These factors limited the penetration of the UV radiation deeper than outer layers, thus preventing composite degradation [61].

Various approaches for the pre-treatment of natural fibers prior to their use as reinforcement in multicomponent systems are extensively presented in literature [67–70].

2.3.3 Natural Protein Fibers (Silk, Wool)

Natural protein fibers, such as silk and wool, were considered for various multicomponent formulations aimed for biomedical applications, apparel industry and electronics, structural applications and civil engineering, although their use in outdoor applications (others than clothing and protective coatings) is rather limited. Therefore, their behavior upon exposure to UV radiation was investigated.

The silk fibroin is known to undergo photochemical degradation after UV exposure, as well as the regenerated fibroin. It was suggested that new products of photodegradation were formed after irradiation, as confirmed by spectroscopy, along with changes in the molecular weight of silk [71]. Coating silk fibers with titania (TiO_2) was considered as UV-protective treatment and the structural and mechanical properties coated silk were studied. Experimental data recorded for raw and treated fibers indicated that tensile strength slightly decreased upon exposure, but the breaking extension decreased significantly for both types of fibers. Most of the tensile characteristics (except for the Young's modulus) of neat and uncoated silk strongly decreased with the increasing time of UV irradiation. FTIR spectroscopy confirmed that photochemical reactions took place in crystalline and in amorphous areas as well. Considering the crystalline:amorphous ratio, it significantly decreased in uncoated silk fibers, while remaining unchanged in titania-treated fibers, even at increased exposure intervals, which may be explained by the enhanced stability of β-sheet microcrystals toward photochemical degradation [72].

Functional intelligent textiles were created to respond to some specific requirements, such as sensing and actuating, pro-active body protection, interactive body–garment interfaces, etc., which embedded various stimuli responsive materials. Illustrative for this trend are the self-cleaning nanocomposites-based textiles. A successful nanocomposite formulation for such applications consists of polyvinyl alcohol, silk and titania [73]. Fibers of 220 nm in diameter and TiO_2 content up to 18 wt% were produced by electrospinning. Submitted to UV irradiation, silk composite fibers underwent a colour change, from brilliant green to colourless.

UV-irradiation of silk fibers was also used as a method to enhance the interfacial interaction between matrix and fibers. Thus, when PP was reinforced with silk fibers (up to 40 wt%) which were previously exposed to UV radiation, the mechanical properties recorded a significant increase due to the better coating of fibers by PP and improvement of the interfacial adhesion [74].

Wool fibers, just like other natural fibers as silk and cotton, have a lower absorption of UV radiation, as compared to synthetic fibers such as PET [75]. Still, wool displays a strong absorption in the range 280–400 nm and even over 400 nm. Most of its applications concern textiles manufacturing, using raw or treated wool fibers [76–78], but it is of recent interest in structural applications as reinforcement in construction materials such as concrete, cementatious or geopolymer formulations [79].

2.4 Thermoplastic Composites with Inorganic and Synthetic Fillers

2.4.1 Inorganic Particulate Fillers

Particulate inorganic fillers, such as clays, mica, talc, calcium carbonate, graphite or graphite oxide, have attracted some attention in the study of photochemical behavior of multicomponent systems as they were considered a possible solution to protect these materials from the UV degradation. In this section, only microscale composites based on thermoplastic matrices were considered.

Studies on the photochemical stability achieved of polymeric materials upon addition of mineral fillers yielded in some unexpected results, since the decrease in properties is, in some cases, more intense after the UV exposure, in comparison with unfilled materials [80–82]. Thus, for PP and PE filled with clays (montmorillonite, MMT) [9] such behavior was explained considering the very small amounts of metal ions (i.e., Fe^{3+}) present as impurities in clays and which can act as catalysts in UV-initiated oxidation reactions that entails accelerated degradation. At the same time, hydroperoxides formed during photo-oxidation are a source of free radical species able to further promote the degradation. Even more, light stabilizers effectiveness was found to be much reduced in polymer-MMT composites, maybe because these molecules were adsorbed on the surface of clay platelets so that their action was diminished or even blocked. Such results were reported for PC/MMT composites [5], as well as for PA6/MMT [83].

The same undesired effect—oxidative degradation catalyzed by impurities—was noticed when talc (hydrated magnesium silicate) was employed in such PP formulations up to 15 wt% [84]. The study concluded that the outdoor conditions concurred to the overall degradation, despite the low energy of the UV radiation as compared with the accelerated test. It was demonstrated that the presence of talc did not inhibit the degradation rate and the photochemical behavior was severely affected by the presence of the acidic rain.

Still, other studies reported opposite effects under certain circumstances. Thus, PP-talc composites showed a different behavior depending on the interval of exposure: UV-initiated degradation was more intense for short terms, but the trend was reversed after 12 weeks of exposure [85]. This behavior may be assigned to the balance between the screening effect of the talc particles and the catalytic activity of the chromophores present as impurities.

Talc was used in PP composites complex formulations mainly to improve their mechanical properties [86, 87], but it was also included in others that contained UV stabilizers in order to compare their behavior under natural weathering conditions. Although it was proved that PP composites containing an UV stabilizer are much less susceptible to undergo photochemical reactions [87], the addition of talc was intended to enhance this effect. Thus, after weathering the considered samples (neat PP, PP-UV stabilizers and PP-talc-UV stabilizers) [61], it was concluded that all underwent photooxidative reactions, as confirmed by Fourier transform infrared (FTIR) spectroscopy which indicated the formation of specific oxidation products. Scanning electron microscopy (SEM) images revealed the in-depth progress of the degradative processes, results corroborated with the decrease in some mechanical properties. The study confirmed that talc exerted only a limited effect on the photostability of composites [88].

HDPE, another polymer extensively used in multicomponent systems for outdoor applications, was also compounded with talc, as well as other inorganic fillers, aiming at limiting the negative effects of UV radiations. It was noticed that filled composites exhibited two opposite behavior trends, depending on the filler. Thus, talc induced a slightly accelerated degradation upon UV exposure, although much less than other filers, in the following sequence: kaolin > diatomite > mica > black mica > talc [89]. Even more, it was proven by pyrolysis gas chromatography–mass spectroscopy (PGC–MS) that volatiles remained after composites photochemical degradation were mostly *n*-alkanes. It was assumed that the crystallinity of HDPE was modified after the UV-initiated oxidation and the relative absorbance of fillers may be a factor that influenced the composites susceptibility.

The opposite behavior was the stabilization of HDPE-inoganic filler composites. In these cases, calcium carbonate ($CaCO_3$) and wollastonite were used, and it was proven that photochemical degradation was much less intense [89].

Calcium carbonate was employed in PP-based composites, either alone or mixed with talc, in order to reciprocally potentiate their UV-protective effect by synergistic action [90]. The experimental data indicated that the mixed filler granted a better retention of mechanical properties after weathering, despite the manifest surface degradation. It was suggested that $CaCO_3$ was responsible for limiting the crack propagation from outer layers to the inner ones, while talc provided the load transfer due to its reinforcing effect.

Another multicomponent system based on PP and $CaCO_3$ included various additives, namely a surfactant (stearic acid) and an UV-stabilizer (Tinuvin 622, a polymeric HALS), and their influence on the composites photochemical behavior was investigated [91]. After weathering, samples containing $CaCO_3$ (30 phr to PP) exhibited a higher susceptibility to degradation and the presence of the surfactant (stearic acid 7 %) added to this effect. A small amount of Tinuvin (1.5 phr to PP) partially balanced the $CaCO_3$ and surfactant action.

Poly(vinyl alcohol) (PVA) is a thermoplastic polymer with properties that make it fit for biomedical applications (solubility in water, hydrophylicity, biocompatibility, lack of toxicity). Therefore, its photochemical behavior was extensively studied, either alone [92–94] or in composites [95, 96] or in complex multicomponent systems [97]. An investigation on the properties of poly(vinyl alcohol)/graphite oxide (PVA/GO) composites [98] submitted to UV irradiation revealed some interesting facts concerning their photochemical behavior. The inorganic filler initially selected was graphite. Due to its incompatibility to PVA given their different nature (PVA is hydrophyl, graphite hydrophob), it was submitted to an oxidation in the presence of a concentrated oxygen acid in order to achieve enhanced hydrophylicity. Experimental data indicated that a small amount of GO (0.1–5 %) slightly stabilized PVA towards UV radiation. FTIR spectra confirmed the photooxidative degradation of composites is slower than in neat PVA. At the same time, the PVA/GO aggregates are destroyed upon irradiation. It was suggested that the protective effect of GO resided in an overall balance between competitive photochemical processes that finally yielded in photo-crosslinked PVA macromolecules at the composite surface. Furthermore, GO amounts of 5 % proved to have a stabilizing effect during the thermal degradation of the UV-irradiated samples.

2.4.2 Synthetic Fibers Reinforced Thermoplastics

Fiber-reinforced thermoplastic composites employed in outdoor applications (mainly in automotive industry) were designed to respond to specific loads over a wide range of stress factors. Thus, it was experimentally confirmed that the presence of glass fiber reinforcement reduced the sensitivity of the material towards the application of tensile stress during UV irradiation [99, 100]. Moreover, tensile stress favoured the diffusion of small radicals formed during polymer degradation.

Since the fiber reinforcement grants composites increased strength, it also make them sensitive to temperature and moisture stress, presence of chemical pollutants, etc. During their service time, addition of tensile stress potentiate the negative effect of the weathering, but compressive stress may reduce the propagation of some degradative processes [101].

Some synthetic fibers used in composites, such as glass and carbon fibers, were designed to be highly resistant to weathering, so, the extent of material susceptibility lays mainly on the polymeric matrix. Other synthetic fibers, such as aramid and polyester fibers, are prone to photochemical degradation, given their structure, but the effects are not always adverse (e.g., polyamides absorbs environmental moisture and this induces an increase in their impact strength; the crystallinity increases upon UV exposure, which may also increase the strength) [102].

Still, the information on the photochemical behavior of these materials is not abundant, although mechanical properties of fiber reinforced thermoplastics are of interest and under study yet now.

PP-based composites reinforced with long E-glass fibers were studied under complex conditions [103, 104]. The effect of the UV radiation and the associated moisture on the composites microstructure and mechanical properties was assessed. Thus, microscopic observation evidenced that the photochemical degradation was limited at the surface of composites, due to the photooxidative degradation of the matrix which cracked and left the fibers exposed [103]. It was proven that PP in composites recorded a large increase in crystallinity compared to neat PP, which entailed a locally increased modulus. This behavior was explained through a chemical crystallization process during which the oligomers formed after chain-scission reactions regrouped into supramolecular structures with higher crystallinity. These changes created local stress concentrators and led to the surface cracking. Despite this and the relatively high amount of fibers (up to 21 %), the overall modulus of composites decreased along with the increasing UV exposure.

Using higher amounts of fibers (30 %) [104], high quantities of photooxidation by-products accumulated along with the increasing exposure time, as confirmed by FTIR. The crystallinity changes also created contraction sites in material surface, which lead to the formation and propagation of cracks, even in the absence of supplemental stress, yielding in an overall embrittlement of the composites. For higher reinforcement load (60 %) [104], the loss of mechanical properties was significant, since the crack propagation favoured the chemical crystallization processes [103].

UV exposure and moisture absorption studies were comparatively performed on long fiber thermoplastics with different matrices, namely PP and PA66 [105], and the first observations were that coloured materials underwent a yellowing process, but all samples displayed cracks at the surface. Depending on time of exposure and the nature of matrix, creep compliance exhibited opposite trends: for PP-based composites, it increased, but PA66-based samples showed a moderate decrease probably due to the low oxidation rate, despite the rather constant crystallinity.

Other composites based on PA6 matrix were reinforced with carbon fibers as fabric. The thermoplastic panels were produced by vacuum assisted resin transfer molding (VARTM) method and, subsequently, exposed to UV and moisture [106]. The samples displayed colour changes and, after 600 h of UV exposure, they displayed an intense yellowing, but no chalking at the surface. SEM confirmed the retention of the fiber–matrix interface integrity and the surface had no evident cracks. It was suggested that the slightly increased crystallinization is due to absorption of the UV-radiation energy. The composites maintained their flexural and impact characteristics.

Thermoplastic polyesters were also considered in such multicomponent systems, reinforced with either short [107] or long [108] glass fibers. For composites with mixed matrices made of PC and PBT and reinforced with short fibers [107] submitted to accelerated weathering, experimental data indicated only a slight modification in yield strength and microhardness, as well as in patterns of fiber fracture. Other tests focused on the long glass fiber reinforced polyesters exposed to UV radiation for long intervals [108]. The composite samples showed a decreased failure strain, ultimate strength, shear and tensile modulus, but significant changes in shear modulus and strength of the polyester, as well. The UV absorbers included in formulations prevented the photooxidative degradation.

3 Thermoplastic Nanocomposites

To suppress the photochemical degradation of multicomponent systems based on synthetic thermoplastics various solutions were designed and tested. The incorporation of nanoparticles in such formulations may yield in higher heat distortion temperature, enhanced flame resistance, significantly increased mechanical and barrier properties, lower thermal expansion coefficient, higher UV-stability, etc. [9]. Some nanofillers were reported to have negative impact concerning the UV-stability of the considered nanocomposites (e.g., clays), whereas others acted as UV absorbers, due to their strong absorption in the UV range, and remarkably limited the photooxidative degradation of the corresponding materials (e.g., carbon nanotubes). The effectiveness of nanoscale fillers as photostabilizers depends on the surface coating of the particles (wettability) and their chemical nature. More precisely, it depends on the compatibility between matrix and filler, the interfacial interaction and, where appropriate, the interphase thickness and properties. All these because at the nanometric scale the interactions follow particular laws.

In the following, some recent data will be summarized according to the nature of the thermoplastic matrix.

3.1 Polyolefins-Based Nanocomposites

Despite expectations that the addition of nanoparticulate layered silicates would improve the photooxidative stability of thermoplastics given their inertness toward UV radiation [109], it was concluded, as resulted from numerous studies, that PE-clays nanocomposites undergo accelerated photochemical degradation as compared to neat PE due to the presence of clay, either raw [110–115] or organically modified [111, 116–119], or in the presence of an oligomeric compatibilizer such as PE-*g*-MA or PP-*g*-MA [114].

In the case of unmodified clays (mostly montmorillonite MMT, but wollastonite, attapulgite, kaolin as well) [110], the enhanced degradation may be explained due to the significantly increased surface area [118] which allow much more numerous reactive sites, and the presence of some ionic species (e.g., Fe^{3+}, Al^{3+}) which may act as redox catalysts for the hydroperoxide decomposition [113]. Aluminum atoms localized at the platelet edges and other transition metals in the higher valency state localized between clay layers are electron acceptor sites, while electron donor sites are transition metals in lower valency state. Thus, clay nanoparticles may act as an electron acceptor and/or donor, contributing to the photooxidation. Thus, the rate of photochemical degradation is significantly higher in PE-MMT than in pure PE.

The presence of the organically modified MMT, typically with octadecylammonium (ODA), yielded in an even more pronounced degradation due to the influence of the ammonium ion which becomes preponderant. It was assumed that NH_4^+ may generate acidic sites in the clay layers and even the complex crystallographic structure of clay may result in some acidic sites after functionalization [116]. Associated to the catalytic effect of transition metal cations *via* the reversible photochemically initiated redox reactions, it induced the formation of free radicals and chain scission upon UV exposure. Therefore, the degradation of these nanocomposites is much faster than the ones with raw MMT.

Even when a compatibilizing agent is added, namely PP-*g*-MA [114], the rate of the photooxidation is significantly greater than in neat PE. Due to the complex dynamics of competing crosslinking and chain scission reactions during photooxidation, changes in the molecular structure of PE occurred and yielded in an altered pseudoplasticity that decreased along with increasing irradiation interval.

The same conclusions are valid for LDPE-silica nanocomposites [117] tested in accelerated weathering and in outdoor conditions, as well. Silica was reported to be a pro-oxidant in oligomeric hydroxyl-terminated PE [120] and to promote free radical formation in biologic systems [121]. The experimental results confirmed its catalytic effect on the considered nanocomposites.

When carbon black (CB) nanoparticles were employed for PE-based nanocomposites, a significant improvement in their UV stability was achieved

[122, 123]. It was proven that, under accelerated weathering conditions and depending on the nanoparticles size, the presence of CB improved the photochemical stability of nanocomposites, probably due to its ability to absorb UV radiation and by limiting or suppressing the Norrish type II scission reactions at the carbonyl moieties formed on the macromolecular chains during irradiation [122]. Even more, the smaller the CB particles (17–60 nm), the larger the surface area (30–210 m^2/g), the more effective the photochemical stabilization.

Associating CB nanoparticles with titania in appropriate ratios [123], an enhanced photochemical stability, along with excellent mechanical properties, was reached due to combined effects of the nanofillers, despite other previous studies that reported on the fast rate of photochemical degradation of PE-titania nanocomposites (17), faster than in pure polymer.

An interesting conclusion was reported on PE-ZnO nanocomposites exposed to accelerated weathering conditions: after an initial degradation, the material reached a stabilization that lasted for the equivalent exposure time of 4 months [117], in good concordance with other related studies on the stabilizing effect of ZnO on PP [125], which is more susceptible to undergo photochemical degradation than PE, and PC [126].

Basically, all findings on PE-clay nanocomposites are valid for PP-based ones [127–136], as well, in either compatibilized formulations—(PP-*g*-MA) [112, 128, 129, 132], $CaCO_3$ [129]—or not. The higher rates of degradation were also explained by the supplemental oxygen trapped inside clays tactoids which increased the oxygen pressure in bulk, decreasing the induction period. The processing seemed to influence the rate of decomposition: samples obtained by injection showed a slight trend to a plateau with higher oxidation rate as compared to extruded ones [127]. The addition of microsized $CaCO_3$ slightly improved the photochemical stability of nanocomposites [129], while nanosized $CaCO_3$ produced a dramatic reduction of photooxidation products related to the carboxylic acid salts formed during the interaction between $CaCO_3$ (basic pH) and PP photooxidation products [133]. This effect—higher degradation rate than microsized $CaCO_3$—may be assigned to an increased nucleation trend and formation of smaller crystals, substantiated by a slight phase segregation evident in SEM images, and catalytic effect as well.

Even when PP nanocomposites were drawn in fibers, the effect of nanofillers (bohemite Disperal40, Cloisite C15A) was the same: the higher the amount of nanofiller, the lower the photochemical stability [134].

Pronounced degradation was reported for polyolefins filled with nanoparticulate titania [118, 119, 123, 124, 134, 137] due to the photocatalytic activity of TiO_2, while opposite effects were recorded for ZnO [125, 126] or HALS and antioxidants [132].

Carbon nanotubes (CNTs, mainly multiwalled carbon nanotubes MWCNT) filled polyolefins fall in the same trend, showing decreased oxidation rates under UV irradiation [134, 138, 139]. Although MWCNT does not act as antioxidants, they work as inner filters limiting the incident UV radiation able to initiate photochemical degradation, without modifying the mechanism of polyolefins photooxidation. Instead, a slight increase in thermal oxidation was noticed as result of the dissipation of the photon energy into thermal energy by MWCNT [138].

3.2 Other Thermoplastics Used in Nanocomposites

Given the practical relevance of materials used in outdoor applications, different synthetic thermoplastic polymers were employed in nanocomposite formulations and submitted to UV irradiation. Thus, PC-clay nanocomposites seemed to display an increased rate of decomposition [110, 140], during which the carbonate scission reactions entailed a yellowing of the material. However, the degradation was less intense than in neat PC [141]. The presence of alumina (Al_2O_3) in PC nanocomposites reduced the overall light transmittance [10, 142].

EVA-CNT nanocomposites stability toward UV radiation was investigated in the presence of oxygen [143] and it was shown that CNTs are antioxidants and act as inner filters, thus reducing the photochemical degradation. At the same time, CNTs absorb UV energy and convert it into thermal energy, which increased local temperature and then initiate thermal degradation. These effects are competing and their balance is influenced by the amount and morphology of CNTs, as well as their functionalization. The antioxidant effect of CNTs in EVA-based formulations, but not only, is assumed to be similar to C60 in terms of electron affinity, so CNTs act as radical traps in chain reactions [143].

Studies of the photochemical behavior of polyamides based nanocomposites were run using MMT [116] or silica [10, 144]. As in other thermoplastics, MMT enhanced the photooxidative processes and accelerated the nanocomposite degradation [116]. In poly(trimethylhexamethylene terephthalamide)-silica nanocomposites, the silica enhanced the light transmittance and the effect decreased gradually when its amount increased over 5 wt% [144].

Polyfluorenes were reported to achieve enhanced photochemical stability and transparency when employed in nanocomposites formulations with ZnO up to 10 wt% [145]. It was assessed that ZnO may shield the material and prevent the photooxidation. Other studies revealed that poly(*p*-phenylene vinylene)-based nanocomposites may reach remarkable photochemical stability when silica-gold [146] or TiO_2 and SiO_2 nanoparticles [147, 148] were employed, although SiO_2 nanoparticles proved to have a better effect.

A polymethyl methacrylate (PMMA)-rutile (TiO_2) nanocomposite was successfully obtained and its photochemical behavior was investigated [149]. The results indicated that rutile nanoparticles are able to screen the UV radiation and, thus, the photochemical stability of the materials is higher than that of neat PMMA. But, when zinc oxide (ZnO) was employed in PMMA formulations [10, 150, 151], the resulting nanocomposites showed an enhanced photochemical stability upon UV exposure, even at very low contents of ZnO (0.017 wt%), but preserved the high transparency characteristic for PMMA.

Poly(vinyl alcohol)-titanium dioxide (PVA-TiO_2) nanocomposites were also studied [152] and proved to have an enhanced absorption in the UV range, hence a remarkable stability to UV degradation, which makes them fit as UV-protective coatings.

A complex multicomponent system based on PMMA and PA6 was developed and processed as fibers by co-axial electrospinning [153]. The co-axial composite nanofibers with a core–shell structure embedded TiO_2 nanoparticles, as well MWCNTs. When TiO_2 nanoparticles were added to the shell, the fibers exhibited a remarkable ability to absorb UV radiation, but a lower transmittance in the visible light range. Using the TiO_2 nanoparticles for the core, the effects were reversed.

As usual in the recent studies on nanocomposites, a serious concern is the release of nanoparticles in the environment as result of the photochemical degradation of materials, their accumulation and impact on nature [154, 155]. Although at this moment the nanoparticles release is reduced and the concentration around the release sites is remarkably low, this is a subject worthy of further investigations.

3.3 Nanocomposite UV-Protective Coatings

Protective coatings made of nanocomposites are a rather new approach in the UV-protection and it mainly refers to outer thin layers of nanocomposites bonded directly to the protected material or by the means of an intermediate layer able to chemically bond to both outer and inner layers. This concept yielded from understanding the effect of nanoparticles on the diffusion rate of volatiles through the nanocomposite. The tortuous path that nanoparticles create makes the oxygen diffusion significantly slower, which provides a barrier against photochemical decomposition. At the same time, given the specific elastic properties of the outer nanocomposite layer, it successfully prevents microcrack propagation, prolonging the lifetime of the materials. This approach appears to be a successful solution for multicomponent systems based on thermoplastic matrices processed in molten state, as well as thermoset ones.

The effect of nanofillers on the photochemical degradation must be considered. Since they are considered UV inert, they are not involved in the chain of reactions of the photochemical decomposition and thus may impart photochemical stability to the material covered by the nanocomposite [4]. Shielding the underlying material from UV irradiation and preventing its subsequent mechanical failure are results of the nanofiller–matrix interactions over a highly expanded interface. Thus, nanoparticles of pigments (rutile TiO_2), carbon nanotubes or fullerenes may be used with good results. It was proven that, at the same content of filler, the opacifiers filled with nanoparticles display an enhanced shielding effect as compared to those containing micro-scale particles [4].

The mechanism of UV shielding of the rutile TiO_2 nanoparticles through which they exert their photochemical stabilization effect was studied [156] and confirmed, unlike anatase TiO_2 which has photocatalytic activity [157] and contributes to the rate of photochemical degradation. It was also proven that coated rutile titania nanoparticles are effective light stabilizers in acrylic and alkydic paints as compared to conventional light stabilizers [4].

The same stabilization effect toward UV radiation was observed in the case of PP filled with ZnO nanoparticles and the effect enhanced along with the increasing amount of filler [138]. ZnO nanofillers proved to be more effective than hindered amine light stabilizers (HALS) when used for UV protection of olefins [158], or even for textiles, in a combination with rutile nanoparticles [159].

Various formulations were designed as coatings and studied in terms of UV protection effectiveness.

Anatase and rutile titania nanoparticles and pigmentary titania particles were used for photooxidation studies on a metallocene PE [4] and proved to be more photochemically active than the pigment particles. On the other hand, titania nanoparticles with enhanced photocatalytic activity (anatase) may be used in nanocomposites coatings having self-cleaning properties.

Titania coatings were used for other polymeric substrates as well, as PET [160], PVC [161] or on poly(ether-ether-ketone) (PEEK) [162]. These applications were performed by radio frequency sputtering in Ar–O_2 plasma. Titania films deposited on PET were amorphous or crystalline, depending on the processing, and their microstructure varied from compact to structured porous, as evidenced by SEM. The rate of the photochemical reactions in PET is influenced by the optical properties of the protective film, which depend on the structure and composition of the titania layer and that is correlated with the processing parameters. For plasticized PVC formulations, the titania coating promoted the photochemical degradation of the plasticizer, thus protecting the PVC [161]. For PEEK protected with ZnO or titania-containing coatings [162], it was noted that ZnO was more efficient given its higher absorption in the UV range, and its activity depended on the film thickness and structure, characteristics that can be optimized by correlating the processing parameters. Zinc oxide coatings proved to be effective protection against UV radiation for PET, too [163]. Depending on the sputtering parameters, ZnO crystallized in würtzite form and its crystallinity decreased with the increasing oxygen partial pressure. Moreover, the optical parameters were affected and, by consequence, the activity as protective coating. It was confirmed that the PET photooxidation reactions rate was significantly reduced when a ZnO film was deposited under optimized processing conditions.

Thin films of ZnO, alumina, silica, ceramers, etc., deposited onto the outer surface of PC, can provide satisfactory levels of photochemical stability, protection toward sunlight, and surface hardness [126, 162, 164–172]. It is known that, upon UV irradiation, PC undergo photochemical oxidation reactions and transformation *via* two successive photo-Fries rearrangements (yielding first in phenylsalycilate and then in 2,2′-dihydroxybenzophenone) by a dual photochemistry, as referred to in literature [173–175]. As consequence, PC undergoes a decrease in mechanical and physical properties, as well as a yellowing along with a loss in clarity.

Coatings made of ZnO deposited by plasma chemical vapor deposition [164] or by sputtering [126] on silicon substrates were used as layers deposited onto bisphenol-A polycarbonate (PC). They effectively reduced the rate of photochemical oxidation and yellowing of the polycarbonate, depending on the ZnO film thickness and properties.

The experiments indicated that an improved compatibility between substrate (PC) and coating is required. Two methods were employed either separately or altogether, depending on the components selected for the coating. Thus: (1) the surface functionalization of the substrate, which may be performed by treatment in various cold plasma media (CO_2, radio frequency, Ar, N_2, air, oxygenated Ar), in order to increase the PC surface energy; this functionalization entails an increase in PC reactivity by the formation of numerous active sites, mainly C=O bonds resulted from breaking of aromatic rings; the most effective media were the oxygenated ones, especially CO_2 plasma [165]; and (2) use of combinations of oxides (ZnO, Al_2O_3, SiO_2, TiO_2), in single or multiple layers (sandwich structures of a SiO_2 film between two TiO_2 layers; Al_2O_3/ZnO/Al_2O_3), or in various succession orders (Al_2O_3/ZnO and ZnO/Al_2O_3).

Deposition of transparent ceramic layers by sputtering is an effective technique, successfully used for UV protection of PCs. In order to document the photocatalytic activity of ZnO at PC surface, a series of different depositions was designed and, subsequently compared: PC/ZnO; PC/Al_2O_3; ZnO/Al_2O_3; PC/ZnO/Al_2O_3; PC/Al_2O_3/ZnO [167]. Upon UV exposure, PC underwent photochemical degradation reactions and characteristic products accumulated on the corroded surface. The phenomenon accentuated along with the increased exposure, but it was significantly reduced by the presence of ZnO layer: the thicker the coating, the more effective the UV protection and the slower the oxygen diffusion. By comparison, the PC/Al_2O_3 system is less effective at the same thickness. The Al_2O_3 coating is a good oxygen barrier and limited the oxygen-mediated reactions, but light initiated reactions were not prevented. When a double ceramic layer was deposited on PC, the results were superior to monolayer coatings. Comparing PC/ZnO/Al_2O_3 and PC/Al_2O_3/ZnO the photocatalytic effect of ZnO was evidenced: PC/ZnO/Al_2O_3 system was less efficient against UV radiation. The better protection offered by the PC/Al_2O_3/ZnO system was explained by the intermediate Al_2O_3 layer that insulated the PC surface from ZnO and thus limited its negative influence, along with preventing oxygen diffusion toward PC.

Optimization of the photochemical protection of PC by ZnO and Al_2O_3 coatings was achieved by sputtering a target, placed on a cooled magnetron cathode, on the PC having the surface previously functionalized in CO_2 cold plasma, in Ar:O_2 atmosphere, under radio frequency [165]. Given the photocatalytic activity of ZnO at the PC surface, and for comparison reasons, it was deposited directly on the substrate and with an intermediate alumina layer. Using a previously functionalized substrate, it was possible to perform the sputtering with good yields at moderate power and the new chemical bonds (Zn–O and Al–O others than those in the originating materials) highly improved the compatibility between components.

Further studies indicated that the best results were obtained for a three-layer coating (Al_2O_3/ZnO/Al_2O_3) that granted the substrate an enhanced protection by an increased oxygen barrier effect, due to the increased coating thickness [168]. An enhanced elasticity was noticed and it was explained by a sum of phenomena: a slight rearrangement of macromolecular chains under the effect of heating and limited photochemical degradation, the formation of new oxygen-containing moieties and hydrogen bonding.

Recently, another type of ZnO coating for UV protection of PC substrates, namely ZnO photonic crystals, was reported to have remarkable inhibiting effect on the photochemical corrosion as compared with nanoparticulate ZnO layers [176].

Other inorganic coatings were also tested for UV protection. Sandwich structures of a SiO_2 film between two TiO_2 layers deposited onto PC substrates proved to have excellent suppressing action on the yellowing associated to photochemical degradation [172]. The SiO_2 film is highly transparent, contributing to the overall clarity of PC, and prevented the peeling of the TiO_2 layer by limiting its photocatalytic effect. The TiO_2 was a multi-task component: absorbed the UV radiation, favoured the decomposition of different organic contaminants upon UV exposure due to its catalytic activity and showed self-cleaning ability granted by the increased hydrophilicity.

Given the differences between substrate, PC, and coating, ZnO, and the technological approaches (gas-phase or vacuum-deposition techniques versus milder technologies), other solutions were suggested. Thus, organic–inorganic hybrid materials, prepared by a sol–gel process starting from poly(ethylene oxide) (PEO) or polycaprolactone (PCL) and tetraethoxysilane (TEOS) as the inorganic network precursor, were considered as UV-protective coatings for PCs [175]. The sol–gel method is advantageous given its versatility and possibility to use it under mild conditions. PEO and PCL were selected because they both have good compatibility with PC and the hybrids adhesion to the substrate surface was enabled. PEO based hybrids were considered only for comparison reasons, despite the high sensitivity of PEO to UV radiation [177]. Furthermore, transparent coatings able to prevent photochemical degradation of PC, while maintaining its transparency and clarity, are of great interest in terms of industrial applications (e.g., in automotive industry). Experimental data confirmed the protective action of hybrids on the PC surface and a further optimization is required.

4 Photochemical Behavior of Thermosetting Matrix Based Composites

The literature is abundant in data on the photochemical behavior of thermosetting matrices based composites. Most of the reported data focused on two main issues: (a) the UV-cured composites and nanocomposites and (b) the photochemical degradation of these materials upon UV exposure. In the first case, the UV radiation is employed in the synthesis of various materials starting from thermoset precursors included in complex formulations that may also contain—besides fillers, whether fibers or particles—initiators, plasticizers, compatibilizing agents, UV absorbers, etc. Concerning case (b), the effects of the UV exposure under specific conditions (natural or accelerated weathering) on the thermosetting-based composites were studied depending on the nature of matrix and filler, type and shape of fillers, processing, etc.

4.1 UV-Cured Thermosetting Matrix Based Composites

The UV curing technology is an advantageous approach on many aspects (low overall energy consumption, reduced thermal stress, short processing intervals, low processing costs) and may be employed in various procedures to obtain composites insofar as the material can be directly exposed to UV radiation. The processing time is reduced to minutes (whereas in thermal curing the processing takes hours), the radiation wavelength may be selected (e.g., mercury arc lamps), but the thickness of the sample is limited in order to ensure the radiation penetration through material. The properties of the composites depend on the crosslinking density which is a function of the type and concentration of the photoinitiator, diluents (if applicable), time and intensity of irradiation, temperature, etc. [11, 178]. The typical photo-curable formulations are basically made of multi-functional monomers and oligomers, as well as certain amounts of photoinitiator able to generate reactive species (radicals or cations) upon exposure. The mechanism is different for different polymers: radical polymerization, as in the case of acrylate-based resins and unsaturated polyesters, cationic polymerization for epoxy- and vinyl ethers [11]. Acrylate-based systems consist of an acrylate functionalized oligomer which will form the backbone of the network, an acrylate monomer which will act as diluent, and a photoinitiator which will enable the free radicals. The resulting networks may be of different structures (polyester, polyether, polyurethane, silicone, etc.). Di-functional monomers that are inert to free radical species, such as vinyl ethers and epoxies, may undergo UV-initiated cationic polymerization. Therefore, the corresponding formulations contain only the photoinitiator, which yield in protonated acids, and the di-functional monomers (ester, ether, urethane and silicone functional groups may be introduced in the system). Aside from typical applications of the UV cured materials (paints and varnishes, protective and/or aesthetic coatings, adhesives, dental prosthetics, etc.), new applications were developed, namely composite thin laminates. In the following, some recent reports on the synthesis of thermosetting matrix-based composites will be considered according to the nature of their corresponding matrix.

4.1.1 Fiber-Reinforced Epoxy Resins

The most common synthetic thermosetting matrices used for composites preparation *via* photochemical crosslinking are epoxy-, epoxy acrylates, siloxanes, vinyl esters.

Glass fibers, just like carbon fibers, tend to be quite inert toward UV radiation [179]. On the other hand, epoxy resins are slightly sensitive to UV exposure, their photochemical degradation being driven mainly by the UVA excitation of carbonyl groups that initiate the chain of reactions either by dissociation or energy transfer [180].

Multi-layers laminate composites made of epoxy resin and glass fibers were produced by UV curing in the presence of various photoinitiators, in order to be used in bullet proof vests [181]. For comparison, a vinyl ester was used. Sandwich structures were produced by lay up process. It was proven that epoxy resin cured rapidly under UV radiation, in opposition to vinyl ester which required 2 days to fully cure, and a 3 min exposure was enough to reach maximum penetration and composite hardness.

A modern approach to produce UV-cured glass fibers-epoxy composites was designed to be used directly on a braiding machine [182, 183] and the resulted materials were compared to corresponding ones obtained by thermal curing. An optimisation study focused on the concentration of photoinitiator and the oxygen quenching phenomenon. Adding diacrylates and triacrylates reduced the effects of oxygen quenching and increased the rate of curing of the acrylated epoxy monomers. It was found that gelation occurred in 10 s upon UV exposure. The samples reached a significantly increased degree of curing when crosslinking agents (imidazole or an anhydride) were added [183].

Adding silane-treated organically modified MMT to a glass fibers reinforced epoxy resin, a complex multi-component system with improved flexural properties was obtained [184].

Novel laminates were obtained by UV-curing after the resin was transferred by infusion between double flexible tooling process (RIDFT) [185]. The production time has significantly shortened and depends on the UV exposure interval. Experimental data confirmed some of the properties (tensile strength) of the new materials were comparable to those obtained by thermal curing, while others (tensile modulus) were higher.

Other synthetic fibers, as well as natural fibers, were employed in epoxy-based multicomponent systems. Thus, a comparative study between polymeric fibers reinforced epoxy resins and natural fibers reinforced ones, all obtained by UV curing, was performed [186]. Isotactic PP (iPP) fibers modified with 20 wt% EVA and hemp fibers were employed, while the matrices used were epoxy acrylate and epoxy methacrylate. Data indicated that relatively regular distribution of fibers was achieved; the addition of fibers caused an increase in Shore hardness of the epoxy methacrylate based composites; the epoxy acrylated composites showed a decrease in hardness when EVA-modified iPP fibers were used, whereas hemp fibers caused an opposite effect. Even more, the iPP fibers reinforced photocurable composites displayed a brittle to ductile fracture transition.

4.1.2 Glass Fiber-Reinforced Polyester Resins

A UV-cured polyester matrix reinforced with glass fibers was tested on both exposed and non-exposed sides in order to establish the relationship between irradiation parameters and surface Barcol hardness [11]. The difference between sides increased along with the distance between irradiation source and laminates.

Specimens made of glass fiber rovings (from the filament winding process) immersed in an unsaturated polyester matrix were UV cured in the presence of a mixture of photoinitiators (BAPO/AHK) in various ratios [187]. For comparison, samples with the same formulation were thermally cured. Data indicated no significant difference between series, with respect to the tensile strength. Still, samples cured in the presence of BAPO (bisacylphosphine oxide)/AHK (α-hydoxy ketone) showed shear strength values higher than all others due, probably, to the better cure throughout the laminate thickness.

Knitted continuous glass fibers were used to produce reinforced polyester composites by UV curing associated with vacuum bag resin infusion (VBRI) technique [188]. This approach allowed a maximum thickness of 13 mm. It was noticed that flexural strength and modulus decreased with the increasing sample thickness due to the reduced crosslinking density.

A study on the UV transmission of various types of glass fiber reinforcements, namely continuous filament random mat, non-crimp fabric and plain weave fabric, was performed using a polyester resin as UV-curable matrix [189]. At the same time, the transmission of the UV radiation was correlated with the curing behavior of the fabric-reinforced laminates. It was already shown that, in the case of woven glass fibers used as reinforcement for UV-curable epoxies, the directions of preferential light propagation are the same as the main reinforcing fiber orientation, and the number of fabric layers influenced the transmission and the dose of radiation [190]. The aforementioned study evidenced that the UV radiation is transmitted through the fiber bundles, when depends on the nature and thickness of fibers, as well as through the voids between the fiber bundles, when depends on the wettability of fibers by the matrix. Reflection at the fibers surface was minimized due to the matching refractive index of both matrix and fibers. Under compression, the sample thickness and, subsequently, the voids volume decreased, and, thus, the transmission increased. On the other hand, the transmission decreases when the fiber packing density and fiber volume fraction increased.

Another comparative study analyzed the effect of different curing strategies on the composite properties, when samples made of same amounts of woven glass fabric and an epoxy-vinyl ester (Derakane 510A—brominated bisphenol A-based vinyl ester) were single-sided UV-cured in the presence of 2 % bisacylphosphine oxide (BAPO) as photoinitiator, or thermally cured in the presence of 1.2 % benzoyl peroxide (BPO) [191]. Data obtained for flexural strength indicated the laminates had comparable characteristics and the procedure may be subjected to further optimization [192].

Using a commercially available UV-curing polyester resin (Sunrez 1401) and orthogonally stitch bonded glass fiber fabric, it was possible to produce laminates of various thickness by VBRI technique and one-sided UV irradiation [188]. The glass transition temperature T_g had lower values as the laminate thickness increased, so thick laminates have a lower crosslinking density. It was found that the hardness on both sides was the same for laminate thickness under 8 mm, while for thicker samples the exposed side was significantly harder than the other. Previous literature reports suggested that a lower crosslinking density grants laminates an enhanced flexibility [11, 193]. But this study showed that this effect is overcompensated by the increased bending strength due to the increased sample thickness, and, even more, the flexural modulus tend to be higher for thin laminates with high crosslinking density than for thick laminates [188].

Chopped glass fibers were used for various polyester resin formulations submitted to UV curing. Thus, photochemically cured samples were obtained using such reinforcement and an unsaturated polyester resin, in the presence of styrene and 0.5 % MAPO (monoacylphosphine oxide) and 0.25 % BAPO, as photoinitiators [193]. Specimens that had been cured with BAPO showed a higher hardness, especially on the non-irradiated side, than those cured with MAPO. The tensile modulus and strength of the post-cured and thermally cured laminates were significantly higher as compared to those of the UV-cured laminates, probably due to the higher crosslinking density in the post-cured and thermally cured specimens.

Other studies focused on the effect of photoinitiators on the cure response of fiber reinforced composites [11, 194]. Thus, BAPO effectiveness (a through-curing photoinitiator) was compared to AHK (a surface-curing photoinitiator) using composites made of unsaturated polyester or vinyl ester resins as matrices reinforced with up to 30 % chopped glass fibers [11]. When a blend of photoinitiators (0.75 % BAPO and 0.25 % AHK) was employed, the best results in terms of time for complete cure were achieved. The value of T_g was higher for composites cured with high-power radiation sources, and the reaction heat during cure and the resin temperature increased along with the increasing irradiation power. It was assessed in literature that a higher temperature is related to a higher crosslinking density and T_g [195], but an excessively high temperature can cause a high loss of styrene and decreased crosslinking density, as observed in thick laminates [11]. The study concluded that intervals for through-cure are shorter for longer wavelength, due to the in-depth penetration of UV radiation.

Using different photoinitiators, namely benzoin alkyl ether, when samples were irradiated on either one side or both sides [11], and benzyl dimethyl ketal, when both sides were exposed [196], experimental data for the latter showed that laminates containing 30–40 % glass fibers and having 2 mm thickness were UV-cured in 15 s at room temperature, in air. Moreover, multifunctional acrylate or acrylether monomers added to the polyester resin improved the tensile and flexural properties of the fiber-reinforced composite to a greater extent than allylic monomers, given the different photocrosslinking mechanisms. Physical properties of the photocrosslinked laminates were correlated with the molecular weight of the resin, the amount of multifunctional monomer, and the glass fiber content [196].

It is of real interest the new trend in thermosetting composites production, namely the use of reneweable epoxidized vegetal oils (EVOs) as matrices (SO,

ELO, AESO, AELO, etc.) able to undergo UV curing and which may represent a viable solution to some specific issues (biodegradability and low environmental impact associated to UV resistance/protection and good mechanical properties), mainly when used as protective coatings [197].

4.1.3 Dual Cure Processing

The dual cure technique has the merit of combining the fast UV curing mechanism with a consecutive thermal cure, especially needed when available incident radiation is not sufficient or some properties must be enhanced (e.g., adhesion, scratch resistance) [178]. The main characteristic of the dual cure systems is the curing occurs in two different stages, unlike the hybrid cure systems where the two curing mechanisms (radical and cationic) occur at the same time. The dual cure systems consist of two basic components, a UV-curable one and another which is thermally curable, that will yield in two independent interpenetrating networks. A successful example is the two-pack polyuretanes (2 K PU) [198] which yield in high scratch and chemical resistant coatings.

This procedure was initially studied on polymeric systems [199–201], then was extended to various composite formulations for general applications [202] or specific ones, such as dentistry and prosthetics [203–208], microelectronics [209–211], photolithography or rapid prototyping [212]. The composites were reinforced mainly with glass fibers [204–207], which granted composites improved mechanical properties and hydrolytic stability when used in humid and acidic media, but carbon fibers proved to be highly appropriate for other envisaged applications [212]. Thus, using a dual curing resin formulation made of an acrylate ester of bisphenol A-epoxy (as primary resin), a photoinitiator (1-hydroxycyclohexyl phenyl ketone) and a thermal initiator (lauroyl peroxide), it was evidenced that the resin present around and inside carbon fibers can be successfully cured by the dual procedure, despite the UV opacity of carbon fibers. The multicomponent system is fit for use in composite photolithography where carbon fibers can allow high electric and thermal conductivity.

4.2 UV-Initiated Degradation of Thermosetting Polymers Based Composites

Most of the reported data in the field were focused on epoxy resins based composites, either filled with microparticles, such as TiO_2 containing pigments [213] and glass microspheres [214], or glass or carbon/graphite fiber reinforced [215–221].

Recent investigations by the means of scanning electron microscopy (SEM) and X-ray photoelectron spectroscopy (XPS) evidenced that epoxy-based resins exhibited cracks (the so-called “silver cracks”) after UV irradiation, which depth and number increased with the exposure time [222]. Thus, reinforcing such resins with UV-inert fibers or particles was considered a viable option in order to extend their field of application and service time.

4.2.1 Particle-Filled Epoxy Resins

Various titania based pigments effect on the photochemical behavior of an amine-cured epoxy resin was studied under simulated weathering conditions [213]. Experimental data confirmed the importance of the pigment type, as its photochemical reactivity is the driving force of all processes. At the same time, the micrometric particles dispersion strongly affected the degradation of the resin. Thus, samples with poor particle dispersion and highly photoreactive pigments showed the most intense degradation, while specimens filled with low photoreactive but well dispersed pigment particles were the most stable.

Another type of particulate filler considered for study was glass microspheres which, technically, are spherical aluminosilicate particles (less than 0.03 mm in diameter) filled with carbon dioxide and nitrogen [214]. Samples of epoxy resin filled with glass microspheres (10 %) were tested for erosive resistance after UV irradiation and experimental results indicated that the filler particles improved the resistance of the composite. Still, after prolonged exposure, the effect diminished due to the photochemical degradation of the matrix. Therefore, a top UV-protective coating is recommended.

4.2.2 Glass/Graphite Fibers Reinforced Epoxy Resins

Glass/carbon fiber-reinforced epoxy composites exhibit high specific strength and stiffness, and good fatigue tolerance, which have made them fit for advanced applications ranging from military and civil aircraft structures to recreational consumer products. These composite materials are typically exposed to multiple environmental stresses during service (UV radiation, moisture, temperature and thermal cycling, etc.), which leads to their synergistic degradation.

A comparative study on the combined effects of the UV exposure and thermal shock on the properties of two selected glass-epoxy and graphite-epoxy systems was conducted in air and “near-vacuum” conditions [215]. The weight loss of the glass-epoxy composites was larger than that of the graphite/epoxy ones in all tests, but for all samples the damages were more severe in air than in “near vacuum” conditions. The tensile strength values were slightly lower upon exposure for both composite types. Microcracks observed only in glass-epoxy specimens were initiated by UV radiation and propagated through the thermal shock, but only on the irradiated surface. All data indicated that graphite-epoxy formulations are more stable toward UV radiation.

But even for such composites it was proved that, upon UV exposure, the interlaminar shear strength and the flexural properties decreased [223, 224]. Other more recent studies, where complex investigations were performed by associating moisture, high vacuum and temperature to the UV radiation, reported slight or no decrease in flexural properties of irradiated composites, although morphology changes—microcracks—were noticed [216, 217].

When high-energy electron radiations in vacuum were combined with UV exposure, it was found that the stability of some graphite-epoxy laminates was rather high to high-energy electron radiations, but relatively low toward UV radiation [218].

Some carbon fiber-reinforced epoxy composites are intended for special applications, such as aerospace and rotorcraft structures, and, therefore, their resistance to UV radiation and/or condensation was required to be investigated [219]. It was concluded that the formulation IM7/997 (a thermoplastic-modified toughened epoxy resin reinforced with polyacrylonitrile (PAN) based carbon fibers having 5 mm in diameter) underwent a synergistic degradation, mainly in the matrix, yielding in a decrease in the mechanical properties due to the limitation of the load transfer. When sequentially submitted to UV radiation followed by condensation, the specimens initially lost weight upon UV exposure and, subsequently, gained weight during the condensation cycle, due to water sorption. Upon further exposure, the samples have lost weight at a steady rate evidencing matrix erosion, void formation, and fiber–matrix interface debonding in the tested composites.

Same conclusions, basically, were reported in a different study [220]: the UV-initiated chemical reactions caused macromolecular chain scission and, subsequently, chain crosslinking; the chain scission lowered the molecular weight of the polymer, yielding in reduced heat and strength, whereas chain crosslinking led to excessive brittleness and caused microcracking; the UV-exposure of carbon fibers-epoxy laminates resulted in microcracks, which led to a decrease in matrix-driven properties due to the embrittlement of the resin by increasing the crosslinking density upon UV exposure. A solution to limit these effects was recommended, namely an epoxy based UV-protective coating made of two different surfacing films and a chromate containing paint primer [220]. After testing the coating, it was concluded that the chromate containing epoxy based paint primer provides protection initially, but the substrate is, however, degraded during extended exposure. Using an epoxy-based film onto the surface to be exposed during the initial stage of the carbon-epoxy laminate cure can limit the UV-induced degradation. In this case, the degradation was either restricted only at the protective layer or slightly penetrated into the carbon-epoxy substrate, since no significant decrease in mechanical strength was noticed.

More recent data on carbon-epoxy composites AS4/3501-6 (an amine cured epoxy resin reinforced with "in-plane fiber tow waviness" AS4 carbon fibers designed for structural applications) were reported [221]. Composites structural details (ply drops, ply joints, etc.) can produce significant fiber deviation. The "in-plane fiber tow waviness" may create areas with high resin content which favoured the high stress concentrators to occur, yielding in a decreased strength ratio and, finally, composite delamination.

It is noteworthy that adding 1–2 wt% MMT to glass fiber reinforced epoxy resins it was possible to improve the composites resistance to UV radiation, even when combined with condensation [225, 226]. The MMT is acting as a compatibilizer between the matrix and glass fibers, contributing to the formation of the interphase and adding to the interfacial interactions. Weight losses and discoloration were observed for all uncompatibilized samples exposed to UV irradiation. These effects were balanced by adding MMT. Even more, SEM images evidenced less delamination, although, after UV irradiation-condensation cycles, the removal of resin from composites was manifest.

4.2.3 Other Thermosetting Resins-Based Composites

Aside from epoxidic-based composites, other resins were also used for various formulations to be tested. Thus, an acrylic-urethane resin was employed in the study of effectiveness of microparticulate titania-based pigments toward UV irradiation [213]. The acrylic-urethane matrix was less susceptible to undergo photoinitiated degradation than the epoxy resin and, thus, was less damaged. The type of pigment and particle dispersion strongly affected the composites photochemical stability, as samples with low reactivity pigments and good particle dispersion showed no significant degradation.

Polysulfone-based laminates were produced using graphite fibers and submitted to UV and high-energy electron radiation [218]. The analysis of volatiles formed during irradiation suggested several free radical mechanisms of photoinitiated degradation. The polysulfone-based laminates proved to be more stable toward high-energy electron radiation, conclusion evidenced by the decrease in mechanical properties after the UV irradiation.

5 Photochemical Behavior of Thermosetting Matrix Based Nanocomposites

Nanocomposites based on thermosetting matrices are heterogeneous materials and their properties, photochemical behavior included, are defined, just as in the case of the thermoplastic-based nanocomposites, by the components nature and compatibility, composition, structure, and interfacial interactions exerted through the interphase. The thickness and properties of the interphase vary and are particular for each system. Even more, properties of the interphase differ from those of the raw components, although it is formed by the adsorption (and, sometimes, by chemical reactions) of the polymer matrix onto the surface of the nanofiller.

If the interphase is formed only by physical or physico-chemical interactions (as in the case of thermoplastics filled with nanoparticulate $CaCO_3$), its thickness is given by the strength of interactions, and its properties are a combination of components properties. When the interphase is formed by chemical reactions (as for the silane-treated fillers used for both thermoplastic and thermosetting matrices), the thickness of the interphase is driven by the amount of silane, while its characteristics are determined by the nature of the reactive functional groups involved [3].

An essential issue for the formation of the interphase and its properties is the compatibility of the components. The higher the compatibility, the higher the amount of interphase, the stronger the interfacial interactions between components. Therefore, many formulations considered one of the compatibilizing options available: (a) the matrix, (b) the filler and (c) both of them. For the case of thermosetting matrices, the most usual approach is the surface treatment of the nanofillers, in order to decrease their surface free energy; this will grant the nanoparticles enhanced wettability by the matrix, along with restricted particle–particle interactions, yielding in limited secondary aggregation of the nanoparticles.

Another aspect of relevance is the micromechanics of such multicomponent systems, as debonding often occurs and leads to breathable layers, layers that develop structural voids during various types of stress, and the voids make the layer pervious toward volatiles.

UV-curable and dual cure nanocomposites are viable elegant solutions for the production of thermosetting resins based nanocomposites, most effective as they are solvent-free, require little energy and low temperatures, evolve in short intervals, and, when components are wisely selected, are environmental friendly [227–231]. But UV reflection, refraction and absorption related to the nanometric fillers must be all considered, as may affect the rate of the photochemical reaction initiation and this may entail longer exposure time and even a modified chemical and supramolecular structure.

In the following, some recent data reported in the field will be presented according to the nanofiller, matrix and processing.

5.1 Clay-Thermosetting Polymer Nanocomposites

The in situ intercalative polymerization associated with the UV curing is a technique which was successfully employed in the synthesis of hybrid films, when fast polymerization of liquid monomers yielded in solid materials with designed properties. It was proved to have high efficiency for epoxy oligomers, vinyl ethers, oxetanes in the presence of onium salts as photoinitiators [232]. The literature is not abundant in reports on the in situ UV-initiated polymerization of epoxides in the presence of layered silicates [233–237], as compared to data on thermally cured or melt compounded nanocomposites. In some studies, the clays were used either unmodified [235, 237] or organically modified [233, 236, 238] or treated with various reagents able to change their surface properties [239–243].

A successful example is the multicomponent system made of an epoxy resin, namely Cyracure® UVR 6110, a bis-cycloaliphatic diepoxy resin 3,4-epoxycyclohexylmethyl-3′,4′-epoxycyclohexyl carboxylate, which was cured in the presence of ethylene glycol (used to increase the photopolymerization rate by the chain transfer mechanism initiated by OH groups and to promote organoclay intercalation) and organically modified MMT (Cloisite 30B—a montmorillonite modified with dihydroxyethyl octadecyl methyl ammonium chloride) [233]. The diol changed the kinetics of the photopolymerization and the properties of the cured films not only by the chemical structure of the matrix, but favouring the organoclay platelets dispersion in the matrix as well.

The use of silanization as treatment for clays prior to their use in UV-curable formulations is based on the availability and reactivity of hydroxyl groups in the clay structure. Thus, the surface modification of MMT by glycidyl-propyl-triethoxysylane (GPTS) allowed new highly dispersed epoxy resin-based nanocomposites with enhanced scratch resistance and low susceptibility to crack formation and propagation [240].

An interesting study on clay-epoxy nanocomposites was focused on the effects of combined exposure to UV radiation and moisture on the nanocomposites properties [244, 245]. It was experimentally proved that clay platelets bearing aliphatic chains as organic modifier (Nanomer I.30P is a montmorillonite modified with octadeclyamine, ODA) had a negative influence on the photochemical degradation of the nanocomposite, due to the presence unexchangeable cations (Fe^{2+} and Fe^{3+} which contributed to the redox reactions) and ODA. On one hand, the material surface exhibited wider cracks than the neat resin, but only at the surface, due to the presence of clay which imparted barrier properties to the nanocomposites, limiting the in-depth penetration of oxygen. On the other hand, the ODA modifier contributed to the photochemical degradation of the matrix due to the free radicals generated. As expected, mechanical properties such as flexural modulus and ductility were negatively influenced.

5.2 *Thermosetting Polymer–Silica Nanocomposites*

Nanocomposites based on thermosetting polymer matrices are increasingly used in outdoor applications given their properties. These polymers have still a high susceptibility to photochemical degradation and, thus, nanoparticles may be released into the service environment. This phenomenon represents a potential risk in terms of ecology and human health. Therefore, studies on the photochemical degradation and weathering of various epoxy-silica nanocomposite formulations were carried out [246–253]. Some reports documented the photochemical degradation, surface morphology changes, total mass loss, and nanoparticle release from epoxy-silica nanocomposites (containing up to 10 % silica nanoparticles) exposed to complex weathering conditions (UV irradiation, 50 °C, 75 % relative humidity) [247, 249, 250, 253], although others evidenced an increased accumulation of silica at the outer surface of nanocomposites, but without any evidence of nanoparticle release [248, 251, 252].

Thus, it was experimentally proven that, under UV exposure, the epoxy matrix undergoes rapid photochemical degradation, which yields in significant weight losses and gradually increased nanosilica concentration onto the exposed surface along with the exposure time.

An analytical method to quantify silica nanoparticles accumulated on the surface of epoxy-based nanocomposites exposed to UV radiation was established [248]. By the chemical extraction of the SiO_2 nanoparticles from the exposed films (containing 5 and 10 % SiO_2, respectively) using hydrofluoric acid (5 %), it was possible to show that the higher the initial content in nanoparticles, the higher the amount extracted after the photochemical degradation of nanocomposites. Even so, no free silica nanoparticles were reported as released into the environment.

The same conclusion was presented in a study on amine-cured epoxy-silica nanocomposite film containing 10 % SiO_2, for which a surface layer containing 75 % silica was determined after long UV exposure intervals [251]. At the same time, no direct evidence was reported for the release of dissociated nanoparticles, although it was suggested that the high accumulation of silica on the surface may, eventually, entail the nanoparticles release.

A comparative study between silica- and CNT-filled amine-cured epoxy nanocomposites [249] revealed the opposite behavior of nanocomposites upon UV irradiation. Thus, silica-filled samples underwent a high rate photochemical degradation followed by a significant accumulation of nanoparticles at the surface, which yielded in nanoparticles release. CNT-filled specimens displayed a high density CNT network at the nanocomposite surface, which limited the in-depth photochemical degradation of the material and prevented the nanoparticles release. The conceptual models for the silica nanoparticles release and CNT preservation upon UV exposure may be used for other multicomponent systems based on epoxy resins in order to assess their potential risks.

Recently, an experimental technique and specific protocols to quantify the release of silica nanoparticles from epoxy-based nanocomposites upon UV exposure were studied [253], but photochemical degradation, weight loss, and surface morphological changes were also investigated. Experimental data indicated a significant amount of silica accumulated on the surface as result of the degradation of the matrix. The relative humidity of the environment added to the degradation of the nanocomposite, evidenced by the higher amount of silica released. The accumulated silica had a catalytic effect on the rates of chain scission and oxidation reactions near the surface, yielding in greater weight losses. It was ascertained that after a period of induction, silica nanoparticles were rapidly released from the material, and their amount increased with the increasing UV irradiation associated with moisture. This set of quantitative data is valuable for developing models able to predict the long-term release of silica nanoparticles from thermosetting polymer based nanocomposites used in outdoor applications.

5.3 CNTs-Epoxy Resins Nanocomposites

Accelerated UV weathering of CNT-epoxy nanocomposites was investigated and most reports did concur in their assessment that it is unlikely for CNTs to be readily released into the environment given their aggregation in high density networks [249, 252, 254–257]. Moreover, an addition of 0.72 or 0.75 wt% [249, 252] of MWCNTs to an amine-cured epoxy-based nanocomposite yielded in an UV radiation resistant material with a surface containing a dense network of nanoparticles that remained unaffected for long terms, without MWCNTs release.

Samples of epoxy-based MWCNTs nanocomposites containing different amounts of nanoparticles (0.25, 0.5, 1 and 2 % MWCNTs) were exposed to accelerated weathering (UV radiation, salt fog and oxygen) [258]. Long exposure intervals yielded in slow degradation rates, reduced weight losses and minimum amount of cracks, evidencing the CNTs contribution to the photochemical and corrosion stability of the nanocomposites. These findings recommended them as viable coatings for aluminum alloy sheets, but such formulations were also tested as nanocomposites coatings for carbon fiber and glass fiber reinforced composites [259]. As expected, CNTs provided a better bonding and an increased strength to the nanocomposite coating, due to their high surface area and aspect ratio. Experimental data indicated that CNTs (2, 4 and 8 wt%) determined a significant decrease in the surface energy of the coating, making it more stable toward weathering.

Amino-functionalized CNTs were employed in epoxy-based nanocomposites in order to improve the compatibility between matrix and filler [254, 260], as they can covalently bond to the epoxy resins [261]. Their behavior toward UV radiation and moisture was compared to untreated CNTs-filled epoxy nanocomposites. Despite the improved dispersion of modified CNTs, the epoxy matrix was less homogenous due, probably, to a higher amount of oligomers resulted from the degradation reactions. The cracks formation was evidenced by SEM images which also revealed CNTs aggregates on the samples surface, as well as inside cracks.

In order to assess the environmental impact of CNT-epoxy nanocomposites degradation under UV irradiation, three types of samples were prepared (neat epoxy resin, unfunctionalized MWCNTs-epoxy nanocomposites with 1.0 wt% filler, and amino-functionalized MWCNTs-epoxy nanocomposites with 1.0 wt% filler), submitted to degradation under UV and elevated humidity conditions, and subsequently tested in vivo on *Drosophila melanogaster* as model for toxicity tests [262]. SEM cross-section images revealed neat CNTs at the surface of the degraded samples, gathered in dense aggregates from which nanoparticles protruded. In contrast, samples with amino-modified CNTs displayed a smoother surface where nanoparticles were embedded. The in vivo tests on *Drosophila melanogaster* showed no indication of increased toxicity for the embedded amino-functionalized CNTs. However, under environmental complex conditions and upon mechanical wear, these materials may undergo a more intense nanoparticle release and, therefore, it is of great interest to assess their toxicological impact.

5.4 Graphene Platelets–Thermosetting Polymer Nanocomposites

Graphene-filled UV-curable nanocomposites are well suited for hi-tech applications (in electronics, energy storage devices, in optoelectronics, coatings) as they combine properties of both basic components. As other graphitic nanoparticles, graphenes, which are two-dimensional single-atom thick layers of carbon atoms organized as a honeycomb lattice, absorb radiation in the UV spectrum. In a UV-curable matrix, a competition between graphene filler and photoinitiator (both UV absorbing) occurs and this may yield into a slow rate curing process and low conversion [263]. Methods to overcome these drawbacks were reported: increasing the irradiation energy, prolonged curing intervals, addition of hyperbranched polymers, replace graphene with functionalized graphene or graphene oxide [263–268].

As compared to non-conductive clays, graphenes have lower density and higher aspect ratio which grant them enhanced gas barrier properties. This strongly limits the in-depth oxygen permeation and reduces the oxidative risk, as proved by some epoxy-graphene nanocomposites used as corrosion protective coatings for cold-rolled steel electrodes [268] prepared using a templating method based on soft lithography [269].

Functionalized graphene nano-sheets were employed in the formulation of some UV-cured epoxy-based nanocomposites [264]. Bis-cycloaliphatic diepoxidic resin 3,4-epoxycyclohexylmethyl-30,40-epoxycyclohexyl carboxylate was submitted to UV curing in the presence of triphenylsulfonium hexafluoroantimonate as photoinitiator (2 wt%) and various amounts of functionalized graphenes (0.5, 1 and 1.5 wt%), yielding in fully crosslinked hybrid network with improved mechanical characteristics.

The effect of UVA radiation on graphene-containing (1 and 2 wt%) UV-cured epoxy nanocomposites were studied [270]. It was found that graphene dispersed into the epoxy matrix granted an increased bearing-load capacity and, simultaneously, drastically reduced the photooxidative degradation.

Considering these aspects, graphene filled UV-curable nanocomposites are cost effective and readily transferred to mass production. Still, their photochemical behavior upon UV exposure and weathering must be seriously evaluated, as well as their environmental risks since graphene may be released from materials along with matrix degradation.

5.5 Organic–Inorganic Hybrid Materials Obtained by the Dual Cure Process

Modern approaches in the field of hybrid materials for advanced applications use dual cure processing which, depending on the nature of components, may be (a) a UV-cure followed by a thermal treatment, or (b) a UV-cure followed by a condensation reaction when the interpenetrated network containing the nanofiller is in situ generated *via* a sol–gel process, especially when silane precursors are used [230].

5.5.1 Epoxy-Based Hybrid Materials

Thus, novel organic–inorganic hybrid epoxy-based nanocomposites containing zirconia were obtained by a cationic UV-initiated polymerization, followed by a thermal treatment when the sol–gel process has completed [271], using zirconium tetrapropoxide as inorganic precursor (up to 5 wt%). Films showed transparency in the visible range that indicated a homogeneous distribution of zirconia, but their increase in the refractive index was correlated with the increasing amount of inorganic precursor and a diminished reflection associated with a gloss loss.

Multicomponent hybrids containing zirconium propoxide, tetraethoxysilane and dimethyl-diethoxysilane, intended for dental restorative or adhesive materials, were successfully obtained by the dual-cure [272]. Zirconium-containing species proved to be highly effective in catalyzing the epoxy polymerization/crosslinking reactions, as compared to those containing Ti, and enhanced mechanical properties, as well as thermal stability of nanocomposites.

Other epoxy-based nanocomposites were obtained by the same dual-cure process using a novel epoxy oligomer, namely 3-isocyanatopropyltriethoxysilane (IPTS)-grafted bisphenol A epoxy resin, and tetraethyl orthosilicate as inorganic precursor [273]. The in situ generated SiO_2 nanoparticles (approx. 40 nm in diameter) were homogeneously dispersed inside the matrix, as evidenced by TEM, and they imparted improved thermal and mechanical properties to nanocomposites.

Transparent photoluminescent hybrids, based on bisphenol A ethoxylatedi-acrylate, silica precursors (tetraethoxysilane and methacryloyloxypropyltrimethoxysilane) and photoluminescent tetracopper iodide clusters [$Cu_4I_4L_4$, where L = $PPh_2(CH2)_2CH_3$], were successfully synthesized by the same technique [274]. The hybrid films showed enhanced transparency and scratch resistance, as well as a yellow-orange bright luminiscence (emission maximum at 565 nm).

5.5.2 Acrylate-Based Hybrid Materials

An interesting study used a poly(ethylene glycol) 600 α,ω diacrylate (PEGDA 600) and another oligomer considered as a modified PEGDA by inserting bisphenol A and terminal α,ω methacrylic groups (BEMA 1400) as organic precursors [275]. Additionally, methacryloyl-oxypropyltrimethoxysilane (MEMO) was employed as organic–inorganic bridging monomer and tetraethoxysilane (TEOS) as inorganic precursor. These multicomponent systems were processed by the dual cure technique and their properties were evaluated. Analytical investigations indicated an almost complete conversion of the reactive functional moieties. The T_g values of the hybrids increased along with the increasing content of TEOS, but were higher for BEMA hybrids than PEGDA ones, at the same amount of TEOS. This may be explained by the presence of methacrylic double bonds and the rigid structure of bisphenol A moieties in BEMA. Hybrid films were transparent and amorphous, and exhibited an improved thermal stability.

Another acrylate resin was employed in a complex multicomponent formulation as a UV-curable matrix. Thus, a novel acid urethane oligodimethacrylate based on poly(ethylene glycol) was synthesized and used to obtain hybrid nanocomposites containing silsesquioxane sequences [3-(trimethoxysilyl)propyl methacrylate (MPTS), 3-(acryloyloxy)-2-hidroxy-propyl methacrylate (MA-OH)] and titania domains (titanium IV butoxide 5–20 wt%) formed through sol–gel reactions, along with silver/gold nanoparticles (Ag/Au NPs) photochemically generated in situ during the UV-curing process [228]. Experimental data indicated a good photochemical reactivity of the novel monomer during the UV-initiated polymerization reactions, even in the presence of the inorganic components. All hybrids showed improved thermal stability, tensile strength and modulus, but moderate toughness. On the other hand, their photocatalytic activity was significant which made them fit for specific applications, such as the water purification by the photocatalytic degradation of organic pollutants.

Perfectly transparent hybrid films were obtained by UV/sol–gel dual cure starting from a fluorinated hydroxyl acrylate monomer, tripropylene glycol diacrylate (TPGDA) and tetraethoxysilane (TEOS) [276]. The hybrids were characterized by an almost complete double bonds conversion, high hydrophobicity and increased surface hardness, T_g values, and tensile modulus. Increased TEOS amounts entailed extensive interfacial interactions through a chemically formed organic–inorganic interphase.

5.5.3 Other Thermosetting Matrices Used for Hybrid Materials

Organic–inorganic hybrids were synthesized by the dual cure process using TEOS as inorganic precursor, triethylene glycol divinyl ether (DVE) as a typical vinyl ether resin, and hydroxybutyl vinyl ether (HBVE) as a coupling agent in order to enhance the interfacial interactions between organic and inorganic phases [277]. These vinyl ether based systems were, thus, reported for the first time in such an application. The resulted transparent films displayed high gel content, increased T_g and storage modulus above T_g along with the increasing amount of TEOS in formulations.

Other hybrid materials were prepared starting from a hyperbranched epoxy functionalized polyester resin, TEOS as an organic precursor of the silica network, and 3-glycidoxypropyltrimethoxysilane (GPTS) as a coupling agent [278]. All formulations yielded in films with high gel content (over 98 %), a high T_g and an increased storage modulus above T_g due to the increased TEOS content. The coupling agent (GPTS) was also involved in the formation of the nanometric filler, as in its absence the SiO_2 particles aggregated in large size inorganic domains and the phase separation occurred. At the same time, GPTS contributed to the formation of the organic–inorganic interphase and enhanced interfacial interactions.

Novel biodegredable, low cost, flame resistant nanocomposites were obtained by combined techniques and using epoxidized vegetable oils as resin precursors and various nanoparticulate fillers [12, 279–283]. Thus, a series of methacrylated and phosphorylated epoxidized soybean oil/silica nanomaterials were prepared by dual cure (photopolymerization and sol–gel process), starting from (TEOS) and acrylated soybean oil which was obtained by reacting the epoxidized soybean oil (ESO) with methacrylic acid and vinyl phosphonic acid [279]. Samples submitted to thermogravimetric analysis showed enhanced thermal stability due to the TEOS incorporation and, subsequently, flame retardancy that recommend these materials as protective coatings for flammable materials. The incorporation of acrylated epoxidized oil, mainly with methacrylic acid, entailed an increase in mechanical and physical properties, as compared to the neat and vinyl phosphonic acid-containing resins. This may be attributed to the higher gel content imparted by the metacrylic acid, as well as the superior thermal stability and more hydrophobic character.

6 Concluding Remarks and Future Developments

Multicomponent systems based on synthetic polymers are competitive materials designed and used in a wide variety of applications, ranging from civil engineering to hi-tech and healthcare. In the case of outdoor applications, materials are exposed to a sum of factors having a high degrading potential, in most cases acting in synergy: UV radiation, temperature and temperature cycles, humidity, pH, pollutants, ozone, microorganisms. The rate of the photochemical reactions increases exponentially with the absolute temperature. The other climatic factors further contribute to an even faster rate of degradation. Therefore, the study of the photochemical behavior of these materials is of high interest for both academic and applicative reasons.

Limiting the negative impact of the UV radiation requires the use of additives with specific functions, such as light stabilizers (as HALS) and UV absorbers (that absorb the most of the damaging UV-B radiation). But since they cannot be added in high amount, other approaches were also considered: the use of organic or inorganic fillers, pigments, nanoparticulate mineral fillers. In most cases, the incompatibility between components may yield in phase segregation which is severely affecting the materials bulk properties. Therefore, compatibilizing agents are a good solution, although other methods were also employed: functionalization and/or surface treatments of the fillers in order to reduce the surface energy, or functionalization of both matrix and filler. This approach favours the formation of the interphase which enhances the interfacial interactions, with positive effects on the materials characteristics and which enables widening the application area.

At the same time, the new high-performance techniques used for the characterization of the UV-initiated degradation have allowed in-depth perceptive correlations. Thus, it was possible to establish the influence of the nature and size distribution of filler particles on the UV-susceptibility of the material. For the same filler, micron size particles have negative effect contributing to the increase of the photochemical reactions rate, while the nanometric particles exhibit an opposite effect. On the other hand, organic fillers (such as wood chips and ligno-cellulosic fibers) enhance the UV-initiated degradation, in contrast to carbon fillers (carbon black, graphene, carbon fibers, CNTs). Even in the case of reinforcing fillers (MWCNT, glass fibers) it was possible to discriminate their particular effect.

One of the key factors in developing these materials is the ability to enhance, in a controlled manner, the interfacial interactions between components, mainly when it comes to complex systems organized at nanometric scale. Given the complexity of formulations, various methods and solutions were employed, in strict dependency with the nature, role and synergy of components. A wisely chosen technology comes to add to the interfacial interactions by contributing to the formation of a well-defined thick interphase layer.

It is quite difficult to envisage the future developments in this very active domain, given the extent of the constant intensive research. But it is reasonably to anticipate that heterogeneous muticomponent materials, such as particulate filled polymers and blends, fiber reinforced composites, etc., will find more applications as the composition-molecular/supramolecular structure-properties relationship will be better understood. At the same time, the improvement of processing technologies will allow lower production costs and an improved control of properties. Even more, the nanotechnology contributes to widening the application area by adding new high performance properties such as flame retardancy, UV-stability, barrier properties, conductivity, etc. The interest in raw materials from renewable resources is expected to further increase as materials with low environmental impact are desired.

A warning note has to be issued: the potential risk of releasing nanoparticles into the environment by the UV degradation of multicomponent materials upon weathering and their negative impact on the nature and human health. This subject is still under debate and opposite opinions, supported by experimental data, were expressed, illustrating a real preoccupation of the academic media.

References

1. Kardos, J.L.: The role of the interface in polymer composites-some myths, mechanisms, and modifications. In: Ishida, H., Kumar, G. (eds.) Molecular Characterization of Composite Interfaces, p. 1. Plenum Press, New York (1985)
2. Eirich, F.R.: Some mechanical and molecular aspects of the performance of composites. J. Appl. Polym. Sci.: Appl. Polym. Symp. **39**, 93–102 (1984)
3. Pukanszky, B.: Interfaces and interphases in multicomponent materials: past, present, future. Eur. Polym. J. **41**, 645–662 (2005)

4. Andrady, A.L., Hamid, H.S., Torikai, A.: Ch. 7. Effects of stratospheric ozone depletion and climate change on materials damage. In: The Environmental Effects Assessment Panel Report for 2006, pp. 185–200. http://ozone.unep.org/pdfs/Final%20UNEP%202006%20Report.pdf
5. Andrady, A.L., Hamid, H., Torikai, A.: Effects of solar UV and climate change on materials. Photochem. Photobiol. Sci. **10**, 292–300 (2011)
6. Allen, N.S., Edge, M., Ortega, A., et al.: Degradation and stabilization of polymers and coatings: nano versus pigmentary titania particles. In: IUPAC Microsymposium on Degradation, Stabilisation and Recycling and Polymers, No. 42, vol. 85, pp. 927–946. Elsevier Science, Oxford (2004)
7. Azwa, Z.N., Yousif, B.F., Manalo, A.C., et al.: A review on the degradability of polymeric composites based on natural fibres. Mater. Des. **47**, 424–442 (2013)
8. Kingston, C., Zepp, R., Andrady, A., et al.: Release characteristics of selected carbon nanotube polymer composites. Carbon **68**, 33–57 (2014)
9. Kumar, A.P., Depan, D., Tomer, N.S., et al.: Nanoscale particles for polymer degradation and stabilization—trends and future perspectives. Prog. Polym. Sci. **34**, 479–515 (2009)
10. Kango, S., Kalia, S., Celli, A., et al.: Surface modification of inorganic nanoparticles for development of organic–inorganic nanocomposites—a review. Prog. Polym. Sci. **38**, 1232–1261 (2013)
11. Endruweit, A., Johnson, M.S., Long, A.C.: Curing of composite components by ultraviolet radiation: a review. Polym. Compos. **27**, 119–128 (2006)
12. Samarth, N.B., Mahanwar, P.A.: Modified vegetable oil based additives as a future polymeric material—Review. Open J. Org. Polym. Mater. **5**, 1–22 (2015). doi:10.4236/ojopm.2015.51001
13. Teaca, C.-A., Rosu, D., Bodarlau, R., Rosu, L.: Structural changes in wood under artificial UV light irradiation determined by FTIR spectroscopy and color measurements—a brief review. BioResources **8**, 1478–1507 (2013)
14. Yousif, E., Haddad, R.: Photodegradation and photostabilization of polymers, especially polystyrene: review. Springer Plus **2**, 398–430 (2013)
15. Bottino, F.A., Cinquegrani, A.R., Di Pasquale, G., et al.: Chemical modification, mechanical properties and surface photooxidation of films of polystyrene. Polym. Test. **12**, 405–411 (2003)
16. Gardella, J.A.: Characterization techniques for polymer blends. Appl. Surf. Sci. **31**, 72–102 (1988)
17. Rabek, J.F.: Polymer Degradation Mechanisms and Experimental Methods. Springer, Stocholm (1994)
18. Gugumus, F.: Light stabilizers. In: Gachter, R., Muller, H. (eds.) Plastic Additives Handbook, p. 129. Hanse Publishers, New York (1990)
19. Jabarin, S.A., Lofgren, E.A.: Photo-oxidative effects on properties and structure of HDPE. J. Appl. Polym. Sci. **53**, 411–423 (1994)
20. Wypych, G.: Handbook of Material Weathering. ChemTech Publishing, Toronto-Scarborough (1995)
21. Tidjani, A.: Comparison of formation of oxidation products during photo-oxidation of LLDPE under different natural and accelerated weathering conditions. Polym. Degrad. Stab. **68**, 465–469 (2000)
22. Stark, N.M.: Photodegradation and photostabilization of weathered wood flour filled PE composites. Ph.D. thesis, Michigan Technological University, USA (2003)
23. Zanoaga, M.: Novel thermoplastic copolyamides based on caprolactam, 1,6-hexane diamine. Synthesis and study of properties. In: Proccedings of 12th Romanian International Conference on Chemistry and Chemical Engineering, September 13–15, 2001, Bucureşti, Romania, pp. 125–130
24. Protection from the Elements. Part IV: radiation exposure. Technical Note, November 2013. https://www.lightingglobal.org/resources/technical-notes/. Accessed 17 Feb 2016

25. Matuana, L.M., Jin, S., Stark, N.M.: Ultraviolet weathering of HDPE/wood–flour composites coextruded with a clear HDPE cap layer. Polym. Degrad. Stab. **96**, 97–106 (2011)
26. Athijayamani, A., Thiruchitrambalam, M., Natarajan, U., et al.: Effect of moisture absorption on the mechanical properties of randomly oriented natural fibers/polyester hybrid composite. Mater. Sci. Eng., A **517**, 344–353 (2009)
27. Joseph, P.V., Rabello, M.S., Mattoso, L.H.C., et al.: Environmental effects on the degradation behaviour of sisal fibre reinforced polypropylene composites. Compos. Sci. Technol. **62**, 1357–1372 (2002)
28. Hu, R.-H., Sun, M.-Y., Lim, J.-K.: Moisture absorption, tensile strength and microstructure evolution of short jute fiber/polylactide composite in hygrothermal environment. Mater. Des. **31**, 3167–3173 (2010)
29. Cappitelli, F., Sorlini, C.: Microorganisms attack synthetic polymers in items representing our cultural heritage. Appl. Environ. Microbiol. **74**, 564–569 (2008)
30. Beg, M.D.H., Pickering, K.L.: Accelerated weathering of unbleached and bleached kraft wood fibre reinforced polypropylene composites. Polym. Degrad. Stab. **93**, 1939–1946 (2008)
31. Fabiyi, J.S., McDonald, A.G., Wolcott, M.P., et al.: Wood plastic composites weathering: visual appearance and chemical changes. Polym. Degrad. Stab. **93**, 1405–1414 (2008)
32. Hyvärinen, M.: Ultraviolet light protection and weathering properties of wood–polypropylene composites. Ph.D. thesis, Lappeenranta University of Technology, Lappeenranta, Finland (2014)
33. Affolter, S.: Long-term behaviour of thermoplastic materials. Materiały Interstate University NTB, Buchs, Szwajcaria. http://www.ntb.ch/fileadmin/Institute/MNT/pdf/NTB_MNT_Polymerics_Ageing_SA.pdf (1999)
34. Biron, M.: Basic criteria for the Selection of thermoplastics. In: Biron, M. (ed.) Thermoplastics and Thermoplastic Composites, 2nd edn., p. 133. Elsevier Ltd., Oxford, UK (2013). doi:10.1016/B978-1-4557-7898-0-0001-9
35. Catto, A.L., Almeida, S.H., Santana, R.M.C.: Abiotic degradation of recycled polymer/wood composites exposed to outdoor applications. J. Mater. Sci. Eng. A **4**, 321–329 (2014)
36. Hon, D.N.S.: Weathering and photochemistry of wood. In: Hon, D.N.S., Shiraishi, N. (eds.) Wood and Cellulosic chemistry, on M. 2nd edn., p. 513. Marcel Dekker Inc., New York (2001), p. 133. Elsevier Ltd., Oxford, UK (2013). doi:10.1016/B978-1-4557-7898-0-0001-9
37. Fabiyi, J.S., McDonald, A.G.: Effect of wood species on property and weathering performance of wood plastic composites. Compos. A Appl. Sci. Manuf. **41**, 1434–1440 (2010)
38. Fabiyi, J.S., McDonald, A.G., Stark, N.M.: Surface characterization of weathered wood–plastic composites produced from modified wood flour. In: Proceedings of the 9th International Conference on Wood and Biofiber Plastic Composites, Forest Products Society, Madison, p. 271 WI (2007)
39. Chaochanchaikul, K., Sombatsompop, N.: Stabilization of molecular structures and mechanical properties of PVC and wood/PVC composites by Tinuvin and TiO_2 stabilizers. Polym. Eng. Sci. **51**, 1354–1365 (2011)
40. Chaochanchaikul, K., Jayamaran, K., Rosarpitak, V., et al.: Influence of lignin content on photodegradation in wood-HDPE composites under UV weathering. BioResources **7**, 38–55 (2012)
41. Fabiyi, J.S., McDonald, A.G.: Degradation of polypropylene in naturally and artificially weathered plastic matrix composites. Maderas. Ciencia y tecnología **16**, 275–290 (2014)
42. Lundin, T.: Effect of accelerated weathering on the physical and mechanical properties of natural-fiber thermoplastic composites. M.S. thesis, University of Wisconsin, Madison, WI (2001)
43. Lundin, T., Cramer, S.M., Falk, R.H., et al.: Accelerated weathering of natural fiber-filled polyethylene composites. J. Mater. Civ. Eng. **16**, 547–555 (2004)

44. Deglise, X., Beatrice, F.: Photodegradation and photostabilization of woods—the state of art. Polym. Degrad. Stab. **88**, 269–274 (2005)
45. Stark, N.M., Matuana, L.M., Clemons, C.M.: Effect of processing method on surface and weathering characteristics of wood-flour/HDPE composites. J. Appl. Polym. Sci. **93**, 1021–1030 (2004)
46. Alexy, P., Kosikova, B., Podstranska, G.: The effect of blending lignin with polyethylene and propylene on physical properties. Polymer **41**, 4901–4908 (2000)
47. Pucciariello, R., Villani, V., Bonini, C., et al.: Physical properties of straw lignin-based polymer blends. Polymer **45**, 4159–4169 (2004)
48. George, B., Suttie, E., Merlin, A., et al.: Photodegradation and photostabilization of wood. Polym. Degrad. Stab. **88**, 268–274 (2005)
49. Muasher, M., Sain, M.: The efficacy of photostabilizers on the color change of wood filled plastic composite. Polym. Degrad. Stab. **91**, 1156–1165 (2005)
50. Basfar, A.A., Ali, K.M.I.: Natural weathering test for films of various formulations of LDPE and LLDPE. Polym. Degrad. Stab. **91**, 437–443 (2006)
51. Borysiak, S., Klapiszewski, Ł., Bula, K., et al.: Nucleation ability of advanced functional silica/lignin hybrid fillers in polypropylene composites. J. Therm. Anal. Calorim. (2016). doi:10.1007/s10973-016-5390-1
52. Li, H., Zhang, Z., Song, K., et al.: Effect of durability treatment on UV resistance, strength and surface wettability of wood plastic composite. BioResources **9**, 3591–3601 (2014)
53. Garcia, M., Hidalgo, J., Garmendia, I., et al.: Wood-plastic composites with better fire retardancy and durability performance. Compos. A Appl. Sci. Manuf. **40**, 1772–1776 (2009)
54. Stark, N.M., Mueller, S.A.: Improving the colour stability of wood–plastic composites through fibre pre-treatment. Wood Fibre Sci. **40**, 271–278 (2008)
55. Stark, N.M., Matuana, L.M.: Influence of photostabilizers on wood flour-HDPE composites exposed to xenon-arc radiation with and without water spray. Polym. Degrad. Stab. **91**, 3048–3056 (2006)
56. Falk, R.H., Lundin, T., Felton, C.: The effect of weathering on wood–thermoplastic composites intended for outdoor applications. In: Proceedings of the 2nd Annual Conference on Durability and Disaster Mitigation in Wood-Frame Housing, Madison, WI, 6–8 November 2000, p. 175 (2000)
57. Butylina, S., Hyvärinen, M., Kärki, T.: A study of surface changes of wood-polypropylene composites as the result of exterior weathering. Polym. Degrad. Stab. **97**, 337–345 (2012)
58. La Mantia, F.P., Morreale, M.: Accelerated weathering of polypropylene/wood flour composites. Polym. Degrad. Stab. **93**, 1252–1258 (2008)
59. Matuana, L.M., Kamdem, D.P.: Accelerated ultraviolet weathering of PVC/wood–flour composites. Polym. Eng. Sci. **42**, 1657–1666 (2002)
60. Campos, A., Marconcini, J.M., Martins-Franchetti, S.M., et al.: The influence of UV-C irradiation on the properties of thermoplastic starch and polycaprolactone biocomposite with sisal bleached fibers. Polym. Degrad. Stab. **97**, 1948–1955 (2012)
61. Abu-Sharkh, B., Hamid, H.: Degradation study of date palm fiber/polypropylene composites in natural and artificial weathering: mechanical and thermal analysis. Polym. Degrad. Stab. **85**, 967–973 (2004)
62. Abu Bakr, A., Hassan, A., Yusof, A.F.M.: Effect of accelerated weathering on the mechanical properties of oil palm empty fruit filled UPVC composite. Iran. Polym. J. **14**, 627–635 (2005)
63. Seldon, R., Nystroem, B., Langstroem, R.: UV aging poly(propylene)/wood–fiber composites. Polym. Compos. **25**, 543–553 (2005)
64. Kaczmarek, H., Oldak, M., Malanowski, P., et al.: Effect of short wavelength UV irradiation on aging of polypropylene/cellulose composition. Polym. Degrad. Stab. **88**, 189–198 (2005)
65. Michel, A.T., Billington, S.L.: Characterization of poly-hydroxybutyrate films and hemp fiber reinforced composites exposed to accelerated weathering. Polym. Degrad. Stab. **97**, 870–878 (2012)

66. Beninia, K.C.C.C., Voorwald, H.J.C., Cioffi, M.O.H.: Mechanical properties of HIPS/sugarcane bagasse fiber composites after accelerated weathering. Procedia Eng. **10**, 3246–3251 (2011)
67. Kalia, S., Kaith, B.S., Kaur, I.: Pretreatments of natural fibers and their application as reinforcing material in polymer composites—a review. Polym. Eng. Sci. **499**, 1253–1272 (2008)
68. Kalia, S., Kaith, B.S., Kaur, I., et al.: Biofiber-reinforced thermoplastic composites. In: Sabu, T., Kuruvilla, J., Malhotra, S.K., Goda, K., Sreekala, M.S. (eds.) Polymer Composites, p. 239. Wiley-VCH Verlag GmbH & Co., Weinheim (2001)
69. Faruk, O., Bledzki, A.K., Fink, H.P., et al.: Progress report on natural fiber reinforced composites. Macromol. Mater. Eng. **299**, 9–26 (2014)
70. Robledo-Ortiz, J.R., Fuentes-Talavera, F.J., González-Núñez, R., et al.: Ch. 26. Wood and natural fiber-based composites (NFCs). In: Saldivar-Guerra, E., Vivaldo-Lima, E. (eds.) Handbook of Polymer Synthesis, Characterization and Processing, p. 493. Wiley, Hoboken (2013). doi:10.1002/9781118480793.ch26
71. Sionkowska, A., Planecka, A.: The influence of UV radiation on silk fibroin. Polym. Degrad. Stab. **96**, 523–528 (2011)
72. Aksakal, B., Koç, K., Yargı, Ö., et al.: Effect of UV-light on the uniaxial tensile properties and structure of uncoated and TiO_2 coated *Bombyx mori* silk fibers. Spectrochim. Acta A Mol. Biomol. Spectrosc. **152**, 658–665 (2016)
73. Wu, M.C., Chan, S.H., Lin, T.H.: Fabrication and photocatalytic performance of electrospun PVA/silk/TiO_2 nanocomposite textile. Funct. Mater. Lett. **8**, 1540013 (2015). doi:10.1142/S1793604715400135
74. Zamana, H.U., Khana, M.A., Khana, R.A.: Comparative experimental studies on the mechanical and degradation properties of natural fibers reinforced polypropylene composites. Compos. Interfaces **19**, 59–70 (2012)
75. Saravanan, D.: UV protection textile materials. AUTEX Res. J. **7**, 53–62 (2007)
76. Treigienė, R., Musnickas, J.R.: Influence of UV exposure on properties of wool fiber pretreated with surfactants solutions. Mater. Sci. (Medžiagotyra) **14**, 75–78 (2008)
77. Coderch, L., Lopez, O., De la Maza-Manich, A.M., et al.: Internal lipid wool structure modification due to a nonionic auxiliary used in dyeing at low temperatures. Text. Res. J. **67**, 131–136 (1997)
78. Musnickas, J., Rupainytė, V., Treigienė, R.: Dye migration influences on colour characteristics of wool fiber dyed with acid dye. Fibers Text. East. Eur. **13**, 65–69 (2005)
79. Alzeer, M., MacKenzie, K.J.D.: Synthesis and mechanical properties of new fibre-reinforced composites of inorganic polymers with natural wool fibres. J. Mater. Sci. **47**, 6958–6965 (2012)
80. Wypych, G.: Handbook of Fillers, 2nd edn. ChemTec Publishing, Toronto (1999)
81. Rothon, R.N.: Particulate-filled polymer composites. In: Rothon, R.N. (ed.) 2nd edn., Rapra Technology, Shrewsbury, UK (2003)
82. Trotignon, J.P., Verdu, J., De Boissard, R., et al.: Polypropylene-mica composites. In: Sedlacek, B. (ed.) Polymer Composites, p. 191. Walter de Gruyter, Berlin (1986)
83. Qin, H.L.: UV photo aging of polyamide 6/montmorillonite nanocomposite. Chem. J. Chin. Univ. **25**, 197–198 (2004)
84. Azuma, Y., Takeda, H., Watanabe, S., et al.: Outdoor accelerated weathering tests for polypropylene and polypropylene/talc composites. Polym. Degrad. Stab. **94**, 2267–2274 (2009)
85. Rabello, M.S., White, J.R.: Photodegradation of talc-filled polypropylene. Polym. Compos. **17**, 691–704 (1996)
86. Denac, M., Smit, I., Musil, V.: Polypropylene/talc/SEBS (SEBS-g-MA) composites. Part 1. Structure. Compos. A Appl. Sci. Manuf. **36**, 1094–1101 (2005)
87. Denac, M., Musil, V., Smit, I.: Polypropylene/talc/SEBS (SEBS-g-MA) composites. Part 2. Mechanical properties. Compos. A Appl. Sci. Manuf. **36**, 1282–1290 (2005)

88. Al-Shabanat, M.: Study of the effect of weathering in natural environment on polypropylene and its composites: morphological and mechanical properties. Int. J. Chem. **3**, 129–141 (2011)
89. Yang, R., Yu, J., Liu, Y., et al.: Effects of inorganic fillers on the natural photo-oxidation of high-density polyethylene. Polym. Degrad. Stab. **88**, 333–340 (2005)
90. Leong, Y.W., Abu Bakar, M.B., Mohd Ishak, Z.A., et al.: Characterization of talc/calcium carbonate filled polypropylene hybrid composites weathered in a natural environment. Polym. Degrad. Stab. **83**, 411–422 (2004)
91. Benavides, R., Gonzalez-Hernandez, R., Gonzalez-Cantu, M.C., et al.: Accelerated degradation of highly loaded polypropylene. J. Vinyl Addit. Technol. **9**, 41–49 (2003)
92. Chen, Y., Sun, Z., Yang, Y., et al.: Heterogeneous photocatalytic oxidation of polyvinyl alcohol in water. J. Photochem. Photobiol. A **142**, 85–89 (2001)
93. Zhang, S.-J., Han-Qing, Y.: Radiation-induced degradation of polyvinyl alcohol in aqueous solutions. Water Res. **38**, 309–316 (2004)
94. Vijayalakshmi, S.P., Madras, G.: Photodegradation of poly (vinyl alcohol) under UV and pulsed-laser irradiation in aqueous solution. J. Appl. Polym. Sci. **102**, 958–966 (2006)
95. Wang, Y., Zhong, M., Chen, F., et al.: Visible light photocatalytic activity of TiO_2/D-PVA for MO degradation. Appl. Catal. B **90**, 249–254 (2009)
96. Khan, M.A., Bhattacharia, S.K., Kader, M.A., et al.: Preparation and characterization of ultra violet (UV) radiation cured bio-degradable films of sago starch/PVA blend. Carbohydr. Polym. **63**, 500–506 (2006)
97. Park, J.S., Yang, J.H., Kim, D.H., et al.: Degradability of expanded starch/PVA blends prepared using calcium carbonate as the expanding inhibitor. J. Appl. Polym. Sci. **93**, 911–919 (2004)
98. Kaczmarek, H., Podgorski, A.: Photochemical and thermal behaviours of poly(vinyl alcohol)/graphite oxide composites. Polym. Degrad. Stab. **92**, 939–946 (2007)
99. Tong, L., White, J.R.: Photo-oxidation of thermoplastics in bending and in uniaxial compression. Polym. Degrad. Stab. **53**, 381–396 (1996)
100. O'Donnell, B., White, J.R.: Stress-accelerated photo-oxidation of polypropylene and glass-fibre-reinforced polypropylene. Polym. Degrad. Stab. **44**, 211–222 (1994)
101. Goel, A., Chawla, K.K., Vaidya, U.K., et al.: Effect of UV exposure on the microstructure and mechanical properties of long fiber thermoplastic (LFT) composites. J. Mater. Sci. **43**, 4423–4432 (2008)
102. Tyler, D.R.: Mechanistic aspects of the effects of stress on the rates of photochemical degradation reactions in polymers. J. Macromol. Sci. Polym. Rev. **C44**, 351–388 (2004)
103. Goel, A.: Fatigue and environmental behavior of long fiber thermoplastic (LFT) composites. Ph.D. thesis, Birmingham, Alabama, USA (2008)
104. Larena, A., Jimenez de Ochoa, S., Dominguez, F.: Dynamic-mechanical analysis of the photo-degradation of long glass fibre reinforced polypropylene: mechanical properties changes. Polym. Degrad. Stab. **91**, 940–946 (2006)
105. Chevali, V.S., Dean, D.R., Janowski, G.M.: Effect of environmental weathering on flexural creep behavior of long fiber-reinforced thermoplastic composites. Polym. Degrad. Stab. **95**, 2628–2640 (2010)
106. Pillay, S., Vaidya, U.K., Janowski, G.M.: Effects of moisture and UV exposure on liquid molded carbon fabric reinforced nylon 6 composite laminates. Compos. Sci. Technol. **69**, 839–846 (2009)
107. Choi, N.S., Takahashi, K., Oschmann, D., et al.: Influence of weathering on unreinforced and short glass fibre reinforced thermoplastic polyester. J. Mater. Sci. **33**, 2529–2535 (1998)
108. Shokrieh, M.M.: Effects of ultraviolet radiation on mechanical properties of glass/polyester composites. J. Compos. Mater. **41**, 2443–2455 (2007)
109. Karger-Kocsis, J.: On the toughness of "nanomodified" polymers and their traditional polymer composites. In: Karger-Kocsis, J., Fakirov, S. (eds.) Nano- and Micromechanics of Polymer Blends and Composites, p. 425. Hanser, Munich (2010)

110. Edraki, M., Keivani, M.B.: Study on the optical and rheological properties of polymer-layered silicate nanocomposites. J. Phys. Theor. Chem. Islam. Azad Univ. Iran **10**, 69–79 (2013)
111. Qin, H., Zhao, C., Zhang, S., et al.: Photo-oxidative degradation of polyethylene/montmorillonite nanocomposite. Polym. Degrad. Stab. **81**, 497–500 (2003)
112. Qin, H., Zhang, S., Liu, H., et al.: Photo-oxidative degradation of polypropylene/montmorillonite nanocomposites. Polymer **46**, 3149–3156 (2005)
113. Kumnayaka, T.O., Parthasarathay, R., Jollands, M., et al.: Molecular weight changes during photo-oxidation of polyethylene nanocomposites. Korea-Aust. Rheol. J. **22**, 173–177 (2010)
114. Lee, J.-H., Daeseung, J., Chang-Eui, H.Y., et al.: Properties of polyethylene-layered silicate nanocomposites prepared by melt intercalation with a PP-*g*-MA compatibilizer. Polym. Degrad. Stab. **65**, 1996–2002 (2005)
115. Hussein, I.A.: Rheological investigation of the influence of molecular structure on natural and accelerated UV degradation of linear low density polyethylene. Polym. Degrad. Stab. **92**, 2026–2032 (2007)
116. Qin, H., Zhang, Z., Feng, M., et al.: The influence of interlayer cations on the photo-oxidative degradation of polyethylene/montmorillonite composites. J. Polym. Sci. B: Polym. Phys. **42**, 3006–3012 (2004)
117. Redhwi, H.H., Siddiqui, M.N., Andrady, A.L., et al.: Accelerated weatherability of the low-density polyethylene nanocomposites with silica, clay, and zinc oxide. J. Nanomater. (2014). doi:10.1155/2014/516406
118. la Mantia, F.P., Dintcheva, N.T., Malatesta, V., et al.: Improvement of photo-stability of LLDPE-based nanocomposites. Polym. Degrad. Stab. **91**, 3208–3213 (2006)
119. Morlat-Therias, S., Fanton, E., Gardette, J.-L., et al.: Photochemical stabilization of linear low-density polyethylene/clay nanocomposites: towards durable nanocomposites. Polym. Degrad. Stab. **93**, 1776–1780 (2008)
120. Sitnikova, O.G., Nazarov, S.B., Shikhanova, I.V., et al.: The study of influence of silica and polyethylene glycols organic–inorganic compounds on free-radical processes *in vitro*. Pharmacol. Pharm. **1**, 69–74 (2010)
121. Shen, H.-M., Zhang, Z., Zhang, Q.-F., et al.: Reactive oxygen species and caspase activation mediate silica-induced apoptosis in alveolar macrophages. Am. J. Physiol. **280**, L10–L17 (2001)
122. Liu, M., Horrocks, A.R.: Effect of carbon black on UV stability of LLDPE films under artificial weathering conditions. Polym. Degrad. Stab. **75**, 485–499 (2002)
123. Salem, M.A.: Mechanical properties of UV-irradiated low-density polyethylene films formulated with carbon black and titanium dioxide. Egypt. J. Solids **24**, 141–150 (2001)
124. Zhao, X., Li, Z., Chen, Y., et al.: Solid-phase photocatalytic degradation of polyethylene plastic under UV and solar light irradiation. J. Mol. Catal. A: Chem. **268**, 101–106 (2007)
125. Zhao, H., Li, R.K.Y.: A study on the photo-degradation of zinc oxide filled polypropylene nanocomposites. Polymer **47**, 3207–3217 (2006)
126. Moustaghfir, A., Tomasella, E., Rivaton, A., et al.: Sputtered zinc oxide coatings: Structural study and application to the photoprotection of the polycarbonate. Surf. Coat. Technol. **180–181**, 642–645 (2004)
127. Diagne, M., Gueye, M., Dasilva, A., et al.: Comparative photo-oxidation under natural and accelerated conditions of polypropylene nanocomposites produced by extrusion and injection molding. J. Appl. Polym. Sci. **105**, 3787–3793 (2007)
128. Morlat, S., Mailhot, B., Gonzalez, D., et al.: Photo-oxidation of polypropylene/montmorillonite nanocomposites. 1. Influence of nanoclay and compatibilizing agent. Chem. Mater. **16**, 377–383 (2004)
129. Morreale, M., Dintcheva, N.T., La Mantia F.P.: On the photo-oxidation behaviour of nano-filled polypropylene: role of the filler type and compatibilizer. In: Proceedings of the 7th International Conference MoDeSt-2012, 28, Prague, Czech Republic (2012)
130. Mailhot, B., Morlat, S., Gardette, J.-L., et al.: Photodegradation of polypropylene nanocomposites. Polym. Degrad. Stab. **82**, 163–167 (2003)

131. Tidjani, A., Wilkie, C.A.: Photo-oxidation of polymeric-inorganic nanocomposites: chemical, thermal stability and fire retardancy investigations. Polym. Degrad. Stab. **74**, 33–37 (2001)
132. Morlat-Therias, S., Mailhot, B., Gonzalez, D., et al.: Photooxidation of polypropylene/montmorillonite nanocomposites. 2. Interactions with antioxidants. Chem. Mater. **17**, 1072–1078 (2005)
133. Morreale, M., Dintcheva, NTz, La Mantia, F.P.: The role of filler type in the photo-oxidation behaviour of micro- and nano-filled polypropylene. Polym. Int. **60**, 1107–1116 (2011)
134. Marcinčin, A., Hricová, M., Ujhelyiová, A., et al.: Effect of inorganic (nano)fillers on the UV barrier properties, photo and thermal degradation of polypropylene fibres. Fibres Text. East. Eur. **17**, 29–35 (2009)
135. Peng, D., Qu, B.: Synthesis of exfoliated PP/LDH nanocomposites via melt-intercalation: structure, thermal properties, and photo-oxidative behavior in comparison with PP/MMT nanocomposites. Polym. Eng. Sci. **46**, 1153–1159 (2006)
136. Chmela, S., Kleinová, A., Fiedlerová, A., et al.: Photo-oxidation of sPP/organoclay nanocomposites. J. Macromol. Sci. Pure Appl. Chem. **42**, 821–829 (2005)
137. Kamrannejad, M.M., Hasanzadeh, A., Nosoudi, N., et al.: Photocatalytic degradation of polypropylene/TiO_2 nano-composites. Mater. Res. **17**, 1039–1046 (2014)
138. Bocchini, S., Di Blasio, A., Frache, A.: Influence of MWNT on polypropylene and polyethylene photooxidation. Macromol. Symp. **301**, 16–22 (2011)
139. Dintcheva, N.T., La Mantia, F.P., Malatesta, V.: Photo-oxidation behaviour of polyethylene/multi-wall carbon nanotube composite films. Polym. Degrad. Stab. **94**, 162–170 (2009)
140. Patterson, P.H., Sloan, J.M., Hsieh, A.J.: Photo degradation mechanisms of layered silicate/polycarbonate nanocomposites. In: Annual Technical Conference of the Society of Plastics Engineering, p. 3936 (2002)
141. Sloan, J.M., Patterson, P., Hsieh, A.: Mechanisms of photo degradation for layered silicate–polycarbonate nanocomposites. Polym. Mater. Sci. Eng. **88**, 354–355 (2003)
142. Chandra, A., Turng, L.S., Gopalan, P., et al.: Study of utilizing thin polymer surface coating on the nanoparticles for melt compounding of polycarbonate/alumina nanocomposites and their optical properties. Compos. Sci. Technol. **68**, 768–776 (2008)
143. Morlat-Therias, S., Fanton, E., Gardette, J.L., et al.: Polymer/carbon nanotube nanocomposites: influence of carbon nanotubes on EVA photodegradation. Polym. Degrad. Stab. **92**, 1873–1882 (2007)
144. Sarwar, M.I., Zulfiqar, S., Ahmad, Z.: Polyamide-silica nanocomposites: mechanical, morphological and thermomechanical investigations. Polym. Int. **57**, 292–296 (2008)
145. Zou, J.P., Le Rendu, P., Musa, I., et al.: Investigation of the optical properties of polyfluorene/ZnO nanocomposites. Thin Solid Films **519**, 3997–4003 (2011)
146. Lim, Y.T., Lee, T.W., Lee, H.C., et al.: Enhanced photo-stability of conjugated polymer nanocomposite doped with functionalized nanoparticles. Opt. Mater. **21**, 585–589 (2002)
147. Yang, B.D., Yoon, K.H., Chung, K.W.: Effect of TiO_2 and SiO_2 nanoparticles on the stability of poly (*p*-phenylene vinylene) precursor. Synth. Met. **143**, 25–29 (2004)
148. Yang, S.H., Nguyen, T.P., Le Rendu, P., et al.: Optical and electrical properties of PPV/SiO_2 and PPV/TiO_2 composite materials. Compos. A **36**, 509–513 (2005)
149. Zan, L., Liu, Z., Fa, W., et al.: A novel polymethyl methacrylate (PMMA)-TiO_2 nanocomposite and its thermal and photic stability. Wuhan Univ. J. Nat. Sci. **11**, 415–418 (2006)
150. Yuwono, A.H., Liu, B., Xue, J., et al.: Controlling the crystallinity and nonlinear optical properties of transparent TiO_2-PMMA nanohybrids. J. Mater. Chem. **14**, 2978–2987 (2004)
151. Khrenov, V., Klapper, M., Koch, M., et al.: Surface functionalized ZnO particles designed for the use in transparent nanocomposites. Macromol. Chem. Phys. **206**, 95–101 (2005)
152. Mallakpour, S., Barati, A.: Efficient preparation of hybrid nanocomposite coatings based on poly(vinyl alcohol) and silane coupling agent modified TiO_2 nanoparticles. Prog. Org. Coat. **71**, 391–398 (2011)

153. Huang, Z.M., Wang, X., Chen, L.: Development of nanofiber reinforced transparent composites. In: Proceedings of the 17th International Conference on Composite Materials, 27–31 July 2009, Edinburgh, UK (2009)
154. Al-Kattan, A., Wichser, A., Vonbank, R., et al.: Characterization of materials released into water from paint containing nano-SiO_2. Chemosphere **119**, 1314–1321 (2015)
155. Al-Kattan, A., Wichser, A., Vonbank, R., et al.: Release of TiO_2 from paints containing pigment-TiO_2 or nano-TiO_2 by weathering. Environ. Sci. Process. Impacts **15**, 2186–2193 (2013)
156. Kemp, T.J., McIntyre, R.A.: Mechanism of action of titanium dioxide pigment in the photodegradation of PVC and other polymers. Prog. React. Kinet. Mech. **26**, 337–374 (2002)
157. Shang, J., Chai, M., Zhu, Y.F.: Solid-phase photocatalytic degradation of polystyrene plastic with TiO_2 as photocatalyst. J. Solid State Chem. **174**, 104–110 (2003)
158. Ammala, A., Hill, A.J., Meakin, P., et al.: Degradation studies of polyolefins incorporating transparent nanoparticulate zinc oxide UV stabilizers. J. Nanopart. Res. **4**, 167–174 (2002)
159. Tsuzuki, T., Wang, X.: Nanoparticles coatings for UV protective textiles. Res. J. Text. Apparel **14**, 9–20 (2010)
160. Ben Amor, S., Baud, G., Jacquet, M., et al.: Photoprotective titania coatings on PET substrates. Surf. Coat. Technol. **102**, 63–72 (1998)
161. Searle, J., Worsely, D.: Titanium dioxide photocatalysed oxidation of plasticizers in thin poly(vinyl chloride) films. Plast. Rubbers Compos. **31**, 329–335 (2002)
162. Giancaterina, S., Ben Amor, S., Baud, G., et al.: Photoprotective ceramic coatings on poly (ether ether ketone). Polymer **43**, 6397–6405 (2002)
163. Bachari, E.M., Ben Amor, S., Baud, G., et al.: Photoprotective zinc oxide coatings on polyethylene terephthalate films. Mater. Sci. Eng. B **79**, 165–174 (2001)
164. Anma, H., Yoshimoto, Y., Tanaka, M., et al.: Preparation of ZnO thin films deposited by plasma chemical vapor deposition for application to ultraviolet-cut coating. Jpn. J. Appl. Phys. **40**, 6099–6103 (2001)
165. Moustaghfir, A., Tomasella, E., Beche, E., et al.: Adhesion of ZnO and Al_2O_3 coatings on polycarbonate. Plasma Process. Polym. **4**, S359–S363 (2007)
166. Moustaghfir, A., Tomasella, E., Ben, Amor S., et al.: Structural and optical studies of ZnO thin films deposited by r.f. magnetron sputtering: influence of annealing. Surf. Coat. Technol. **174–175**, 193–196 (2003)
167. Moustaghfir, A., Tomasella, E., Jacquet, M., et al.: ZnO/Al_2O_3 coatings for the photoprotection of polycarbonate. Thin Solid Films **515**, 662–665 (2006)
168. Mailhot, B., Rivaton, A., Gardette, J.L., et al.: Enhancement of the photoprotection and nanomechanical properties of polycarbonate by deposition of thin ceramic coatings. J. Appl. Phys. Am. Inst. Phys. **99**, 104310–254102 (2006)
169. Mailhot, B., Rivaton, A., Gardette, J.-L., et al.: Enhancement of polycarbonate lifetime and surface properties by deposition of ceramic coatings. In: Proceedings of the 40th International Symposium on Macromolecules, World Polymer Congress, Macro 2004, 4–9 July 2004, Paris, France (2004)
170. Moustaghfir, A., Rivaton, A., Tomasella, E., et al.: Photostabilisation of polycarbonate by ZnO coatings. J. Appl. Polym. Sci. **95**, 380–385 (2005)
171. Collin, S., Bussière, P.-O., Thérias, S., et al.: Physicochemical and mechanical impacts of photo-ageing on bisphenol a polycarbonate. Polym. Degrad. Stab. **97**, 2284–2293 (2012)
172. Won, D.S., Lee, W.G.: Improvement of polycarbonate properties by coating of TiO_2 and SiO_2 thin film. Appl. Chem. Eng. **25**, 41–46 (2014)
173. Rivaton, A., Mailhot, B., Soulestin, J., et al.: Comparison of the photochemical and thermal degradation of bisphenol-A polycarbonate and trimethylcyclohexane-polycarbonate. Polym. Degrad. Stab. **75**, 17–33 (2002)
174. Scheirs, J., Gardette, J.L.: Photo-oxidation and photolysis of poly(ethylene naphthalate). Polym. Degrad. Stab. **56**, 339–350 (1997)

175. Fabbri, P., Leonelli, C., Messori, M., et al.: Improvement of the surface properties of polycarbonate by organic–inorganic hybrid coatings. J. Appl. Polym. Sci. **108**, 1426–1436 (2008)
176. Meng, S., Li, D., Zheng, X., et al.: ZnO photonic crystals with enhanced photocatalytic activity and photostability. J. Mater. Chem. A **1**, 2744–2747 (2013)
177. Morlat, S., Gardette, J.L.: Phototransformation of water-soluble polymers. I: photo-and thermooxidation of poly(ethylene oxide) in solid state. Polymer **42**, 6071–6079 (2001)
178. Schwalm, R.: UV Coatings—Basics, Recent Developments and New Applications. Elsevier Science, Amstardam (2006). ISBN-10: 0444529799; ISBN-13: 978-0444529794
179. Maxwell, A.S., Broughton, W.R., Dean, G., et al.: Review of accelerated ageing methods and lifetime prediction techniques for polymeric materials. NPL Report DEPC MPR 016, Teddington, Middlesex, UK, p. 22 (2005)
180. Karumuri, S., Hamim, S.U., Singh, R.P., et al.: Photodegradation of epoxy polymers. In: Proceedings of the XXV-th AIAA/ASME Oklahoma Symposium, Oklahoma State University, Stillwater, USA (2014)
181. Ramli, J., Jeefferie, A.R., Mahat, M.M.: Effects of UV curing exposure time to the mechanical and physical properties of the epoxy and vinyl ester fiber glass laminates composites. ARPN J. Eng. Appl. Sci. **6**, 104–109 (2011)
182. Adanur, S., Arumugham, Y.: Characteristics of ultraviolet cured glass–epoxy textile composites. Part 1: experimental procedures and testing. J. Ind. Text. **32**, 93–106 (2002)
183. Adanur, S., Arumugham, Y.: Characteristics of ultraviolet cured glass–epoxy composites. Part 2: results and discussion. J. Ind. Text. **32**, 107–118 (2002)
184. Zulfli, N.H.M., Shyang, C.W.: Flexural and morphological properties of epoxy/glass fibre/silane-treated organo-montmorillonite composites. J. Phys. Sci. **21**, 41–50 (2010)
185. Nwabuzor, A.O., Okoli, O.I.: Preliminary assessment of the ultra violet curing of composites manufactured by the resin infusion between double flexible tooling process. Polym. Compos. **27**, 417–424 (2006)
186. Andrzejewska, E., Andrzejewski, M., Jeczalik, J., et al.: Composites of photocured acrylic resins with natural and polymeric fibers. Polimery **54**, 382–385 (2009)
187. Lackey, E., Inamdar, K., Worrel, L., et al.: Demonstration of filament winding using photoinitiated resins. RadTech Rep. **15**, 36–42 (2001)
188. Yuan, Q., Yang, M.B., Mai, Y.W.: Ultraviolet curing of glass fibre reinforced polyester composites. Adv. Compos. Lett. **9**, 341–346 (2000)
189. Endruweit, A., Ruijter, W., Johnson, M.S., et al.: Transmission of ultraviolet light through reinforcement fabrics and its effect on ultraviolet curing of composite laminates. Polym. Compos. **29**, 818–829 (2008)
190. Crivello, J.V., Narayan, R., Sternstein, S.S.: Fabrication and mechanical characterization of glass fiber reinforced UV-cured composites from epoxidized vegetable oils. J. Appl. Polym. Sci. **64**, 2073–2087 (1997)
191. Narayanan, V., Scranton, A.B.: UV curing of composites. RadTech Rep. **11**, 25–30 (1997)
192. Coons, L.S., Rangarajan, B., Godshall, D, et al. Photopolymerization of vinyl ester-glass fiber composites. In: Scranton, A.B., Bowman, C.N., Peiffer, R.W. (eds.) Photopolymerization: Fundamentals and Applications ACS Symposium Series, vol. 673, p. 203 (1997)
193. Jung, T., Koehler, M., Wostratzky, D.: Photocuring of glass-reinforced polyester composites. SAMPE J. **34**, 40–44 (1998)
194. Sitzmann, E.V., Wostratzky, D.A., Bramer, D.A., et al.: Impact of photoinitiators on the light cure response of composites. RadTech Rep. **14**, 28–41 (2000)
195. Lange, J.: Viscoelastic properties and transitions during thermal and UV cure of a methacrylate resin. Polym. Eng. Sci. **39**, 1651–1660 (1999)
196. Shi, W., Ranby, B.: UV curing of composites based on modified unsaturated polyester. J. Appl. Polym. Sci. **51**, 1129–1139 (1994)
197. Mahendran, A.R., Wuzella, G., Aust, N., et al.: Photo crosslinkable modified vegetable oil based resin for wood surface coating application. Prog. Org. Coat. **74**, 697–704 (2012)

198. Fey, T., Muhle, J.: UV-dual cure systems for automotive applications. RadTech Eur. **2**, 741–747 (2003)
199. Takeshige, F., Kinimoto, Y., Torii, M.: Additional heat-curing of light-cured composite resin for inlay restoration. Adhes. Dent. **35**, 59–66 (1995)
200. Ito, K., Shigeru, Y., Kawata, Y., et al.: Photo-initiated and thermal curing of epoxides by the use of photo-base generators bearing acyloxyimino groups. Can. J. Chem. **73**, 1924–1932 (1995)
201. Schultz, J.W., Bhatt, J., Chartoff, R.P., et al.: Polymerization and viscoelastic behavior of networks from a dual-curing liquid crystalline monomer. J. Polym. Sci. B: Polym. Phys. **37**, 1183–1190 (1999)
202. Sipani, V., Scranton, A.B.: Cationic photopolymerization. In: Kroschwitz, J.I. (ed.) Encyclopedia of Polymer Science and Technology, vol. 10, 3rd edn, p. 838. Wiley, New York (2004)
203. Corvey, D.A., Tahaney, S.R., Davenport, J.M.: Mechanical properties of heat-treated composite resin restorative materials. J. Prosthet. Dent. **68**, 458–461 (1992)
204. Kennedy, K.C., Kusy, R.P.: Investigation of dual-staged polymerization and secondary forming of photopultruded, fiber-reinforced, methacrylate-copolymer composites. J. Biomed. Mater. Res. **41**, 549–559 (1998)
205. Kennedy, K.C., Chen, T., Kusy, R.P.: Behavior of photopolymerized silicate-glass-fiber-reinforced dimethacrylate composites subjected to hydrothermal aging. Part 1: steady-state sorption characteristics. J. Mater. Sci. Mater. Med. **9**, 243–248 (1998)
206. Kennedy, K.C., Chen, T., Kusy, R.P.: Behavior of photopolymerized silicate-glass-fiber-reinforced dimethacrylate composites subjected to hydrothermal aging. Part 2: hydrolytic stability of mechanical properties. J. Mater. Sci. Mater. Med. **9**, 651–660 (1998)
207. Kennedy, K.C., Kusy, R.P.: UV-cured pultrusion processing of glass reinforced polymer composites. J. Vinyl Addit. Technol. **1**, 182–186 (1995)
208. Sipani, V., Coretsopoulos, C.N., Scranton, A.B.: Photopolymerization of composite materials. Trends Photochem. Photobiol. **7**, 169–175 (2001)
209. Baikerikar, K.K., Scranton, A.B.: Viscosity characterization of highly filled photopolymerizable liquid encapsulants for microelectronic devices. Polym. Compos. **21**, 297–304 (2000)
210. Baikerikar, K.K., Scranton, A.B.: Photopolymerizable liquid encapsulants for microelectronic devices. Polymer **42**, 431–441 (2001)
211. Sipani, V., Coons, L.S., Rangarajan, B., et al.: Photopolymerization of composites: recent developments in glass-fiber and silica-filled systems. RadTech Rep. **17**, 22–26 (2003)
212. Gupta, A., Ogle, A.A.: Dual curing of carbon fiber reinforced photoresins for rapid prototyping. Polym. Compos. **23**, 1162–1170 (2002)
213. Wang, D.L., Watson, S.S., Sung, L.P., et al.: Effect of TiO_2 pigment type on the UV degradation of filled coatings. J. Coat. Technol. Res. **8**, 19–33 (2011)
214. Kotnarowska, D.: Influence of ultraviolet radiation on erosive resistance of modified epoxy coatings. Solid State Phenom. **113**, 583–588 (2006)
215. Liau, W.B., Tseng, F.P.: The effect of long-term ultraviolet light irradiation on polymer matrix composites. Polym. Compos. **19**, 440–445 (1998)
216. Phelps, H.R., Long, E.R.: Property changes of a graphite/epoxy composite exposed to nonionizing space parameters. J. Compos. Mater. **14**, 334–341 (1980)
217. Phelps, M.L.: In-service inspection methods for graphite-epoxy structures on commercial transport aircraft. NASA Technical Report NTRS 1981, document ID 19820004269. http://ntrs.nasa.gov/archive/nasa/casi.ntrs.nasa.gov/19820004269.pdf.
218. Giori, C., Yamauchi, T.: Effects of ultraviolet and electron radiations on graphite-reinforced polysulfone and epoxy resins. J. Appl. Polym. Sci. **29**, 237–249 (1984)
219. Kumar, B.G., Singh, R.P., Nakamura, T.: Degradation of carbon fiber-reinforced epoxy composites by ultraviolet radiation and condensation. J. Compos. Mater. **36**, 2713–2733 (2002)

220. Tipton, B.: Prevention of environmentally induced degradation of carbon/epoxy composite material via implementation of a polymer based coating system. M.Sc. Dissertation, University of Central Florida, Orlando, Florida, USA (2008)
221. Pottavathri, S.B., Nair, R., Asmatulu, R.: Effect of in-plane fiber tow waviness upon the tensile strength characteristics of fiber reinforced composites of carbon/epoxy AS4/3501-6. In: Proceedings of the ASME 2015 International Mechanical Engineering Congress and Exposition (V009T12A059-V009T12A059), American Society of Mechanical Engineers (2015)
222. Al-Turaif, H.A.: Surface morphology and chemistry of epoxy-based coatings after exposure to ultraviolet radiation. Prog. Org. Coat. **76**, 677–681 (2013)
223. Bassewitz, H.V.: Non semiconducting materials. In: Proceedings of the International Conference on Evaluation of the Effect of the Space Environment on Materials, Toulouse, France, p. 525 (1974)
224. Hancox, N.L., Minty, D.C.: Materials qualification and property measurements of carbon fiber reinforced composites for space use. Br. Interplanet. Soc. J. **30**, 391–399 (1977)
225. Zainuddin, S., Hosur, M.V., Tcherbi-Narteh, A., et al.: Study on the degradation of dynamic compressive properties of nanophased E-glass/epoxy composites by ultraviolet (UV) radiation and condensation. In: Proceedings of the Society of Experimental Mechanics Annual Conference, Albuquerque, New Mexico, USA (2009)
226. Zainuddin, S., Hosur, M.V., Zhou, Y., et al.: Durability study of neat/nanophased GFRP composites subjected to different environmental conditioning. Mater. Sci. Eng. A **527**, 3091–3099 (2010)
227. Jeong, Y.G., An, J.E.: UV-cured epoxy/graphene nanocomposite films: preparation, structure and electric heating performance. Polym. Int. **63**, 1895–1901 (2014)
228. Chibac, A.L., Melinte, V., Buruiana, T., et al.: Preparation of photocrosslinked sol-gel composites based on urethane-acrylic matrix, silsesquioxane sequences, TiO_2, and Ag/Au nanoparticles for use in photocatalytic applications. J. Polym. Sci. A: Polym. Chem. **53**, 1189–1204 (2015)
229. Kim, D., Jeon, K., Lee, Y., et al.: Preparation and characterization of UV-cured polyurethane acrylate/ZnO nanocomposite films based on surface modified ZnO. Prog. Org. Coat. **74**, 435–442 (2012)
230. Malucelli, G.: Hybrid organic/inorganic coatings yhrough dual-cure processes: state of the art and perspectives. Coatings **6**, 10 (2016). doi:10.3390/coatings6010010
231. Liu, K., Miao, S., Su, Z., et al.: Castor oil-based waterborne polyurethanes with tunable properties and excellent biocompatibility. J. Lipid Sci. Technol, Eur (2016). doi:10.1002/ejlt.201500595
232. Fouassier, J.P., Rabek, J.F.: Radiation Curing in Polymer Science and Technology, vol. I. Elsevier, London (1993)
233. Ceccia, S., Turcato, E.A., Maffettone, P.L., et al.: Nanocomposite UV-cured coatings: organoclay intercalation by an epoxy resin. Prog. Org. Coat. **63**, 110–115 (2008)
234. Benfarhi, S., Decker, C., Keller, L., et al.: Synthesis of clay nanocomposite materials by light-induced crosslinking polymerization. Eur. Polym. J. **40**, 493–501 (2004)
235. Uhl, F.M., Webster, D.C., Davuluri, S.P., et al.: UV curable epoxy acrylate–clay nanocomposites. Eur. Polym. J. **42**, 2596–2605 (2006)
236. Decker, C., Keller, L., Benfarhi, S., et al.: Synthesis of composite materials by photopolymerization. Polym. Prepr. **45**, 18–19 (2004)
237. Bongiovanni, R., Mazza, D., Ronchetti, S., et al.: The influence of water on the intercalation of epoxy monomers in Na-montmorillonite. J. Colloid Interface Sci. **296**, 515–519 (2006)
238. Keller, L., Decker, C., Zahouily, K., et al.: Synthesis of polymer nanocomposites by UV-curing of organoclay-acrylic resins. Polymer **45**, 7437–7447 (2004)
239. Malucelli, G., Bongiovanni, R., Sangermano, M., et al.: Preparation and characterization of UV-cured epoxy nanocomposites based on o-montmorillonite modified with maleinized liquid polybutadienes. Polymer **48**(7000–7007), 240 (2007)

240. Di Gianni, A., Amerio, E., Monticelli, O., et al.: Preparation of polymer/clay mineral nanocomposites *via* dispersion of silylated montmorillonite in a UV curable epoxy matrix. Appl. Clay Sci. **42**, 116–124 (2008)
241. Herrera, N.N., Letoffe, J.M., Putaux, J.M., et al.: Aqueous dispersions of silane-functionalized laponite clay platelets. A first step toward the elaboration of water-based polymer/clay nanocomposites. Langmuir **20**, 1564–1571 (2004)
242. Park, K.W., Jeong, S.Y., Kwon, O.Y.: Interlamellar silylation of H-kenyaite with 3-aminopropyl-triethoxysilane. Appl. Clay Sci. **27**, 21–27 (2004)
243. Park, M., Shim, I.K., Jung, E.Y., et al.: Modification of external surface of laponite by silane grafting. J. Phys. Chem. Solids **65**, 499–501 (2004)
244. Woo, R.S.C., Chen, Y., Zhu, H., et al.: Environmental degradation of epoxy–organoclay nanocomposites due to UV exposure. Part I: photo-degradation. Compos. Sci. Technol. **67**, 3448–3456 (2007)
245. Woo, R.S.C., Zhu, H., Leung, C.K.Y., et al.: Environmental degradation of epoxy-organoclay nanocomposites due to UV exposure: part II residual mechanical properties. Compos. Sci. Technol. **68**, 2149–2155 (2008)
246. Işin, D., Kayaman-Apohan, N., Güngör, A.: Preparation and characterization of UV-curable epoxy/silica nanocomposite coatings. Prog. Org. Coat. **65**, 477–483 (2009)
247. Pellegrin, B., Nguyen, T., Mermet, L., et al.: Degradation and nanoparticle release of epoxy-nanosilica composites exposed to UV radiation. NSTI-Nanotech 1:94–97 (2009). ISBM:978-1-4398-1782-7
248. Rabb, S.A., Yu, L.L., Bernard, C., et al.: An analythical mathod to quantify silica nanoparticles accumulated on the surface of polymer nanocomposites exposed to UV radiation. NSTI-Nanotech 1:728–731 (2010). ISBN:978-1-4398-3401-5
249. Nguyen, T., Pellegrin, B., Bernard, C., et al.: Fate of nanoparticles during life cycle of polymer nanocomposites. J. Phys.: Conf. Ser. **304**, 012060 (2011). doi:10.1088/1742-6596/304/1/012060
250. Nguyen, T., Pellegrin, B., Bernard, C., et al.: Characterization of surface accumulation and release of nanosilica during irradiation of polymer nanocomposites by ultraviolet light. J. Nanosci. Nanotechnol. **12**, 6202–6215 (2012)
251. Gorham, J., Nguyen, T., Bernard, C., et al.: Photo-induced surface transformations of silica nanocomposites. Surf. Interface Anal. **44**, 1572–1581 (2012)
252. Froggett, S.J., Clancy, S.F., Boverhof, D.R., et al.: A review and perspective of existing research on the release of nanomaterials from solid nanocomposites. Part. Fibre Toxicol. 11:17 (2014). http://www.particleandfibretoxicology.com/content/11/1/17
253. Sung, L., Stanley, D., Gorham, J.M., et al.: A quantitative study of nanoparticle release from nanocoatings exposed to UV radiation. J. Coat. Technol. Res. **12**, 121–135 (2015)
254. Schlagenhauf, L., Nuesch, F., Wang, J.: Release of carbon nanotubes from polymer nanocomposites. Fibers **2**, 108–127 (2014). doi:10.3390/fib2020108
255. Petersen, E.J., Lam, T., Gorham, J.M., et al.: Methods to assess the impact of UV irradiation on the surface chemistry and structure of multiwall carbon nanotube epoxy nanocomposites. Carbon **69**, 194–205 (2014)
256. Chin, J., Byrd, E., Embree, N., et al.: Accelerated UV weathering device based on integrating sphere technology. Rev. Sci. Instrum. **75**, 4951–4959 (2004)
257. Nguyen, T., Pellegrin, B., Mermet, L., et al.: Network aggregation of CNTs at the surface of epoxy/MWCNT composite exposed to UV radiation. NSTI-Nanotech 1:90–93 (2009). ISBN:978-1-4398-1782-7
258. Asmatulu, R., Mahmud, G.A., Hille, C., et al.: Effects of UV degradation on surface hydrophobicity, crack, and thickness of MWCNT-based nanocomposite coatings. Prog. Org. Coat. **72**, 553–561 (2011)
259. Asmatulu, R., Khan, W.S., Ghaddar, M.H.: Investigating the surface free energy changes of UV-exposed nanocomposite coatings on carbon fiber- and glass fiber-reinforced composites. J. Res. Appl. Sci. **2**, 119–130 (2015)

260. Orlov, A., Ramakrishnan, G., Ging, J., et al.: Evaluating safety and stability of CNT nanocomposites exposed to environmental conditions. In: Proceedings of the Technical Proceedings of the 2012 NSTI Nanotechnology Conference and Expo, Santa Clara, CA, USA, pp. 335–337 (2012)
261. Fiedler, B., Gojny, F.H., Wichmann, M.H.G., et al.: Fundamental aspects of nano-reinforced composites. Compos. Sci. Technol. **66**, 3115–3125 (2006)
262. Ging, J., Tejerina-Anton, R., Ramakrishnan, G., et al.: Development of a conceptual framework for evaluation of nanomaterials release from nanocomposites: environmental and toxicological implications. Sci. Total Environ. **473–474**, 9–19 (2014)
263. Bastani, S., Darani, M.K.: Ch. 8. Graphene-based UV-curable nanocomposite coatings. In: Milne, W.I., Cole, M. (eds.) Carbon Nanotechnology. Publisher One Central Press, pp. 187–209 (2015). ISBN eBook: 978-1-910086-11-7. http://www.onecentralpress.com/carbon-nanotechnology/
264. Martín-Gallego, M., Verdejo, R., Lopez-Manchado, M., et al.: Epoxy-graphene UV-cured nanocomposites. Polymer **52**, 4664–4669 (2011)
265. Guo, Y., Bao, C., Song, L., et al.: Radiation cured epoxy acrylate composites based on graphene, graphite oxide and functionalized graphite oxide with enhanced properties. J. Nanosci. Nanotechnol. **12**, 1776–1791 (2012)
266. Das, T.K., Prusty, S.: Graphene-based polymer composites and their applications. Polym.-Plast. Technol. Eng. **52**, 319–331 (2013)
267. Martin-Gallego, M., Lopez-Manchado, M.-A., Calza, P., et al.: Gold-functionalized graphene as conductive filler in UV-curable epoxy resin. J. Mater. Sci. **50**, 605–610 (2015)
268. Chang, K.C., Hsu, M.H., Lu, H.I., et al.: Room-temperature cured hydrophobic epoxy/graphene composites as corrosion inhibitors for cold-rolled steel. Carbon **66**, 144–153 (2014)
269. Peng, C.W., Chang, K.C., Weng, C.J., et al.: UV-curable nanocasting technique to prepare bio-mimetic super-hydrophobic non-fluorinated polymeric surfaces for advanced anticorrosive coatings. Polym. Chem. **4**, 926–932 (2013)
270. Lopes da Cruz, D.T., Ferreira Avila, A.: Mechanical characterization of nano-modified single lap bonded joints: the aging effect investigation. In: The Proceedings of the IV International Symposium on Solid Mechanics—MecSol 2013, Porto Alegre, Brazil (2013). http://www.ufrgs.br/mecsol2013/articles/mecsol2013-015.pdf
271. Sangermano, M., Voit, B., Sordo, F., et al.: High refractive index transparent coatings obtained *via* UV/thermal dual-cure process. Polymer **49**, 2018–2022 (2008)
272. Copuroglu, M., Sen, M.: Synthesis and characterization of a Zr-containing silicate-based epoxy-functional polymer nanocomposite system. Polym. Eng. Sci. **55**, 792–798 (2015)
273. Bi, Y.T., Li, Z.J., Liang, W.: Preparation and characterization of epoxy/SiO_2 nanocomposites by cationic photopolymerization and sol–gel process. Polym. Adv. Technol. **25**, 173–178 (2014)
274. Roppolo, I., Messori, M., Perruchas, S., et al.: Multifunctional luminescent organic/inorganic hybrid films. Macromol. Mater. Eng. **297**, 680–688 (2012)
275. Malucelli, G., Priola, A., Sangermano, M., et al.: Hybrid nanocomposites containing silica and PEO segments: preparation through dual-curing process and characterization. Polymer **46**, 2872–2879 (2005)
276. Sangermano, M., Bongiovanni, R., Longhin, M., et al.: Hybrid organic/inorganic UV-cured acrylic films with hydrophobic surface properties. Macromol. Mater. Eng. **294**, 525–531 (2009)
277. Sangermano, M., Amerio, E., Epicoco, P., et al.: Preparation and characterization of hybrid nanocomposite coatings by cationic UV-curing and the sol–gel process of a vinyl ether based system. Macromol. Mater. Eng. **292**, 634–640 (2007)
278. Amerio, E., Sangermano, M., Malucelli, G., et al.: Preparation and characterization of hyperbranched polymer/silica hybrid nanocoatings by dual-curing process. Macromol. Mater. Eng. **291**, 1287–1292 (2006)

279. Basturk, E., Inan, T., Gungora, A.: Flame retardant UV-curable acrylated epoxidized soybean oil based organic–inorganic hybrid coating. Prog. Org. Coat. **76**, 985–992 (2013)
280. Tsujimoto, T., Uyama, H., Kobayashi, S.: Synthesis of high-performance green nanocomposites from renewable natural oils. Polym. Degrad. Stab. **95**, 1399–1405 (2010)
281. Wold, C.R., Soucek, M.D.: Viscoelastic and thermal properties of linseed oil-based ceramer coatings. Macromol. Chem. Phys. **201**, 382–392 (2000)
282. Deffar, D., Teng, G.: Soucek M D. Comparison of titanium-oxo-clusters derived from sol-gel precursors with TiO_2 nanoparticles in drying oil based ceramer coatings. Macromol. Mater. Eng. **286**, 204–215 (2001)
283. Mosiewicki, M.A., Aranguren, M.I.: A short review on novel bocomposites based on plant oil precursors. Eur. Polym. J. **49**, 1243–1256 (2013)

Stabilization of Polymers Against Photodegradation

Eduard Marius Lungulescu and Traian Zaharescu

1 Introduction

Due to their organic nature the most of polymers currently used in various applications in different areas of industry or household are characterized by a certain degree of stability on the environmental natural conditions leading to their degradation. By this reason the functional properties of these aged materials become improper for long term operation. The accumulation of oxidation products by various chemical reactions during light absorption quantifies this instability progress.

The absorption of natural UV radiation (from solar light) or artificial UV beam (from different UV light sources) generates excited electrons on unstable high energy entities initiating physical and chemical damage processes with the consequences on the structure changing of polymeric materials, i.e. photodegradation [1]. The photodegradation process is based on the cleavage of polymer chains followed by the formation of radicals and the building up of lower molecular weight compounds, the crosslinking (the formation of high molecular weight compounds) or the increase in unsaturation level and photo oxidation.

Certain polymeric structures, such as the configurations containing benzene rings, double bonds (C=C, C=O), hydroxyl or carboxyl groups may themselves absorb UV radiation [2]. The conjugated structures show also remarkable efficiency

E.M. Lungulescu (✉) · T. Zaharescu
INCDIE ICPE CA, 313 Splaiul Unirii, P.O. Box 139, Bucharest, Romania
e-mail: marius.lungulescu@icpe-ca.ro; lungulescu_eduard@yahoo.com

T. Zaharescu
e-mail: traian.zaharescu@icpe-ca.ro; traian_zaharescu@yahoo.com

D. Rosu and Visakh P. M. (eds.), *Photochemical Behavior of Multicomponent Polymeric-based Materials*, Advanced Structured Materials 26,
DOI 10.1007/978-3-319-25196-7_6

in the absorption of UV radiation. The groups of this structural type known as chromospheres groups often induce an advanced tendency of polymers to achieve photo-induced transformations. The photodegradation of saturated polymers, such as polyethylene or polypropylene, which do not contain such UV sensitive groups in their structure, is explained by the photo-sensitizing action of impurities—double bonds, hydroperoxides, traces of catalyst, the structural irregularities (like chain branches) [1]. These impurities appear during the synthesis process of polymers. Subsequently, the processing, storage and service conditions add new and different imperfections that increase the susceptibility to photodegradation of polymeric materials.

The combined effect of sunlight and oxygen occurred during outdoor service is accompanied by altering useful properties of polymeric material. The photodegradation effect can be observed as discoloration, the decrease in molecular weight due to the chain scission, the crosslinking of susceptible substrate, if formulation allows it, and the hardening of material, the formation of oxygenated chemical groups (hydroxyl, carbonyl, carboxyl, etc.) leading to the deterioration of mechanical and dielectric properties [3–10].

2 Mechanisms of Polymer Photodegradation

The mechanisms of photochemical degradation depend peculiarly on the type of polymer, as well as on the specificity of environmental factors. Usually, whereas the applications of polymeric materials run in the presence of air and of atmospheric oxygen, the polymer degradation lead to a variety of physical and chemical effects. This damaging process is practically an oxidative photodegradation (photo-oxidation).

The main macroscopic effects induced into polymeric material by the photodegradation processes [11, 12] are related to:

- *mechanical properties*. The modification of coloration accompanies the material damage during photodegradation. For example, polystyrene exposed to sunlight turns to yellow, polyvinyl chloride becomes dark yellowish or black reddish, the change degree being proportional with the exposure time. The discoloration is often accompanied by the formation of cracks or brittle regions on material surface. In general, the mechanical properties of polymeric materials are declined as the result of photo energy absorption;
- *chemical transformations*. Besides chain scission and modification in material structuring, photodegradation produces oxygen-containing functional groups such as: ketones, carboxylic acids, peroxides, alcohols, etc.;
- *electrical properties*. Photochemical degradation alters the main electric properties such as dielectric constant, volume resistivity, breakdown tension or any other dipole concentration dependent in the sense of their deterioration by the accumulation of polar groups [13].

Table 1 The characteristics of some single chemical bond in polymers

Chemical bond	Binding energy (kcal mol^{-1})	Wavelength (μm)
O–H	110.6	259
C–F	105.4	272
C–H	98.8	290
N–H	93.4	306
C–O	84.0	340
C–C	83.1	342
C–Cl	78.5	364
C–N	69.3	410

The photochemical transformations require certain transferred radiation energy which transports equal or greater energy than the dissociation energy of molecules for the efficient absorption of solar radiation. The UV radiation processes such suitable photonic energy [14]. When a molecule absorbs a photon it is raised onto photo-activated energetic level and, consequently, it passes into an electronic excitation state. The energy of incidental UV radiation is higher enough to dissociate some chemical bonds from polymer structure, whose main features are listed in Table 1.

The energy retained polymeric molecules in the excited state can cause several chemical reactions called primary photochemical reactions i.e. free radical formation, ionization, cyclization, addition, intermolecular regrouping and fragmentation followed by a number of secondary reactions through which the former intermediates are decayed. In the case of polymers with mobile hydrogen atoms the transferred UV radiation energy is consumed for the expelling such types of hydrogen atoms and the formation of specific polymeric macroradicals which create new molecular structures.

During the oxidative degradation of the most industrial engineering polymers it was found that the specific inhomogeneities including ketones or other products containing carbonyl groups, hydroperoxide, catalyst residues become the start points for the reactions with primary fragments produced by damage initiators. The photolysis of polymers occurred in processed materials increases the sensitivity to UV radiation, the initiation threshold and the acceleration of the photodegradation process depend on the spectral structure. In Table 2 the wavelengths of UV radiation to which the sensitivity of polymers attains maximum are presented.

Basically, the photooxidation mechanism involves the creation of some excited molecules species by the interactions between photons and irradiated polymeric material, as well as with the oxygen molecules, when reactive excited species appear, such as oxygen singlet. The photochemical reactions are chained reactions of free radicals accompanied by different photo-induced physical processes, such as: fluorescence, internal conversion of exciting energy or the intermolecular energy transfer (phosphorescence). The oxygen singlet, $^1O_2$, is involved in many reactions due to its high chemical reactivity. The nature of the formed chemical species depends on the polymeric structure, environmental conditions, temperature, etc.

The photooxidation mechanism of polymers is a process similar to thermooxidation process [15]. The difference between the two oxidative degradation

Table 2 UV wavelengths to which the polymers are the most sensitive

Polymer	Wavelength (μm)
Polyester	325
Polystyrene	318.5
Polyethylene	300–320
Polypropylene	310–370
Polycarbonate	295–305 and 350–360
Polyvinyl chloride	310–320
EVA copolymer	322 and 364
Polymethylmethacrylate	290–315
Polyamide	400
Styrene-acrylonitrile copolymer	290 and 325
Polyvinyl acetate (film)	<280

processes consists in the difference of initiation stage, when photooxidation is much faster due to the photo activation and the higher volume density of intermediates.

Photooxidation takes place at ambient temperatures and it can also be produced by different artificial light sources.

The rate of photooxidation depends on the nature of polymer, on the polymeric layer thickness, on oxygen solubility and diffusion rate, as well as on the crystallinity degree or on the branching degree of macromolecular chain [11] and on the level of irradiance [16].

A scheme describing photooxidation processes accomplished in polymers can be represented by a certain sequence of stages [2, 15, 17]:

(a) *Chain initiation*

- Primary

$$\mathrm{RH + O_2 \rightarrow (RH^+ - O_2) \rightarrow R\cdot (free\,radical)}$$

- *secondary* (from polymers with irregular structure);

$$\left.\begin{array}{l}\text{Carbonyl groups } >\mathrm{C{=}O}\\ \text{Hydroperoxides ROOH}\\ \text{Catalyst residues}\\ \text{Charge transfer complex } (\mathrm{RH, O_2})\end{array}\right\} \xrightarrow[\mathrm{Me^{++},\ Me^{+++}}]{h\nu} \text{free radicals}(\mathrm{R^\bullet, RO^\bullet, HO^\bullet}) + \text{products}$$

(b) *Chain propagation*

- primary:

$$\mathrm{R\cdot + O_2 \xrightarrow{h\nu} RO_2\cdot}$$

$$\mathrm{RO_2\cdot + RH \xrightarrow{h\nu} ROOH + R\cdot}$$

– chain branching:

$$ROOH \xrightarrow{h\nu} RO\cdot + \cdot OH$$

$$2ROOH \xrightarrow{h\nu} RO_2\cdot + RO\cdot + H_2O$$

$$RO^{\cdot} + RH \xrightarrow{h\nu} ROH + R\cdot$$

$$HO^{\cdot} + RH \xrightarrow{h\nu} H_2O + R\cdot$$

(c) *Chain Termination*

$$ROO\cdot + RO_2\cdot \rightarrow \text{non radical products}$$

$$\left.\begin{array}{l} ROO\cdot + \text{Radical} \rightarrow \\ ROO\cdot + \text{Inhibitor} \rightarrow \end{array}\right\}\text{Inactive products}$$

$$R\cdot + R\cdot \rightarrow \text{crosslinked inactive products}$$

During to the UV exposure of polyolefin the most important chromophores species are the carbonyl groups formed in presence of air [18]. These species contribute to the initiation of photodegradation reactions, produce excite states of singlet and triplet and initiate Norrish type I reactions (1), which lead to the scission of the main polymeric chain [19]

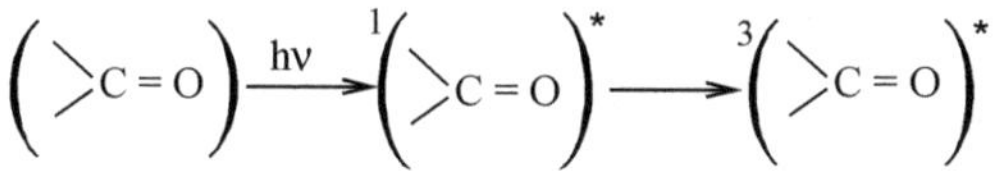

(1)

In the case of polyethylene it has been shown that about two-thirds of carbonyl species are involved in a Norrish type II reaction (2) without radical intermediates and about one third of those gives a Norrish type I reaction [19]

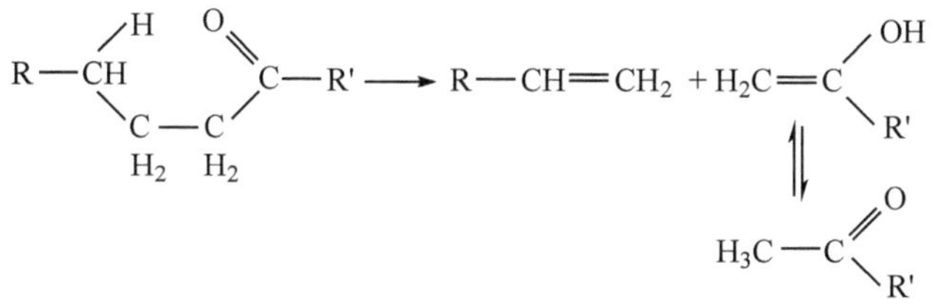

(2)

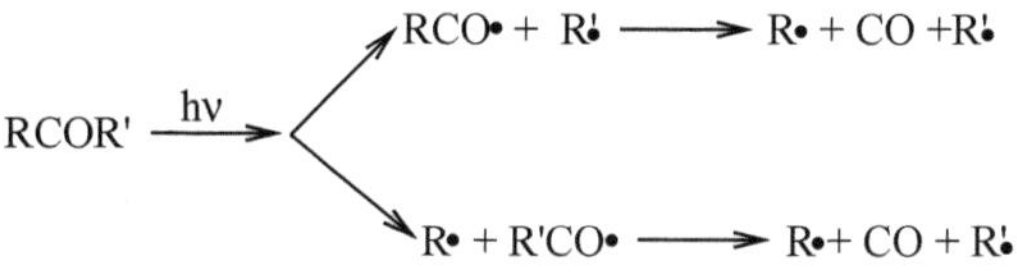

(3)

Besides carbonyl groups, hydroperoxides formed during the polymer processing play an important role in the increasing polymer sensibility on the UV radiation [20]. Other impurities such some traces of catalyst (titanium compounds) decompose the hydroperoxides and affect the thermal and photostability of polymer over time [21].

In the case of polymers the photostabilization methods are related to the mechanisms that describe the harmful effects of photonic UV radiation energy on polymer structures and they illustrate how additives (photostabilizers or antioxidants) can achieved the protection against oxidation.

The highest efficiency against the photooxidation process is obtained, if the used stabilizers fulfil certain conditions:

- To absorb efficient UV light and to present good dissipation properties of energy;
- To be stable to UV light action;
- To present low optical absorption on the visible domain.

The most known photostabilizers can be classified by their action mechanisms: UV reflectors, UV absorbers, quenchers of excited states, hydroperoxide decomposers, radical scavengers [22].

3 Photostabilization

3.1 Screening of UV Radiation

The light screening means the protection against the degrading action of solar radiation by the preventing penetration of UV radiation into polymer, the degradation process being limited only on the surface of polymer. An efficient screening of light can be carried out by using some protective coatings (compound with cellulose materials, appropriate paints and lacquerers, a. s. o., but, in time, themselves are subjected to photooxidative degradation).

Another method of photoscreening is the use of some substances to reflect UV radiation generically called UV light reflectors. The action mechanism of such substances consists in the reflection of UV light. The incidental light interacts with polymer macromolecules only in a lower proportion. The most important method used to reflect the light are: polymer surface metallization [23], coating with pigmented fluoropolymers or dispersing TiO_2 particles [24] or aluminium which can integrally reflect UV light. The pigmentation is limited by many factors such as: product colouring, dispersion issues, negative influences on functional electrical and mechanical properties. The efficiency of pigments in the protection of polymers against the progress in photodamage depends also on the nature of substrate. The iron oxide can be used as the protector polyolefin playing the role of light screener [25], but, in the same time, it catalyses the decomposition of polymeric substrate, like polyvinyl chloride exposed to UV radiations [26]. TiO_2 could have a photocatalytic effect for fibres, polyethylene insulation materials [27] and films of polypropylene [28] or epoxy resins [29]. A high amount of this TiO_2 pigment accelerates the degradation of ABS polymer.

Another method to screen the UV light is the use of ***UV absorbers***. The photostabilizers from this category can be fluorescent compounds which absorb the UV light and re-emit immediately these photons at higher wavelengths as fluorescent light. Compounds such as cumarins, naphtalimide, benzimidazolyl, diaminostilbenedisulphonate are efficient photoprotectors as UV absorbers, even at concentrations lower than 1 % [30].

One of the most efficient stabilizers as UV absorber is carbon black. The disadvantage of using black carbon is its disadvantageous color due which polymeric materials becomes nontransparent. Additionally, the black carbon generates heat which causes a thermal degradation of polymer [31, 32].

Besides organic UV absorbers already presented, inorganic compounds such as TiO_2, CeO_2, ZnO, could be used as UV absorbers [33–35].

The protection mechanism involving such kind of substances used as UV absorbers is followed by many action systems:

(a) *Intersystem crossing molecules*

In these systems the stabilizer absorbing the UV light is excited to triplet state T_1 (π-π*) which either produces a photochemical rearrangement or becomes relatively inert in respect to photochemical reactions. The efficient compounds for this system can be phenol ester of benzoic acid, salicylates, butyrophenones and their substitution compounds.

Some examples of the rearrangement systems under photoexposure are presented in the following reactions:

- The rearrangement of the esters of 4-hydroxy-3-5-di-*t*-butyl benzoic acid to *o*-hydroxybenzophenone:

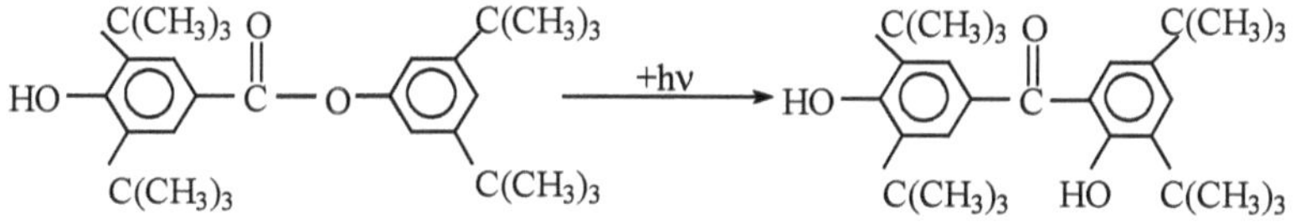

(4)

These esters have a good efficiency to the UV light stabilization of polyolefin.

- The rearrangement of the phenyl salicylate onto 2,2′-dihydroxybenzophenone and 2,4′-dihydroxybenzophenone:

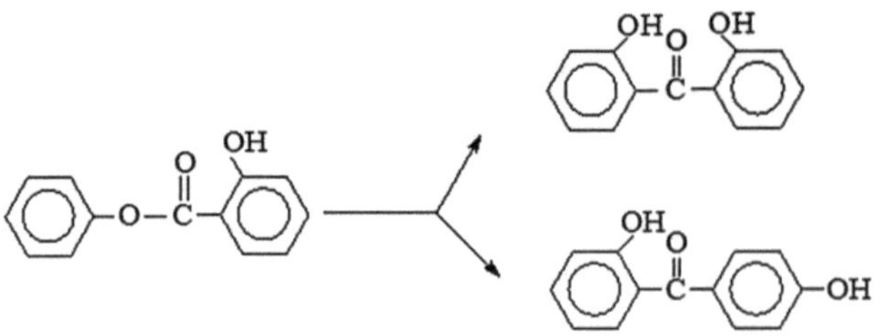

(5)

The product of 2,6-dimethylphenyl salicylate is successfully used to stabilize polypropylene. The addition of *p-t*-octylphenyl salicylate to polyethylene significantly increases its resistance against aging [36].

(b) *Internal conversions*

In these circumstances, the energy of absorbed light is converted into vibration energy by a radial process without spin changing i.e. S_1–S_0, S_2–S_1, the returning to the nonionic stage is possible with a presumable transformation of radiation energy into heat. The frequently used photostabilizers are *o*-hydroxybenzophenone compounds, whose photostabilization mechanism is based on a rapid tautomerization of excited states.

Another class of compounds is represented by benzotriazoles i.e. *o*-hydroxyphenylbenzotriazoles, which have a photostabilisation mechanism similar to *o*-hydroxyphenones, by rapid tautomerization of the excited states.

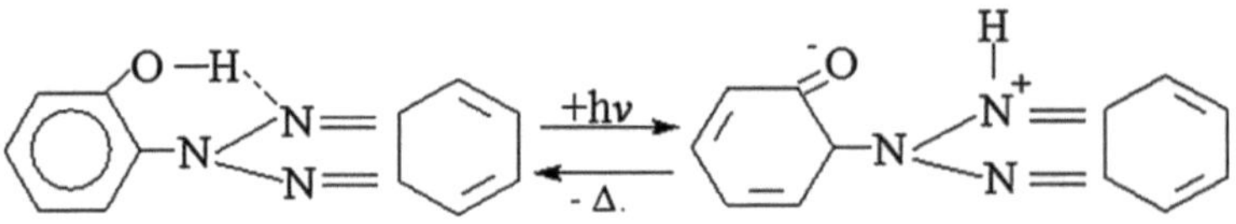

(6)

- S-triazine, such hydroxyphenyl-*s*-triazines, act by a mechanism similar to *o*-hydroxyphenones with the following considerations:
- compounds with high number of *o*-hydroxyphenile groups provide a better protection and a good absorption of UV radiations with higher wavelengths;
- various substituents, which may reduce the basicity of the *s*-triazine ring, can increase the light resistance of the compounds.

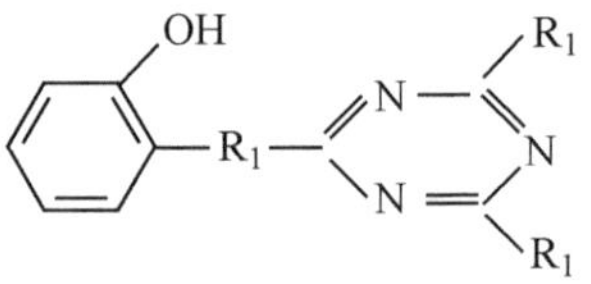

(7)

(c) *Deactivation*

This process is based on the deactivation of the different energy stages for polymer (singlet, triplet) produced by the photonic energy during solar radiation exposures. On this way, the destructive action of those energies on the polymer structure is avoided. This manner of energetic decay can be considered as the energy donation when an excitation of a stabilizer molecule (quencher) is produced:

$$P \xrightarrow{h\nu} P_s \longrightarrow P_t \xrightarrow{S} P + S_t \longrightarrow P + S + \text{heat}$$

where

- P is a polymer molecule which contains chromophore groups such as [C=O];
- S is an acceptor stabilizer molecule which can have excited energy levels;
- S and T are singlet excitation stages and triplet excitation states, respectively
- The stabilizer should possess the capacity to return to the underground state $S_t \rightarrow S$.

(d) *Formation of a complex in excited stages by photo physical processes (fluorescence, internal conversion, etc.)*

$$P^* + S \rightarrow [P + S]^*$$

The stabilizers used for these mechanisms are:

- *Organometallic compounds* such as: copper acetylacetonate, zinc diethylthiocarbamate, nickel dibutyldithiocarbamate, nickel acetylacetonate, metal chelates, and many others,
- *hindered amines*
- *derivatives of piperidine*
- *derivatives offerrocene* such as benzyl ferrocene, *o*-hydroxybenzoilferoccene, etc.

3.2 Quenchers of the Excited States

The quenchers of the excited states accept energy from an excited molecule polymer before this excess of energy produces its degradation. The energy transfer can be achieved by dipole-dipole interaction (Förster mechanism [37]). Energy quenchers must have excited stages with energies lower than the excited state of polymer molecule. Once energy is transferred, the quencher returns to its ground state by emissions of fluorescence, phosphorescence and heat [38].

As additives belonging to this category of stabilizers, organometallic compounds of nickel [39] can be used such as nickel dibuthyldithiocarbamate, nickel-2,2,2-thio-bis (4-octophenolate)-*n*-buthylamine, as well as some phenolic stabilizers [40].

In principle, the difference between the UV absorbers and energy quenchers consists of the protection of absorbers in polymer by absorbing UV light, while the quenchers deactivate the excited states of the polymer molecules. In reality, it was shown that the absorbers can function also as quenchers [1].

3.3 *Hydroperoxides Decomposers*

Hydroperoxides represent, as it is generally accepted, ones of the most reactive species in the polymer photodegradation [41]. They can be induced during the processing of polymer, as well as during the subsequent exposure of polymers to light and heat in the presence of air. The present transitional metals accelerate the decomposition by radicalic process of the accumulation of hydroperoxides. In the case of polyolefin, the hydroperoxides photolysis leads to the formation of some carbonyl compounds and/or alcoxy radicals.

The efficient compounds in respect to hydroperoxide decomposers include nickel and sulphur—containing complexes [42]. Such substances can also act as UV absorbers and quenchers of excited states [43].

3.4 *Radical Scavengers*

3.4.1 Hindered amine (hindered amine light stabilizers, HALS)

The hindered amine represent the most recently type of UV light stabilizers. These products have been industrial produced since 1980, the first stabilizer being HALS-1 (Tinuvin 770, Sanol LS 770) with the chemical formula bis (2,2,6,6-tetramethyl-4-piperidinyl) sebacate. Other important HALS compounds are: bis (1,2,2,6,6-pentamethyl-4-piperinidyl) sebacate (known as Tinuvin 292), the polymeric form of the dimethyl ester of butyric acidwith 4-hydroxy-2,2,6,6-tetramethyl pyridine (HALS-2), poly[[6-[1,1,3,3-tetramethylbutyl) amino]-1,6-hexandiyl[(2,2,6,6-tetramethyl-4-piperidinyl)imino]] (HALS 3).

The HALS compounds are largely used for polyolefin photostabilisation due to their competitive costs and high performances, which permit the utilization of lower amounts than UV absorbers such as benzophenone and benzotriazole compounds. The mixtures of HALS compounds with benzophenones present some synergic photostabilization effects [11], being reported also an antirad efficiency of them [44, 45].

Other hindered amines used with good results in the stabilization promotion of polyolefin, polyurethanes, polystyrene, ABS, acrylonitrile styrene copolymers, are derivatives of piperidine such as: 2,2,6,6-tetramethylpiperidine, 2,2,6,6-tetramethyl-4-hydroxy-phosphonopiperineor by stable free radicals such as: 4-[cyan (phenylmethylene)-2,2,6,6-tetramethyl piperidinoxil], 4-(ureidoimino)-2,2,6,6,-tetramethyl-piperidinoxil, 2′,2′,6′,6′ tetramethylspiro(benzimidazoline)-2,4-piperi dinoxil.

Taking into account the experimental results obtained with the hindered amines, it was possible to formulate appropriate stabilization mechanisms. They were reported as several academic approaches. Unfortunately, the mechanism of photoprotection is not fully elucidated.

The photostabilization mechanism of hindered amines consists of the scavenging free radicals [46, 47], i.e. through Denisov cycle. It was observed that these compounds absorb UV light beams with wavelengths lower than 290 nm. Several compounds such as tetramethylpiperidines act either as destroyers of free radicals reacting directly with peroxyl units or as a catalyst which forms nitroxyl radical under the action of UV light by the regeneration process:

$$>N{-}O^{\bullet} \xrightarrow{R^{\bullet}} >N{-}O{-}R \longrightarrow RO_2^{\bullet} + >N{-}O^{\bullet} \qquad (8)$$

It has been reported that other photooxidation mechanisms are also reliable concerning the complexation of transitional metals, the complexation and decomposition of hydroperoxides, the quenching of the polymer-oxygen charge transfer complexes [38].

In order to assess the efficiency of a light stabilizer in its coupling with a phenolic antioxidant, multifunctional hindered amine light stabilizer or some complexes of tetramethylpiperidine with nickel were synthesized [48]. These products were successfully used for the stabilization of monofilament polypropylene and the usage range was expanded to low density polyethylene. For the PVC stabilization, some stilbene compounds such as 4-chloro-α-cyan-3-methyl-stilben (9) were also efficiently used [49].

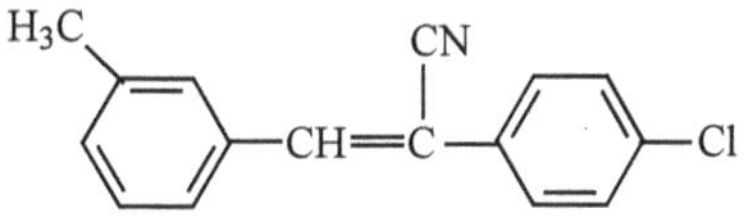

(9)

An interesting solution for photostabilisation is the use of some substances containing structural units which combines both the properties of UV absorbers and HALS such as derivatives of 1,8-naphtalimide which contain fragments of 2-(2-hydroxyphenyl)-1,3,5-triazine and 2,2,6,6-tetramethylpiperidine previously included in the structures capable to polymerize via vinyl double bond (Scheme 1) [50]. These substances present yellow-green fluorescent emission due to the triazine

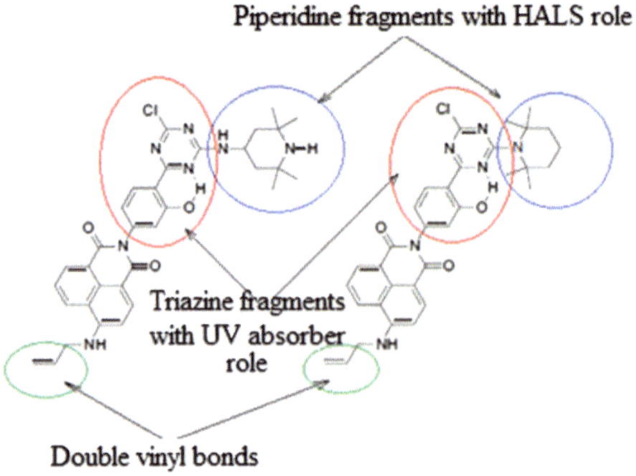

Scheme 1 Derivatives structure of 1,8-naphtalimide containing traces of HALS and UV absorber [50]

group having UV absorber role. They are efficient in the photostabilization simultaneously acting by absorption mechanism of UV radiation and by blocking oxidation of photoinduced free radicals by means of hindered amine group.

Besides the above mentioned methods, there are other photostabilization methods of polymeric materials such as:

- **the use of organometallic compounds of nickel, copper and transitional metals** such as those presented in the stabilised polymer formulas (10–17):

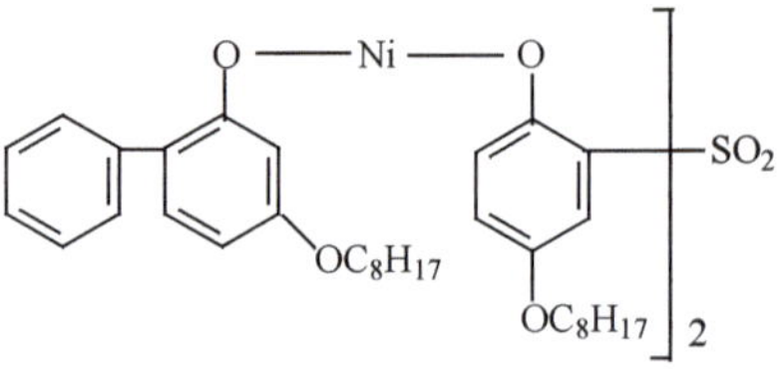

(10)

Nickel complex with bis-sulphonyl for polyolefins polyethylene

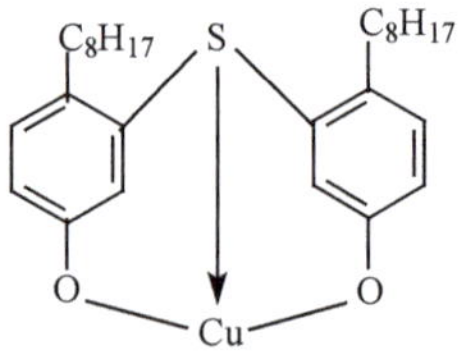

(11)

Copper complex for polyethylene

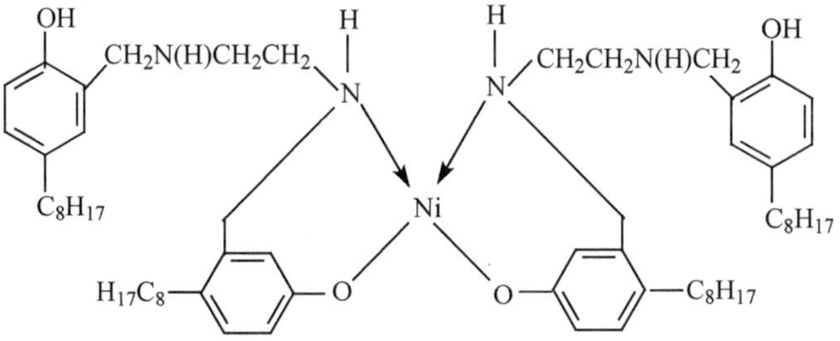

Nickel salts with N-(alkyl-hydroxybenzyl) alkylen-polyamine for polypropylene

(12)

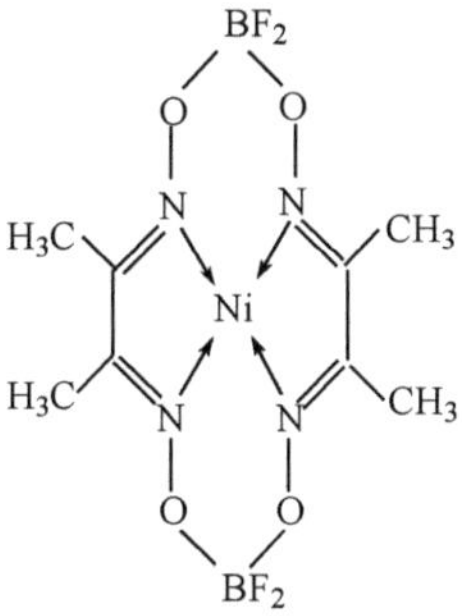

(13)

Complex of nickel-boron bifluorine-N, N-bis (dimethylglyoxime)

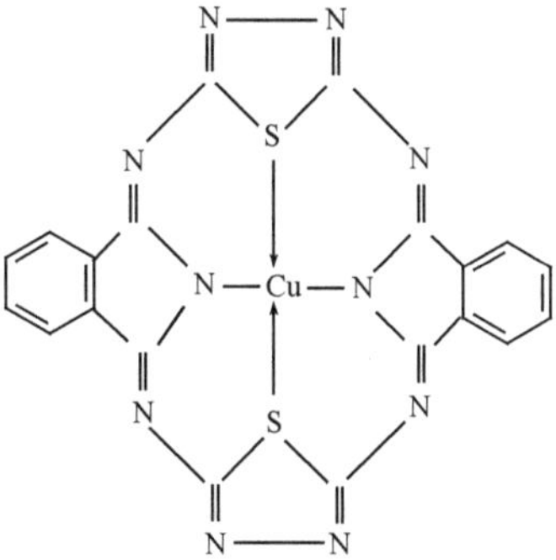

(14)

Copper complex with thiodiazoleindiole for polyamides

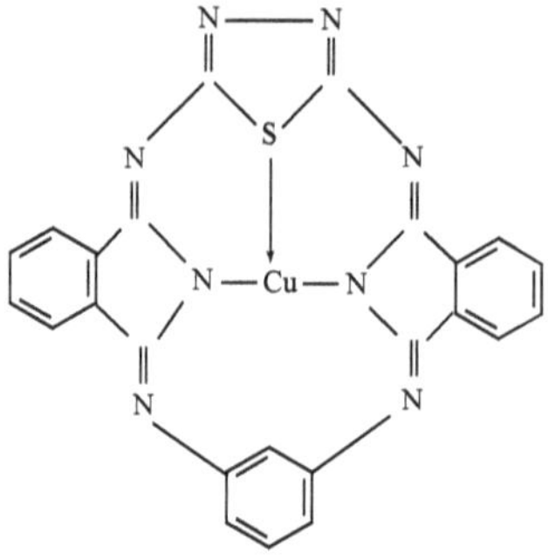
(15)

Copper complex with thidiazoleizo-
indolephenylene for polyamides

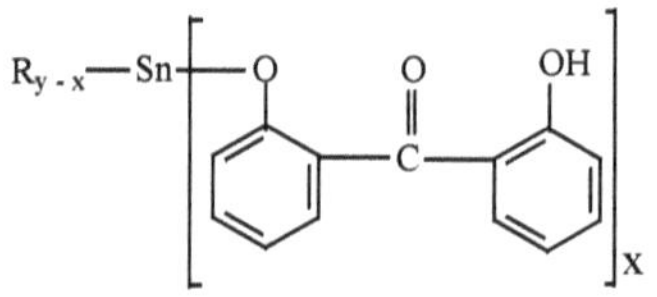
(16)

o-hydroxybenzophenone – organo – tin
compounds

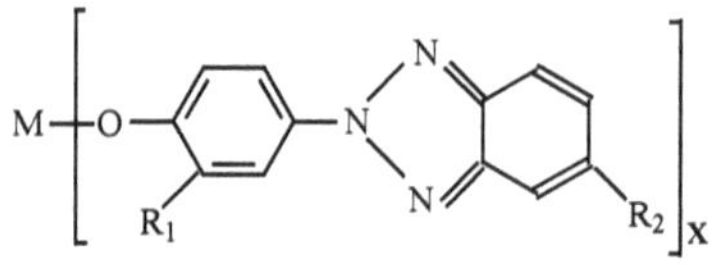
(17)

Benzotriazole with metal

- **synergistic pairs of photostabilizers with antioxidants.**

These couples of oxidation protectors have the main role of delay on the progress of photochemical degradation by the oxidation inhibition of free radicals and by hydroperoxides decomposition. This type of compounds includes antioxidants (either for the inhibition of radical decay onto peroxyl radicals or for the decomposition of hydroperoxides), carbon black, ZnO (which is activated by UV radiation) or other compounds which are involved in the oxidation/reduction reactions.

- **polymer light resistance improving by purification and modification**

The increase of polymer light resistance can be achieved by many ways such as:

(a) *The assurance of high purity of polymer and additives*

– It is well known that the polymers can incorporate molecules or parts of other trace materials or impurities during synthesis which contribute to the decrease of

their resistance against photostabilisation: metallic impurities or compounds which catalyse the photooxidation such manganese traces.
- It is also known the catalytic effect of copper and other transition metals to the thermooxidation of many polymers [51]. The traces of other foreign groups in the polymer structure may drop down the resistance to photooxidation of polymers.

(b) *The convenient modification of polymer structure by the incorporation of some stabilizers during polymerization.*

This process eliminates the inconvenient brought about by the photostabilizer evaporation during the polymer processing or its migration during the storage and it can be achieved by:

- the attachment of photostabilizer by introducing some reactive or polymerizable groups such vinyl or allyl group by a chemical or physical process;
- the preparation of polymerizable UV absorbers which lead to the preparation of polymers containing photostabilization groups by polymerization or copolymerization of certain monomers;
- the use of the photo-rearrangement of polymers with additives. For example, there were obtained homopolymers and copolymers of phenyl 5-acryloxymethylsalicylate and phenyl 5-methacryloxymethyl-salicylate with vinyl acetate, vinylidene chloride and vinyl chlorine.

4 Stabilization Characterization of Various Polymers

4.1 Polyethylene

Polyethylene is the most common thermoplastic polymer with multiple applications: insulating materials in electrical cables, pipes, packaging (bags, film, containers, a. s. o.), medical devices and other fields because of its good chemical and stress-crack resistance besides high strength.

In order to increase the product durability under environmental conditions, especially to the UV exposure, polyethylene is stabilized by different types of additives.

There are many ways to assess the level of photodegradation in PE such: electron spin resonance (ESR), UV and FTIR absorption [52–54], gel fraction, mechanical properties measurements [52], different thermal analysis techniques (TGA [55, 56], DSC [27, 56, 57], CL [55, 58, 59]).

The most used photostabilizers for PE includes HALS compounds. These compounds are up to four times more effective than nickel chelates and up to ten times than usual UV absorbers [53, 60].

According to Malik [61] an efficient stabilizer must comply with three basic requirements: higher stability or compatibility with polymer matrix, minimal diffusion and high degree of homogeneity of active species. It is well known that in the case of polymers it is very difficult to obtain a good homogeneity of the stabilizers inside the polymer matrix. In order to decrease the diffusion of stabilizer out of polymer matrix it is necessary to use stabilizers with high molecular weight or, the currently used technique in the UV stabilization of polyolefin, the grafting of stabilizer on the polymer backbone [53, 62].

Kaci et al. [53], using a PBH-3 compound, commercially known as Sanduvor PR31, to stabilize the PE under natural weathering conditions. It was demonstrated that the grafting of the stabilizer on polymer is very fast and occurs on few days. Due to the presence of methylenic double bond on the PBH-3 structure, the grafting process occurs by a photochemical reaction.

The UV radiation (λ = 210 nm) have been used the photoinduced grafting various diazo-compounds and methyl azidocarboxylate, bearing extended mono-HAS (Hindered Amine Stabilizer) [62] on polyethylene structures. It can be noticed that the grafting process increases the polyethylene photostability, the high stability against photo-oxidation being achieved for those films supposed to bear the largest amount of bonded HAS groups close to surface.

Yang et al. [54] studied the photodegradation of linear low density polyethylene (LLDPE) modified with inorganic nanofillers and/or light stabilizers compounds. They observed by FTIR spectroscopy measurements that the used nano-fillers (Al_2O_3, SiO_2 and ZnO) had positive effect on the photostabilization of LLDPE, the fillers lowering the degradation rate. However, the presence of ZnO and TiO_2 nanoparticles induces degradation more rapidly in polyethylene exposed to UV radiation. This effect was demonstrated by the increase in the rates of both carbonyl group build up and of CO_2 generation in LDPE substrate with nanoparticles of ZnO [63].

The chemiluminescence (CL) is an appropriate method to assess the effectiveness of some UV stabilizers on polyethylene exposed to UV radiation. Jipa et al. [64] studied by isothermal and non-isothermal CL the contribution to the photostability of polyethylene brought about by: an UV absorber *2-(3-methyl-6-hydroxyphenyl) benzotriazole* (Tinuvin P), and two HALS UV stabilizers: *(2,2,6,6-tetramethyl-4-pyperidyl-succinate) polyester* (Tinuvin 622) and *[(4-methoxyphenyl)-methylene]-bis-(1,2,2,6,6-pentamethyl-4-pyperidinyl)propanedioate* (Sanduvor PR31). The CL provides reliable information about the improvement in the life time of polyethylene by the stability qualification measuring oxidation induction time (OIT) by isothermal CL and onset temperature oxidation (OOT) by non-isothermal CL. An indicative parameter for photodegradation of polymers is the integrated non-isothermal CL signal (Σ_{CL}) which is proportional to the concentration of peroxides intermediates built up during the UV exposure. The minimum values of Σ_{CL} are related to the significant efficiency in oxidation stability [64].

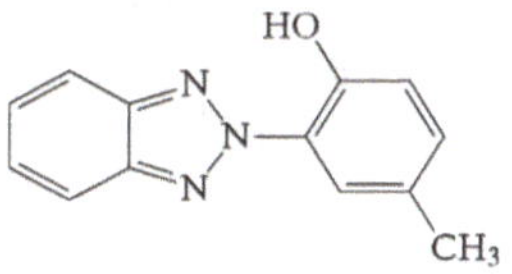

(18)

Tinuvin P

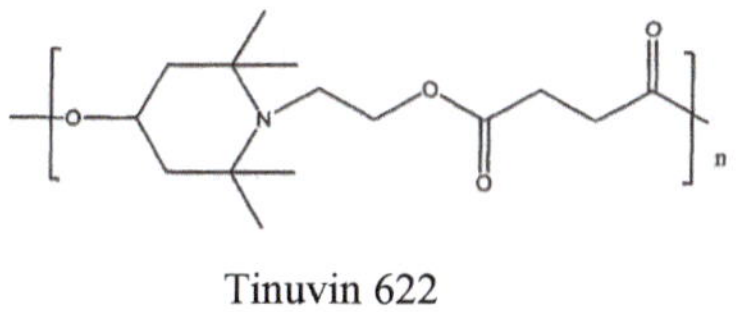

(19)

Tinuvin 622

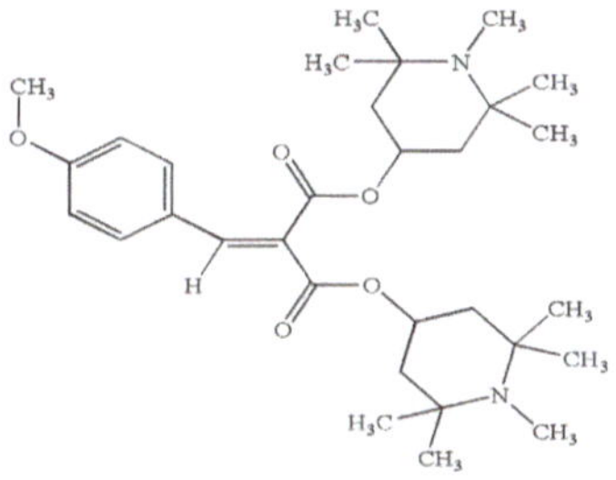

(20)

Sanduvor PR31

In Fig. 1 the decreasing order of UV stability provided by various additives is illustrated: Tinuvin 622 > Tinuvin P > Sanduvor PR 31.

The addition of HALS compounds on the PE resin can be efficiently extended to the protection against surface photo-oxidation of XLPE onto the photo-crosslinking process of PE. The FTIR and XPS data gave the evidence that the surface photo-oxidation of XLPE increase as the UV-irradiation time increases, the main formed products being identified as hydroperoxides and carbonyl groups [65].

It was proved that some compounds of transitional metals such copper stearate could retard the photo-degradation of HDPE [66] and PP [67]. Cooper stearate can acts both as UV light absorber and as radical scavenger, respectively. The stability effect of copper stearate on polyolefin was explained by two directions [67]:

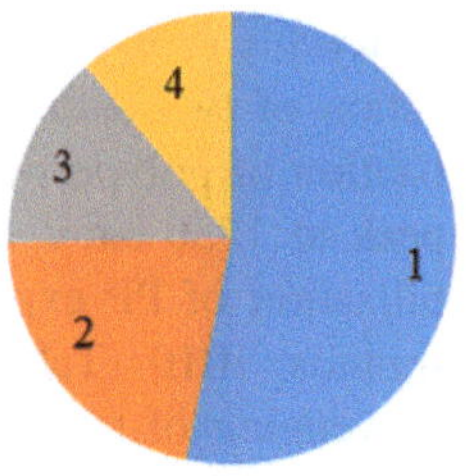

Fig. 1 Comparison of Σ_{CL} for LDPE containing various UV stabilizers: *1* unstabilized, *2* Sanduvor PR 31, *3* Tinuvin P, *4* Tinuvin 622 [64]

- the photo-chemical reaction of the copper stearate takes place before of the degradation of the polymeric substrate;
- the radicals formed by Norrish type I cleavage of ketone groups react with the copper ion to form inactive groups such as terminal carbon-carbon double bonds.

Many studies carried out on polyethylene/nanoclay systems [68–70] showed that the polymer as the major component of nanocomposites degrades faster than the pristine material, when it is exposed to UV radiation. This feature was explained on the basis of interaction of organo-clays and the antioxidant used for the protection of polymer substrate. Morlat-Therias et al. observed in the study on the photo-chemical stabilization of LLDPE/clay nanocomposites by infrared spectroscopy [71] that the presence of organo-clay promotes a lower durability of the nanocomposites exposed to artificial or natural UV light. By the comparison of the effectiveness provided by various UV absorbers (Cyasorb UV-1164, Cyasorb UV-2337, Cytec THT 6460) and by a metal deactivator compound (i.e. Irganox MD-1024), it was observed that the use of metal deactivator is an efficient solution for the stabilization of nanocomposite. In the presence of the metal deactivator, the active metal ions are scavenged by chelation forming an inactive complex [71].

Various compounds such as thiobisphenol were reported to exhibit a pro-oxidant effect in the early period of polyethylene photooxidation, which is accelerated as their concentration is enhanced, while on the later irradiation period the carbonyl build up was retarded [72]. The retardation of the carbonyl accumulation was obtained with a compound based on dodecyl-3,3′-thiodipropionate over the whole period of ultraviolet exposure. Kinetic data showed [72] that the concentration of 0.1 w/w% of thiobisphenols determines the most effective retardation for the accumulation of carbonyl group in LDPE films.

4.2 Polypropylene

The photooxidation of polypropylene involves the initiation of free radical chain reaction by the photolysis of hydroperoxides, producing peroxyl radicals as well as alkoxy radicals. In the last decades many studies dealing with the types of stabilizers used to delay the polypropylene photodegradation were reported as well as several studied on the experimental methods for assessing their photostability effectiveness [73–78].

Rychlý et al. used infrared and UV spectroscopies to estimate and to compare the action of two antioxidants, Irganox HP 136 and Irganox 1010, by the UV stabilization of polypropylene films [78]. It was observed by the evolution of carbonyl peak at 1730 cm^{-1} that the high efficiency of Irganox Hp 136 is attained in stabilization of PP exposed to UV radiation, whose activity is quite competitive to Irganox 1010 (Table 3). The stabilization mechanism is mainly based on the scavenging of free radicals formed by UV initiation. It can be observed that the

Table 3 Values of induction time from infrared spectroscopy measurements for evolution of carbonyl absorption [78]

Samples	Induction time (h)
Pure polypropylene	100
PP + 0.10 % Irganox 1010	885
PP + 0.15 % Irganox HP 136	1214
PP + 0.30 % Irganox HP 136	1243

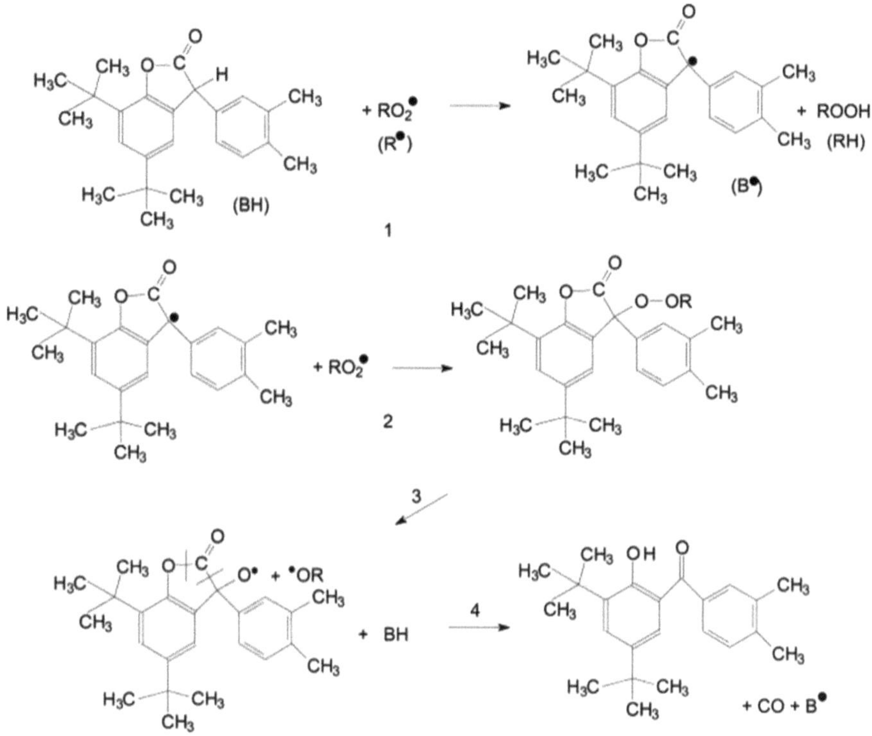

Fig. 2 The stabilization mechanism of Irganox HP 136 in polypropylene [78]

value of induction time of Iganox HP 136 at 0.15 % is similar with the oxidation induction time measured at the loading of 0.30 %.

The authors proposed a stabilisation mechanism of Irganox HP 136 by which benzofuranone lactone is converted in the oxidized PP into 2-hydroxy benzophenone derivatives, which is efficient against the effect of light (Fig. 2) [78].

Because the chemiluminescence method is a tool frequently used to qualify the stabilizing effect in polypropylene of various hindered amine, the analysis of CL kinetic parameters (oxidation induction time, oxidation rate, activation energy of oxidation, CL intensity) could provide reliable information related to the photodegradation state of polypropylene modified with different HALS compounds [76–78].

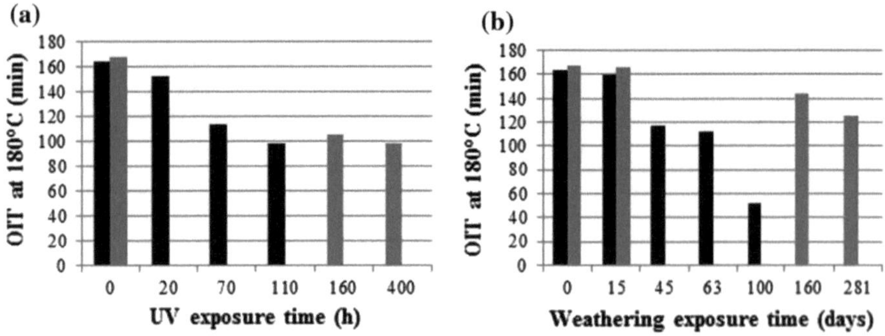

Fig. 3 Activation energy versus exposure time for iPP unstabilized (*black*) and stabilized with 0.3 % Sanduvor PR31 (*Grey*): **a** artificial UV light, **b** weathering exposure time [77]

In Fig. 3 the values of oxidation activation energy obtained from CL measurements for isotactic PP film stabilized with 0.3 % (w/w) Sanduvor PR31, exposed to UV light from both artificial and natural weathering, respectively, are presented [77].

It is generally accepted that higher values of activation energy correspond to higher stability of tested material. It can be observed that the presence of Sanduvor PR31 induces higher values of activation energy as compared to unstabilized samples.

Jipa et al. [76] studied by isothermal chemiluminescence the behaviour to photooxidation of isotactic polypropylene in the presence of some HALS (H2-H4) compounds and compared them with mercapto-1,3,5-triazinic phenolic stabilizer (antioxidant H5) and Tinuvin 770 (H1). The values of oxidation induction time (OIT) (Fig. 4) obtained from CL curves are related to the photostability of PP induced by each antioxidant. The longer the induction time, the higher the

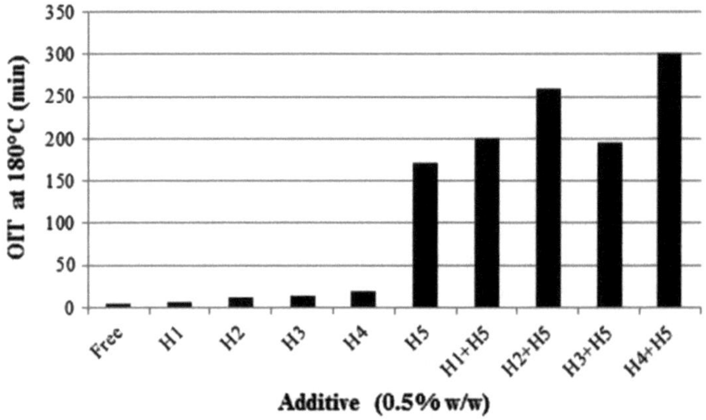

Fig. 4 CL OIT values for PP modified with various HALS compounds

stabilisation efficiency. It can be observed that H1-H4 compounds present lower photooxidative protection of PP. The most effective antioxidant is the mercapto-triazine compound. It can be also observed the synergistic activity between the H1-H4 compounds and H5 which lead to higher OIT values. This effect is due to the regeneration of phenoxy radicals provided by the hydroxylamine derivatives resulting from oxidation of piperidine [76].

The exposure to accelerated photodegradation of polypropylene plaques improved with various systems based on hindered amine stabilizers (Tinuvin 327, Tinuvin 123), UV absorbers (Tinuvin 327, Tinuvin 120) and phenolic antioxidant Irganox 1010 revealed by ESRI (Electron spin resonance imaging) a modification of nitroxide concentration profile (U-shape profile) as a function of the additives type and the manner of exposure [79]. The U shape profiles mean that almost the same nitroxide concentrations on both sides of irradiated and unexposed PP plaque surfaces were detected, but small nitroxide concentration in the inner layers was revealed. The use of a couple including UV absorber (Tinuvin 120) and HAS compound (Tinuvin 327) delays efficiently the PP photodegradation. The lack of evidence for any modification in nitroxide concentration, similarly proved in the samples without UV absorber [79] indicates the proper involvement of additives in the diminution of oxidation effects. Accordingly, the ATR/FTIR spectra recorded on the surface of PP plaque shown an increase of the absorption bands corresponding to carbonyl and hydroxyl groups with the irradiation time, the height of carbonyl peak being the same on both exposed and unexposed PP sample [79].

The use of some metal oxides i.e. rare earth oxides [33], TiO_2 [27, 29], in a proper concentration may improve the photodegradation resistance of the polymer matrix to UV exposure due to light screening effect provided by these particles. The addition of rare earth oxides, for example CeO_2 at 0.1 w/w% in PP, showed a minor effect on the changes in mechanical and thermal properties when the photodegradation of PP was accomplished [33]. The increase of the concentration of CeO_2 to 1 w/w% improves the thermal stability and mechanical properties of PP matrix. The FTIR, DSC, ATG analyses and mechanical characterization evidenced simultaneously that CeO_2 particles improve in a significant extent the resistance of PP to photodegradation [33]. The CeO_2 particles act as UV filter [27, 29, 33–35] by absorbing part of UV irradiation, reducing the UV intensity which can promote the oxidation of the PP chains.

4.3 Other Polymers

Poly(vinyl chloride) **(PVC)** is one of the most extensively investigated polymers since many years, due its large range of utilization in many application areas: automotive, pipes, electrical cables, window frames, etc.

Nevertheless, PVC suffers from poor light stability. Upon light exposure undergoes a rapid autocatalytic dehydrochlorination, which forms conjugated polyene, responsible for PVC discoloration [80–82]. The photodehydrochlorination is accompanied by modification of the mechanical properties of the PVC [82, 83]. In order to increase the PVC resistance to photodegradation are used many compounds which act as UV absorbers, HALS compounds, UV screeners a. s. o.

Rabie et al. [80] investigated the stabilizing efficiency of some diamide derivatives as photostabilisers for PVC containing dioctyl phthalate and dibutyl phthalate by evaluating the weight loss of the polymer. The unadditivated PVC weight loss, resulted from evolution of HCl during the photodegradation, increased with the UV radiation exposure time. The photostabilising efficiency of diamide derivatives was indicated by the lowering of the weight loss, reflecting a lower rate of dehydrochlorination [80].

The modification of commercial PVC by introducing of benzothiazole and benzimidazole compounds as pending groups into the repeating unit of PVC, evidenced an increase of the photostability of modified PVC, as compared to unmodified PVC [84]. The photostabilization of films of PVC could be achieved by using of some diorganotin (IV) complexes. These additives stabilize the PVC films through HCl scavenging, UV absorption or screening, peroxide decomposer and radical scavenger mechanisms [85].

The photostability of PVC-based materials for outdoor application can be substantially improved by means of some protective coatings such as: photocured urethane-acrylate clearcoats containing a UV absorber and a HALS radical scavenger which can filter the most harmful solar radiation, delaying the onset of PVC degradation [86], photo cross-linked epoxy-acrylate resins [87]. An additional benefit of this method of photostabilization subsists in the improvement of some surface properties of the coated polymer by making it more resistant to organic solvents, chemicals, abrasion and scratching [86].

Zhang et al. investigated the photostabilizing efficiency of different light stabilizers (TiO_2, Tinuvin 234, Tinuvin XT 833, Tinuvn 531, Chimasorb 944) in poly (vinyl chloride) by means of discoloration, Fourier transform infrared spectroscopy (FTIR), and scanning electron microscopy (SEM). The results show that the addition of light stabilizers can slow down discoloration of PVC, the most effective being TiO_2, due to its UV light reflective properties. However, the FITR analysis showed high carbonyl content for pure PVC and PVC modified with TiO_2 and Chimassorb 944, indicating that the photo-oxidation reactions of these UV irradiated samples are relatively serious [88].

Besides stabilizing methods presented above, there are many other compounds which can be used to delay de photodegradation of PVC among which we mention: Cadmium-stearate (II) [89], organotin compounds [90], dibutyltin maleate (DBTM) mixed with trisnitro (1,3-dihydroxyl-2-hydroxylmethyl-2-nitropropane) compounds as stabilisers [91], Schiff's bases [92], etc.

Copolymers cover the large part of polymer market, because they gather the most interesting features required by long term application. Luengo et al. [93] studied the photostabilization of poly(styrene-b-ethylene-co-butylene-b-styrene) (SEBS) modified with some hindered phenols and their combination with phosphite antioxidants by using a variety of spectroscopic methods including FTIR, UV, and luminescence spectroscopies coupled with crosslinking and hydroperoxide analysis.

The addition of a hindered phenol photostabilized the SEBS by inhibiting of discoloration, and the formation of hydroperoxides, acetophenone, and oxidation products, as well as chain scission and disaggregation of the styrene units. The combination hindered phenol-phosphite exposes high synergistic effect in the inhibition of SEBS yellowing, the reduction in the concentrations of chromophore groups and the decrease in the probability of chain scission. Consequently, higher phosphite concentration inhibits crosslinking reactions in the photo-oxidative degradation of SEBS as well as the formation of hydroperoxides groups [93].

Recently, a new method for attaining improved resistance against photodegradation of SBS copolymers (polystyrene-block-polybutadiene-block-polystyrene) was proposed [94]. The reported method consists in the modification of SBS copolymer with multiwall carbon nanotubes (MWCNT) and carbon nanotube functionalised with carboxylic groups (CNT-COOH). The ATR-FTIR spectra recorded on pristine SBS, SBS/CNTs and SBS/CNTs-COOH samples at different exposure times showed the accumulations of oxygen containing products, i.e. both carbonyl and hydroxyl species at the sample surface is considerably lower for CNTs containing nanocomposites in comparison to pristine SBS. The CNTs improve the mechanical properties of SBS nanocomposite as compared to pristine SBS: elastic modulus higher with 20 % for SBS/CNT and 35 % from SBS/CNT-COOH, ductility of the nanocomposites decreases in the same amount. This result proves that the presence of CNTs and CNTs-COOH have a beneficial effect on the photo-oxidative resistance of SBS nanocomposites by hindering the crosslinking reactions and slowing down the oxidized species formation. The CNTs stabilizing effect is attributed to their UV-light shielding properties and their radical scavenging activity [94]. The modification of EVA copolymers with carbon nanotubes evidenced the same stabilising effect of CNT, especially at higher CNT content (3–5 %) [95].

5 Conclusions

The stabilization of polymers is one of the most important questions related to the product lifetime. The proper solution for the delay of oxidation during the inevitable exposure to the joint actions of light and oxygen is the addition of compounds which present efficient activities in the scavenging and blocking reaction tendency of free radicals. The outdoor services of polymeric items require the improvement in the oxidation resistance avoiding accidents and frequent maintenance. The delay efficiency of oxidation depends strongly on the protector efficiency, whose action is macroscopically measured by the diminishing oxidation induction time and by the slowing down the progress of oxidation ageing.

The analysis of stabilizer activity must be accomplished by the incorporation of increasing amount of involved compound for the determination of concentration threshold where functional properties are not altered. The correlation between the oxidation protection and the materials durability must be considered as the key of the integrity persistence in the polymer material science.

References

1. Rabek, J.F.: Comprehensive Chemical Kinetics, Vol. 1, Degradation of Polymers, Ed. Bamford CH, Tipper CHF. Elsevier, Amsterdam (1975)
2. Yousif, E., Haddad, R.: Photodegradation and photostabilization of polymers especially polystyrene: review. SpringerPlus **2**, 398 (2013). doi:10.1186/2193-1801-2-398
3. Searle, N.D., McGreer, M., Zielnik, A.: Weathering of Polymer Materials. Encyclopedia of Polymer Science and Engineering. Wiley, New York (2010). doi:10.1002/0471440264.pst401.pub2
4. Davis, A., Sims, D.: Weathering of Polymers. Applied Science Publishers, London (1983)
5. Kockott, D.: Natural and artificial weathering of polymers. Polym. Degrad. Stab. **25**, 181–208 (1989)
6. Gijsman, P., Hennekens, J., Janssen, K.: Polymer durability, stabilization and lifetime prediction. In: Clough, R.L., Billingham, N.C., Gillen, K.T. (eds.) Advances in Chemistry Series, vol. **249**, pp. 621–649 (1996). ISBN13: 9780841231344
7. Gonzalez, A.V., Cervantes-Uc, J.M., Veleva, L.: Mineral filler influence on the photo-oxidation of high density polyethylene: I. Accelerated UV chamber exposure test. Polym. Degrad. Stab. **63**, 253–260 (1999)
8. Mendes, L.C., Rufino, E.S., de Paula, F.O.C., Torres Jr., A.C.: Mechanical, thermal and microstructure evaluation of HDPE after weathering in Rio de Janeiro City. Polym. Degrad. Stab. **79**, 371–383 (2003)
9. Wu, G., Lu, C., Cai, X., Xiancheng, R.J.: Mechanical properties and solid-state structure of photodegraded polyoxymethylene and effect of UV stabilizers modification. J. Macromol. Sci. Phys. Part B **50**, 1521–1534 (2011)
10. Eve, S., Mohr, J.: Study of the surface modification of the PMMA by UV-radiation. Procedia Eng. **1**, 237–240 (2009)
11. Andrei, C., Drăguţan, I., Balaban, A.T.: Polyolefin Photostabilization by Hindered Amine Stabilisers (In Romanian). Romanian Academy Printing House, Bucharest (1990)
12. Guillet, J.E., Dhanraj, J., Golemba, F.J., Hartley, G.H.: Stabilization of polymers and stabilizer processes. In: Gould, R. F. (ed.) Fundamental Processes in the Photodegradation of Polymers. Adv. Chem. Series, vol. 85, ACS, Washiongton, DC. Ch. **19**, pp. 272–286 (2008)
13. Lungulescu, E.M., Zaharescu, T., Podina, C.: Thermal and radiation stability of polyolefins modified with silica nanoparticles. J. Opt. Adv. Mater. **16**, 719–725 (2014)
14. Jipa, S., Setnescu, R., Setnescu, T., Zaharescu, T., Gorghiu, L.M.: Chemical effects induced by ultraviolet and nuclear radiation. Electra, Bucharest (2004)
15. Bolland, J.L.: Kinetics of olefin oxidation. Q. Rev. Chem. Soc. **3**, 1–21 (1949)
16. Martin, J.W., Ryntz, R.A., Chin, J., Dickie, R.A.: Service Life Prediction of Polymeric Materials: Global Perspectives, pp. 79–81. Springer, Berlin (2009)
17. Rabek, J.F.: Polymer Degradation Mechanisms and Experimental Methods. Springer, Stocholm (1994)
18. François-Heude, A., Richaud, E., Desnoux, E., Colin, X.: A general kinetic mechanism for the photothermal oxidation of polypropylene. J. Photochem. Photobiol. A Chem. **296**, 48–65 (2015)

19. Herdan, J.M., Giurginca, M., Meghea, A.: Antioxidants (In Romanian). Technical Printing House, Bucharest (1995)
20. Mellor, D.C., Moir, A.B., Scott, G.: The effect of processing conditions on the UV stability of polyolefins. Eur. Polym. J. **9**, 219–225 (1973)
21. Allen, N.S. (ed.): Degradation and Stabilization of Polymers. Applied Science Publishers Ltd., London (1983). ISBN 13: 978-94-010-8034-7
22. Harper, C.A. (ed.): Handbook of plastics technologies. The complete guide to properties and performances. McGraw Hill eBooks, (2006)
23. Broasca, G., Borcia, G., Dumitrascu, N., Vrinceanu, N.: Characterization of ZnO coated polyester fabrics for UV protection. Appl. Surf. Sci. **279**, 272–278 (2013)
24. Yang, H., Zhu, S., Pan, N.: Studying the mechanisms of titanium dioxide as ultraviolet-blocking additive for films and fabrics by an improved scheme. J. Appl. Polym. Sci. **92**, 3201–3210 (2004)
25. Tolinski, M.: Additives for Polyolefins: Getting the Most out of Polypropylene and TPO, 2nd edn, p. 34. Elsevier, Amsterdam (2015)
26. Girois, S.: Effect of Iron compounds on PVC (Thermal and photochemical stability). In: Conference Proceedings ANTEC'99, Volume III—Special Areas, New York City, May 2–6, 3604 (1999)
27. Ilie, S., Setnescu, R., Lungulescu, E.M., Marinescu, V., Ilie, D., Setnescu, T., Mares, G.: Investigations of a mechanically failed cable insulation used in indoor conditions. Polym. Test. **30**, 173–182 (2011)
28. Kamrannejad, M.M., Hasanzadeh, A., Nosoudi, N., Mai, L., Babaluo, A.A.: Photocatalytic degradation of polypropylene/TiO_2 nano-composites. Mater. Res. **17**, 1039–1046 (2014)
29. Crăciun, E., Ioncea, A., Jitaru, I., Covaliu, C., Zaharescu, T.: Nano oxides UV protectors for transparent organic coatings. Rev. Chim. **62**, 21–26 (2010)
30. Zakrzewski, J., Szymanowski, J.: 2-Hydroxybenzophenone UV-absorber containing 4,4,5,5-tetramethylimidazolidine fragment. Polym. Degrad. Stab. **72**, 109–113 (2001)
31. Kurzböck, M., Wallner, G.M., Lang, R.W.: Black pigmented polypropylene materials for solar absorbers. Energy Procedia **30**, 438–445 (2012)
32. Liu, M., Horrocks, A.R.: Effect of Carbon Black on UV stability of LLDPE films under artificial weathering conditions. Polym. Degrad. Stab. **75**, 485–499 (2002)
33. Bezati, F., Massardier, V., Balcaen, J., Froelich, D.: A study on the dispersion, preparation, characterization and photo-degradation of polypropylene traced with rare earth oxides. Polym. Degrad. Stab. **96**, 51–59 (2011)
34. Podbrščček, P., Dražić, G., Anžlovar, A., Crnjak, O.Z.: The preparation of zinc silicate/ZnO particles and their use as an efficient UV absorber. Mater. Res. Bull. **46**, 2105–2111 (2011)
35. Fonseca de Lima, J., Figueredo-Martins, R., Neri, C.R., Serra, O.A.: ZnO:CeO_2-based nanopowders with low catalytic activity as UV absorbers. Appl. Surf. Sci. **255**, 9006–9009 (2009)
36. Singh, B., Sharma, N.: Mechanistic implications of plastic degradation. Polym. Degrad. Stab. **93**, 561–584 (2008)
37. Feron, K., Belcher, W.J., Fell, C.J., Dastoor, P.C.: Organic solar cells :understanding the role of förster resonance energy transfer. Int. J. Mol. Sci. **13**, 17019–17047 (2012)
38. Malik, J., Ligner, G., Avar, L.: Polymer bound HALS—expectations and possibilities. Polym. Degrad. Stabil. **60**, 205–213 (1998)
39. Abdel-Bary, E.M., Sarhan, A.A., Abdel-Razik, E.A.: Studies of the photo-oxidation of polydienes: Part 1—Photo-stabilising effect of some derivatives of diacetyl-monooxime benzoylhydrazone nickel (II) chelates on cis-1,4-polybutadiene. Polym. Degrad. Stab. **18**, 145–155 (1987)
40. Soltermann, A.T., de la Peña, D., Nonell, S., Amat-Guerri, F., García, N.A.: Phenolic-type stabilizers as generators and quenchers of singlet molecular oxygen ($O2(1\Delta g)$). Part I: methyl salicylate, salicylic acid and some related compounds. Polym. Degrad. Stab. **49**, 371–378 (1995)

41. Carlson, D.J., Wiles, D.M.: The photooxidative degradation of polypropylene. Part I. Photooxidation and photoinitiation processes. J. Macromol. Sci. **C-14**, 65–106 (1979)
42. Salamone, J.C. (ed.): Concise polymeric materials encyclopedia. CRC Press, Boca Raton (1999)
43. Carlsson, D.J., Wiles, D.M.: Photostabilization of polypropylene. II. Stabilizers and hydroperoxides. J. Polym. Sci. Polym. Chem. Ed. **12**, 2217–2233 (1974)
44. Setnescu, R., Kaci, M., Jipa, S., Setnescu, T., Zaharescu, T., Hebal, G., Benhamida, A., Djedjelli, H.: Chemiluminescence study on irradiated low-density polyethylene containing various photo-stabilisers. Polym. Degrad. Stab. **84**, 475–481 (2004)
45. Jia, H., Wang, H., Chen, W.: The combination effect of hindered amine light stabilizers with UV absorbers on the radiation resistance of polypropylene. Radiat. Phys. Chem. **76**, 1179–1188 (2007)
46. Kattas, L., Gastrock, F., Levin, I., Cacciatore, A.: Plastic additives Chap. 4. In: Harper CA (ed.) Modern Plastics Handbook. McGraw-Hil, New York (1999)
47. Xie, H., Lai, X., Zhou, R., Li, H., Zhang, Y., Zeng, X., Guo, J.: Effect and mechanism of N-alkoxy hindered amine on the flame retardancy, UV aging resistance and thermal degradation of intumescent flame retardant polypropylene. Polym. Degrad. Stab. **118**, 167–177 (2015)
48. Ryts, G., Slongo, M.: Unsaturated derivatives of 2,2,6,6-tetramethylpiperidine. US 4942238 A (1989)
49. Plotnikov, V.G., Efimov, A.A.: Polymer photostabilisers. Photophysical properties and stabilizing efficiency. Russ. Chem. Rev. **59**, 792–806 (1990)
50. Bojinov, V.B., Panova, I.P., Simeonov, D.: Design and synthesis of polymerizable, yellow-green emitting 1,8-naphthalimides containing built-in s-triazine UV absorber and hindered amine light stabilizer fragments. Dyes Pygments **78**, 101–110 (2008)
51. Jipa, S., Setnescu, R., Zaharescu, T., Setnescu, T., Gorghiu, L.M., Bancuta, I., Chelarescu, E. D.: Copper diffusion in cable-insulating materials by chemiluminescence and DSC techniques. J. Therm. Anal. Calorim. **122**, 251–259 (2015). doi:10.1007/s10973-015-4668-z
52. Lukáč, I., Hrdlovič, P., Guillet, J.E.: Synthesis and photochemical properties of poly(1-(4-carboethoxyphenyl)-2-propen-1-one). Polym. Photochem. **7**, 163–244 (1986)
53. Kaci, M., Hebal, G., Benhamida, A., Boukerrou, A., Djidjelli, H., Sadoun, T.: Kinetic study of hindered amine light stabilizer photografting in low density polyethylene films under natural weathering conditions. J. Appl. Polym. Sci. **84**, 1524–1532 (2002)
54. Yang, R., Li, Y., Yu, J.: Photo-stabilization of linear low density polyethylene by inorganic nano-particles. Polym. Degrad. Stab. **88**, 168–174 (2005)
55. Corrales, T., Catalina, F., Peinado, C., Allen, N.S., Fontan, E.: Photooxidative and thermal degradation of polyethylenes: interrelationship by chemiluminescence, thermal gravimetric analysis and FTIR data. J. Photochem. Photobiol. A Chem. **147**, 213–224 (2002)
56. Allen, N.S., Edge, M., He, J., Chen, W., Kikkawa, K., Minagawa, M.: Thermal and photooxidative behaviour of 2-hydroxybenzophenone stabilisers in polyolefin films: Effect of 4-butoxy-4-amino-tetramethylpiperidine substitution. Polym. Degrad. Stab. **42**, 293–306 (1993)
57. Roy, P.K., Surekha, P., Rajagopal, C., Chatterjee, S.N., Choudhary, V.: Studies on the photo-oxidative degradation of LDPE films in the presence of oxidised polyethylene. Polym. Degrad. Stab. **92**, 1151–1160 (2007)
58. Jipa, S., Zaharescu, T., Setnescu, R., Gorghiu, L.M., Dumitrescu, C., Oros, C.: Chemiluminescence study on HALS antioxidant activity in LDPE. Polym. Bull. **57**, 545–552 (2006)
59. Kaci, M., Touati, N., Setnescu, R., Setnescu, T., Jipa, S.: Characterization by Chemiluminescence of Unstabilized and HALS-Stabilized LDPE Films Exposed to Natural Weathering Conditions Int. J. Polym. Anal. Charact. **9**, 275–287 (2004)
60. Law, W.Y., Pan, J.Q.: Photostabilizing effectiveness of new HALS produced by isocyanation of hindered piperidine derivatives. Polym. Degrad. Stabil. **60**, 459–464 (1998)

61. Malik, J., Ligner, G.: Hindered amine light stabilizers: introduction. In: Pritchard, G. (ed.) Plastic Additives. Polymer Science and Technology Series, vol. 1, pp. 353–359. Springer, Dordrecht (1998)
62. Mosnacek, J., Bertoldo, M., Kosa, C., Cappelli, C., Ruggeri, G., Lukáč, I., Ciardelli, F.: Modification and photostabilization of low density polyethylene film by photodecomposition of various diazo-compounds and methyl azidocarboxylate. Polym. Degrad. Stabil. **92**, 849–858 (2007)
63. Yang, R., Christensen, P.A., Egerton, T.A., White, J.R.: Degradation products formed during UV exposure of polyethylene-ZnO nano-composites. Polym. Degrad. Stab. **95**, 1533–1541 (2010)
64. Jipa, S., Zaharescu, T., Setnescu, R., Setnescu, T., Brites, M.J.S., Silva, A.M.G., Marcelo-Curto, M.J., Gigante, B.: Chemiluminescence study of thermal and photostability of polyethylene. Polym. Int. **48**, 414–420 (1999)
65. Wu, Q., Qu, B., Xu, Y., Wu, Q.: Surface photo-oxidation and photostabilization of photocross-linked polyethylene. Polym. Degrad. Stab. **68**, 97–102 (2000)
66. Osawa, Z., Kurisu, N., Nagashima, K., Nakano, K.: The effect of transition metal stearates on the photodegradation of polyethylene. J. Appl. Polym. Sci. **23**, 3583–3590 (1979)
67. Osawa, Z., Kobayashi, K., Kayano, E.: Mechanism of inhibition by copper stearate of the photo-degradation of polyolefins. Polym. Degrad. Stab. **11**, 63–75 (1985)
68. Qin, H., Zhao, C., Zhang, S., Chen, G., Yang, M.: Photo-oxidative degradation of polyethylene/montmorillonite nanocomposite. Polym. Degrad. Stab. **81**, 497–500 (2003)
69. La Mantia, F.P., Tzankova Dintcheva, N., Malatesta, V., Pagani, F.: Improvement of photo-stability of LLDPE-based nanocomposites. Polym. Degrad. Stab. **91**, 3208–3213 (2006)
70. Qin, H., Zhang, Z., Feng, M., Gong, F., Zhang, S., Yang, M.: The influence of interlayer cations on the photo-oxidative degradation of polyethylene/montmorillonite composites. J. Polym. Sci. Part B Polym. Phys. **42**, 3006–3012 (2004)
71. Morlat-Therias, S., Fanton, E., Gardette, J.-L., Dintcheva, N.T., La Mantia, F.P., Malatesta, V.: Photochemical stabilization of linear low-density polyethylene/clay nanocomposites: towards durable nanocomposites. Polym. Degrad. Stab. **93**, 1776–1780 (2008)
72. Jiráčková-Audouin, L., Bory, J.F., Farrenq, J.F., Verdu, J., Pospíšil, J.: Influence of thiobisphenols on the photo-oxidation of low density polyethylene. Polym. Degrad. Stab. **6**, 17–29 (1984)
73. Allen, N.S., Parkinson, A., Gardette, J.-L., Lemaire, J.: Thermal and photo-oxidative behaviour of hindered piperidine compounds in polypropylene: Importance of hydroxylamine in stabilization. Polym. Degrad. Stab. **5**, 135–144 (1983)
74. Allen, N.S., Fatinikun, K.O., Henman, T.J.: Influence of transition metal ions on the photo-behaviour of BHT in polypropylene: UV-study. Polym. Degrad. Stab. **17**, 81–88 (1987)
75. Chmela, S., Carlsson, D.J., Wiles, D.M.: Photo-stabilizing efficiency of N-substituted hindered amines in polypropylene: Effects of processing conditions and exposure to a protonic acid. Polym. Degrad. Stab. **26**, 185–195 (1989)
76. Jipa, S., Zaharescu, T., Setnescu, R., Setnescu, T., Wayne, W.Y., Pau, J.-Q.: Kinetic effect of photoexposed isotactic polypropylene in the presence of some hindered-amine light stabilisers. Polymer **41**, 6949–6953 (2000)
77. Jipa, S., Setnescu, R., Zaharescu, T., Setnescu, T., Kaci, M., Touati, N.: Chemiluminescence of isotactic polypropylene induced by photo-oxidative degradation and natural weathering. J. Appl. Polym. Sci. **102**, 4623–4629 (2006)
78. Rychlý, J., Mosnáčková, K., Rychlá, L., Fiedlerová, A., Kaszab, G., Nádorb, A., Osváth, S., Stumphauser, T., Szarka, G., Czaníková, K., Chmela, Š., Iván, B., Mosnáček, J.: Comparison of the UV stabilisation effect of commercially available processing stabilizers Irganox HP 136 and Irganox 1010. Polym. Degrad. Stab. **118**, 10–15 (2015)

79. Marek, A., Kaprálková, L., Schmidt, P., Pfleger, J., Humlíček, J., Pospíšil, J., Pilař, J.: Spatial resolution of degradation in stabilized polystyrene and polypropylene plaques exposed to accelerated photodegradation or heat aging. Polym. Degrad. Stab. **91**, 444–458 (2006)
80. Rabie, S.T., Khalil, A.M., Nada, A.A.: Diamide derivatives as photostabilizers for plasticized poly(vinyl chloride). J. Vinyl. Addit. Technol. **14**, 191–196 (2008)
81. Anton-Prinet, C., Mur, G., Gay, M., Audouin, L., Verdu, J.: Photoageing of rigid PVC—III. Influence of exposure conditions on the thickness distribution of photoproducts. Polym. Degrad. Stab. **60**, 283–289 (1998)
82. Torikai, A., Hasegawa, H.: Accelerated photodegradation of poly(vinyl chloride) Polym. Degrad. Stab. **63**, 441–445 (1999)
83. Yu, Q., Zhu, S., Zhou, W.: Peroxide induced crosslinking and degradation of polyvinyl chloride. J. Polym. Sci. Polym. Chem. **36**, 851–860 (1998)
84. Yousif, E., Hameed, A., Rasheed, R., Mansoor, H., Farina, Y., Graisa, A., Salih, N., Salimon, J.: Synthesis and photostability study of some modified poly(vinyl chloride) containing pendant benzothiazole and benzimidozole ring. Int. J. Chem. **2**, 65–80 (2010)
85. Yousif, E., Salimon, J., Salih, N.: New photostabilizers for PVC based on some diorganotin (IV) complexes. J. Saudi Chem. Soc. **19**, 133–141 (2015)
86. Decker, C.: Photostabilization of poly (vinyl chloride) by protective coatings. J. Vinyl. Addit. Technol. **7**, 235–243 (2001)
87. Decker, C.: Surface protection of poly (vinyl chloride) by photografting of epoxy–acrylate coatings. J. Appl. Polym. Sci. **28**, 97–107 (1983)
88. Zhang, X., Pi, H., Guo, S.: Photostabilizing efficiency of ultraviolet light stabilizers for rigid poly(vinyl chloride) against photo-oxidation. Polym. Eng. Sci. **53**, 378–388 (2013)
89. Vymazal, Z., Švorčik, V.: Photodegradation of PVC stabilized with cadmium stearate-II. Eur. Polym. J. **21**, 1035–1038 (1985)
90. Vymazal, Z., Vymazalová, Z.: Photodegradation of PVC stabilized by organotin compounds. Eur. Polym. J. **27**, 1265–1270 (1991)
91. Toroti, M., Olayemi, J.Y., Adeniyi, J.B., Peters, O.: The photooxidative degradation of poly (vinylchloride)—2. The stabilising action of dibutyltin maleate and tris-nitro (1,3-dihydroxyl-2-hydroxymethyl-2-nitropropane) on PVC from ultraviolet light radiation. Polym. Degrad. Stab. **61**, 297–302 (1998)
92. Yousif, E., Salih, N., Salimon, J.: Improvement of the photostabilization of PVC films in the presence of 2-N-salicylidene-5-(Substituted)-1,3,4-Thiadiazole. J. Appl. Polym. Sci. **120**, 2207–2214 (2011)
93. Luengo, C., Allen, N.S., Wilkinson, A., Edge, M., Parellada, M.D., Barrio, J.A., Santa, R.: Photostabilization of styrene–ethylene–butylene–styrene block copolymer by hindered phenol and phosphite. J. Vinyl. Addit. Technol. **12**, 2–7 (2006)
94. Dintcheva, N.Tz., Arrigo, R., Catalanotoo, F., Morici, E.: Improvement of the photo-stability of polystyrene-blockpolybutadiene-block-polystyrene through carbon nanotubes. Polym. Degrad. Stab. **118**, 24–32 (2015)
95. Marlat-Therias, S., Fanton, E., Gardette, J.L., Peeterbroeck, S., Alexandre, M., Dubois, P.: Polymer/carbon nanotube nanocomposites: Influence of carbon nanotubes on EVA photodegradation. Polym. Degrad. Stab. **92**, 1873–1882 (2007)

Weathering Degradation of Polymers

Traian Zaharescu and Eduard Marius Lungulescu

1 General Considerations

The structures of polymers based on several kinds of bonds are always subjected to the modifications, when they are placed in the specific conditions. The activities where polymer items are involved depend entirely on the material strength to the energy transfer from the surrounding space. The answers of products to the destructive actions of environmental factors must be known by the manufacturers and customers, because the life time is strongly connected to the modifications occurred at the molecular levels. The great interest paid to the functional features of materials is fruitfully applied to the characterization of durability, as well as to the extension of application areas.

The polymeric products are degraded by an accelerated process which occurs by bond cleavage, diffusion of radicals, oxidation of intermediates and distribution of final products along the depth of penetration. The main factors that contribute by synergetic activities are heat, light, humidity mechanical charge. The intimate mechanism of structural changes can be found elsewhere [1, 2]. The consequences of chemical modifications may be found in the shifting materials onto the cracked outer layers, which allow the facile penetration of water and pollutants completing the advanced ageing. The competition between various radical intermediates for the oxygen consumption determines the degradation rate [3]. The degree of crosslinking, the molecular size and entanglement, the presence of inorganic filler and the history of samples are important factors that controls the diffusion by penetration of oxygen and the diffusion by spreading of degradation products [4, 5].

T. Zaharescu (✉) · E.M. Lungulescu
INCDIE ICPE CA, 313 Splaiul Unirii, P.O. Box 139, Bucharest, Romania
e-mail: traian.zaharescu@icpe-ca.ro; traian_zaharescu@yahoo.com

E.M. Lungulescu
e-mail: marius.lungulescu@icpe-ca.ro; lungulescu_eduard@yahoo.com

D. Rosu and Visakh P. M. (eds.), *Photochemical Behavior of Multicomponent Polymeric-based Materials*, Advanced Structured Materials 26,
DOI 10.1007/978-3-319-25196-7_7

The indoor accelerated and outdoor weathering of polymers occurs similarly, but not identical. The great difference between the two degradation procedures consists of the local radical concentrations and their migration in the material bulk [6]. The intensity of irradiation beam induces different degrees of degradation and the ageing rate reflects the level of energy density received by the surface [7]. The oxidation takes place with peculiar rates, because the concentrations of radicals and the mobilities of intermediates are interconditioned and the overall processes are determined by the involved activation energies [8].

The degradation profile of materials subjected to simultaneous actions of environmental stressors follows the photo ageing effects, which is visually monitored by coloring and photolytic contributions to structural modifications. The artificial weathering investigation is the prediction testing whose results become the selection criteria for long term service in appropriate conditions of outdoor durability. The lifetime features complete the understanding material answer to the formulation reliability by which inspected compositions are designed for specific applications. The apparent activation energy required for the degradation of materials by weathering has as much components as the number of factors are. However, it is seldom noticed that the overall behavior illustrates the synergistic effects of degrading agents.

The sequence of modification steps during material oxidation involves the fragmentation of polymer chains on the weakest sites simultaneously occurred with the diffusion of oxygen, the formation of hydroperoxides, the decomposition of these intermediates during their supply and the accumulation of final and stable products as inert compounds.

The structural modifications have correspondingly effects on the functional characteristics in direct relation with the duration and intensity of exposures [9]. It can give the example—the protection of outdoor products like statue stones [10]. The reference light (λ = 340 nm) induces an oxidation process with significant rate, which simulates real exposure. However, there is a parallelism between indoor and outdoor processing, which is illustrated in Table 1.

Consequently, the errors of each kind of procedure are determined by the absence of simultaneously acting factors or supra estimation of applied conditions. The synergistic effects of environmental stressors determine correct results, which must be correlated with the chemistry of material. The repeatability of artificial weathering tests provides somewhat similar information, which is influenced by the material history. The outdoor exposure does not ever create identical intensities of degrading agents. The meaningful results are obtained by the simulation of natural

Table 1 The comparison between indoor and outdoor weathering

Indoor exposure	Outdoor exposure
Xenotest type procedure	Frame positioning in well defined area
Controlled investigation	Aleatoric exposure
Artificial lighting	Sunlight irradiation
Selected appropriate exposure	Synergistic experiment

parameters, if the main picture is considered during the proper application of reliable variants. The faithful studies are regular accomplished as cycling exposures, which must be designed in relation with the rate of oxygen penetration. The bond strength in macromolecules, the morphology, the molecular distribution, the crystallinity, the compacting feature, the applied mechanical charge, the presence of stabilizers like HALS, the addition of filler in material formulation, the processing conditions are the main peculiarities that characterize the resistance and the degradation rates during climatological examination [11, 12]. The planning of the test concerning the manner through which nature and hazard affect the durability of polymers can be conceptualized in direct relation with the proportions in which various factors are combined [13].

In fact, the key factor that determines the bond scissions is UV light, which transfer the energy required for the formation of free radicals. The geographical location of testing represents a confluence of factors with several uncertainties [14] through which the place identity is expressed. The most important exhibiting condition that is very often negligible is the angle with which the sun rays drop on material. It can receive the certain coefficient in the statistics of this kind of studies. The specific temperature, rainfalls, permanent humidity, ozone concentration become synergistic parameters for sunlight action. Their intensities and regularities must be considered as particular variables in the accumulation of degradation products, in the modifications of physical and mechanical features. The calculation of product lifetime is seldom affected by the wrong estimation of regional climatic profile and by its transfer into indoor experiments [15–17].

As it was demonstrated [18], the rate of photodegradation follows an exponential relationship, where the certain intensity of applied light acts partially in the direct dependence of experimental conditional including exposure time (Eq. 1), where k represents the oxidation rate, K means the rate constant, I denotes the exposure intensity, a is a material constant that describes the physical and chemical peculiarities and t is exposure time.

$$k = K.I^{a.t} \tag{1}$$

The light wavelength whose energy determines the specific degradation rate must be included through the intensity exponent, as well as the contribution of each wavelength from exposure spectrum. In the indoor experiments, the irradiance range must be correctly selected in the area taken into account. An example is Miami (USA) region, where the natural irradiance varies between 0.35 W m^2 for normal sun shining and 0.74 W m^2 for intense sun activity [18].

The life times of materials subjected to climatic ageing are controlled by the received energy, which presents highest values for the wave lengths between 290 and 400 nm [19]. The photo-degradation is a complex process which progress as a chain mechanism—a sequence of reaction producing radicals. The peroxides must be considered as oxidation initiators (Fig. 1). The abundances of final and stable products depend on the material structure and the presence of stabilizer. The entirely formed amount of products are oxygen-containing configuration appeared

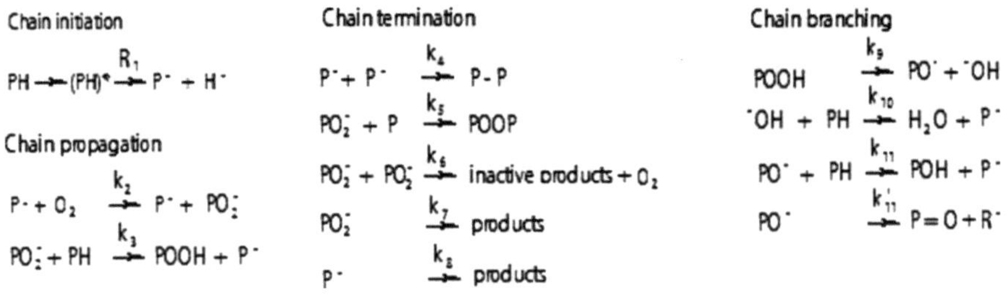

Fig. 1 The photo-degradation of the majority of polymers by UV exposure

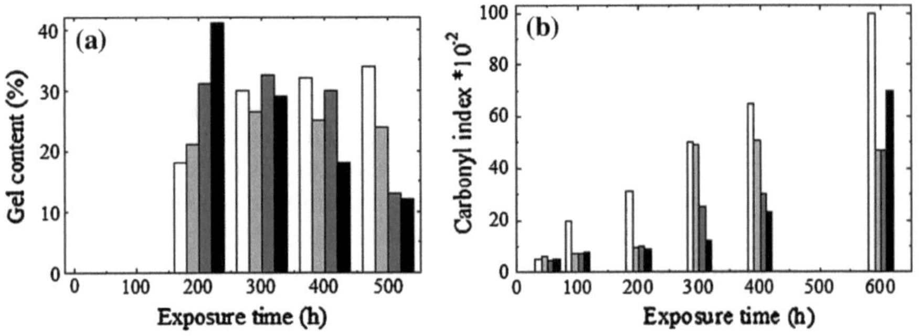

Fig. 2 Variation of gel content (**a**) and carbonyl index (**b**) in the systems consisting of LDPE and EOP in different loadings. (*white*) 0 %; (*light grey*) 5.8 %; (*dark grey*) 8.8 %; (*black*) 10.3 %

by the collateral reactions (chain branching) or by self-decaying of peroxyl radicals (chain termination).

The evolution of systems containing double bonds, for example ethylene–octene copolymer in low density polyethylene blends, is illustrative for the materials with unsaturation orient the behavior during UV exposure (Fig. 2). The contribution of this copolymer is evident. If the gelation of samples does not occur for the first 100 h of exposure [20], the accumulation of radicals allow a sudden increase at 200 h. Then the degradation affects the gel fraction by the cleavage of inter-molecular bridges.

The carbonyl indexes increase smoothly in the blends containing EOP in comparison with neat polyethylene, because the consumption of double bonds feeds the crosslinking. The final amounts of oxygenated products reach specific levels according to the various routes of hydroperoxides decomposition [20].

The temperature that is the parameter characterizing the minute energetic state determines the movement speed of molecules and, consequently, their availability for degradation [8, 21]. The variation in activation energy required for outdoor weathering depicts the progress in the photooxidation of polymers. Indeed, the rate of oxidation is enhanced, when the material temperature is augmenting (Fig. 3).

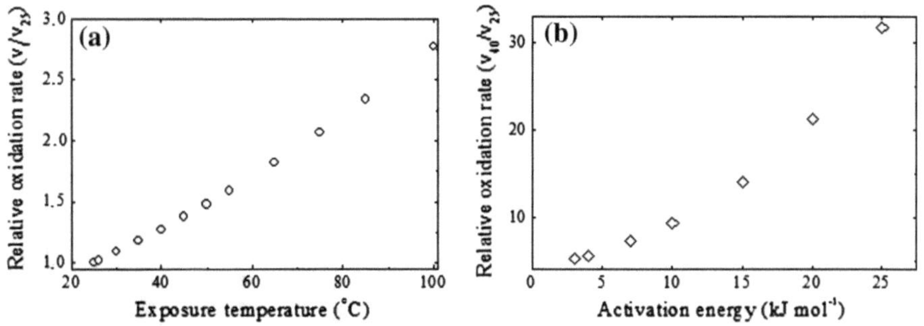

Fig. 3 Modification in relative rates of oxidation by weathering of ABS. **a** Calculated for an activation energy of 12 kJ mol^{-1}; **b** calculated at 40 °C

The exponential increase of relative oxidation rate during weathering ageing emphasizes the contribution of stabilization factors like higher crosslinking or the presence of a proper antioxidant.

The interpretation of property changes occurred in different polymer materials is always based on the existence of weaker bonds, which are preferentially cleaved. The resulting radicals receive energy either from incidental radiation or by heating. The pathway of degradation is demonstrated by the predominant manner of scissions initiated by the structural peculiarities. According to the radical mechanism of photooxidation, the simultaneity of thermally and photochemically induced degradation causes an accelerated process based on the synergetic effects of acting factors [22, 23].

The material life time can be calculated by means various experimental procedures, which provide relevant information obtained by structural modifications. The long-term stability is estimated for the preservation of upper-limit performances and longest operation periods over several degradation cycles with high irradiation fluxes. In fact, the durability is expressed by the reliability of material behavior, which is described by the failure rate. The definition of durability must also include the condition of testing: irradiance, duration, complementary environmental factors, physical properties (crystallinity, hardness, viscosity, diffusion constant), homogeneity, presence of antioxidant, especially for engineering polymers [24].

The achievement of faithful studies on the material strength under indoor or outdoor climatic degradation depends strongly by the surface thermal properties. In the relationship that depicts the modeling of thermal conduction of outer layers, the contribution factors of light intensity and spectrum, wind, humidity, rain frequency, pollutants, heat convection must be foreseen [25]. The rate of gas diffusion into material is dependent on the pressure according with the Eq. (2) [26].

$$S = \frac{(M_{\infty} - M_o)/m}{\left(\frac{M_o}{\rho}\right) \cdot P} \left[\frac{\mathrm{cm^3 CO_2} STP}{\mathrm{cm^3} sample \cdot \mathrm{cm\ Hg}}\right] \quad (2)$$

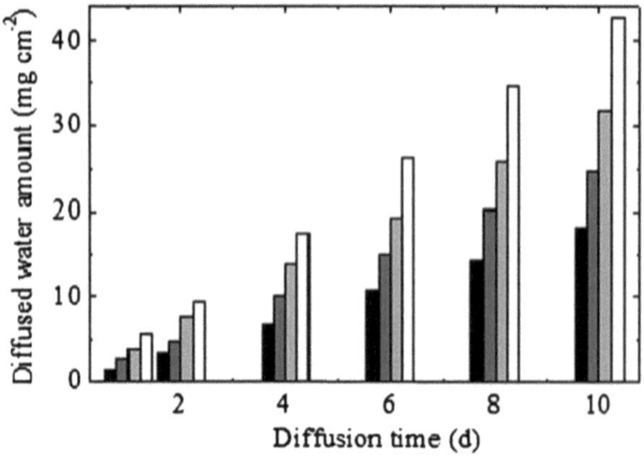

Fig. 4 Evolution of untaken water by weathered LDPE exposure: (*black*) non, (*dark grey*) 15 days, (*light grey*) 17 days (*white*) 25 days

where S is diffusion coefficient, M_∞ and M_o are the sample weights at saturation and in initial state, m is the molecular mass of diffusing gas, ρ is material density and P is the outer gas pressure. It is evident that the principal parameters influencing the diffusion process are the molecular mass of the gas and applied pressure. In the case of oxygen, the partial pressure in air is one fifth from atmospheric pressure, but its penetration becomes important in the air exposing samples. For water vapors that are generically nominated humidity penetrate sample sufficient easy because the molecular weight and volume are fitted for a facile migration into the inner layer of polymer. The internal distribution of degradation products gets a parabolic profile, because the abundance of oxygen-containing molecules obeys an exponential law as it is suggested by Fick's laws of diffusion.

The consequence of photodegradation of polymers is the easier movement of foreign molecules into the material bulk. Figure 4 illustrates the accelerated degradation caused by the water vapor penetration [26]. The prolonged exposure makes possible the fragmentation of molecules by the incidental radiation, while the resulting units allow the stranger molecules to pass by towards the inner free space.

The penetration of gases and water brings about the modification of mechanical properties, because the interaction between molecules in the vicinity of diffused molecules is turned away from origin. In the Eq. (3, [25]), σ(t), σ_o and σ_∞ represent the property at moment t, the start of ageing and at completely relaxed material, respectively, τ is relaxation time, β is a non-exponential factor that is placed between 0 and 1 indicating the availability to degradation.

$$\sigma(t) = \sigma_o . \exp\left[\left(-\frac{t}{\tau}\right)^{\beta}\right] + \sigma_\infty \qquad (3)$$

The functionality of any polymer subjected to the combined action of environmental factors induces a deviation in the substrate features. The amplitudes of either the penetrability or mechanical properties are influenced by the temperature range where the ageing process is provoked. The grass transition temperature is also

Table 2 The Tg values for weathered polyurethane and epoxy resin [27]

Polymer	Glass transition temperature (°C)				
	0 h	250 h	500 h	1000 h	200 h
Polyurethane	37	47	46	45	50
Epoxy resin	76	70	68	66	65

influenced by the material damaging [27]. The variation of Tg for the weathered polyurethane and epoxy resin is presented in Table 2.

The correlation between the effects of similar experiments, artificial or natural exposure can be managed by the Eq. (4) as the result of rearrangement of Arrhenius relationship:

$$\frac{k_1}{k_2} = \exp\left[\frac{E_a}{R} \cdot \frac{T_2 - T_1}{T_1 . T_2}\right] \tag{4}$$

where k_1 and k_2 are the rate constants of the process for the both situations, E_a is activation energy required by the process, R is gas constant and T_1 and T_2 are the temperatures at which the process occurs. By contrary if the same material is subjected to artificial and, separately, natural ageing, the difference between then is enlarging as the exposure becomes longer and longer [28]. In Fig. 5 the studies on oxidation by the exposure of bisphenol A polycarbonate to indoor and outdoor ageing emphasize the investigation result discrepancy due to the preference in the selection of experimental procedure.

During the studying or operation of polymer-based products, the most accelerated modifications take place in the thin outer layer of material, because the absorption of light prevails. The correlation between oxidation level and exposure time (Fig. 6) illustrated by the accumulation of carbonyl structures and the contact angle at the surface of ethylene–propylene–diene terpolymer demonstrates the proportionality between the concentration of collected oxidation products and the physical behavior of specimen [7]. The main carbonyl structures (Fig. 7) appeared

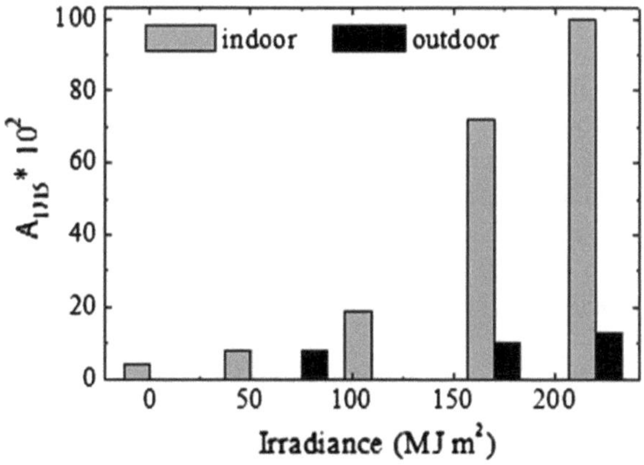

Fig. 5 The accumulation of ketonic carbonyls during the weathering of bisphenol A polycarbonate [28]

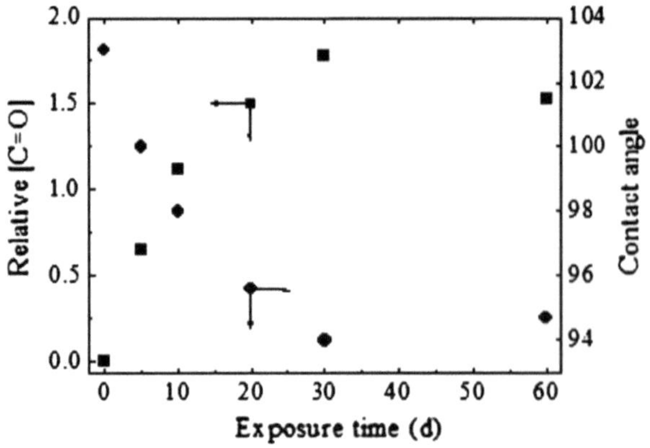

Fig. 6 Dependencies of carbonyl concentration and contact angle on the exposure time of ethylene–propylene–diene terpolymer [7]

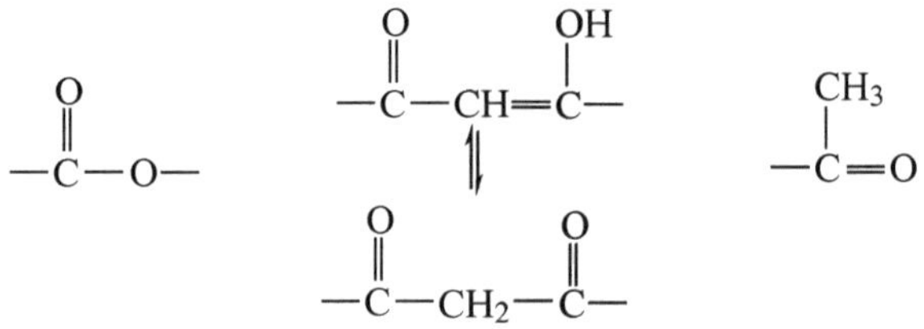

Fig. 7 Various carbonyl structures appeared during weathering treatment of polymers

simultaneously with alcohols are formed starting from the same source (alkoxy radicals, RO˙) according with the mechanism presented in Fig. 1. Their relative concentrations depend on the branching degree, the presence of stabilizer and primary level of unsaturation [29].

A comprehensive survey on the factors that affects the photostability of polymers was reported [30], where a significant aspect of basic studies defaulted the effects of specific characteristics of place preliminary integrated in a certain geographical region. Other aspects that are dismissed regard the severity of environmental factors, the limit of failure, the inherent sensitivity of material, the correlation between the irradiation conditions and the size of samples. Sometimes this kind of investigations are achieved on damps or sufficient thick (~1 mm), other time the samples are films, where the penetration is effective onto the whole length. The parabolic distribution of degradation products influences the probability of chemical decay or mechanical failure.

Other important energetic problem that is usually omitted by the majority of researchers is the reflectance of sample surface that diminishes the received energy [31]. The Eq. (5) describes the error that alters the results due to the discarding the problem of incidental angle of exposure.

$$\frac{R_t}{R_0} = \exp\left[-\left(\frac{4\pi(\cos\varphi)\sigma}{l}\right)^2 kt\right] \tag{5}$$

where R_t and R_0 are the spectral reflectance of sample after irradiation and of control (black body), respectively, ϕ is the incidental angle, σ represents the surface roughness, l is sample thickness, k is Boltzmann's constant and t is exposure time. For reliable results and considerations specific standards are required.

The general practice related to the use of plastics in various areas, predominantly for the fabrication of consumer goods, food packaging, medical and sanitary applications, roof and building materials asks the certification of products durability by the description of polymer capacity to retard degradation during environmental exposures. The modeling of degradation impact on the materials resistance during the property alterations in environmental conditions would be made starting from the Eq. (6) [1], which takes into account by the energetic of process:

$$v = A\cdot\exp\left(\frac{\Delta G - B\delta}{RT}\right) \tag{6}$$

where, v represents the degradation rate, ΔG denotes the Gibbs potential that is characteristic for involved process, δ is an energetic factors that depends on the stress susceptibility of material, A and B are process constants that depend on the collision frequency and constitutive bond energies, respectively, R is gas constant and T means the degradation temperature.

The far better understanding of rescue consequences caused by the deterioration of functional features of materials is directly related to the practice of industrial applications and the durability studies accomplished on multipurpose products. The synergetic actions of various factors leads to the conclusion that the simple addition of individual effects is possible, if a predominant factor directs degradation of a specific way. One illustrative kinetic parameter that characterizes the diminution of certain property is half life, defined by Eq. (7) [32]:

$$t_{1/2} = C + \frac{D}{T} \tag{7}$$

where $t_{1/2}$ means the time when the considered property reaches the half value of property in respect with initial one, T is absolute temperature at which degradation takes place, C and D are material constants depicting the resistance under standard conditions. They can be determined by measurements carried out in two different circumstances. These constants, C and D, can be assimilated with the straight line coefficients from logarithmic from of Arrhenius relationship. This approach allows the evaluation of critical conditions, when material becomes improper for a certain service.

The durability of building materials exposed to the direct action of light discarding the thermal effects, which are continuously irradiated by sun light can be evaluated by Eq. (8) [33, 34] degraded by outdoor factors:

$$D(t) = \int_{0}^{t} \int_{l_{min}}^{l_{max}} E(l,t).\left(1 - e^{-k(l,t)}\right) F(l) dl\, dt \qquad (8)$$

where, D(t) is material durability, E(l, t) is the UV spectrum applied for irradiation, l_{min} and l_{max} are the wave length of the inferior and superior, respectively, ends of range, k(l, t) is the spectral absorption at wavelength l over t time, F(l) is the spectral quantum efficiency assuming that the processed material has a very narrow thickness. However, there is large discrepancies between the fitting of experimental results obtained during outdoor and indoor exposure, the first ones reveal a less intensive ageing [35].

The typical weathering degradation studies disregard the contributions of several other factors that affect the magnitudes of permeabilities of oxygen and water vapors penetrating materials [36, 37]. The first example is the degradation of ethylene–propylene–diene terpolymer (EPDM) which was previously subjected to a gamma irradiation to form free reactive radicals. The existence of a certain concentration of oxidation promoters, polymer is easily degraded in the presence of ozone [38]. The amount of carbonyl groups increases because the material availability to oxidation allows evidently the promotion of oxidative ageing. This situation is representative not only for the mountain zones, where the concentrations of ozone are significant, but also in the indoor experiments, when moderate and high intensity UV light is used for photo-ageing of weathered products.

Other aspect to which the majority of researchers do not pay attention, especially in the studies on chemical weathering of insulating materials destined to electrical cables and wires is the evaluation of altered functional characteristics by electrical field. This aspect is quite important for the security of aerial and buried electric transport networks. The channels appeared as the result of the orientation of dipoles in oxidizing polymers die to electrical field allow the further penetration of water which forms electrical trees [39, 40]. The length, the density and the size extension of these defects influence deeply the durability of electrical insulator. Moreover, the continuous action of electrical field amplifies the degradation, which increases proportionally the failure of outdoor operating cabled.

The most frequent stressor acting during weathering ageing is mechanical charge. The pressure that is temporary or permanently applied on polymer products causes the modification in the intermolecular distances. Even though the diffusion coefficients for oxygen and water penetration are lowering down, the new formed peroxyl radicals have higher probability to be involved in propagation of oxidation which is a chain process [41]. If there are taken into consideration the supplementary action of other degrading factors simultaneously present, the effects of mechanical compression. The degradation rate is therefore augmented due to the restructuration of material morphology.

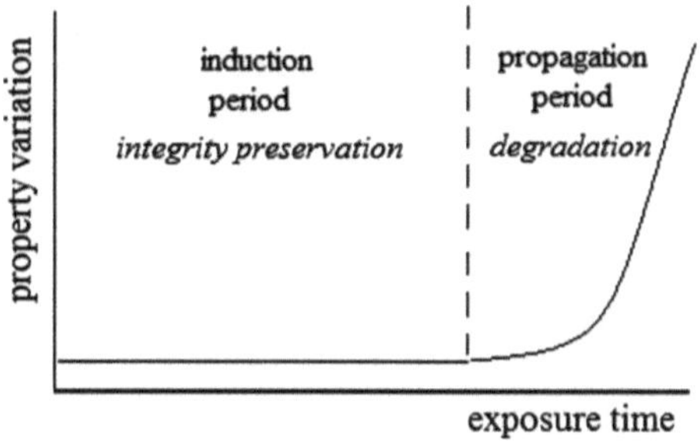

Fig. 8 The evolution of degradation during oxidation by weathering

One factor that is also discarded is the weathering material testing by immersion in water. The continuous contact with this fluid creates the constant penetration of water onto inner layers of polymer, but it allows the depletion of degradation products by their diffusion towards outer limits of sample and their dissolution in degradation environment. This scenario can be applied to any material, if the migration of stabilizers is the main process through which specimens are exposed to the fast chemical alterations [42–44].

The weathering process that implies profound variations in the material properties can be graphically represented by an ascendant curve with certain characteristics (Fig. 8).

The two well defined regions which compose the graph of degradation versus exposure are temporally separated by the oxidation induction time. It delimitates the period over which the material resistance keeps the polymer in the nonoxidized state and the conservation of properties is the main feature in the development of weathering treatment or exposure. The duration of this period as well as the rate of degradation on the propagation stage depend on the formulation (the lack of weak bonds), the crystallinity degree (the minimum amount of amorphous phase), the branching level and the polymerization degree of macromolecules, the presence of stabilizer (it is preferred a hindered amine stabilizer—HAS, a hindered amine light stabilizer—HALS or a couple of two additives for attaining a synergetic effect), the diffusion coefficients for oxygen and water, the material homogeneity and the absence of degradation initiator traces like metallic oxides (from manufacture catalyst) or peroxides (the excess from polymerization process), the lack of mechanical tension, the material history including storage period, the intensities of degrading factors, the relative contributions of environmental factors that describe the sustained co-operation actions in the generation of free radicals as the first step of degradation chain process, the concentration and particle size distribution of filler, the spectral distribution, the temperature of testing/exposure conditions.

The oxidation induction time (OIT) in the isothermal processing, the onset oxidation temperature (OOT) determined in the nonisothermal testing experiments and oxidation rate (v_{ox}) are the fundamental parameters that describe the kinetic ageing process. They may be completed with other peculiarities like glass transition temperatures (Tg), swelling degree (Q), coloring or discoloring level, type and density of cracks and bubbles, carbonyl and hydroxyl indexes, a.s.o.

In fact, the life time of materials subjected to the action of the environmental degradation factors is generally described by the oxidation induction time, because during this operation period the constancy in the material properties is assured by the formation of negligible amounts of oxidation initiators. The longest the oxidation induction time, the more advanced stability the polymer presents. Of course, the oxidation rate is correlated to OIT, because the increasing concentrations of each oxygen-containing radical unit promote faster progress of propagation step.

The qualification of materials by the evaluation of kinetic parameters and attributes is based on the degradation mechanisms, which combine the contribution of each sequential step to the product accumulation and the modification of material behavior [45–48].

2 Weathering of Engineering Polymers

The diversity of plastics products and, of course, the multitude of service conditions ask the certification of durability and the evaluation of property modifications happened while they are in operation. The weathering degradation tests as well as the imposed functional requirements provide the useful and complementary considerations by which the stability studies are conducted. Unfortunately, the geographic criteria are not generally applicable, but they offer some guidelines by which the customers can select suitable materials.

2.1 Polyethylene

The most used polymer due to the pertinent properties under a large variety of applications, polyethylene is manufactures as different sorts. They are selectively concerned and their long term performances are based on the specific chemical resistance. Starting from the influence of contact with metals which promotes oxidation in the large extent [49] the reduction in the functional characteristics results from the efficient attack of oxygen accompanied by the simultaneous energy transfer by photo-irradiation. The diffusion of small molecules like oxygen and water control the degradation profiles and imposes a certain approach of degradation rate. The penetration depth is described by Eq. (9) [50]:

$$TOL = \left(\frac{D_{O_2}[O_2]}{r_{ox}}\right)^{1/2} \tag{9}$$

where TOL expresses the distance to which oxygen goes in, D_{O2} denotes the characteristic diffusion coefficient of material, $[O_2]$ is local concentration of diffusing oxygen and r_{ox} is the oxidation rate. It can be remarked that the reverse proportionality of oxidation progress being the oxygen consuming process allows the advanced penetration to the deeper layers of material. The reported model for polyolefins describes the possibility to restrict degradation by following the specific roles of product manufacture [51]. The troublesome problems appeared as the consequence of oxidative degrading can be surpassed by the application of proper manufacture parameters [52].

The course of oxidation of polyethylene can be tracked by the spectral measurements of accumulation of ketone, acid and hydroperoxide concentrations versus absorbance which is proportionally changed with exposure time [53]. If the humidity conditions are similar, the accumulation of oxidation products depends lightly on the sun exposure, expressed by the spectral absorbance of certain organic function appeared during weathering exposure. The elimination of irradiance as independent factor allows the good comparison between results obtained at different geographic sites.

The outdoor exposure to sun light in Sicily during November period has deep modifications in the properties of LDPE (Fig. 9) [54]. The benefit of stabilization is evident, even though the highest oxidation degree is attained in the first microns from the outer layers [55].

The formation of oxidation products is reported by the ways through which radicals are involved in the weathering exposures and treatments of polyolefins [53].

The presence of nanoparticles in the compositions of polymer products causes negative effects on the chemical stability of host polyethylene. The promotion of degradation was revealed by nanoparticles of natural montmorillonite (Cloisite 15A), whose contribution to the increasing carbonyl index is enhanced with its concentration (Fig. 10). The molecular weights of polyethylene drops drastically, when nanoparticles of montmorillonite are included in the formulations [56]. The hybrid nanocomposites consisting of LDPE and Cloisite 20A are degraded with three times higher oxidation rates, which would be explained by the greater

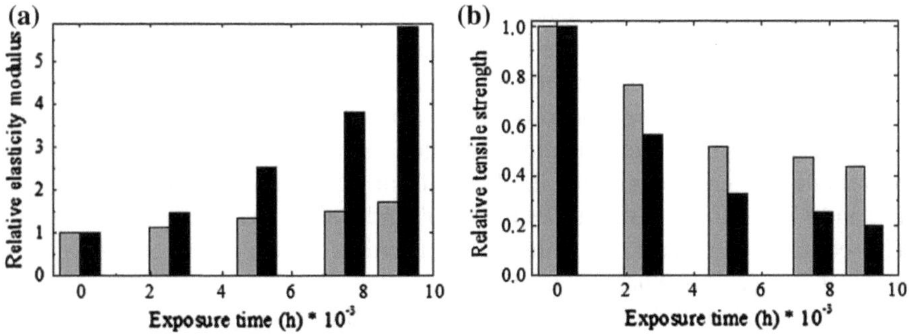

Fig. 9 Modifications of elasticity modulus and tensile strength by exposure of LDPE to sun light in outdoor irradiation. (*black*) free of stabilizer, (*grey*) in the presence of Irganox 2001 + Chemisorb 81 (0.5 % for each of stabilizers)

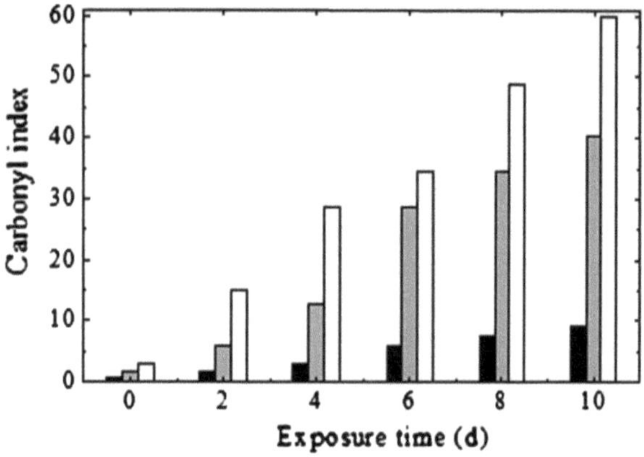

Fig. 10 Changes in carbonyl index of UV irradiated LDPE modified with Cloisite 15A. (*black*) neat LDPE, (*grey*) LDPE modified with 2 % filler, (*white*) LDPE modified with 7 % filler

diffusion coefficient [57]. The filler allows the gaseous oxygen to penetrate more easily due to the larger free spaces.

The consequence of UV irradiation is the progressive decrease in the polyethylene viscosity [58]. This essential element must be correlated with the modification of diffusion property of polymer, by which the degradation is significantly accelerated. The type of nanofiller (multi-walled carbon nanotubes, fumed silica, neat Cloisite and modified Cloisite) influences differently the behavior of pristine HDPE [59]. The increase order of tensile strength measured at Yield point places the contribution of studied nanoparticle phases for the first 100 h of UV exposure describes promotion of a crosslinking process involving the radicals formed by photolysis.

$$\text{Neat HDPE} < \text{HDPE} + \text{MWCNT} < \text{HDPE} + \text{neat MMT} < \text{HDPE} + \text{SiO}_2 \\ < \text{HDPE} + \text{modified MMT}$$

The continuation of UV exposure brings about the alteration of property, which can be explained by the multiple scissions and the accentuation of photo-oxidation.

The Fig. 11 presents the ways through which oxidation products appear. Of course, Norrish I and II complete mechanism. The consumption of hydroperoxides is screened by the continuous formation during photolysis.

The incidental photons deposit their energy on the molecules till they are completely absorbed. The oxidation induction time for the photolytic oxidation of polyethylene decreases with the irradiance from 125 min for unexposed sample to 112 and 97 min for exposure times of 60,000 and 100,000 kJ m^2, respectively [60]. This comparison reveals the contribution of tertiary carbon atoms on the vulnerability of hydrocarbon chains (Fig. 12).

In the case of the photooxidation of polyethylene/polyamide-6 blends, the mechanism reported for these blends shows the involvement of polyethylene, the component with the lower stability under the action of UV light. The degradation starts in polyamide fraction and the diffusion of radical into polyethylene phase

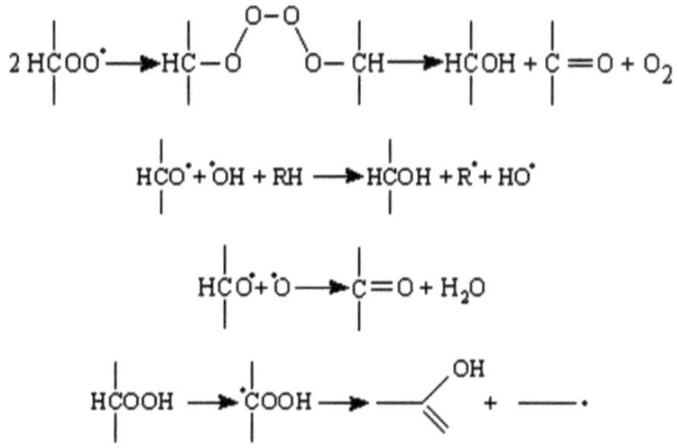

Fig. 11 The summary scheme on the genesis of oxidation products appear in polyethylene

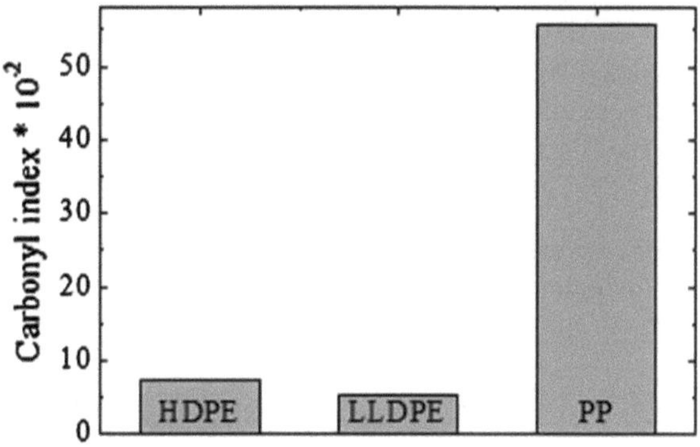

Fig. 12 The comparison between carbonyl indexes for three linear polyolefins

brings about oxidation which is propagated faster in this blend component [61]. A synergistic consequence in these blends is explained by the formation of unsaturation in the polyethylene part. If the accumulation of unsaturation in neat polyethylene is compared with the upgraded amount of the same structure in LDPE/PA 6 blends it can be noticed that the former process occurs less faster that the last one [62].

The most intensive investigations on the resistance of various sorts of polyethylene are related to the sheets used for green-houses [63–66].

2.2 *Polypropylene*

Polypropylene is another polymer materials which has large areas of applications. The presence of propyl moieties, where the tertiary carbon atoms represents the possible places for degradation is the main factor taken into consideration for estimation of durability. The comparison between polyethylene, polypropylene and ethylene–propylene–diene monomer was reported and the criterion of stability is the comparison point of view [67]. The blends consisting of polyethylene and polypropylene under natural and artificial degradation are more resistant when the polyethylene load is higher. The severe absence of UV irradiation during the exposure in the open field minimized the rate of degradation [68], but the day light caused a progressive coloring as the indication of advanced depreciation.

Starting from the degradation studied on labeled polypropylene the scheme presented in Fig. 13 could be conceived [69]. The studies on the UV exposure of polypropylene confirm this mechanism [70]. However, the differences in the oxidation product abundances was noticed [70] because of the unlike energetic conditions and the specific contributions of degrading factors that act in weathering investigations (Fig. 14).

It was stated that the low crystalline polypropylene (LCP) is easily oxidized than the high crystalline polypropylene (HCP). The greater degree of ordering is mirrored by the lower diffusion rate. For example, the final content of ketones in LCP after 21 days with the irradiance of 14,000 kJ m^2 exceeds with 25 % the similarly degraded HCP, while acid content is more abundant in LCP with about 57 % than it was accumulated in HCP [70].

The seasonal ageing of polypropylene is characterized by some specific features (Fig. 15) [71]:

- the increase in the carbonyl indexes in the summer time is more accentuate than the enhance in winter time,

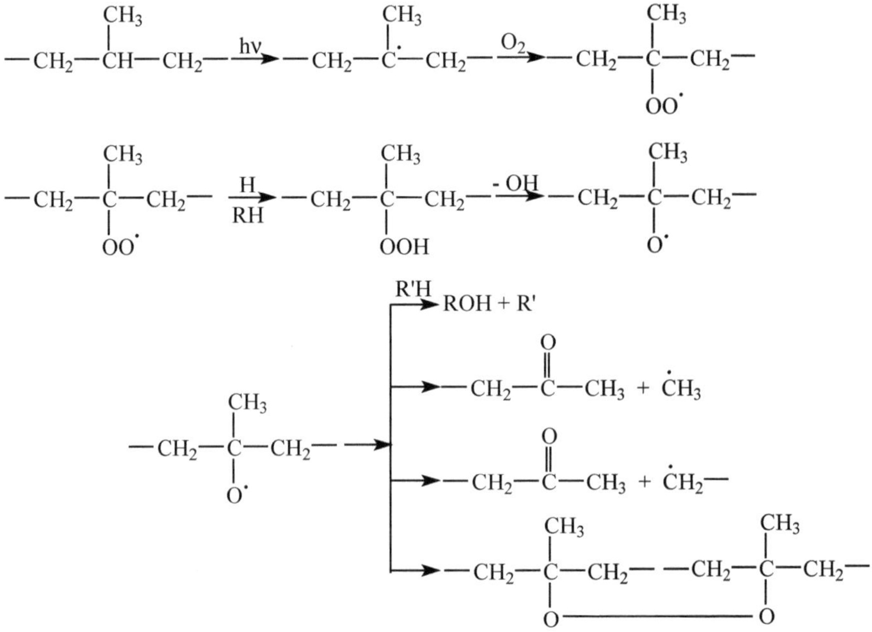

Fig. 13 Decay scheme for the formation of oxidation products in UV exposed polypropylene

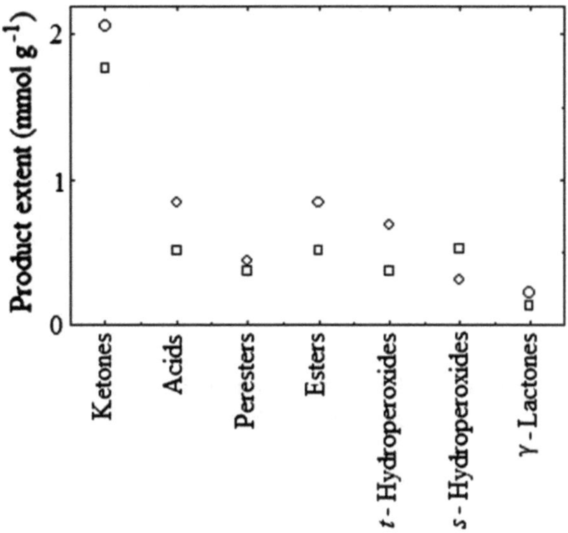

Fig. 14 Distribution of oxidation products in low-crystalline polypropylene exposed to weathering degradation. (*open circle*) outdoor oxidation, (*open square*) accelerated oxidation

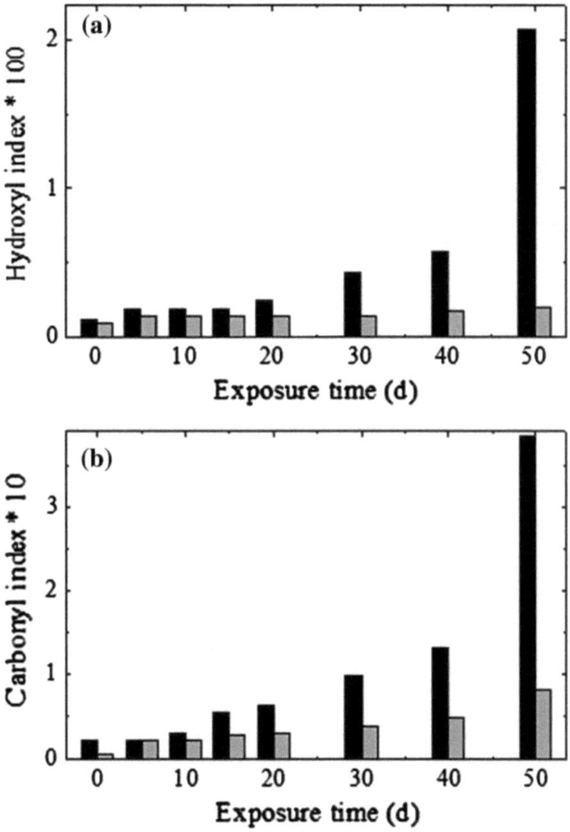

Fig. 15 Variations of **a** hydroxyl index and **b** carbonyl index for polypropylene exposed to weathering treatment on two opposite seasons. (*black*) summer, (*grey*) winter

- the hydroxyl indexes have present constant values over large time interval indicating that the formation of carbonyl units is preferentially done by the conversion of hydroperoxides maintaining OH-containing structures at a minimal level and their consumption is not compensated by suddenly generated peroxyl radicals,
- the sharp increase in carbonyl-containing products is occurred either in summer time or in winter time after 40 days of exposure. It means that a critical concentration is reached at this long period. It may be assumed that the higher concentration of hydroperoxides allows the bimolecular process [72] in a good agreement with the proposed scheme of Fig. 13.

The molecular configuration indicating the spatial arrangement of atoms and the existence of substituted carbon atoms is a decisive factor for the description of material oxidation resistance. The two kinds of polypropylene, isotactic and syndiotactic present different responses to the degrading activities of environmental aggressive factors (Fig. 16) [73]. The certain ordering degrees of molecules in our examples lead to the decrease in the susceptibility of polymer to oxidation.

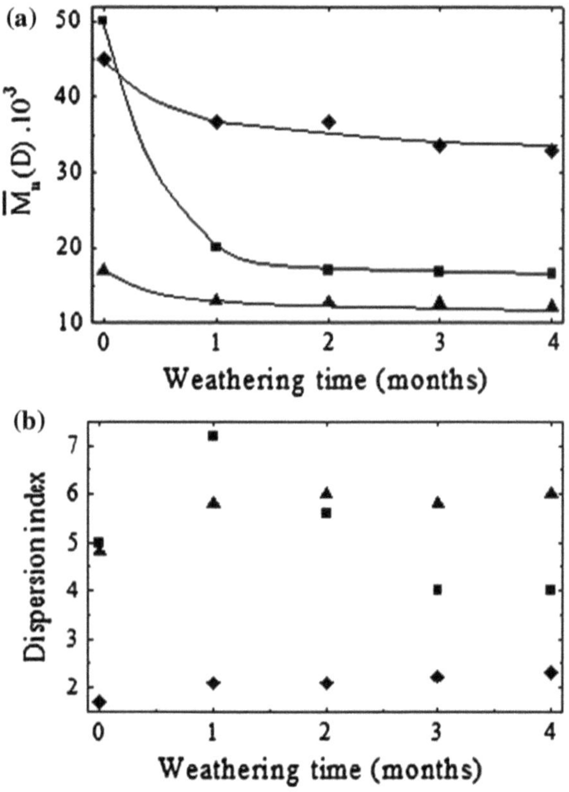

Fig. 16 The modification in **a** average numerical molecular weights and **b** dispersion indexes for (*filled square*) isotactic polypropylene, (*filled diamond*) syndiotactic polypropylene, (*filled triangle*) high density polyethylene

The FTIR investigation on the carbonyl accumulation confirms the discrepancies that exist between the oxidation rates of i-PP, s-PP and HDPE. The stability ranking places these polymers in the following series

$$\text{HDPE} > \text{i-PP} > \text{s-PP}$$

and their physical properties as well as durabilities are modified accordingly. Each incidental wave length induces the increase in the concentration of carbonyl units, but this amount becomes smaller and smaller as the wave length is increased, because the deposited energy on the macromolecules is higher at lower wave length.

The similar concept regarding the stability of various structured polypropylenes, the activation energies required for the oxidative degradation depends evidently on the nucleation of substrates [74, 75]. For three sorts of polypropylene: homopolymer (h-PP), random-copolymer (rc-PP) and impact-copolymer (ic-PP), the activation energies of photodegradation present a relevant sequence which particularizes the involvement of molecular packaging in the polymer volume:

$$\text{h-PP}\left(37.9\ \text{kJ mol}^{-1}\right) < \text{rc-PP}\left(42.4\ \text{kJ mol}^{-1}\right) < \text{ic-PP}\left(57.8\ \text{kJ mol}^{-1}\right)$$

The hybrid nanocomposites with various applications in the manufacture of improved products are evaluated by weathering tests [76, 77]. Polypropylene exposed to UV irradiation for 340 h do not present grooves, microcracks or microcavities [76]. This attribute illustrates the increased resistance to the action of degradation products, which are accumulated during photolysis. Polyethylene doped with multi-wall carbon nanotubes presents peculiar times after which the elongation at break becomes half of initial value (Fig. 17) [77].

The attempts for the description of cracks formation in UV-irradiated polypropylene emphasized the dependence of local oxidation state which determines the appearance of material discontinuities (Fig. 18) [78]. This variation is caused by the difference in the energies of each photo-beam, whose penetration is

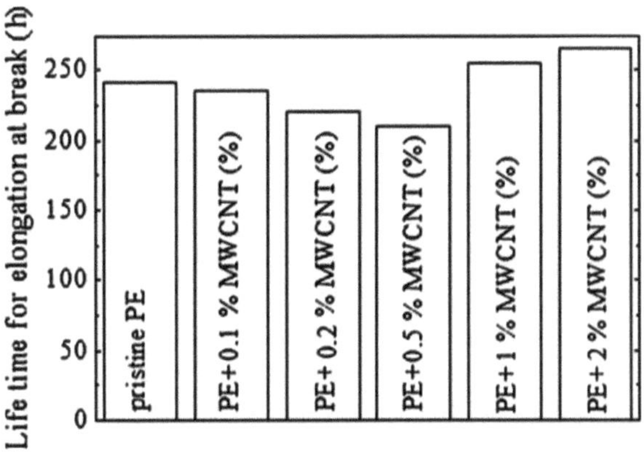

Fig. 17 Half life of polyethylene/MWCNT composites for irradiation with UV-B lamps

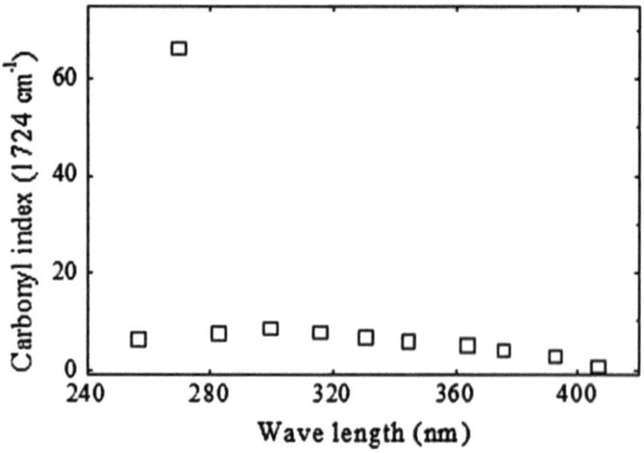

Fig. 18 Dependence of cracking availability expressed by carbonyl index on irradiation wave length

accompanied by the energy transfer onto polypropylene molecules. The distribution of cracks is inhomogeneous and propagation of these cracks is possible due to the oxidation process that accompanies UV irradiation as the main component of weathering ageing. The Arrhenius representation of the variation of carbonyl indexes at various temperatures provides a reliable picture of material durability.

The easy degradation of polypropylene by UV irradiation was explained by the mechanistic bases, where the generation of peroxyl radicals as the initiators of loop oxidation [80, 81] is done aleatory in the bulk, but they become the degradation start points in the blends with other more stable polymers. The decomposition contribution of polypropylene fraction loadings on polyamide-6 host polymer in their blends is important being demonstrated by thermal analysis of these systems [82].

2.3 *Ethylene–Polypropylene Elastomers*

The resistance of this class of polymers was reported in detail because they have multiple applications in the various economical branches [7, 83–85]. The natural ageing was also intensively studied because of the outdoor usage as cables, gaskets, membranes in various equipments. Because the permeability of EPDM membranes depends on their oxidation states, the prediction of oxygen diffusion is not enturely possible during or after weathering treatments [86].

The composition of these elastomers plays an important role in the degradation kinetics. The relative ration of ethylene content to propylene loading influences the progress in the oxidation process by the involvement of various values of activation energy (Fig. 19). It was reported [85] the higher content of ethylene increases the thermal stability of EPDM.

The resistance of EPDM to the oxidation process during weathering was studied by the variation in the number of crosslinks [87, 88]. The dependence of tensile strength on the number of crosslinks is shown in Fig. 20a. These results indicate the

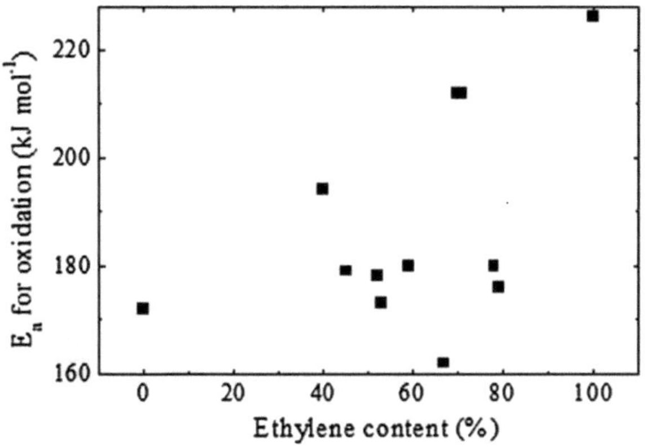

Fig. 19 The distribution of activation energy values required for thermal oxidation versus ethylene content in EPDM

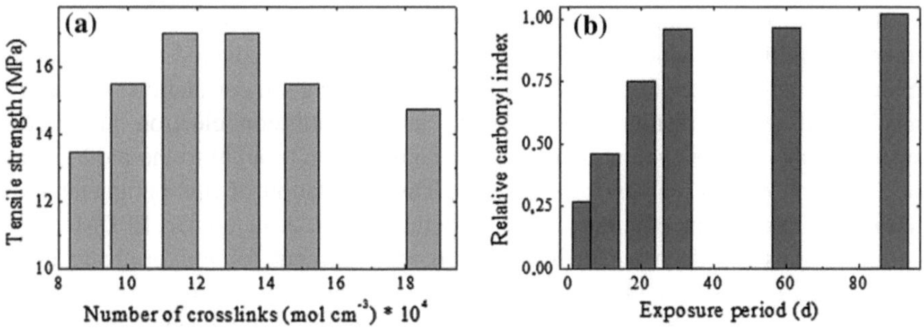

Fig. 20 **a** The modification in mechanical behavior versus number of crosslinks and **b** the accumulation of carbonyl functions versus exposure time

slight curing of elastomers due to the UV irradiation and the mechanical properties are improved on the first stage of treatment. The further exposure brings about the start of degradation and the integrity of material is spoiled. However, the build-up of carbonyl units occurs (Fig. 20b), even the degree of crosslinking is simultaneously enhanced.

The diffused oxygen during degradation, even it occurs naturally [89] or under accelerated conditions [90] is consumed by the reactions with free radicals. The oxygen uptake is quite significantly increased because of the abundance of free radicals that consume diffused oxygen (Fig. 21) [91]. If it is taken into consideration that the fragmentation of molecules happens simultaneously and the viscosity of exposed material is dropping down [92], the acceleration of degradation is promoted due to the increase in the diffusion coefficient.

The complex environmental conditions that describe the majority of applications impose the analysis of the defined influence of blending on the stability of EPDM

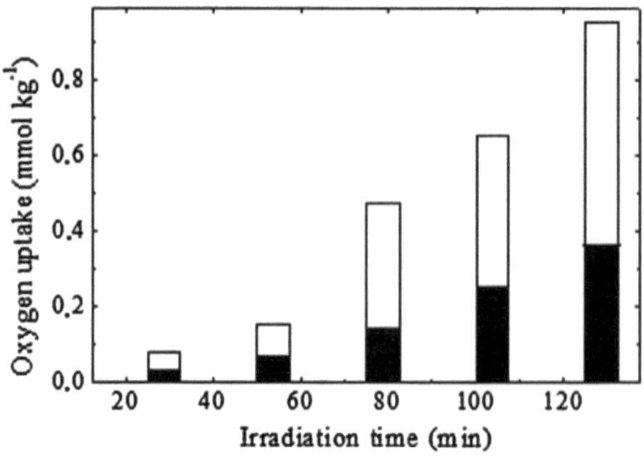

Fig. 21 The oxygen consumption during photolysis of EPDM: (*black*) neat material, (*white*) material exposed to UV irradiation (λ > 300 nm) testing temperature: 100 °C

over time. The molecular scissions occurred in each component phase change the physical behavior of basic materials, but the amplitude of variation is determined by the difference extent between materials. The polyurethane coating sheets get the modification in the molecular distribution, so it is more probable the faster penetration of fluids. The shielding effect is diminished and the covered bulk is exposed more direct to the intensive attack of oxygen [93]. The good correlation between the mechanical testing results and the XRD and transmission electron microscopy sustained also by the evaluation of coating contribution brought to the modification of activation energy values for oxidation. The calculation of this parameter illustrating the protection efficiency of polyurethane thin sheet on the EPDM degradation reveals the decrease in the energetic requirement for polyurethane in the favor of elastomers. The observed valley in the dependency of activation energy for oxidation on polyurethane loading presents a minimum at a decomposition degree of 50 %. The further increase in the energetic condition seems to be due to the recombination of radicals belonging to the both types of materials.

In fact, one of the most important factors that determine the stability of EPDM is not only the C2/C3 ratio, but also the type and content of the thirds component included in the macromolecules of this polymer [94]. The accumulation of degradation products follows the principle that the greater propylene groups, the more accelerated degradation is occurred (Fig. 22a, b).

The oxygen consumption is accelerated by the increasing of diene (5-ethylidene-2-norbornene) content, because it provides the higher volume concentration of unsaturation sites. The microhardness and Young modulus present maximum values at 100 h of irradiation in Xenotest device. The promotion of crosslinking by the incidental UV light is confirmed being followed by an accelerated degradation due to the presence of molecular oxygen in augmented amounts.

It was proved that the expanded ethylene–propylene–diene monomer is a good component for the manufacture of high quality roofing spare parts [95]. It minimizes the harsh characteristics under the conditions of permanently subjected plates to the activities of natural factors. The thermal conductivity of EPDM/NR/wood

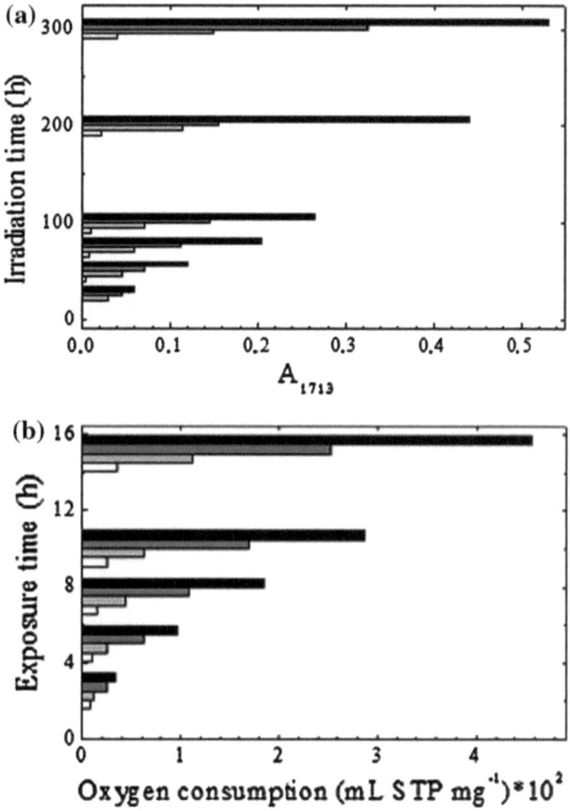

Fig. 22 Oxidative degradation by weathering monitored by **a** FTIR absorbance and **b** oxygen uptake for EPDM with different EBN content: (*white*) non (C2/C3 = 50:50), (*light grey*) 2 % (C2/C3 = 56.5:43.5), (*dark grey*) 5 % (C2/C3 = 54:46), (*black*) 9 % (C2/C3 = 48:52)

laminates presents a minimum value after 30 days of ageing, but, at an advanced exposure, this property increases slightly. The explanation of this feature is the accumulation of degradation products in the expanded elastomers. The wood fraction of these blends retains humidity, so the synergic effect water vapor/oxygen cannot be considered as ongoing action. The gapping effect of bubbles in EPDM fraction is also considered for the accumulation of gasses formed during photolysis. The manufacture procedure influences the behavior of EPDM products. The absorption of oxygen is more quantitatively uptaken as the amount of dicumyl peroxide (DCP) used for elastomers crosslinking is higher (Fig. 23) [96]. The explanation of this fact is the existence of peroxide trace remained incorporated in processed material, which initiates degradation under UV irradiation. The similar effect was noticed when the traced metals originated from polymerization catalyst initiate depreciation of product by self or accelerated oxidation in low density polyethylene [97].

The simple comparison between the degrading effects of heating and UV irradiation has pointed out the acceleration of degradation by the incidental light. While the irradiation of EPDM with a light beam (340 nm) for 80 days (2500 kJ m^2) the formation of absorption peaks at 1715 and 3382 cm^{-1} characterizing the

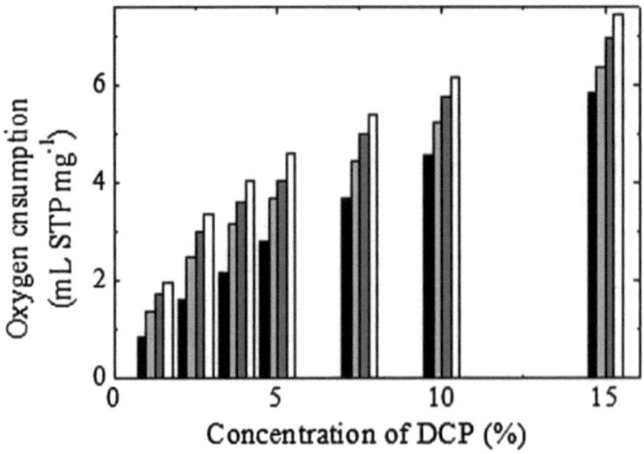

Fig. 23 The dependence of oxygen consumption on the concentration of DCP used for EPDM crosslinking (ENB content: 9.0 %) (*black*) 0 %, (*light grey*) 1.5 %, (*dark grey*) 2.5 %, (*white*) 4.0 %

accumulation of carbonyl units and hydroperoxides, respectively, the heat treatment (90 °C for 7 days) of the same kind of specimens does not induced any degradation revealed by the whitening measurements [98].

The majority of nanofillers added in the formulations of polymer products act as adsorbers of free radicals formed during ageing [99]. This process marks the delay of oxidation that accompanies the other modification such as the changes in material morphology, the susceptibility to oxidation, the hydrophilicity, the functional features determined by the structural alterations in polymer phase [100]. An illustrative example is EPDM/carbon black nanocomposites. The type of degradation has a decisive influence on the evolution of degradation (Fig. 24), because the rate of accumulation for oxidation products does not follow similar development. The induced difference brings the dissimilarities between the alternatives of testing [101].

2.4 Other Polymers

Polyamides are resistant polymers to degradation because of their low diffusion coefficients for gases and water [102, 103]. Regardless of destructive action photoirradiation stimulates an improvement in the stability of polyamides by crosslinking [104]. The association of polyamides with other polymers produces blends whose resistance to oxidation varies with the percentage loadings. Polypropylene in the mixtures with polyamide-6 is the free radical supplier for the whole of material [79]. The generation of oxidation promoters initiates the material ageing, but during the UV exposure the crystallinity increases and the oxidation process is hindered.

The oxidation of two types of polyethylene in the mixtures with polyamide-6 exhibit shows opposite behavior in the respect with the addition of clay nanoparticles [105]. While low density polyethylene/polyamide-6 modified with montmorillonite nanoparticles presents inferior values of carbonyl content in comparison

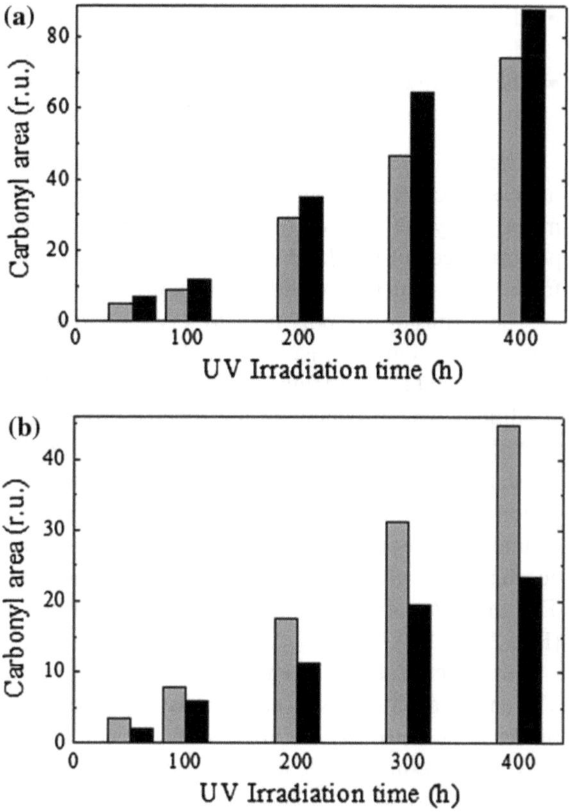

Fig. 24 Evolution of carbonyl build up in **a** LDPE/PA-6 and **b** HDPE/PA-6 caused by UV exposure. (*grey*) neat blends, (*black*) blends modified with nanoparticles of montmorillonite

with neat blends, high density polyethylene/polyamide-6 with montmorillonite shows superior values of carbonyl accumulation in respect with unmodified systems (Fig. 24). This unlike response to the weathering treatment would be caused by the difference in the diffusion characteristics, which modifies the penetration of oxygen.

The mechanism of the thermal decomposition of polyamide 6 reveals the priority of scissions for C–C bonds in respect with C–N bonds [106–108].

Rubbers are polymers with large engineering application areas. The weathering oxidation progresses in the elastomers according with the nature of substrate. A comparison between styrene-butadiene rubber (SBR), ethylene–propylene–diene monomer (EPDM), chloroprene rubber (CR) and natural rubber (NR) [109] allowed stating some relevant outlines:

- the changes of mechanical properties are influenced by the polymer structure which characterizes the durability under climatologically induced ageing,
- the correlation between material resistance and the degrading conditions can be reliable after medium or long exposure, because some irregularities in the evolution of property can be notices on the starting stage,

- the geographical site where the experiments are accomplished must be well defined due to the integration of measured effects,
- the formulation and the sequence of molecular constitutive moieties, as well, are the predominant factors that indicate the mechanism of degradation.

The stability study on styrene-butadiene-styrene rubber [110] established that the increasing content of butadiene improves the mechanical strength of UV exposed material. By the addition of convenient amounts of nanotubes (0.25 and 1 phr), the performances of this kind of elastomers are enhanced because this filler has satisfactory consequences on the scavenging free radicals against oxidation.

The exposure of polymers to UV radiation causes a depletion of stabilizer, because it is either consumed by thermal reactions with radicals, or expelled my migration [111]. Evidently, the decrease in antioxidant content restores the bad conditions which reduce the life time of products.

Polyurethanes subjected to weathering degradation show the variation of resistance in the function of constitutive segments [112–114]. The FTIR investigations evidences the urethane structure is lost, while other configurations (for example, hydroperoxides) are accumulated. The unsaturation which initiates accelerated degradation is also formed by the fragmentation of macromolecules. The integrity of polyurethanes can be described by the variation of roughness (Fig. 25, [112]).

Similar consideration regarding the stability of *epoxy resins* were reported [115]. The curing agent offers a certain resistance, which is modified by the degradation of binder simultaneously occurred with scission of macromolecules.

The interesting applications of polymers are the protection foils and the polymeric components of solar cells. The permanent exposure to the action of weathering agents, the foils are continuously degraded [116, 117]. The studying papers must be focused on the durability of these membranes because the quantum yield of material appoints the performances of electrical generators based by sun light.

Fig. 25 The modification of roughness during exposure of polyurethane at weathering test

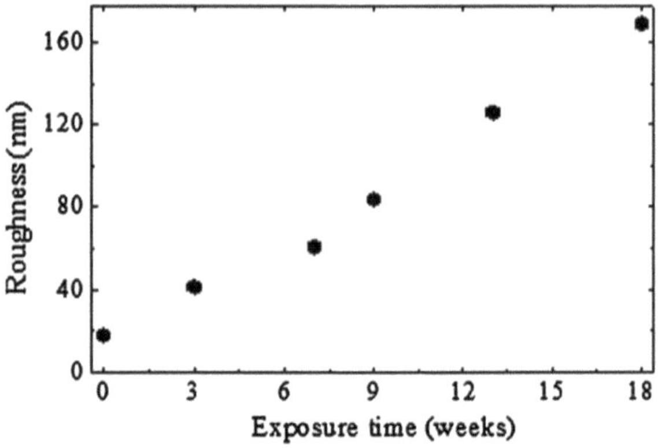

3 Conclusion

The applications of polymer systems are define by the chemical resistance against the action of degrading factors. The environmental stressors are elements that act independently on the human willing. The outdoor service of any polymeric product is subjected to degradation, even though the weather is more or less favorable to the perfect conditions of operation. The most important component of weathering ageing is the UV exposure, which causes chemical modifications by the depositing of energy on macromolecules, they being fragmented according to the bond energy spectrum. The formation of oxidation products determines the changes in physical properties, such as fluid diffusion coefficient, crystallinity degree, softening temperature range, hardness, mechanical and electrical characteristics, as well as the chemical attributes like induction period and rate of oxidation, duration of oxidation, and, in the end, the durability. These all features are interconnected, because the accumulation of degradation products changes the intimate interaction between various structures existing in the polymer bulk. The intensities and the time extension of degradation are correlated with the initial free radical abundance and with the competition between the formation rate for hydroperoxides and their decay into the degrading products.

There are many similar attributes of thermal-, photo- and radiation-degradation. They progress on the similar mechanisms based on the homolytic or heterolytic scissions and the formation of degradation initiators for loop process of oxidative degradation. The discrepancies between these three ageing ways consist of the concentration of primary free radicals and their distribution in the material. The depth of degradation is only some microns, but the diffusion of chain promoters, hydroperoxides, towards the inner layers of materials leads to a parabolic distribution of oxidation products around the symmetry axe.

The accelerated weathering offers a plausible description of outdoor degradation. The absence of synergistic effects between the natural ageing factors: sun light, heat, humidity, wind blowing is a basic dissimilarity between indoor and outdoor exposures. The predominance of UV irradiation which is the manner through which material is seriously altered completes the comparison. However, the option of one of these degradation procedures is dependent on the purpose of materials, on the material sensitivity, on the required duration that characterizes the service conditions, on the extending range of applications.

The conservation of initial properties of materials can be achieved either by the improvement of chemical resistance after additional crosslinking, or by the addition of light protectors, which delay oxidation in direct relation of their antioxidant activities. The transformations induced by weathering treatments are minimized, if the polymer is efficiently screened with other protective layer.

The weathering studies describing the oxidation state of polymer wastes can be used for the design of recycling purposes, because the pre-existing oxidation products in the raw materials imposes a certain strategy for the waste recovery goals.

References

1. White, J.R., Turnbull, A.: Weathering of polymers: mechanisms of degradation and stabilization, testing strategies and modeling. J. Mater. Sci. **29**, 584–613 (1994)
2. Searle, N.D., McGreer, M., Zielnik, A.: Weathering of polymer materials. Encycl. Polym. Sci. Technol. (2010). doi:10.1002/0471440264.pst401.pub2
3. Ojeda, T., Freitas, A., Birck, K., Dalmolin, E., Jacques, R., Bento, F., Camargo, F.: Degradability of linear polyolefins under natural weathering. Polym. Degrad. Stab. **96**, 703–707 (2011)
4. Tocháček, J., Vrásničková, Z.: Polymer life-time prediction: the role of temperature in UV accelerated ageing of polypropylene and its copolymers. Polym. Test. **36**, 82–87 (2014)
5. Zhang, X.-F., Zhou, L.-C., Pi, H., Guo, S.-Y., Fu, J.-W.: Performance of layered double hydroxides intercalated by a UV stabilizer in accelerated weathering and thermal stabilization. Polym. Degrad. Stab. **102**, 204–211 (2014)
6. White, C.C., Tan, K.T., Hunston, D.L., Nguyen, T., Benatti, D.J., Stanley, D., Chin, J.W.: Laboratory accelerated and natural weathering of styrene–ethylene–butylene styrene (SEBS) copolymer. Polym. Degrad. Stab. **96**, 1104–1110 (2011)
7. Zhao, Q.-L., Li, X.-G., Gao, J.: Aging of ethylene–propylene–diene monomer (EPDM) in artificial environment. Polym. Degrad. Stab. **92**, 1841–1846 (2007)
8. Pickett, J.E., Gibson, D.A., Rice, S.T., Gardner, M.M.: Effects of temperatures on the weathering of engineering thermoplastics. Polym. Degrad. Stab. **93**, 684–691 (2008)
9. Kockott, D.: Natural and artificial weathering of polymers. Polym. Degrad. Stab. **25**, 181–208 (1889)
10. Favaro, M., Mendichi, R., Ossola, F., Russo, U., Simion, S., Tomasin, P., Vitago, P.A.: Evaluation of polymers for conservation treatments of outdoor exposed stone monuments. Part I: photo-oxidative weathering. Polym. Degrad. Stab. **91**, 3083–3096 (2006)
11. Wypych, G. (ed.) Artificial weathering versus natural exposure. Handbook of Material Weathering, 5th edn, pp. 231–244, Chem. Tec. Publishing (2013). ISBN 978-1-895198-62-1
12. Nichols, M.E.: Paint weathering tests. In: Kutz, M. (ed.) Handbook of Environmental Degradation of Materials, 2nd edn, ch. 20, pp. 597–619. Elsevier, Amsterdam (2016). ISBN 978-1-437734-56-0
13. Dorsey, J., Rushmeier, H., Sillion, F. (eds.): Ageing and weathering. Digital Modeling of Material Appearance, Morgan Kaufmann Publisher, 193–225 (2008). ISBN 978-0-12-221181-2
14. Heikkilä, A., Tanskannen, A., Kärhä, P., Hanhi, K.: adjusting timing of weathering test for seasonial variation in UV exposure. Polym. Degrad. Stab. **92**, 675–683 (2007)
15. Sampers, J.: Influence of weathering factors other than UV radiation and temperature in indoor exposure. Polym. Degrad. Stab. **76**, 455–465 (2002)
16. Vink, P., Wisse, J.D.M.: The photo-oxidation of polyolefin containing hindered piperidine compounds. Polym. Degrad. Stab. **4**, 51–57 (1982)
17. Pilař, J., Michálková, D., Šlouf, M., Vacková, T.: Long-term accelerated weathering of HAS stabilized PE and PP plaques. Compliance the ESRI, IR, and microhardness data characterizing heterogeneity of photooxidation. Polym. Degrad. Stab. **120**, 114–121 (2015)
18. Pickett, J.E., Gibson, D.A., Gardner, M.M.: Effects of irradiation conditions on the weathering of engineering thermoplastics. Polym. Degrad. Stab. **93**, 1597–1606 (2008)
19. Dilara, A.P., Briassoulis, D.: Standard testing methods for mechanical properties of low density polyethylene (LDPE) film used as greenhouse covering materials: a critical evaluation. Polym. Degrad. Stab. **17**, 549–585 (1998)
20. Liu, Z., Chen, S., Zhang, J.: Photodegradation of ethylene–octene copolymers with different octane contents. Polym. Degrad. Stab. **96**, 1961–1972 (2011)
21. Pickett, J.E., Gibson, D.A., Gardner, M.M.: Effect of irradiation conditions on the weathering engineering thermoplastics. Polym. Degrad. Stab. **93**, 1597–1606 (2008)

22. Larché, J.-F., Bussière, P.-O., Thérias, S., Gardette, J.-L.: Photooxidation of polymers. Relating material properties to chemical changes. Polym. Degrad. Stab. **97**, 25–34 (2012)
23. Haillant, O., Dumbleton, D., Zielnik, A.: An Arrhenius approach to estimating organic photovoltaic module weathering acceleration factors. Sol. Energy Mater. Sol. Cells **95**, 1889–1895 (2011)
24. Haillant, O.: Accelerated weathering testing principles to estimate the service life of organic PV modules. Sol. Energy Mater. Sol. Cells **95**, 1284–1292 (2011)
25. Bijl, P., Heikilä, A., Syrjälä, S., Aarva, A., Poikonen, A.: Modelling of sample surface temperature in an indoor weathering test. Polym. Test. **30**, 485–492 (2011)
26. Nakatsuka, S., Andrady, A.: Studies on the enhanced degradable plastics. III. The effect of weathering on polyethylene and (ethylene–carbon monoxide) copolymer on moisture and carbon dioxide permeability. J. Environ. Degrad. Polym. **2**, 161–167 (1994)
27. Croll, S.C., Shi, X., Fernando, B.M.D.: The interplay of physical ageing and degradation during weathering of two crosslinked coatings. Prog. Org. Coat. **61**, 136–144 (2008)
28. Diepens, M., Gijsman, P.: Outdoor and accelerated studies of bisphenol A polycarbonate. Polym. Degrad. Stab. **96**, 649–652 (2011)
29. Zaharescu, T., Giurginca, M., Mihalcea, I.: Effects of some antioxidants on the stability of ethylene–propylene elastomers. Polym. Degrad. Stab. **48**, 151–154 (1995)
30. Pospíšil, J., Pilař, J., Billingham, N.C., Marek, A., Horák, Z., Nešpůrek, S.: Factors affecting accelerated testing of polymer photostability. Polym. Degrad. Stab. **91**, 417–422 (2006)
31. Croll, S.C., Hinderliter, B.R., Liu, S.: Statistical approaches of predicting weathering degradation and service life. Prog. Org. Coat. **55**, 75–87 (2006)
32. Maxwell, A.S., Broughton, W.R., Dean, G., Sims, G.D.: Review of accelerated ageing methods and lifetime prediction techniques for polymer materials. NPL Report DEPC MPR 016 (2005)
33. Yang, T.C., Wu, J.H., Noguchi, T., Ishiki, M.: Methodology of accelerated weathering test through physicochemical analysis of polymeric materials in building construction. Mater. Res. Innov. **18**, S3–S5 (2014)
34. Adrady, A.L.: Wavelength sensitivity in polymer photodegradation. Adv. Polym. Sci. **128**, 47–94 (1997)
35. Martin, J.W., Nguyen, T., Byrd, E., Dickens, B., Embree, N.: Relating laboratory and ourdoor exposure of acrylic coatings. I. Cumulative damage model and laboratory exposure apparatus. Polym. Degrad. Stab. **75**, 193–210 (2002)
36. Kimball, W.H., Munir, Z.A.: The effect of accelerated weathering on the degradation of polymeric films. Polym. Eng. Sci. **18**, 230–237 (1978)
37. Gijman, P., Diepens, M.: Photolysis and photooxidation in engineering plastics. In: Celina, M.C, Wiggins, J.S., Billingham, N.C. (eds.) Polymer Degradation and Performances. ACS Symp, Series vol. 1004, ch. 24, ACS, Washington (2009)
38. Giurginca, M., Zaharescu, T., Meghea, A.: Degradation of ethylene–propylene elastomers in the presence of ozone. Polym. Degrad. Stab. **50**, 45–48 (1995)
39. Andrei, L., Vlad, I., Ciuprina, F.: Electrical field distribution in power cable insulation affected by water tree. In: 9th International Symposium on "Advanced Topics in Electrical Engineering", Bucharest, 7–9 May 2015
40. Ciuprina, F., Teissèdre, G., Filippini, J.C.: Polyethylene crosslinking and water treeing. Polymer **42**, 7841–7846 (2001)
41. Caruso, M.M., Davis, D.A., Chen, Q.L., Odom, S.A., Sottos, N.R., White, S.R., Moore, J.S.: Mecanochemically-induced chemical changes in polymer materials. Chem. Rev. **109**, 5755–5798 (2009)
42. Joseph, P.V., Rabello, M.S., Mattoso, L.H.C., Joseph, K., Thomas, S.: Environmental effects on the degradation behavior of sisal fiber reinforced polypropylene composites. Compos. Sci. Technol. **62**, 1357–1372 (2002)
43. Rashdi, A.A.A., Supuan, S.M., Ahmad, M.M.H.M., Kholina, A.: Combined effects of water absorption due to water immersion, soil buried and natural weather on mechanical properties of kenaf unsaturated polyesters. Int. J. Mech. Mater. Eng. **1**, 11–17 (2010)

44. Setnescu, R., Kaci, M., Dehouche, N., Setnescu, T., Nasri, L., Zaharescu, T.: Hydrothermal ageing of metallocene polyethylene films in presence of grafted amine stabilizers. Arab. J. Sci. Eng. **40**, 69–80 (2015)
45. Merah, M., Bazoune, A., Fazal, A., Khan, Z.: Weathering degradation mechanism of chlorinated PVC. Int. J. Plast. Technol. (2013). doi:10.1007/s12588-013-9052-x
46. Jipa, S., Zaharescu, T., Setnescu, R., Setnescu, T., Brites, M.J.S., Silva, A.M.G., Marcelo-Curto, A.J., Gigante, B.: Chemiluminescence study on thermal and photostability of polyethylene. Polym. Int. **48**, 414–420 (1999)
47. Feldman, D.: Polymer weathering—photo-oxidation. J. Polym. Environ. **10**, 163–173 (2002)
48. Zaharescu, T., Jipa, S., Setnescu, R., Wurm, D., Brites, M.J.S., Silva, A.M.G., Marcelo Curto, A.J., Gigante, B.: Effects of some secondary amines on the oxidation of ethylene–propylene elastomers. Polym. Degrad. Stab. **68**, 83–86 (2000)
49. Gorghiu, L.M., Jipa, S., Zaharescu, T., Setnescu, R., Mihalcea, I.: The effect of metals on thermal degradation of polyethylenes. Polym. Degrad. Stabil. **84**, 7–11 (2004)
50. Audouin, L., Langlios, V., de Verdu, J.: Bruijn. J. C. M. Role of oxygen diffusion in polymer ageing. Kinetic and mechanical aspects. J. Mater. Sci. **29**, 569–583 (1994)
51. Rincon-Rubio, L.M., Fayolle, B., Audouin, L., Verdu, J.: A general solution of the closed-loop kinetic scheme for the thermal oxidation of polypropylene. Polym. Degrad. Stab. **74**, 177–188 (2001)
52. Hancox, N.L.: Thermal effects on the polymer matrix. Part 2. Thermal degradation. Mater. Des. **19**, 93–97 (1998)
53. Tidjani, A.: Comparison of formation of oxidation products during photo-oxidation of linear low density polyethylene under different natural and accelerated weathering conditions. Polym. Degrad. Stab. **68**, 465–469 (2000)
54. La Mantia, F.P.: Natural weathering of low density polyethylene. III. Mechanical properties. Eur. Polym. J. **20**, 993–995 (1984)
55. Blais, P., Carlsson, D.J., Wiles, D.M.: Surface changes during polypropylene photo-oxidation: a study by infrared spectroscopy and electron microscopy. J. Polym. Sci. Part A1 **10**, 1077–1092 (1972)
56. Kumanayaka, T.O., Parthsarathy, R., Jollands, M.: Accelerating effect of montmorillonite on oxidative degradation of polyethylene nanocomposites. Polym. Degrad. Stab. **95**, 672–676 (2010)
57. Sánchez-Valdés, S., Martínez Colunga, J.G., López-Quintanilla, M.L., Yañez Flores, L., García-Salazar, M.L., Gonzáles Cantu, C.: Preparation and UV weathering of polyethylene nanocomposites. Polym. Bull. **60**, 829–836 (2008)
58. Dintcheva, N.T., Al-Malaika, S., La Mantia, F.P.: Photo-oxidation of polyethylene/clay nanocomposites. Polym. Degrad. Stab. **94**, 1571–1588 (2009)
59. Grigoriadou, I., Paraskeopoulos, K.M., Chrissafis, K., Pavlidou, E., Stamakopoulos, T.G., Bikiaris, D.: Effect of different nanoparticles on HDPE UV stability. Polym. Degrad. Stab. **96**, 151–163 (2011)
60. Maria, R., Rode, K., Brüll, R., Dorbath, F., Baudrit, B., Bastian, M., Boendlé, E.: Monitoring the influence of different weathering conditions on polyethylene pipes by IR-microscopy. Polym. Degrad. Stab. **96**, 1901–1910 (2011)
61. Ojeda, T., Freitas, A., Birck, K., Dalmolin, E., Jacques, R., Bento, F., Camargo, F.: Durability of linear polyolefins under natural weathering. Polym. Degrad. Stab. **96**, 703–707 (2011)
62. Therias, S., Dintcheva, N.T., Gardette, J.L., La Mantia, F.P.: Photooxidation behavior of polyethylene/polymide-6 blends. Polym. Degrad. Stab. **95**, 522–526 (2010)
63. Dehbi, A., Mourad, A.-H.I.: Durability of mono-layer versus tri-layers LDPE films used as greenhouse cover: comparative study. Arab. J. Chem. (2011). doi:10.1016/j.arabjc.2011.04.010
64. Guenachi, K., Hamou, A., Hassini, N., Saiter, J.-M., Marais, S., Bucher, E.: Influence of sandwind on the durability of polyethylene greenhouse aged under sub-saharian climatic conditions. Mater. Res. Innov. **6**, 174–178 (2002)

65. Conti, A., Sudhakar, M., Chellini, E.: Assessment of the whole environmental degradation of oxo-biodegradable linear low density polyethylene (LLDPE) film designed for mulching applications. J. Polym. Environ. **20**, 1007–1018 (2012)
66. Russo, P., Aciermo, D., Marinucci, L., Greco, A., Frigone, M.: Influence of natural and accelerated weathering on performances of photoselective greenhouse films. J. Appl. Polym. Sci. **127**, 2213–2219 (2013)
67. Zaharescu, T., Plesa, I., Jipa, S.: Kinetic effects of silica nanoparticles on thermal and radiation stability of polyolefins. Polym. Bull. **70**, 2981–2994 (2013)
68. Philip, M, Attwood, J., Hulme, A., Williams, G., Shipton, P.: Evaluation of weathering in mixed polyethylene and polypropylene products. PLA Report 0007-008 (2004). ISBN 1-84405-092-0
69. Bernstein, R., Thomberg, S.M., Irwin, A.N., Hochrein, D.K., Derzon, D.K., Klamo, S.B., Clough, R.L.: Radiation oxidation mechanism: Volatile organic degradation products from polypropylene having selective C-13 labeling studied by GC–MS. Polym. Degrad. Stab. **93**, 854–870 (2008)
70. Tidjani, A.: Photodegradation of polypropylene under natural and accelerated conditions. J. Appl. Polym. Sci. **64**, 2494–2503 (1997)
71. Rajakunar, K., Sarasvathy, V., Chelvan, A.T., Chitna, R., Vijayakumas, C.T.: Natural weathering study of polypropylene. J. Polym. Environ. **17**, 191–202 (2009)
72. Azuma, Y., Takeda, H., Watanabe, S., Hakatani, H.: Outdoor and accelerated weathering tests for polypropylene and polypropylene/talc composites: a comprehensive study of their weathering behavior. Polym. Degrad. Stab. **94**, 2267–2274 (2009)
73. Kato, M., Tsurita, A., Kuroda, S., Osawa, Z.: Effect of stereoregularity on the photo-oxidation degradation and weathering of polypropylene. Polym. Degrad. Stab. **67**, 1–5 (2000)
74. Výchopňová, J., Černák, R., Obadal, M., Raab, M., Verney, V., Commereuc, S.: The role of specific nucleation in polypropylene photodegradation. Polym. Degrad. Stab. **92**, 1763–1768 (2007)
75. Tocháček, J., Vrátničková, Z.: Polymer lift-time prediction. The role of temperature in UV accelerated ageing of polypropylene and its compounds. Polym. Test. **36**, 83–87 (2014)
76. Guadagno, L., Naddao, C., Raimondo, M., Gorrasi, G.: Effect of carbon nanotubes in the photo-oxidative durability of syndiotactic polypropylene. Polym. Degrad. Stab. **95**, 1614–1626 (2010)
77. Dintcheva, N.T., La Mantia, F.P., Malatesta, V.: Photo-oxidation behavior of polyethylene multi-wall carbon nanotube composite films. Polym. Degrad. Stab. **94**, 162–170 (2009)
78. Geburtig, A., Wachtendorf, V.: Determination of the spectra; sensitivity and temperature dependence of polypropylene crack formation caused by UV-irradiation. Polym. Degrad. Stab. **95**, 2118–2123 (2010)
79. Chevali, V.S., Dean, D.R., Janowski, G.M.: Effect of environment weathering on flexural creep behavior of long fiber-reinforced thermoplastic composites. Polym. Degrad. Stab. **95**, 2628–2640 (2010)
80. Gillen, K.T., Wise, J., Clough, R.L.: General solution for the basic autoxidation scheme. Polym. Degrad. Stab. **47**, 149–161 (1995)
81. Philippart, J.-L., Sinturel, C., Arnaud, R., Gardette, J.-L.: Influence of the exposure parameters on the mechanism of photooxidation of polypropylene. Polym. Degrad. Stab. **64**, 213–225 (1999)
82. Mouffok, S., Kaci, M.: Artificial weathering effect on the structure and properties of polypropylene/polyamide-6 blends compatibilized with PP-*g*-MA. J. Appl. Polym. Sci. **132**, 41722 (2015)
83. Jipa, S., Zaharescu, T., Kappel, W., Setnescu, R., Oros, C.: RTL investigation on the radiochemical oxidation of polyolefins. Macromol. Symp. **242**, 87–92 (2006)
84. Liu, J., Wang, L.-W., Zhu, L., Luo, T.-Y.: Thermal ageing behavior of ethylene propylene diene monomer (EPDM) rubber. Appl. Mech. Mater. **727–728**, 47–50 (2015)

85. Gamlin, C.D., Dutta, N.K., Choudhury, N.R.: Mechanism and kinetics of the isothermal thermodegradation of ethylene–propylene–diene (EPDM) elastomers. Polym. Degrad. Stab. **80**, 525–531 (2003)
86. Rutherford, S.W., Kuntz, R.E., Smith, M.G., Honnell, K.G., Coons, J.E.: Measurement and correlation of sorption and transport properties of ethylene–propylene–diene monomer (EPDM) elastomers. J. Membr. Sci. **263**, 57–65 (2005)
87. Duek, E.R., Juliano, V.F., Guzzo, M., Kascheres, C., De Paoli, M.A.: Photo-oxidation of EPDM rubber. Part II—the photo-initiation process. Polym. Degrad. Stab. **28**, 235–248 (1999)
88. Zhao, Q.L., Li, X.G., Gao, J.: Ageing of ethylene–propylene–diene monomer (EPDM) in artificial weathering environment. Polym. Degrad. Stab. **97**, 1841–1846 (2007)
89. Fraïsse, F., Kumar, A., Commereuc, S., Veney, V.: Photo-oxidation of polymers. Validation of oxygen uptake and relationship with extent of hydroperoxides. J. Appl. Polym. Sci. **99**, 2238–2244 (2008)
90. Zaharescu, T., Podină, C.: Radiochemical stability of EPDM. Polym. Test. **20**, 141–149 (2001)
91. Gugumus, F.: Re-examination of the role of hydroperoxides in polyethylene and polypropylene: chemical and physical aspects of hydroperoxides in polyethylene. Polym. Degrad. Stab. **49**, 29–50 (1995)
92. Kumar, A., Commereuc, S., Verney, V.: Ageing of elastomers: a molecular approach based on rheological characterization. Polym. Degrad. Stab. **85**, 751–757 (2004)
93. Giric-Marcovici, M., Choudhury, N.R., Dimploulos, M., Matisons, J.G.: Weathering of coated EPDM rubber compound by controlled UV irradiation. Polym. Degrad. Stab. **69**, 157–168 (2000)
94. Snijers, E.A., Boersmer, A., van Baarle, B., Noordermeer, J.: Effect of the third monomer type and content on the UV stability of EPDM. Polym. Degrad. Stab. **89**, 200–207 (2005)
95. Yamsaengsung, W., Sombatsompop, N.: Use of expanded-EPDM as protecting layer for moderation of photo-degradation of wood/NR composites for roofing applications. J. Appl. Polym. Sci. **117**, 335–342 (2010)
96. Snijers, E.A., Boersmer, A., van Baarle, B., Gijman, P.: Effect of dicumyl peroxide crosslinking on the UV stability of ethylene–propylene–diene (EPDM) elastomers containing 5-ethylidene-2-norbornene (ENB). Polym. Degrad. Stab. **89**, 484–491 (2005)
97. Roy, P.K., Surekha, S., Raman, R., Rajagopal, C.: Investigating the role of metal oxidation state on the degradation behavior of LDPE. Polym. Degrad. Stab. **94**, 1033–1039 (2009)
98. Choi, S.-S., Chung, H.-S., Joo, Y.-T., Lee, S.-H.: Analysis of the whitening phenomenon of a thermopleastic elastomers article by UV weathering. Polym. Test. **30**, 415–419 (2011)
99. Zaharescu, T., Jipa, S., Supaphol, P.: Thermal stability of isotactic polypropylene modified with $CaCO_3$ nanoparticles. Polym. Bull. **64**, 783–790 (2010)
100. Singh, B., Sharma, N.: Mechanistic implications of plastic degradation. Polym. Degrad. Stab. **93**, 561–584 (2008)
101. Delor-Jestin, F., Lacoste, J., Barrois-Oudin, N., Cardinet, C., Lemaire, J.: Photo-, thermal and natural ageing of ethylene–propylene–diene monomer (EPDM) rubber used in automotive applications. Influence of carbon black, crosslinking and stabilizing agents. Polym. Degrad. Stab. **67**, 467–477 (2000)
102. Valko, E.I., Chiklis, CK: Effect of the thermal exposure on the physicochemical properties of polyamides. J. Appl. Polym. Sci. **9**, 2855–2877 (1965)
103. Forsström, D., Terselius, B.: R Thermo oxidation stability of polyamide 6 films. I. Mechanical and chemical characterization. Polym. Degrad. Stab. **67**, 69–78 (2000)
104. He, D.M., Susanto, H., Ulbricht, M.: Photo-irradiation for preparation, modification and stimulation of polymer membranes. Prog. Polym. Sci. **34**, 62–98 (2009)
105. Dintcheva, N.Tz., Filippone, G., La Mantia, F.P., Acierno, D.: Photo-oxidation behaviour of polyethylene/polyamide 6 filled with organomodified clay: improvement of the photo-resistance through morphology modification. Polym. Degrad. Stab. **95**, 527–535 (2010)

106. Tan, K.T., White, C.C., Benatti, D.J., Hunston, D.L.: Effects of ultraviolet radiation, temperature and moisture on ageing of coatings and sealants—a chemical and rheological study. Polym. Degrad. Stab. **95**, 1551–1556 (2010)
107. Pramoda, K.P., Liu, T.-X., Liu, Z.-H., He, C.-B., Sue, S.-H.: Thermal decomposition behavior of polyamide-6/nanocomposites. Polym. Degrad. Stab. **81**, 47–56 (2003)
108. Levchik, S.V., Weil, E.D., Lewin, M.: Thermal decomposition of aliphatic nylons. Polym. Int. **48**, 532–557 (1999)
109. Aimura, Y., Wada, N.: Reference materials for weathering tests on rubber products. Polym. Test. **25**, 166–175 (2006)
110. Costa, P., Ribeiro, S., Botelho, G., Machado, A.V., Mendes, S.L.: Effect of butadiene/styrene ratio, block structure and carbon nanotube content on the mechanical and electrical properties of thermoplastic elastomers after UV ageing. Polym. Test. **42**, 225–233 (2015)
111. Santos, R.M., Pimenta, A., Botelho, G., Machado, A.V.: Influence of the testing conditions on the efficiency and durability of stabilizers against ABS photo-oxidation. Polym. Test. **32**, 78–85 (2013)
112. Yang, X.F., Vang, C., Tallman, D.E., Bierwagen, G.P., Croll, S.G., Rohlik, S.: Weathering degradation of a polyurethane coating. Polym. Degrad. Stab. **74**, 341–351 (2001)
113. Rosu, D., Rosu, L., Cascaval, C.N.: IR-change and yellowing of polyurethane as the result of UV irradiation. Polym. Degrad. Stab. **94**, 591–596 (2009)
114. Yang, X.F., Tallman, D.E., Bierwagen, G.P., Croll, S.G., Rohlik, S.: Blistering and degradation of polyurethane coatings under different accelerated weathering tests. Polym. Degrad. Stab. **77**, 103–109 (2002)
115. Delor-Jestin, F., Drouin, D., Cheval, P.-Y., Lacoste, J.: Thermal and photochemical ageing of epoxy resin—influence of curing agent. Polym. Degrad. Stab. **91**, 1247–1255 (2006)
116. Bet-moushoul, E., Mansourpanaah, Y., Farhadi, K.M., Tabatabai, M.: TiO_2 nanocomponents based polymeric membranes. A review on performance improvement for various applications in chemical engineering processes. Chem. Eng. J. **283**, 29–46 (2016)
117. Parlak, E.A.: The blend ration effect on the photovoltaic performances and stability of poly-(3-hexylthiophene);[6:6]-phenyl-C_{61} butilic acid methylester (PCBM) and poly (3-octylthyophene): PCBM solar cells. Sol. Energy Mater. Sol. Cells **100**, 174–184 (2012)

Life-Time Prediction of Multicomponent Polymeric Materials

Cristian-Dragos Varganici, Dan Rosu and Liliana Rosu

Abstract Since the continuous emerging of new technologies and outdoor applications of polymers and polymer based materials, and thus a dire need in increasing their lifetime service quality, the number of methods predicting aging has also experienced growth. Most of such methods are reported on natural and artificial aging and may be based on accelerated aging predictions, different damage models and empirical formulas. The present chapter is a systematization of literature studies focused on discovering new means for accurate linking of laboratory and field exposure results, which represent the actual and future main challenges in estimating and assessing quality and service lifetime of polymeric materials.

1 Methods of Life-Time Prediction of Multicomponent Polymeric Materials

Polymers, whether selfstanding or in the form of polymer blends or as composites and/or multicomponent materials, encompass a wide palette of applications of which, a substantial part are destined for direct outdoor exposure. During their service life, outdoor exposed polymeric materials encounter direct contact with a series of environmental factors (i.e. light, humidity, oxygen, microorganisms and others) which cause gradual and irreversible structural modifications, either chemical and/or physical, and thus leading to performance reduction until unevitable disposal of the certain material. It is due to this aspect that manufacturers are increasingly demanded to assure products service lifetime. For efficient mainte-

C.-D. Varganici (✉) · D. Rosu · L. Rosu
Centre of Advanced Research in Bionanoconjugates and Biopolymers, "Petru Poni" Institute of Macromolecular Chemistry, 41A Grigore Ghica-Voda Alley, 700487 Iasi, Romania
e-mail: varganici.cristian@icmpp.ro

D. Rosu
e-mail: drosu@icmpp.ro

L. Rosu
e-mail: lrosu@icmpp.ro

D. Rosu and Visakh P. M. (eds.), *Photochemical Behavior of Multicomponent Polymeric-based Materials*, Advanced Structured Materials 26,
DOI 10.1007/978-3-319-25196-7_8

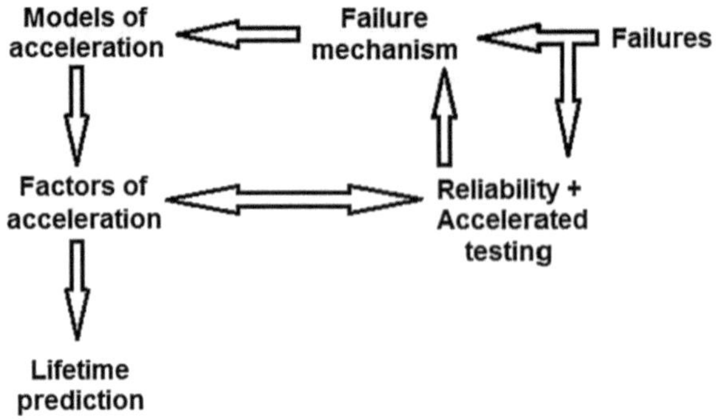

Scheme 1 General representation of a lifetime prediction process

nance of a polymeric material in a specific application, decomposition prediction studies under service conditions must first be undertaken before use. A definition of service life consists of the time from the beginning of material installation up to the point in which a crucial property determining its performance meets a certain minimum accepted standard. Assessment of material service lifetime is usually based either on data obtained from practice or from durability tests conducted prior to use. The service lifetime of a material may be quantitatively expressed as the evolution of a property as a function of time until infliction of structural damage is gradually initiated by environmental factors. Scheme 1 depicts a general diagram of lifetime prediction.

Due to the increasing demand for new polymeric materials for outdoor applications and enhancement of their service quality and life, literature has abounded in a variety of methods developed for predicting aging of polymeric materials exposed outdoors [1–30], as specified by Liu et al. [31], and correlation studies between natural and artificial aging [1–4].

Such methods include predictions from tests of accelerated aging [7–23] and from weatherability based on empirical formulas (extrapolation, ageing kinetics model derivation) [5, 6], while others exploit new ways for detection of service lifetime [24–26]. Some methods are based on monitoring of cumulative damage [27–30], etc. [31].

Service lifetime prediction of polymers and/or polymer based materials may be undertaken from different types of tests, such as: creep behavior tests (*linear and non-linear creep, physical aging, time-dependent plasticity*), fatigue behavior tests (*stress transfer* and *normalized life prediction models, empirical fatigue theories, fracture mechanics theory* and *strength degradation*) and standard accelerated aging tests (*chemical resistance, thermal stability, liquid absorption*) [32].

However, there exists a major drawback of numerous such methods, since many theoretical models and empirical formulas are restricted only to specific materials and environmental conditions. Another disadvantage resides in the involvement of abundant data of mathematical statistics from aging results and the requirement of deep knowledge on different aging mechanisms of polymers [31].

The standard aging tests also encompass *weathering test methods*, both under *natural* and *artificial exposure*. Such tests possess the advantage of not being as restrictive as the above mentioned ones, thus allowing testing of a wider range of polymers and polymer based materials under their conditions [32].

The first issue which arises concerning reproducing of *natural exposure* consists in finding the laboratory conditions which exactly match the environmental ones. Another issue resides in the fact that the exposure period and that of product life expectance must be at least of the same duration. For this purpose *accelerated natural weathering* is made under harsher conditions than the ones expected for product service. Therefore, a series of standard test procedures for polymer weathering under different daylight conditions (natural daylight, concentrated sunlight via Fresnal mirrors and glass-filtered daylight) were developed over the years and named: ISO 877 [33], ASTM D4364-02 [34], ASTM D1435 [35] and ASTM G24-05 [32, 36, 37].

During *artificial weathering*, the tested materials are placed in a chamber and exposed to a source of artificial light. Humidity and temperature are controlled during testing. Since all parameters in the aging chamber are controlled, the ease in accelerating weathering tests represents an advantage of the method. On the other hand, one must know how these parameters interact during such a test, in order to efficiently set them. The main and most important component of a system for accelerated weathering is the lamp, which must be adequately chosen in order to simulate the solar radiation with high fidelity [32].

Table 1 depicts the main light sources used in artificial weathering tests and their characteristic wavelengths. Table 2 shows typical photochemical terms and their corresponding measurement units. Table 3 gives a detailed global radiation spectral distribution.

Table 1 Main light sources used for acceleration weathering [37]

Light source type	Wavelength (nm)
Xenon arc	270–800
Mercury vapour	200–400
Fluorescent lamp	295–400
Carbon arc	300–800

Table 2 Specific terms and definitions

Name	Symbol	Unit
Wavelength	λ	nm
Irradiance	E	$W\ m^{-2}$
Radiant intensity	I	$W\ sr^{-1}$
Radiant energy	Q	J
Radiant exposure	H	$J\ m^{-2}$
Radiant flux	Φ	W
Radiance	L	$W\ m^{-2}\ sr^{-1}$

Table 3 Detailed global radiation spectral distribution

Spectral domain	Range (nm)	Irradiance (W m^{-2})/(%)
UV–B	280–320	5/0.4
UV–A	320–360	27/2.4
	360–400	36/3.2
VIS	400–440	56/5
	440–480	73/6.5
	480–520	71/6.4
	520–560	65/5.8
	560–640	121/10.8
	640–680	55/4.9
	680–720	52/4.6
	720–780	67/6
IR	780–1000	176/15.7
	1000–1200	108/9.7
	1200–1800	138/70
	1800–3000	12.3/6.3
Total	280–3000	1120/100

Selecting the most appropriate light source for artificial weathering must be made based on the sought application, targeted environment and decomposition mechanism(s) assessment criteria. The increase of acceleration effect of an applied artificial method is inversely proportional to its increase in accuracy regarding wavelength and intensities matching to real outdoor solar radiation, thus requiring longer exposure times [37].

E_0 represents a solar constant defined as the irradiance value corresponding to the mean of distance between earth and sun, without including terrestrial atmosphere. This solar constant influences irradiance from solar spectra at the earth surface and its total value is 1120 W m^{-2} when the sun reaches zenith, according to CIE (Table 3) [37].

The best spectral accuracy to that of natural sunlight was proven by the use of *xenon long-arc lamps*, since the spectral irradiance, with total value of 1120 W m^{-2}, is specified within the whole wavelength range of 280–3000 nm. Specific filters are applied for removing wavelengths in the UV domain, found in the xenon-arc spectrum, which are shorter than the ones in natural solar light. An issue which must be taken into account consists in the high infrared radiations generated by the xenon long-arc lamps, which cause sample overheating, therefore requiring removal. Other disadvantages of these lamps include high purchase costs and expensive maintenance. Standard procedures for the use of xenon lamps are given in ASTM D2565-99 [38] and ISO 4892: part 2 [32, 39].

Mercury lamps possess deficiencies in red and infra-red spectra regions and strong spectral line, therefore being generally not recommended [37, 40].

The same as mercury lamps, *fluorescent lamps* also have deficiencies in red and infra-red spectra regions and are regarded as a significantly cheaper alternative of the xenon-arc lamps [32]. Almost the whole irradiation of this type of lamps manifests in the critical UV spectra domain (below 400 nm), this making them suitable for materials properties evaluation in indoor applications. During outdoor exposure short wavelength values accelerate reactions in polymers leading to different structural modifications, such as oxidation, cross-linking, lowering of mechanical and thermal stability, molecular weight modifications etc. [37].

A fluorescent lamp generates radiation by re-emission of long wavelength spectra produced by low pressure vapor source based mercury lamp, due to a phosphor layer coating the bulbs interior surface. UVB tubes are generally applied in accelerated degradation processes (intense peak at 313 nm). As a disadvantage, UVB fluorescent tubes also generate intense radiation at wavelengths lower than 270 nm, not within solar radiation, thus needing elimination. UVA tubes spectra in the UV region are close to ones corresponding to solar radiation, because they generate radiation at seemingly higher wavelengths. Since the majority of polymer degradation phenomena take place within the UV region, this may not be an issue for most polymers. However, there exist exceptions, such as in the case of nylon-6 which significantly decomposes at 365 nm [32].

The *carbon-arc* spectrum consists of a sum of line emission series continuously overlapping on to a background and not matching solar radiation, especially at wavelength values below 350 nm. Carbon-arc light sources emit high energy radiations at short wavelengths, leading to intense structural damage of the exposed polymer or polymer based material, thus manifesting a very slight similitude to natural weathering processes. Despite this aspect, such devices are commercially available since the 1960s and still used for plastics testing (standards ASTM D1499–99 [41] and ISO 4892: part 4 [42]) [32].

2 Accelerated and Natural Aging of Multicomponent Polymeric Materials

As previously mentioned, UV radiation is a component of the electromagnetic spectrum and, due to its effects varying with wavelength, it is divided into three regions: UVA (400–315 nm), UVB (315–280 nm) and UVC (290–100 nm). UV radiation is generated either naturally, by reaching the incandescent temperature of a body through heating, as the case of solar UV, or artificially, by inserting electric current into a gas (e.g. vaporized mercury). In many experiments, particularly the ones impling photobiology, it is practical to use artificial sources which simulate the UV component from natural sunlight [43].

It is well known that the vast majority of polymers are immiscible, this leading to the occurence of phase separation phenomenon. In physical terms this aspect is described by a positive value of the free energy, thus by an increased mixing heat, an issue which may be overcomed if solubility parameters difference between components ideally disappear [44]. Phase separation phenomenon generates a decline in material performances, particularly mechanical and thermal properties. In surpassing this problem, a series of compatibilization methods have been developed over the years, detailed in the literature [45].

Among these, obtaining of interpenetrating (IPNs) and semi-interpenetrating polymer networks (SIPNs) showed high efficiency. An IPN contains two intertwined chemically crosslinked polymers, whilst a SIPN presents a linear component entangled in a crosslinked polymer [45]. During SIPNs synthesis a forced phase compatibilization occurs, leading to a synergetic effect upon individual components properties, thus a good components miscibility being also reflected by the presence of a single glass transition temperature (T_g).

Before further discussions on studies of accelerated and natural aging, one must first comprehend basic theoretical and practical approaches to laboratory testing of life-time prediction and the included different properties assessment. One class of SIPNs consists of crosslinked epoxy matrices and linear semi-aromatic polyurethanes. Epoxy resins are applied in construction industry as coating materials, because they possess excellent mechanical properties and adhesion to metals. On the other hand, epoxy resins exhibit a series of disadvantages, such as after curing stress, brittleness, they crack under impact, which is why they are blended with elastomers, like polyurethanes, into SIPNs. Polyurethanes improve epoxy resin previously mentioned disadvantages due to their good elastic, non-abrasive and damping properties [46].

Rosu et al. [47] conducted accelerating laboratory aging tests of SIPNs comprised of a semi-aromatic polyurethane, as linear component, and epoxy resin, as crosslinked component, in varying quantities of 5, 10, 15, 20, 30 and 40 % (SIPNs 1–6) (Scheme 2—reproduced with kind permission from Elsevier—License No. 3723021063514). Miscibility and thermal characterization studies were previously undertaken [48, 49], the SIPNs exhibiting a single increasing T_g up to and including 30 % crosslinked epoxy resin and two T_gs at 40 % crosslinked epoxy resin, thus manifesting phase separation. The good miscibility of the components was also reflected in the reducing of the melting/crystallization transitions of polyurethane with crosslinked epoxy resin concentration.

For the UV irradiation studies there was used a medium pressure mercury lamp, experiments being conducted in the UVA region ($\lambda > 300$ nm) [47]. Color studies are normally a steping stone in predicting different UV aging mechanisms. Such studies indicate if chromophores accumulation occurs at the surface and their stability and general coloration during irradiation. CIEL $a^* b^*$ is a system in which color modifications determination are made according to Eq. 1, where ΔE is total color modification, L^* is the lightness factor, varying from 0 (pure black) to 100 (pure white), and a^*, b^* are chromatic coordinates of greenness ($-a^*$)—redness (a^*) and blueness ($-b^*$)—yellowness (b^*), respectively. Numbers 1 and 2 are indexes before and after irradiation [50].

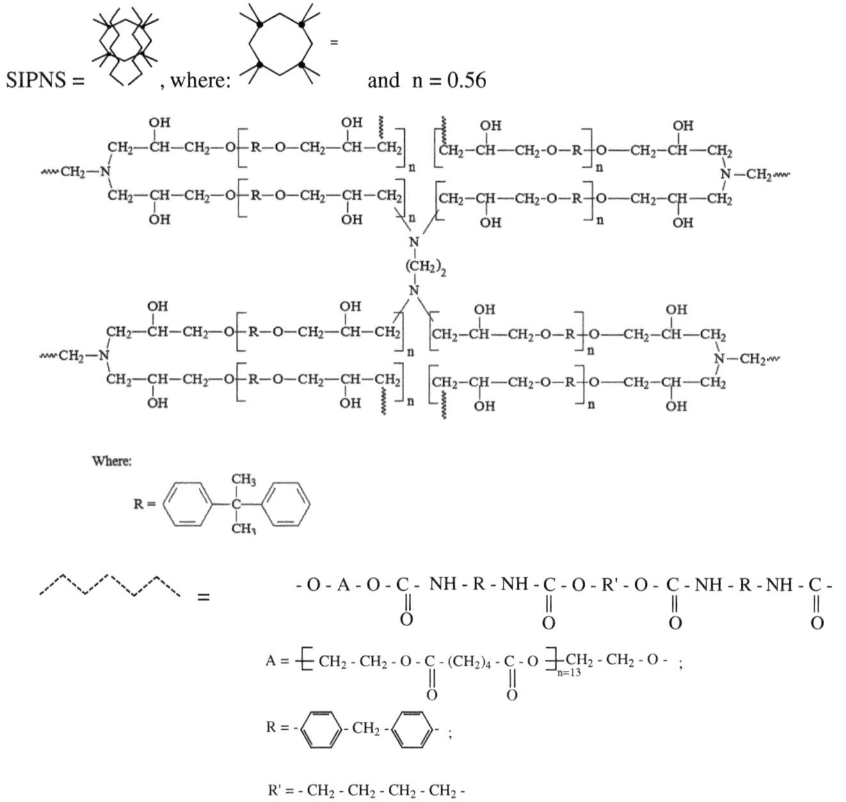

Scheme 2 Structure of the studied SIPNs [47] (reproduced with kind permission from Elsevier—License No. 3723021063514)

$$\Delta E = \left[\left(L_2^* - L_1^*\right)^2 + \left(a_2^* - a_1^*\right)^2 + \left(b_2^* - b_1^*\right)^2\right]^{1/2} \tag{1}$$

An increase in total color change (Fig. 1—reproduced with kind permission from Elsevier—License No. 3723021063514), confirmed by an increase in a^* and b^* values, and a decrease in lightness factor values (Fig. 2—reproduced with kind permission from Elsevier—License No. 3723021063514) are generally correlated with accumulation of cromophores, and darkening of the surface, respectively. These observations were confirmed by gloss retention decrease with irradiation time [47].

ATR–FTIR technique presents a crucial role in the assessment of structural modifications during UV irradiation of polymer films. Authors introduced a difference spectrum C (Fig. 3—reproduced with kind permission from Elsevier—License No. 3723021063514) between initial sample spectrum and that of sample after irradiation [47].

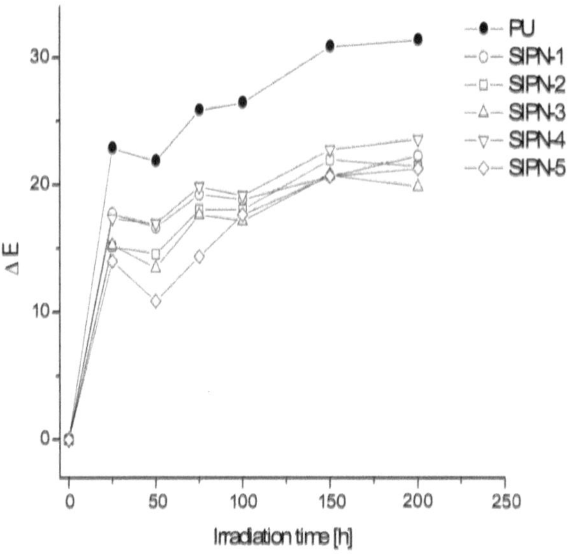

Fig. 1 Total color modification of SIPNs with irradiation time (reproduced with kind permission from Elsevier—License No. 3723021063514)

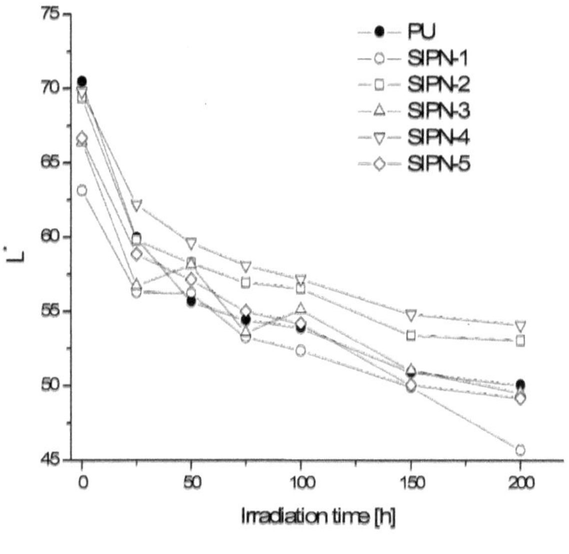

Fig. 2 Lightness factor variation with irradiation time [47] (reproduced with kind permission from Elsevier—License No. 3723021063514)

Based on the difference spectrum C, where negative signals indicate newly formed entities and positive ones show moieties lost during photodegradation, authors found two different mechanisms of photodegradation for polyurethane in the networks. One of these mechanisms is represented by Norrish type reactions through aliphatic ester linkages decomposition and carbon dioxide elimination (Scheme 3) (reproduced with kind permission from Elsevier—License No. 3723021063514). The photo-Fries rearrangement occurs in urethane bond and

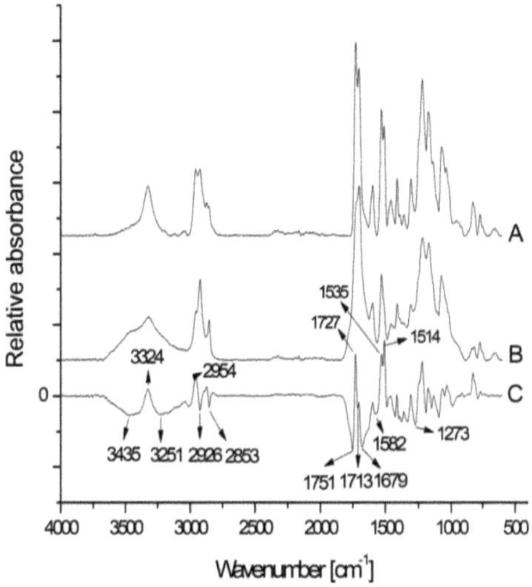

Fig. 3 ATR–FTIR spectra of sample SIPN–3: *A* before irradiation; *B* after 200 h irradiation time and *C* difference spectrum between spectra *A* and *B* [47] (reproduced with kind permission from Elsevier—License No. 3723021063514)

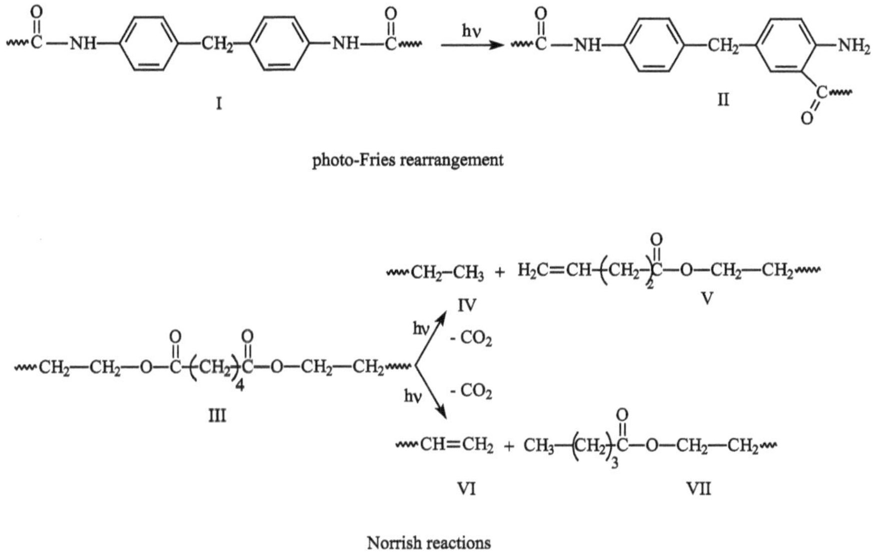

Scheme 3 Photodecomposition mechanisms of semiaromatic polyurethane [47] (reproduced with kind permission from Elsevier—License No. 3723021063514)

leads to formation of ortho-amino ester structures (Scheme 3). The epoxy network partially protected the transparent polyurethane from photo-oxidation by absorbing some of the UV radiations and increasing surface opacity [47].

In similar conditions Varganici et al. [51] conducted controlled photo-aging experiments in order to study a potential photo-protective effect of PVA over cellulose in PVA and cellulose based cryogels with 10, 30 and 50 % cellulose

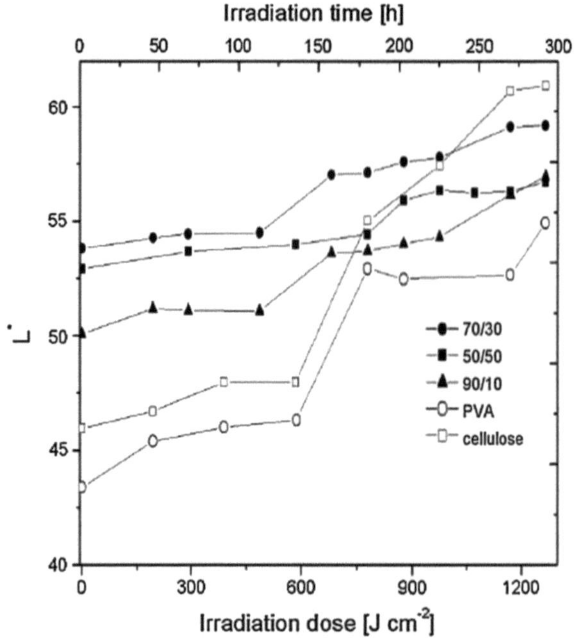

Fig. 4 Lightness factor variation with irradiation time [51] (reproduced with kind permission from Elsevier—License No. 3723030695158)

content. Increase in L^* values was correlated with surface discoloration (Fig. 4—reproduced with kind permission from Elsevier—License No. 3723030695158). Instable red chromophores were formed during pure PVA and cryogels photo-oxidation (Fig. 5—reproduced with kind permission from Elsevier—License No. 3723030695158), while cellulose exhibited continuous surface yellowing by stable chromophores formation (Fig. 6—reproduced with kind permission from Elsevier—License No. 3723030695158). For the cryogels containing up to 70 % PVA, no new chromophores and/or formation of small concentrations of instable red ones were found during irradiation, due to absence of negative signals in the FTIR difference spectrum in Fig. 7c (reproduced with kind permission from Elsevier—License No. 3723030695158), as compared to presence of negative signals in difference spectra from Fig. 7a, b (reproduced with kind permission from Elsevier—License No. 3723030695158). This was clearly an indication of cellulose chains protection against UV radiation by PVA, which is known for its general behavior as an oxygen barrier. Carbonyl peak area variation with irradiation time and dose in Fig. 8 (reproduced with kind permission from Elsevier—License No. 3723030695158), from which one may observe significant photooxidation phenomena occurrence, similar to those for cellulose, at higher than 70 % PVA concentration (90 %) supported this aspect. PVA mostly decomposed by depolymerization processes (Fig. 7d—reproduced with kind permission from Elsevier—License No. 3723030695158), leading to the photodecomposition mechanism proposed in Scheme 4 (reproduced with kind permission from Elsevier—License No. 3723030695158). By applying the second derivative to the carbonyl absorbtion

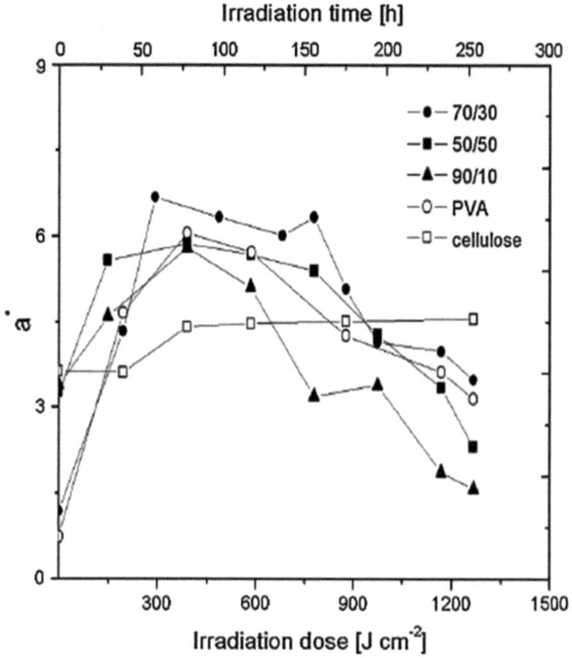

Fig. 5 Variation of chromatic coefficient a* with irradiation time [51] (reproduced with kind permission from Elsevier—License No. 3723030695158)

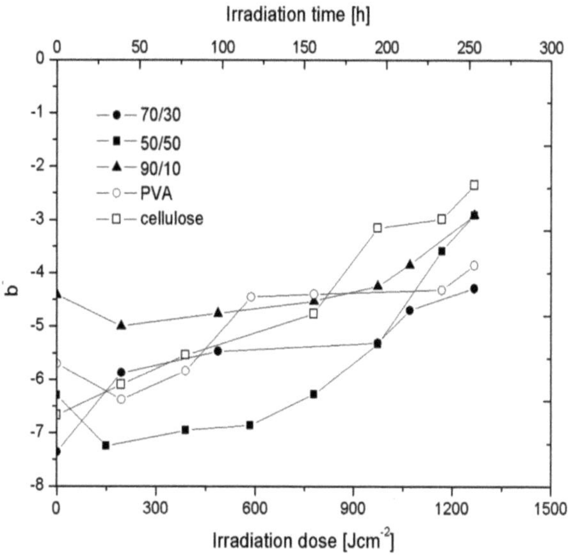

Fig. 6 Variation of chromatic coefficient b* with irradiation time [51] (reproduced with kind permission from Elsevier—License No. 3723030695158)

region corresponding to cellulose, new peaks were identified, leading to the decomposition mechanism which authors proposed in Scheme 5 (reproduced with kind permission from Elsevier—License No. 3723030695158). All observations were supported by by UV–Vis, FTIR, SEM, XPS and TG–MS techniques [51].

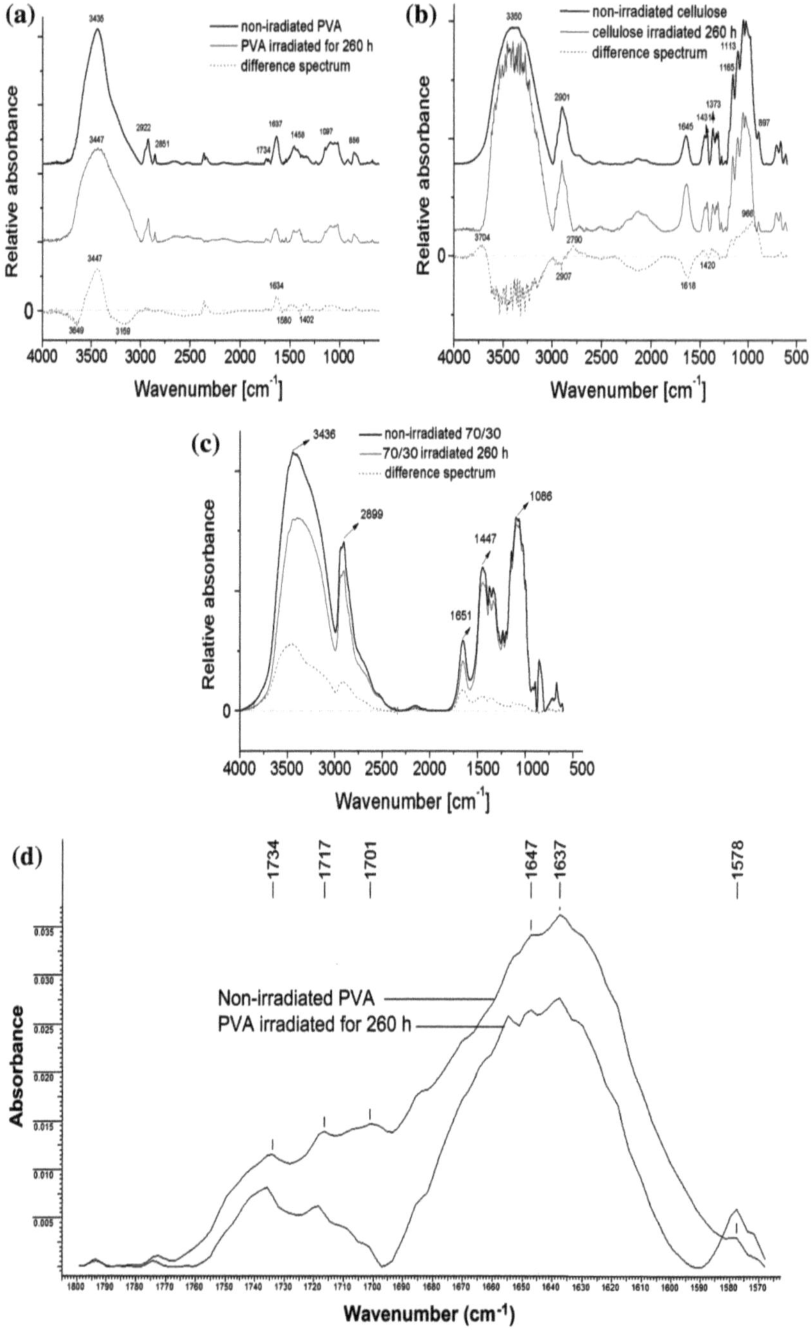

Fig. 7 **a** FTIR spectra of PVA [51] (reproduced with kind permission from Elsevier—License No. 3723030695158). **b** FTIR spectra of cellulose [51] (reproduced with kind permission from Elsevier—License No. 3723030695158). **c** FTIR spectra of cryogel 70/30 [51] (reproduced with kind permission from Elsevier—License No. 3723030695158). **d** Carbonyl region in the PVA FTIR spectra before and after irradiation [51] (reproduced with kind permission from Elsevier—License No. 3723030695158)

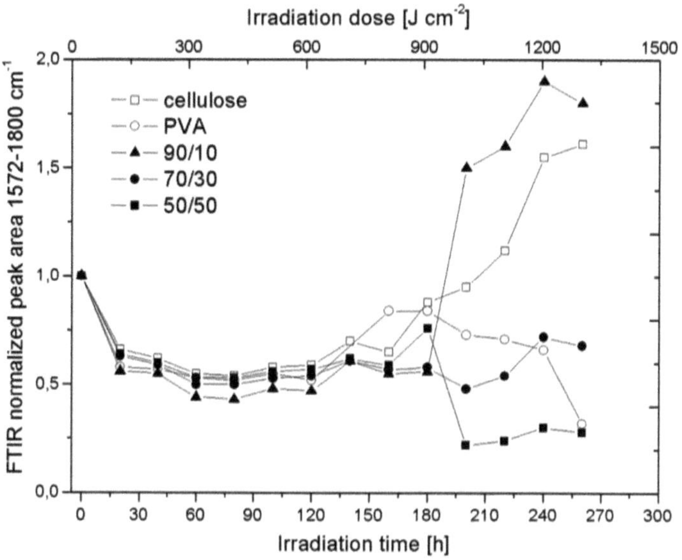

Fig. 8 Carbonyl normalized peak area as a function of irradiation time and dose [51] (reproduced with kind permission from Elsevier—License No. 3723030695158)

In similar conditions Rosu et al. [52] conducted controlled photo-aging experiments of chemically modified surface of *Abies alba L.* softwood species. Chemical modification was undertaken in two stages (Scheme 6—reproduced with kind permission from Elsevier—License No. 3723031483298). The former stage consisted in the reaction of wood components with succinic anhydride (SA) to yield monoester entities with final carboxyl groups, followed by crosslinking with epoxidized soybean oil (ESO) in the latter stage. Success of the chemical modification reactions was confirmed by FTIR.

A general increasing trend was observed in color differences up to almost one day exposure time, followed afterwards by decrease, this being an indication of wood surface protection by the applied chemical modifications (Fig. 9—reproduced with kind permission from Elsevier—License No. 3723031483298). Also, unprotected wood surface manifested the highest darkening tendency (Fig. 10—reproduced with kind permission from Elsevier—License No. 3723031483298). Increase in chromatic coefficients values with irradiation time (dose) is an indication of wood surface reddening, due to generation of quinoid moieties from lignin, and yellowing, due to formation of new carbonyl entities during UV irradiation. (Fig. 11—reproduced with kind permission from Elsevier—License No. 3723031483298). Authors confirmed structural modifications in terms of colour changes by FTIR (Fig. 12—reproduced with kind permission from Elsevier—License No. 3723031483298).

A confirmation of color studies was made by the decrease in the carbonyl/carbohydrate ratio of treated wood surfaces compared to that of

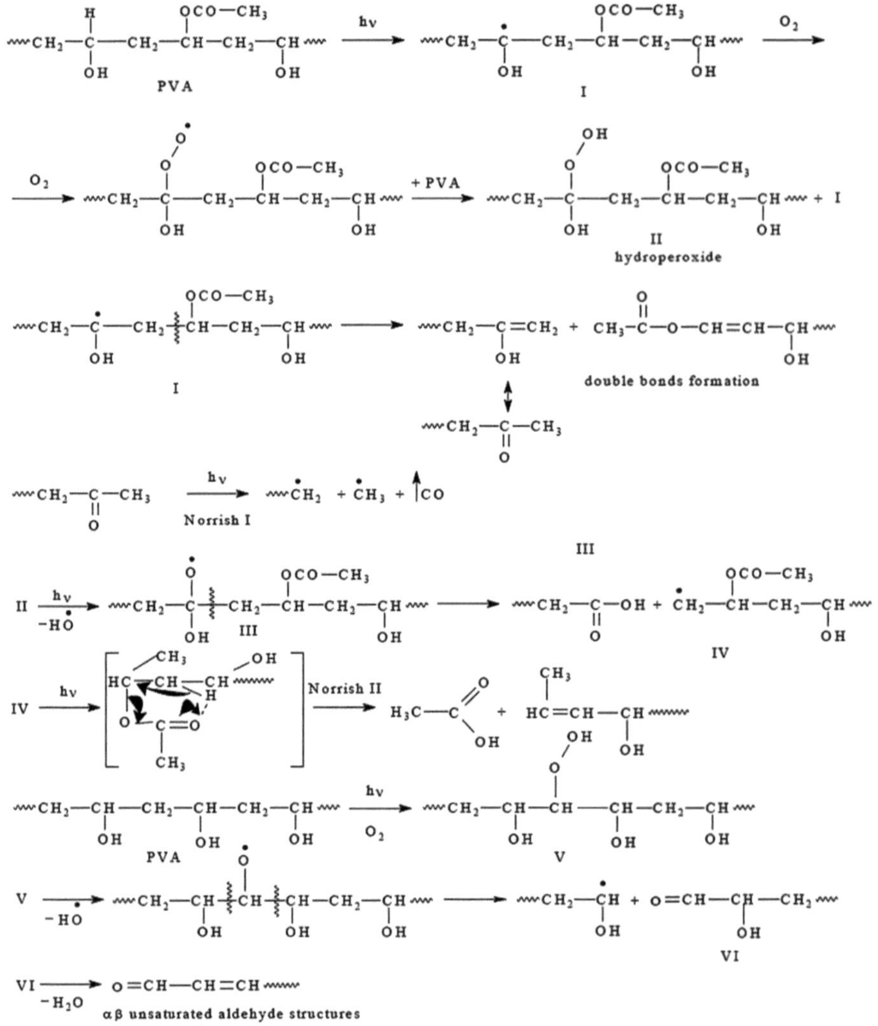

Scheme 4 Proposed photodecomposition mechanism of PVA in the cryogels [51] (reproduced with kind permission from Elsevier—License No. 3723030695158)

unmodified wood surface (Fig. 13—reproduced with kind permission from Elsevier—License No. 3723031483298). Another aspect resided in the decrease of the lignin/carbohydrate ratio of the of treated wood surfaces compared to that of unmodified wood surface, lignin being the most susceptible component to photodegradation (Fig. 14—reproduced with kind permission from Elsevier—License No. 3723031483298) [52].

Tocháček and Vrátníčova [53] linked six months of outdoor exposure data to accelerated ageing by means of total UV radiation energy needed in sample photodecomposition at four different temperatures. In both cases there was used the carbonyl index onset increase and its same relative increase point as polymer decomposition measures. Total UV radiation energy values were calculated at different induction periods for the studied temperatures. Authors found that for direct comparison, accelerated aging must be conducted at temperatures between 32 and 36 °C, since the same quantity of total UV radiation generates the same polymer degradation extent under these conditions in both aging types. This temperature range may be applied for obtaining a reliable service life-time prediction.

Gu et al. [54] have managed to correlate results from laboratory and field exposures for an epoxy system designed as coating material through three different strategies, since the obtaining of accurate service lifetime data of environmentally exposed polymers, in general, remains to this day an issue, perturbing and delaying new product achievement. Frequent errors arising in successfully establishing the linkage between laboratory and outdoor conditions include: aging of the artificial light source, absence of an optimum control over temperature and relative humidity inside the aging device, lack of uniformity in irradiance, distribution differences between natural and artificial light source in terms of spectral emissions [55–57].

The first strategy consists in correlating laboratory simulated data and that obtained from outdoor exposure via a predictive model by a reliability-based methodology. The parameters for this model, in terms of weathering, include reciprocity and additivity law, total effective dosage and accumulated damage and are obtained exclusively from laboratory experiments. Data resulted from outdoor exposure experiments are applied in verifying laboratory obtained parameters, using as input the temporal field values of moisture, temperature and spectral irradiance throughout their entire variation range. Basically, laboratory experiments parameters values are regarded as standards to which outdoor exposure results are compared. This is possible, knowing that, although the lack of control over individual environmental factors, each individual variable can be monitored with the same accuracy for outdoor tests as laboratory ones.

According to Martin et al. [58] the total effective dosage model, given in Eq. (2), correlates the absorbed radiation energy in the studied material to damage via the quantum yield. The model includes all UV-visible spectral radiation through all stages of material exposure: surface contact, absorbtion and degradation. Hence, it is a flexible model for environmental factors, i.e. radiant flux, relative humidity and temperature.

$$D_{eff}(t) = \int_{0}^{t} \int_{\lambda \min}^{\lambda \max} E_c(\lambda_{uv}, t)(1 - 10^{-A(\lambda_{uv},t)})(\phi\lambda)d\lambda dt \tag{2}$$

in which D_{eff} (t) represents the effective dosage; $\lambda_{\max}$, $\lambda_{\min}$ are the maximum and minimum wavelengths; $A(\lambda_{uv},t)$ is the absorbtion at a certain wavelength and time value; $E_c(\lambda_{uv},t)$ is the irradiance exposure at a certain time t and ϕ (λ_{uv}) is the

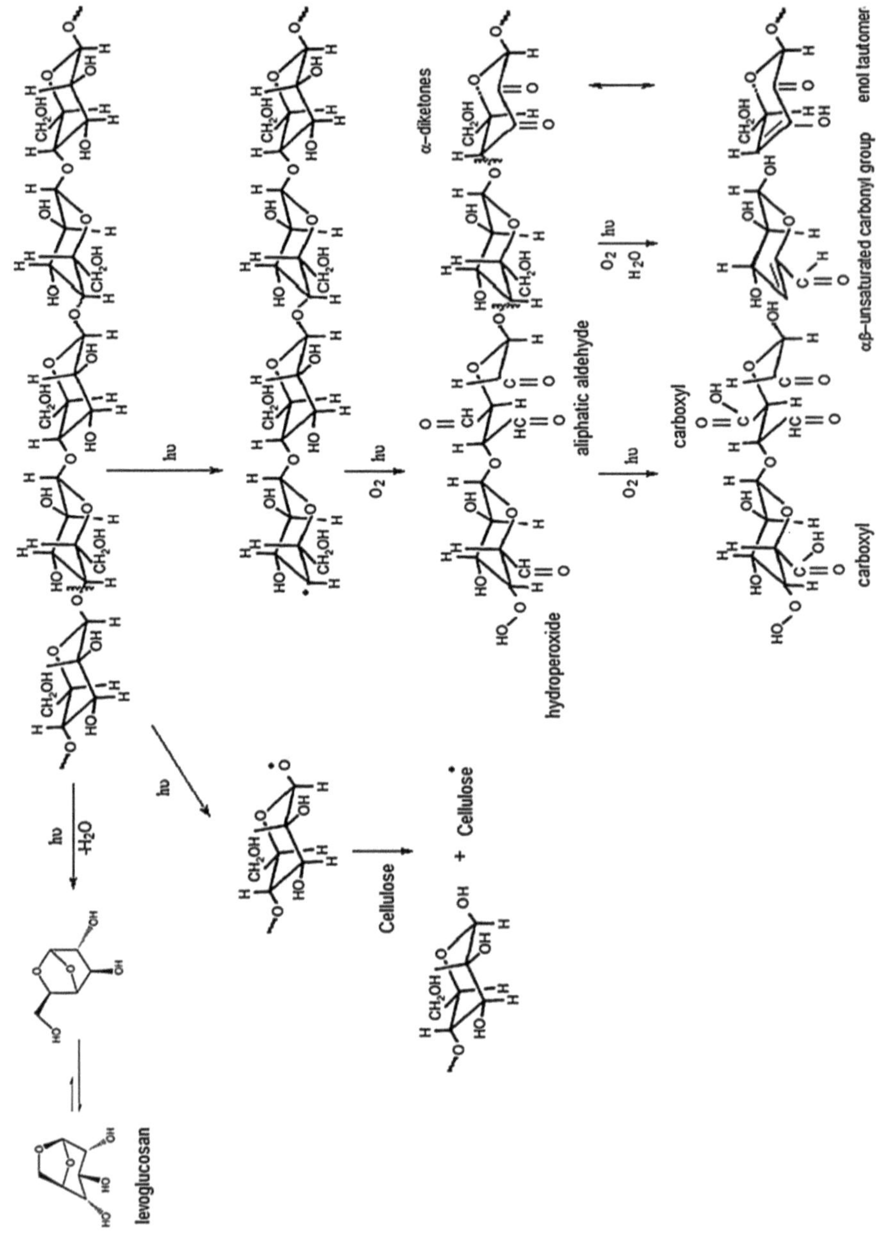

Scheme 5 Proposed photodecomposition mechanism of cellulose in the cryogels [51] (reproduced with kind permission from Elsevier—License No. 3723030695158)

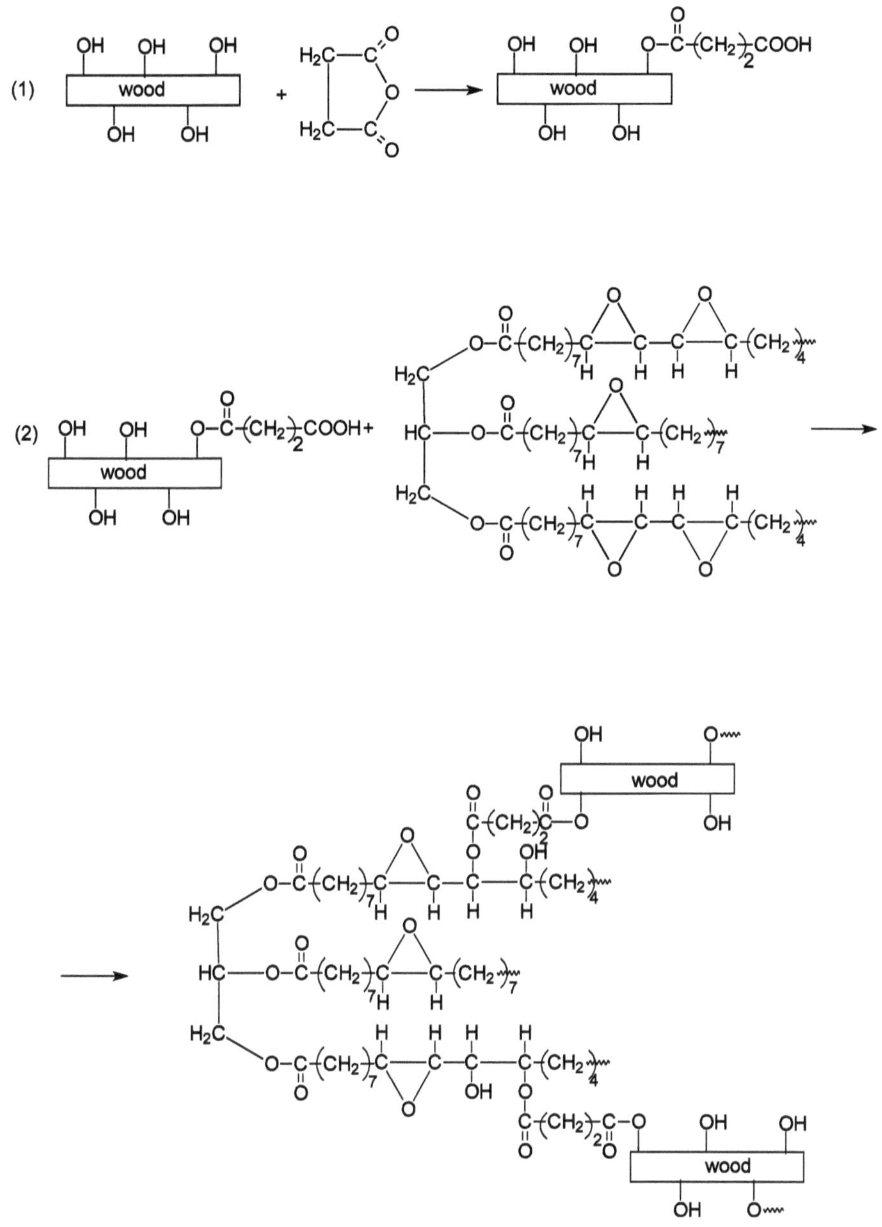

Scheme 6 Stages of wood surface modification [52] (reproduced with kind permission from Elsevier—License No. 3723031483298)

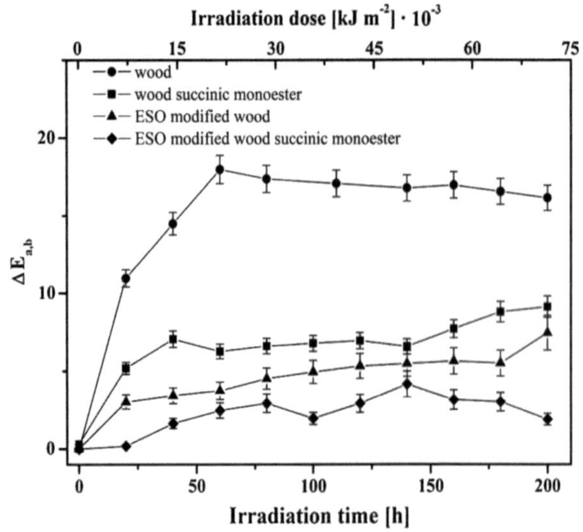

Fig. 9 Total color changes during irradiation [52] (reproduced with kind permission from Elsevier—License No. 3723031483298)

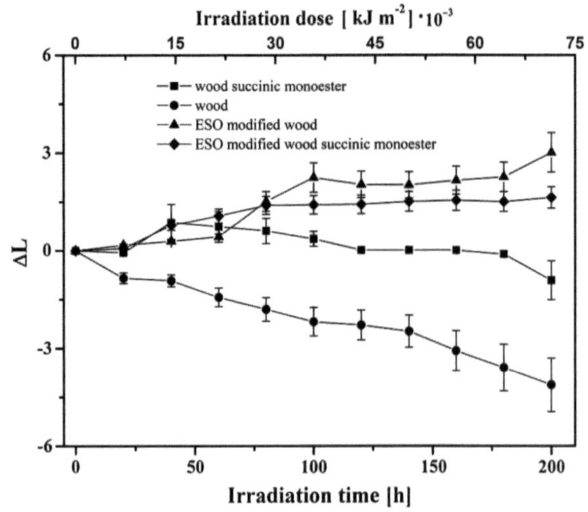

Fig. 10 Lightness factor variation during irradiation [52] (reproduced with kind permission from Elsevier—License No. 3723031483298)

apparent quantum yield, which links photolytic dosage to damage, D_{damage}, through Eq. (3), where $E_c(\lambda_{uv},t) \cdot A(\lambda_{uv},t) = D_{dosage}$, defined as the absorbed incident radiation fraction.

$$D_{damage} = \phi(\lambda_{uv}) D_{dosage}(\lambda_{uv}) \tag{3}$$

$E_c(\lambda_{uv},t)$ and $A(\lambda_{uv},t)$ may be obtained from UV–visible spectroscopy and D_{damage} from FTIR transmission spectroscopy from the degradation versus absorbed dosage or time plot. Based on such determinations one may estimate the

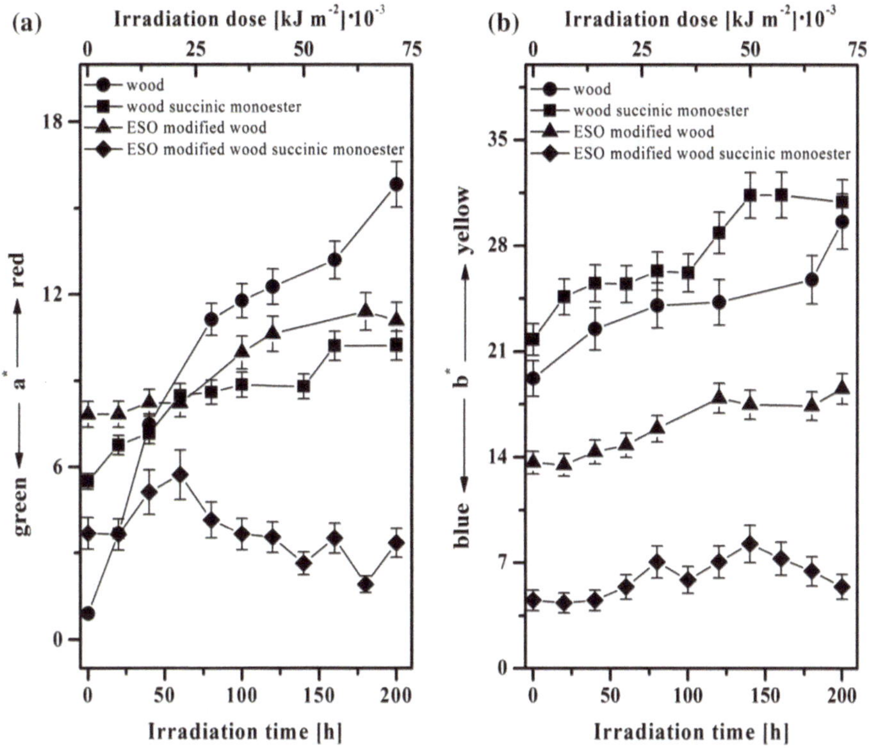

Fig. 11 Chromatic coordinates variation during irradiation [52] (reproduced with kind permission from Elsevier—License No. 3723031483298)

Fig. 12 Difference FTIR spectra before and after irradiation [52] (reproduced with kind permission from Elsevier—License No. 3723031483298)

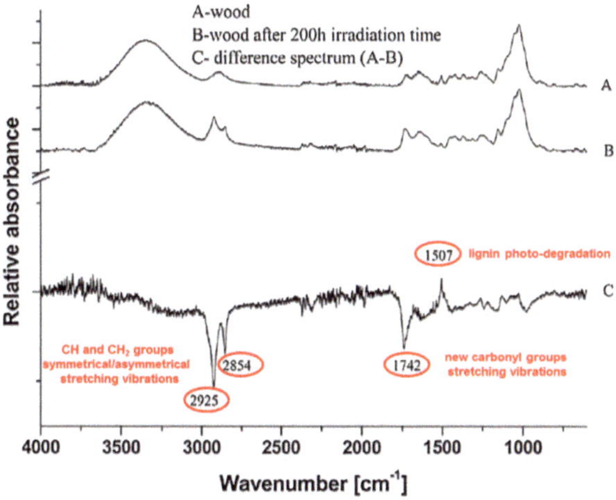

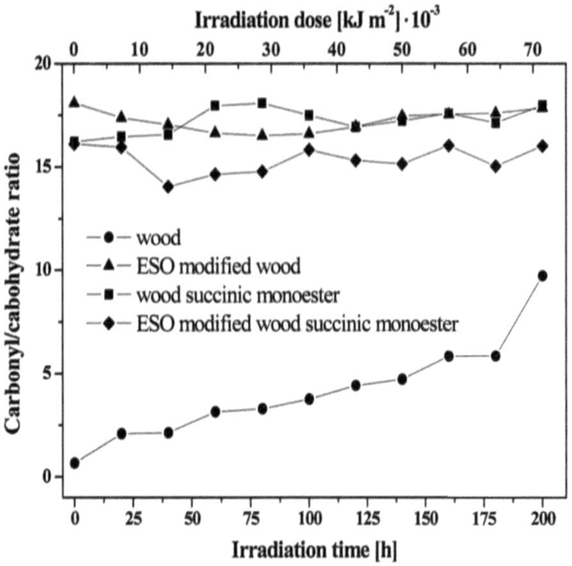

Fig. 13 Carbonyl/carbohydrate variation during irradiation [52] (reproduced with kind permission from Elsevier—License No. 3723031483298)

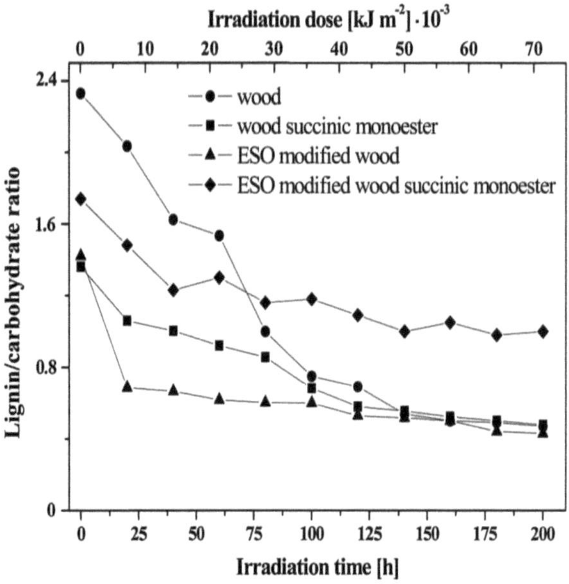

Fig. 14 Lignin/carbohydrate variation during irradiation [52] (reproduced with kind permission from Elsevier—License No. 3723031483298)

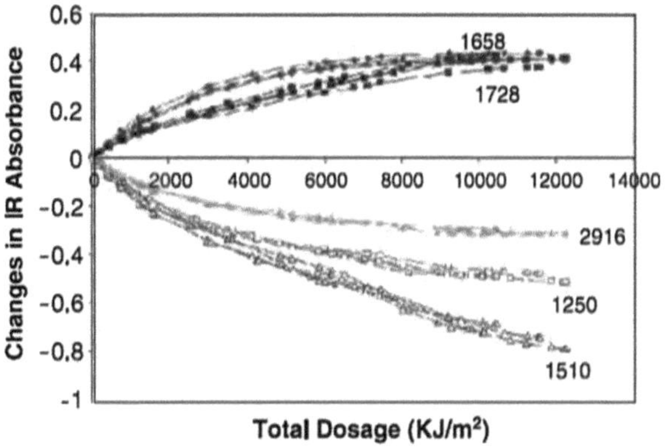

Fig. 15 Changes in FTIR bands intensities at 1250, 1510, 1658, 1728 and 2916 cm^{-1} with dosage increase for an amine cured epoxy inside aging chamber with two filters at 35 °C and 25 % RH. Data collected from four replicates [54] (reproduced with kind permission from Springer Science + Business Media—License No. 3835920446669)

apparent quantum yield as the initial slope from the degradation as function of absorbed dosage plot.

Since degradation during photochemical exposure is generated solely by the amount of absorbed incident energy, the total effective dosage model thus possesses the highest scientific accuracy.

In terms of laboratory exposure, experiments are usually undertaken in aging chambers. FTIR changes in IR absorbance corresponding to baseline corrected band peaks which exhibit significant modifications in intensities at different exposure times are usually plotted as a function of dosage. The plotting of dosage is preferred over exposure time due to the high UV radiation energy, making it impossible to monitor device components and optical parameters at a same level during extended exposure times, thus irradiation intensity variation is used. From such a plotting one finds the UV irradiance effect on decomposition, dependent neither on irradiation source, nor on total radiation [54]. As previously stated, the dosage is determined from UV–visible spectra and dose. Using custom made mathematical software calculus, one may determine the total effective dosage and plot the changes in IR absorbance (chemical changes) as a function of it (Figs. 15 and 16—reproduced with kind permission from Springer Science + Business Media—License No. 3835920446669). From the initial slope the apparent spectral quantum yield is determined and further used in initial radiation efficiency evaluation at different wavelengths. Further information on the assessment of photochemical decomposition mechanism may also be obtained from the slope shape. For example, the decreasing of slopes with total dosage is an indication of a screening effect generated by the decomposition products.

The overlapping of all curves measured at different irradiance levels in the form of a master curve, such is presented in Fig. 16, indicates that a certain damage level

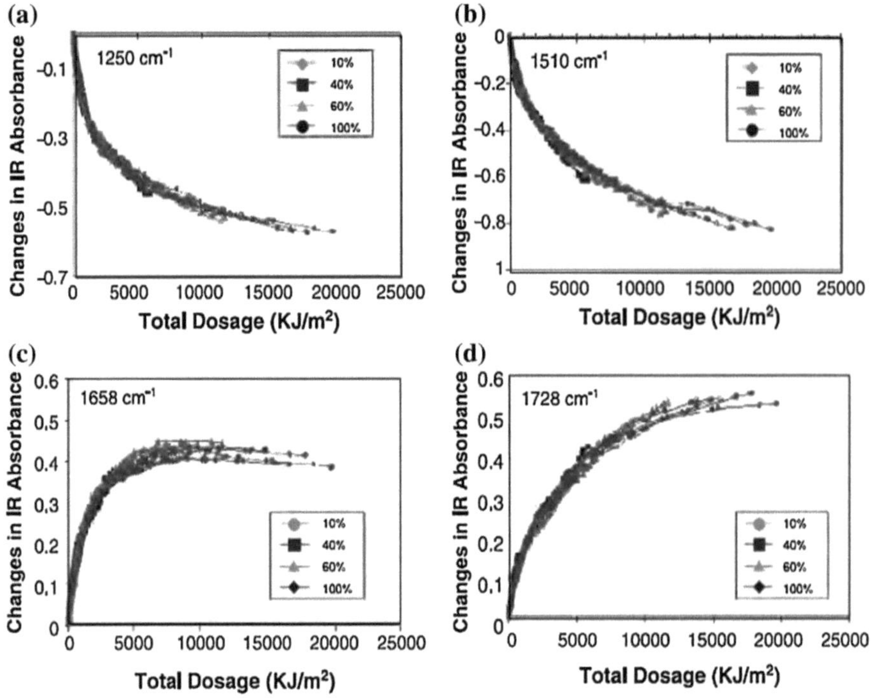

Fig. 16 Changes in FTIR bands intensities at 1250, 1510, 1658, 1728 and 2916 cm^{-1} with dosage increase for an amine-cured epoxy inside aging chamber recorded at four irradiance levels and with a 326 nm filter. Data collected from four replicates [54] (reproduced with kind permission from Springer Science + Business Media—License No. 3835920446669)

is generated by a dosage not dependent on the incident intensity. If other parameters obtained in different conditions and at different wavelengths exhibit similar trend, then the additivity law is respected throught the spectral irradiance, relative humidity and radiance intensity ranges.

A second strategy of linking photodecomposition mechanisms obtained for outdoor exposure to the ones resulted under laboratory exposure conditions consists in plotting IR bands corresponding to photochemical damage one as a function of the other (Fig. 17—reproduced with kind permission from Springer Science + Business Media—License No. 3835920446669). Thus similar peak variation trends reflect similar photochemical decomposition mechanisms, regardless of different kinetic paths and with respect to the different roles of the exposure conditions to total chemical damage. Basically, an overlapping of curves reflect similar photodecomposition mechanisms at different wavelengths (different quantum efficiencies) up to their ends where scattering occurs due to severe decomposition resulted from higher amounts of products, such as carbonyls in Fig. 17, from secondary reactions.

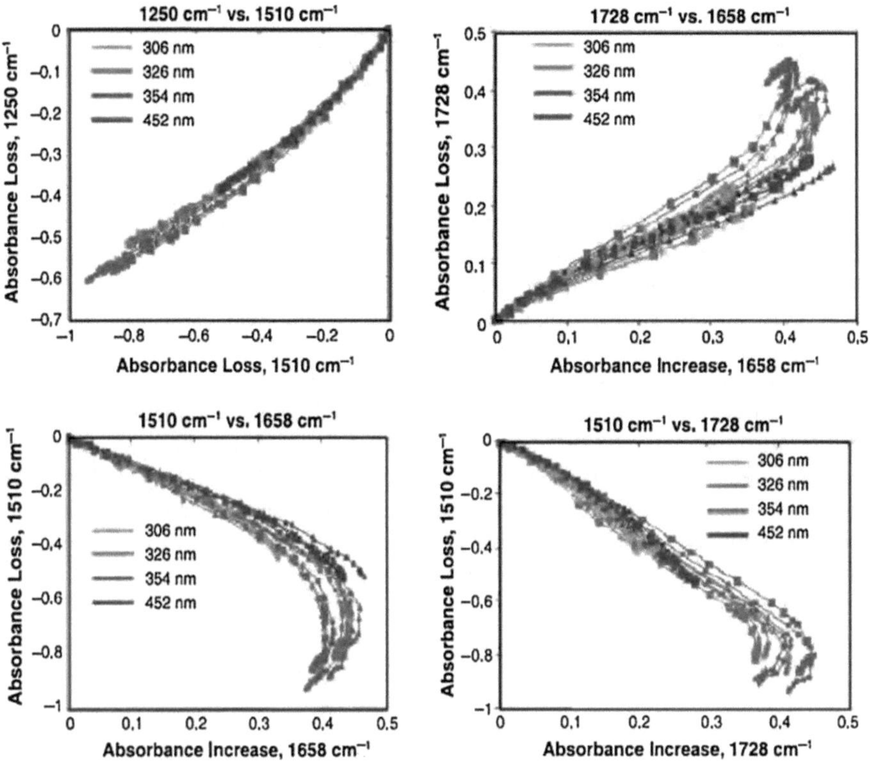

Fig. 17 Effect of wavelength of filters on different chemical modifications ratios inside aging chamber equipped with four filters (306, 326, 354, and 452 nm) and an irradiance level of 100 % at 45 °C and 25 % RH. Data collected from four replicates [54] (reproduced with kind permission from Springer Science + Business Media—License No. 3835920446669)

The cumulative damage model [54, 59] is given in Eq. (4):

$$g'(t) = \frac{d\text{Đ}[\tau,\xi(\tau)]}{d(t)} = \frac{1}{d(t)\cdot\sigma}[\text{Đ}(\infty) - \text{Đ}(0)][\frac{\exp(z)}{(1+\exp(z))^2}] \tag{4}$$

where: $\xi(\tau) = [D_{\text{inst}}(\tau), \text{temp}(\tau), \text{RH}(\tau)]$, Đ(0) represents the standardized level of damage at time 0, Đ(0) is the long term asymptote and z is given in Eq. (5):

$$z = \frac{\log[d(t)] - \mu}{\sigma} \tag{5}$$

where *d(t)* is total effective dosage, μ and 0 describe the location and steepness of the damage curve.

The cumulative amount of damage at time *t*, based on dosage incremental values is predicted by Eq. (6):

$$Dosage_{CUM}(t) = \sum_{i=0}^{t} \Delta d(i)$$
$$Damage_{CUM}(t) = \sum_{i=0}^{t} \Delta Đ(i) \tag{6}$$

where $\Delta d(t) = d(t) - d(t-1)$ and $\Delta Đ(t) = g'(t)\cdot\Delta d(t)$.

Specific parameters are taken from laboratory aging experiments. The predictive damage model is tested by comparing the observed damage to the calculated one during outdoor exposure with respect to the total dosage.

A third an last strategy in linking laboratory and field exposure test results consists in the so-called model free heuristic approach [54] where damage levels of outdoor exposure are estimated through a computer program from dosage or damage curves obtained during laboratory artificial aging. Parameters from latter testing are input into the program. In Fig. 18 (reproduced with kind permission from Springer Science + Business Media—License No. 3835920446669) there are indicated some examples in which observed outdoor damage and that calculated with the predictive model are relatively in a good agreement. However there is still much work in developing this prediction model in the future.

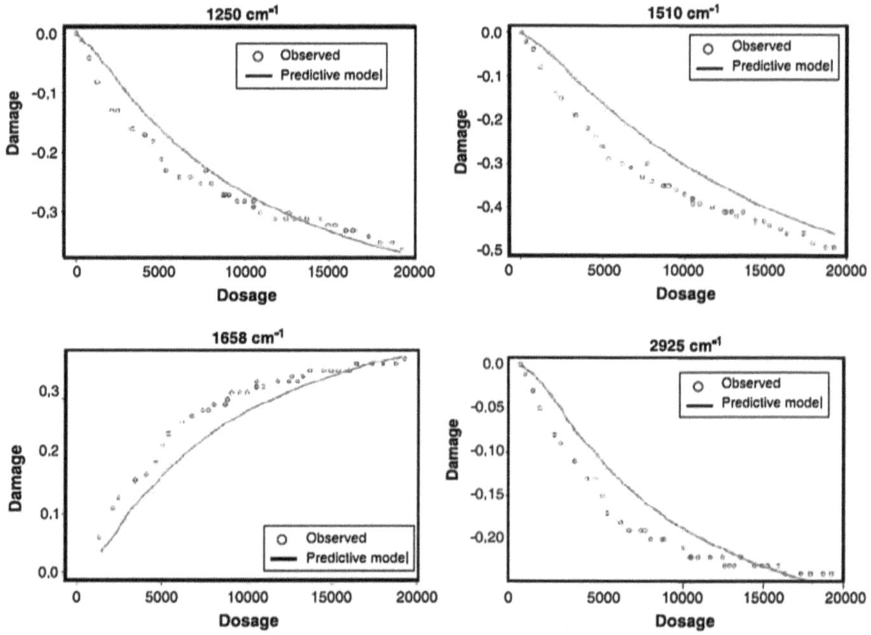

Fig. 18 Comparison between observed and predicted damage for IR bands at four signals for a single outdoor exposed specimen [54] (reproduced with kind permission from Springer Science + Business Media—License No. 3835920446669)

2.1 The Reciprocity Law

Bunsen and Roscoe [60] reported the first reciprocity law based experiments, according to which all mechanisms of photochemical processes are exclusively dependent on total absorbed energy and independent of exposure time, t, and radiant intensity, I, and expressed as:

$$I \cdot t = \text{constant} \tag{7}$$

Obeying of the reciprocity law consists in the same photoresponse to the same received integrated total dosage absorbed by a sample regardless of different exposure conditions i.e. low or high radiant flux and long or short exposure times, flash photolysis and intermittency experiments, ramping the radiant flux higher, maintaining it for a specific period of time and ramping it afterwards at lower levels [61]. Failures of reciprocity law were identified at very low or very increased radiant fluxes, thus papers doubting the validity of the reciprocity law started being published shortly after [62–66]. This led to the modification of Eq. (7) by astronomer Schwarzschild [67] in order to fit at low intensities stellar data. Schwarzschild's empirical law is given by either Eq. (8.1) or (8.2):

$$It^p = \text{constant} \tag{8.1}$$

$$I^p t = \text{constant} \tag{8.2}$$

In the above mentioned equations p denotes the Schwarzschild coefficient, which was initially equal to 0.86 and varies within different materials, in the same material and even with variations in radiant flux. By plotting log(photoresponse) as a function of log(I) from equations (8) one may test the validity of the reciprocity law [68–70]. If the line slope is equal to 1 data points overlap with the line, the reciprocity law is validated. In the same conditions but with a slope value different than 1, Schwarzschild's law is validated with p being the slope. If data does not fit the line, then the p value varies and is dependent on radiant flux. Other models were also reported after the reciprocity and Schwarzchild's laws [71, 72], mostly proposed in the photographic field, however few being widely accepted.

2.2 Effect of Temperature on Photochemical Accelerated Aging

It is important to predict the aging effects of photochemical exposure on properties of materials before their use in different applications. In many applications, mostly ones occurring in the natural environment, exposure to UV irradiation must be taken into account at certain elevated test temperatures as means of accelerated aging simulation, especially concerning polymers for medical devices [73]. One typical approach consists in assuming that aging rate is increased by the factor [74]:

$$f = 2^{\Delta T/10} \tag{9}$$

where $\Delta T = T - T_{\text{reference}}$ in which T depicts a specific accelerating temperature [73, 75] and $T_{\text{reference}}$ is a temperature value of reference at which aging effects are determined. As an example, for storing of materials, $T_{\text{reference}}$ is the value of storage temperature and T is room temperature. For future implant biomaterials, $T_{\text{reference}}$ is the the body temperature (37 °C). Since Eq. (9) is no more than an empirical expression, its validity was challenged by stating that results obtained from accelerated aging studies should be rather validated through kinetic studies of chemical processes and not via empirical laws [74, 76].

From a kinetic standpoint, supposing no secondary effects occur during chemical modifications in a polymer structure, it is considered that such modifications follow a first order reaction [74, 77], where C_{A} represents the reactant molar concentration and k is defined in Eq. (11):

$$-\frac{dC_A}{dt} = kC_A \tag{10}$$

$$k = K\exp\left(\frac{-E}{RT}\right) \tag{11}$$

In Eq. (11) E represents activation energy of the aging process, R is the universal gas constant (8.314 J mol^{-1}) and K is an empirical parameter describing the probability of other molecules interactions with the material.

At a time $T = T_{\text{reference}}$, if $k = k_{\text{reference}}$, then Eq. (11) may be rewritten as Eq. (12), which, in turn, may be expressed in the form of Eq. (13):

$$f = \frac{k}{k_{reference}} = \exp\left(\frac{R\Delta T}{E}\right) \tag{12}$$

$$\log\left(\frac{k}{k_{reference}}\right) = \frac{R\Delta T}{E} \tag{13}$$

E is determined from the slope of the plot $\log(k/k_{\text{reference}})$ versus ΔT (Eq. 13), that being R/E, after previously determining k over a range of ΔT values, and inserted into Eq. (12) for obtaining the factor of elevated temperature between $T_{\text{reference}}$ and T which increases the application rate [74, 77]. Although this is a better approach, since it relies on experimental data, a single issue arises which consists in the supposition that the studied aging process is described by first order kinetics, thus the validity conditions of Eq. (9) must be taken into account and comprehended. Debates arose concerning the T values range over which Eq. (9) is valid [73–75].

If ΔT = 10 °C, then $f = 2$, according to Eq. (9), which further substituted into Eq. (13) results in Eq. (14), in which $E' = E$ when the ten degree rule is respected. By further introducing the value of R into Eq. (14), then $E' = 120\ \text{J mol}^{-1}$, presumed with respect to the ten degree rule. By inserting E' from Eq. (14) into Eq. (12) there results the expression for f given in Eq. (15).

$$E' = \frac{10R}{\log_e 2} \tag{14}$$

$$f' = \exp\left(\frac{\Delta T \log_e 2}{10}\right) \tag{15}$$

By plotting f as a function of ΔT, one may check the ten degree rule predictions validity. The predictions are valid if the values of Eqs. (9) and (15) superimpose onto a single curve. The temperature must be below a value which initiates other chemical processes or physical transitions such as glass transition, melting or other phenomena not associated with normal aging processes [73, 74]. Initiations of any of such processes would also not validate the first order kinetic assumption.

The general issue is that a ten degree temperature increases doubles the aging rate, issue which may be surpassed only by calculating E of aging from Eq. (13). This aspect also brings forth the issue of ambient temperature value, which must be selected in such a manner as not to initiate previously mentioned chemical or physical processes, thus a generalization of the empiric law is required.

In this sense one may consider an increment of temperature noted μ for increasing n times the reaction rate. As a consequence, the raising by ΔT of ambient temperature will lead to a rating age increase shown in Eq. (16). Based on the previous discussions, this equals to the obtaining of an activation energy E'' value expression given in Eq. (17).

$$f'' = n^{\frac{\Delta T}{\mu}} \tag{16}$$

$$E'' = \frac{\mu R}{\log_e n} \tag{17}$$

The lifetime of a polymer may be expressed through Eq. (18), in which t_{fail} is the failure time of the material property, P is relative property, $f(P)$ reflects the property change function, T is the absolute temperature and the rest of the parameters were previously explained [78].

$$t_{fail} = \frac{f(P)}{A} \exp\left(\frac{E}{RT}\right) \tag{18}$$

The decomposition rate of a monitored property of interest was established via the equation proposed by Dakin et al. [79, 80] (Eq. 19), in which $g(P)$ is the differential property degradation function [78]:

$$-\frac{dP}{dt} = g(P)A\exp\left(-\frac{E}{RT}\right) \quad (19)$$

Integration of Eq. (19) leads to the obtaining of Eq. (18) where $f(P)$ is the integral function of a property degradation and its expression is given in Eq. (20) [78].

$$f(P) = -\int_{1}^{P} \frac{dP}{g(P)} \quad (20)$$

3 Conclusions

Most polymers and different types of polymer based materials, such as composites, blends, interpenetrating and semi-interpenetrating networks, are implied in a wide variety of applications, many of which occur outdoors, more or less directly exposed into the environment. Consequences of outdoor exposure reside in either irreversible physical and/or chemical structural modifications inflicted by different environmental factors, gradually determining material performance reduction until disposal. It is why materials service lifetime assessment studies simulating different environmental conditions must first be conducted based on data obtained from simulated practice or durability tests. Such tests consist in the monitoring of a desired property during exposure time until structural damage is gradually generated by environmental factors, thus increasing the lifetime and service quality of polymers has always constituted a major challenge in obtaining of new polymeric materials with predefined properties and life cycle assessment.

Due to this aspect and the continuous development of new technologies and outdoor applications of polymers, increasing number of methods on the matter have been also reported, many of which debate the challenge that is to accurately link data from natural and artificial aging experiments and rely on either accelerated aging predictions, empirical formulas, different damage models or new ways for estimating service lifetime of polymers.

References

1. Perrin, F.X., Merlatti, C., Aragon, E., Margaillan, A.: Degradation study of polymer coating: improvement in coating weatherability testing and coating failure prediction. Prog. Org. Coat. **64**, 466–473 (2009)
2. Merlatti, C., Perrin, F., Aragon, E., Margaillan, A.: Natural and artificial weathering characteristics of stabilized acrylic-urethane paints. Polym. Degrad. Stab. **93**, 896–903 (2008)

3. Rajakumar, K., Sarasvathy, V., Thamarai, C.A., Chitra, R., Vijayakumar, C.: Natural weathering studies of polypropylene. J. Polym. Environ. **17**, 191–202 (2009)
4. Larché, J.F., Bussière, P.O., Gardette, J.L.: Characterisation of accelerated ageing devices for prediction of the service life of acrylic-melamine/urethane thermosets. Polym. Degrad. Stab. **96**, 1530–1536 (2011)
5. Wang, Q.H., Zhang, G.Y., Zhang, X.L., Wang, X.M., Li, H.: PC out-door aging relationship research in different typical climates. Eng. Plast. Appl. **21**, 49–51 (1993)
6. Davis, A.: Effect of climate on the weathering of polyacetal. Polym. Degrad. Stab. **3**, 187–198 (1981)
7. Khelidj, N., Colin, X., Audouin, L., Verdu, J.: A simplified approach for the lifetime prediction of PE in nuclear environments. Nucl. Instrum. Methods B **236**, 88–94 (2005)
8. Colin, X., Monchy, L.C., Audouin, L., Verdu, J.: Lifetime prediction of polyethylene in nuclear plants. Nucl. Instrum. Methods B **265**, 251–255 (2007)
9. Pinel, B.: A methodology to predict the life duration of polymers used in nuclear power stations. Industrial needs and their approach. Nucl. Instrum. Methods B **151**, 471–476 (1999)
10. Bellenger, V., Ganem, M., Mortaigne, B., Verdu, J.: Lifetime prediction in the hydrolytic ageing of polyesters. Polym. Degrad. Stab. **49**, 91–97 (1995)
11. Pickett, J.E., Coyle, D.J.: Hydrolysis kinetics of condensation polymers under humidity aging conditions. Polym. Degrad. Stab. **98**, 1311–1320 (2013)
12. Brown, R.: Predictive techniques and models for durability tests. Polym. Test **14**, 403–414 (1995)
13. Celina, M., Gillen, K., Assink, R.: Accelerated aging and lifetime prediction: review of non-Arrhenius behaviour due to two competing processes. Polym. Degrad. Stab. **90**, 395–404 (2005)
14. Gillen, K., Bernstein, R., Clough, R., Celina, M.: Lifetime predictions for semicrystalline cable insulation materials: I. Mechanical properties and oxygen consumption measurements on EPR materials. Polym. Degrad. Stab. **91**, 2146–2156 (2006)
15. Kahlen, S., Wallner, G., Lang, R.: Aging behavior and lifetime modeling for polycarbonate. Sol. Energy **84**, 755–762 (2010)
16. Pickett, J.E., Sargent, J.R., Blaydes, H.A., Babbie, N.: Effects of temperature on the weathering lifetime of coated polycarbonate. Polym. Degrad. Stab. **94**, 1085–1091 (2009)
17. Ram, A., Zilber, O., Kenig, S.: Life expectation of polycarbonate. Polym. Eng. Sci. **25**, 535–540 (1985)
18. Mehr, Y.M., Driel, W.D.V., Jansen, K.M.B., Deeben, P., Zhang, G.Q.: Lifetime assessment of Bisphenol-A Polycarbonate (BPA-PC) plastic lens, used in LED-based products. Microelectron. Reliab. (2014). doi:10.1016/j.microrel.2013.09.009
19. Gillen, K., Celina, M.: The wear-out approach for predicting the remaining lifetime of materials. Polym. Degrad. Stab. **71**, 15–30 (2000)
20. Gillen, K., Celina, M., Bernstein, R., Shedd, M.: Lifetime predictions of EPR materials using the wear-out approach. Polym. Degrad. Stab. **91**, 3197–3207 (2006)
21. Malik, J., Tuan, D., Spirk, E.: Lifetime prediction for hindered amine light stabilizer-stabilized low-density polyethylene and polypropylene. Polym. Degrad. Stab. **47**, 1–8 (1995)
22. Guseva, O., Brunner, S., Richner, P.: Service life prediction for aircraft coatings. Polym. Degrad. Stab. **82**, 1–13 (2003)
23. Evans, M.: A statistical degradation model for the service life prediction of aircraft coatings: with a comparison to an existing methodology. Polym. Test **31**, 46–55 (2012)
24. Hoàng, E.M., Lowe, D.: Lifetime prediction of a blue PE100 water pipe. Polym. Degrad. Stab. **93**, 1496–1503 (2008)
25. Celina, M., Trujillo, A., Gillen, K., Minier, L.: Chemiluminescence as a condition monitoring method for thermal aging and lifetime prediction of an HTPB elastomer. Polym. Degrad. Stab. **91**, 2365–2374 (2006)
26. Hakkarainen, M., Albertsson, A.: Indicator products: a new tool for lifetime prediction of polymeric materials. Biomacromolecules **6**, 775–779 (2005)

27. Croll, S., Hinderliter, B., Liu, S.: Statistical approaches for predicting weathering degradation and service life. Prog. Org. Coat. **55**, 75–87 (2006)
28. Croll, S., Hinderliter, B.: A framework for predicting the service lifetime of composite polymeric coatings. J. Mater. Sci. **43**, 6630–6641 (2008)
29. Croll, S., Hinderliter, B.: Estimating service lifetimes in weathering: an optimistic view. J. Coat. Technol. Res. **4**, 217–230 (2007)
30. Shah, C.S., Patni, M.J., Pandya, M.V.: Accelerated aging and life time prediction analysis of polymer composites: a new approach for a realistic prediction using cumulative damage theory. Polym. Test **13**, 295–322 (1994)
31. Liu, H., Zhou, M., Zhou, Y., Wang, S., Li, G., Jiang, L., Dan, Y.: Aging life prediction system of polymer outdoors constructed by ANN. 1. Lifetime prediction for polycarbonate. Polym. Degrad. Stab. **105**, 218–236 (2014)
32. Maxwell, A.S., Broughton, W.R., Dean, G., Sims, G.D: Review of accelerated ageing methods and lifetime prediction techniques for polymeric materials, NPL Report DEPC MPR 016, National Physics Laboratory, Teddington, Middlesex, United Kingdom, Queen's Printer for Scotland (2005)
33. ISO 877: Plastics—methods of exposure to direct weathering, to weathering using glass-filtered daylight, and to intensified weathering by daylight using Fresnel mirrors (1994)
34. ASTM D4364-02 Standard practice for performing outdoor accelerated weathering tests of plastics using concentrated sunlight
35. ASTM D1435-99 Standard practice for outdoor weathering of plastics
36. ASTM G24-05 Standard practice for conducting exposures to daylight filtered through glass
37. Robinson, J., Poulsen, K.V., Högström, P.A., Linder, A., Burkhard, H., Kobilsek, M., Gemmel, A.: Comparison of standard and UV test methods for the ageing of cables. In: Proceedings of the 60th IWCS conference, Charlotte (NC) (2011)
38. ASTM D2565-99 Standard practice for xenon arc exposure of plastics intended for outdoor applications
39. ISO 4892 part 2: Plastics—methods of exposure to laboratory light sources—Part 2: xenon-arc sources (2004)
40. IEC 60068-2-9 Environmental testing, Part 2: tests—guidance for solar radiation testing
41. ASTM D1499-99 Standard practice filtered open-flame carbon-arc type exposures of plastics
42. ISO 4892 Part 4: Plastics—methods of exposure to laboratory light sources—part 4: open-flame carbon-arc lamps (2004)
43. Diffey, B.L.: Sources and measurement of ultraviolet radiation. Methods **28**, 4–13 (2002)
44. Skiest, I.: Handbook of Adhesives, 2nd edn, pp. 12–17. Litton Educational Publishing, New York (1977)
45. Benson, L.M.: Polymer Blends: A Comprehensive Review, pp. 65–108. Carl Hanser Verlag, Munich (2007)
46. Rosu, D., Varganici, C.-D., Rosu, L.: Thermal degradation of thermosetting blends. In: Visakh, P.M., Arao, Y. (eds.) Thermal Degradation of Polymer Blends, Composites and Nanocomposites, pp. 21–23. Springer International Publishing, Switzerland (2015)
47. Rosu, D., Rosu, L., Mustata, F., Varganici, C.-D.: Effect of UV radiation on some semiinterpenetrating polymer networks based on polyurethane and epoxy resin. Polym. Degrad. Stab. **97**, 1261–1269 (2012)
48. Varganici, C.D., Rosu, L., Rosu, D., Simionescu, B.C.: Miscibility studies of some semiinterpenetrating polymer networks based on an aromatic polyurethane and epoxy resin. Compos. B Eng. **50**, 273–278 (2013)
49. Rosu, D., Rosu, L., Varganici, C.-D.: The thermal stability of some semi-interpenetrated polymer networks based on epoxy resin and aromatic polyurethane. J. Anal. Appl. Pyrol. **100**, 103–110 (2013)
50. DIN 6174:1979-01 Farbmetrische Bestimmung von Farbabständen bei Körperfarben nach der CIELAB-Formel

51. Varganici, C.-D., Rosu, L, Mocanu, O.M., Rosu, D.: Influence of poly(vinyl alcohol) on cellulose photochemical stability in cryogels during UV irradiation. J Photochem. Photobiol. A Chem. **297**, 20–30 (2015)
52. Rosu, D., Bodîrlău, R., Teacă, C.-A., Rosu, L., Varganici, C.-D.: Epoxy and succinic anhydride functionalized soybean oil for wood protection against UV light action. J. Clean. Prod. **112**(1), 1175–1183 (2015)
53. Tocháček, J., Vrátníčova, Z.: Polymer life-time prediction: the role of temperature in UV accelerated ageing of polypropylene and its copolymers. Polym. Test **36**, 82–87 (2014)
54. Gu, X., Stanley, D., Byrd, W.E., Dickens, B., Vaca-Trigo, I., Meeker, W.Q., Nguyen, T., Chin, J.W., Martin, J.W.: Linking accelerated laboratory test with outdoor performance results for a model epoxy coating system. In: Martin, J.W., Ryntz, R.A., Chin, J., Dickie, R.A. (eds.) Service Life Prediction of Polymeric Materials. Global Perspectives, pp. 1–26. Springer Science + Business Media, New York (2009)
55. Kinmonth, R.A., Norton, J.E.: Effect of spectral energy distribution on degradation of organic coatings. J. Coat. Technol. **49**(633), 37 (1977)
56. Fischer, R.M., Ketola, W.D., Murray, W.P.: Inherent variability in accelerated exposure methods. Prog. Org. Coat. **19**, 165–179 (1991)
57. Searle, N.D., Giesecke, P., Kinmonth, R., Hirt, R.C.: Ultraviolet spectra distributions and aging characteristics of xenon arcs and filters. Appl. Opt. **3**, 963–966 (1964)
58. Martin, J.W., Nguyen, T., Byrd, E., Embree, N.: Estimation of absorbed photolytic dosage and quantum yield using UV–Vis and FTIR spectroscopies. In: Proceedings of the American Chemical Society, PSME, vol. 82, Spring Meeting, San Francisco, CA, 26–30 March 2000
59. Vaca-Trigo, I., Meeker, W.Q.: A statistical model for linking field and laboratory exposure results for a model coating. In: Proceedings of the International Symposium on Service Life Prediction: Global Perspectives, Key Largo, FL (2006)
60. Bunsen, R.W., Roscoe, H.E.: Photochemische untersuchungen. Ann. Phys. **108**(2), 193–273 (1859)
61. Martin, J.W., Chin, J.W., Nguyen, T.: Reciprocity law experiments in polymeric photodegradation: a critical review. Prog. Org. Coat. **47**, 292–311 (2003)
62. Scheiner, J.: Application de la photographie à la determination des grandeurs stallaire. Bull Comité Permanent Int Execut Photogr Carte Ciel **1**, 227–246 (1889)
63. de Abney, W.W.: On a failure of the law in photography that when the products of the intensity of the light acting and of the time of exposure are equal, equal amounts of chemical action will be produced. Proc. R. Soc. Lond. **54**, 143–147 (1893)
64. de Abney, W.W.: The failure in a photographic law with very intense light. Photogr. J. **18**, 302 (1894)
65. Englisch, E.: Über die wirkung intermittierender belichtungen auf bromsilbergelatine. Photogr. Korresp. **36**, 109 (1899)
66. de Abney, W.W.: Chemical action and exposure. Photogr. J. **18**, 56 (1893)
67. Schwarzschild, K.: On the law of reciprocity for bromide of silver gelatin. Astrophys. J. **11**, 89 (1900)
68. Vohl, P.: Surface photovoltage effects in titanium dioxide films. Photogr. Sci. Eng. **13**(3), 120 (1969)
69. Egerton, T.A., King, C.J.: The influence of light intensity on photoactivity in TiO_2 pigmented systems. J. Oil Colour Chem. Assoc. **62**, 386–391 (1979)
70. Tollin, G., Kearns, D.R., Calvin, M.: Electrical properties of organic solids. I. Kinetics and mechanism of conductivity of metalo–free phthalocyanine. J. Chem. Phys. **32**(4), 1013–1019 (1960)
71. Mees, C.E.K., James, T.H.: The Theory of the Photographic Process. Macmillan, New York (1966)
72. Mees, C.E.K.: The Theory of the Photographic Process, p. 249. Macmillan, New York (1942)
73. Hemmerich, K.J.: Accelerated aging. General aging theory and simplified protocol for accelerated aging of medical devices. Med. Plast. Biomater **5**, 16–23 (1998)

74. Hukins, D.W.L., Mahomed, A., Kukureka, S.N.: Accelerated aging for testing polymeric biomaterials and medical devices. Med. Eng. Phys. **30**, 1270–1274 (2008)
75. ASTM: Standard practice/guide for the mechanical characterization of lumbar nucleus devices: Report WK4863 (2005)
76. Verdu, J., Colin, X., Fayolle, B., Audouin, L.: Methodology of lifetime prediction in polymer aging. J. Test. Eval. **35**, 1–8 (2007)
77. Laidler, K.J.: Reaction kinetics. In: Homogeneous Gas Reactions, vol. 1, pp. 1–41. Pergamon, Oxford
78. Budrugeac, P.: Lifetime prediction for polymers via the temperature of initial decomposition. J. Therm. Anal. Calorim. **65**, 309–312 (2001)
79. Dakin, T.W.: Electrical insulation deterioration treated as a chemical rate phenomenon. AIEE Trans. **67**, 113–122 (1948)
80. Dakin, T.W.: Electrical insulation deterioration. Electrotechnology **66**, 124 (1960)

Photochemical and Photophysical Behavior of Azopolymer Compounds

Elnaz Bagherzadeh-Khajehmarjan, Sohrab Ahmadi-Kandjani and Shahla Golghasemi-Sorkhabi

1 Introduction

Two phenyl rings separated by an azo (–N=N–) bond known as Azobenzene, functions as the base molecule for a wide class of aromatic azo compounds. Azobenzenes almost cover the full rainbow, and up to and almost 70 % of the world's commercial dyes are still azobenzene-based. These chromophores are adaptable molecules and have received much attention, in both fundamental and applied research. Azobenzenes can be categorized into three spectroscopic classes: azobenzene-type molecules, aminoazobenzene-type molecules, and pseudo-stilbenes.

The rigid mesogenic shape of these molecules is suitable, for spontaneous organization into liquid crystalline (LC) phases. One of the most attractive methodologies for inserting azobenzene into functional materials is through covalent attachment to polymers. Incorporation of azobenzenes into matrix of polymer is an adequate inclusion technique. Both side-chain and main-chain azobenzene polymers have been prepared. Also, Azobenzenes can be used in a "guest-host" system. These guest–host systems can be cast or spin-coated from solution mixtures

E. Bagherzadeh-Khajehmarjan · S. Ahmadi-Kandjani (✉) · S. Golghasemi-Sorkhabi
Research Institute for Applied Physics and Astronomy (RIAPA),
University of Tabriz, Tabriz, Iran
e-mail: s_ahmadi@tabrizu.ac.ir

E. Bagherzadeh-Khajehmarjan
e-mail: e.bagherzadeh7@gmail.com

S. Golghasemi-Sorkhabi
e-mail: sh.golghasemi@tabrizu.ac.ir

S. Golghasemi-Sorkhabi
Laboratoire Moltech Anjou Université D'Angers/UMR CNRS 6200 2,
Bd Lavoisier, 49045 Angers, France

D. Rosu and Visakh P. M. (eds.), *Photochemical Behavior of Multicomponent Polymeric-based Materials*, Advanced Structured Materials 26,
DOI 10.1007/978-3-319-25196-7_9

of polymer and azo small molecules, where the azo content in the thin film is easily adjustable, through concentration. In addition to the unusual photoresponsive behavior of the azo moieties, resulting materials profit from the inherent stability, rigidity and processability of polymers.

The most intriguing properties of these materials are orientation of azobenzenes in direction perpendicular to the electric field of light and the readily induced, and reversible, isomerization of azo bond, between the two geometric isomers: a more stable trans state and meta-stable cis. Upon absorbing a photon (with a wavelength in the trans absorption band), the azobenzene converts into the cis isomer, with high efficiency. A second wavelength corresponding to the cis absorption band can cause the back-conversion, to the trans state. Or in the absence of second wavelength, alternately, azos will thermally reconvert from the cis to trans state, with a timescale ranging from milliseconds to hours, depending on the substitution pattern and local environment. This photoisomerization can be exploited as a photos witch to orient the chromophore (which induces birefringence), or even to perform all-optical surface topography patterning. The geometrical change upon absorption of light is the fundamental molecular photomotion in azobenzenes.

A large-scale surface mass transport occurs when the Azopolymer films are irradiated with light which leads to surface patterning, i.e., a surface relief grating (SRG). This mass transport occurs at room temperature, which is well below the Tg of polymer. It should be noted that, this phenomenon is a reversible mass transport, not irreversible material ablation. The flat film, with the original thickness, could be recovered upon heating above Tg. Several models of surface evolution upon illumination have been discussed but none provides a satisfactory explanation for the mass transport at a molecular level. Nevertheless, it is widely accepted that multiple trans–cis and cis–trans photoisomerizations of the chromophores trigger the inscription process. These photochemical and photophysical nature of chromophores and their photomotions enable many interesting applications, ranging from optical components and lithography to sensors and smart materials, photoswitches, photoprobes, optical data storage and nonlinear optics.

2 Photochemical Behaviors of Azopolymers

As a result of smart materials' ability to perform specific functions in response to changes in stimuli, they have transformed material science. Therefore, they have potential applications in many fields such as; actuators, sensors and micro-pumps. The capability to respond in a precise way, to slightest variation in the neighboring environment like; temperature, pH, light, magnetic or electric field or the existence of biological molecules is a feature that makes these materials "smart". Fast macroscopic changes occurring in their structure is one of the reasons for exceptionality of these materials. The other distinct property is that some of these shape changes are reversible [1]. As a result of constant interest in remotely controllable

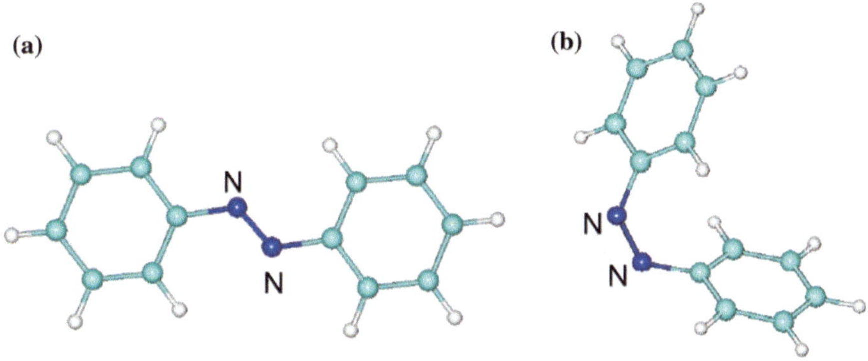

Fig. 1 Azobenzene molecule: **a** trans-form and **b** cis-form

devices and high speed and capacity in optical processes, development of photosensitive organic materials has been one of the top subject of intense research over the past years [2]. Synthesizing photoresponsive smart polymers can be done via functionalizing the material with photosensitive molecules for instance cinnamic acid (CA), cinnamylidene or azo compounds. Azo dyes are, considerably, the most important and versatile class, accounting for over 50 % of all commercial dyes. Due to this significance, azo dyes have been studied more than any other class and a lot of studies has been done to understand their properties. Azo dyes can contain two, three, or more rarely, four azo groups (–N=N–), but they should contain at least one azo group. The nitrogen atoms are Sp^2 hybridized. The azo group is attached to two radicals, which one or more usually both are aromatic [3].

Azobenzene is a compound where an azo linkage joins the two aromatic phenyl rings. Azobenzenes possess two isomeric forms, as shown in Fig. 1; a thermally stable trans state and a meta-stable cis form. Each form is associated with various physicochemical features such as; absorption spectrum, dipole moment, refractive index and so on, as a result of change in molecular geometry [1]. In equilibrium, the material mostly consists of trans azobenzenes in the lower energy, i.e. trans conformational state. The trans isomers have rod like shape, no dipole moment and are elongated. Whereas, the cis configurations have bended geometry and a dipole moment of 3.0 D [4, 5].

2.1 Spectroscopic Properties of Azobenzenes

Through substituting an aromatic ring with various substituents, in order to change geometry and electron donating/withdrawing mechanism, different types of azo materials can be obtained. This class of chromophores share numerous spectroscopic and photo-physical properties [1]. As described by Rau [6], azobenzenes can

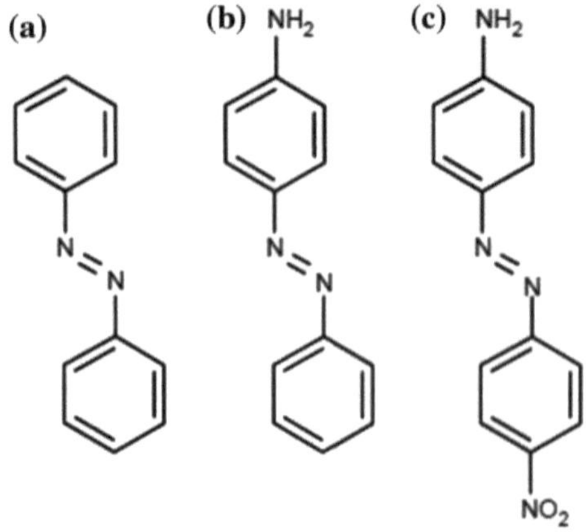

Fig. 2 Different types of azobenzenemolecules, **a** azobenzene, **b** aminoazobenzene, **c** pseudo-stilbene [6]

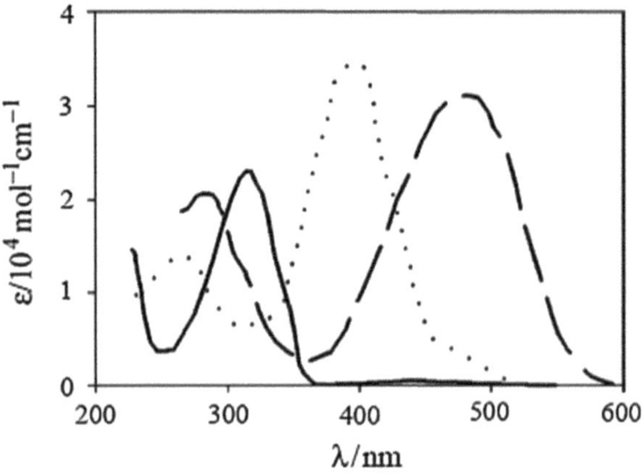

Fig. 3 Absorption spectra of azobenzene molecules in trans form, azobenzene, aminoazobenzene and pseudo-stilbene shown in *solid line*, *dot line* and *dash line*, respectively

be categorized into three spectroscopic classes: azobenzene-type molecules, aminoazobenzene-type molecules and pseudo-stilbenes (refer to Fig. 2). The particularity of absorption spectra for these three groups (shown in Fig. 3) give rise to their prominent colors: yellow, orange, and red, respectively. Absorption characteristics of many azos are similar to those of the unsubstituted azobenzenes. In the spectra, these compounds show a low-intensity $n \rightarrow \pi^*$ and a much stronger $\pi \rightarrow \pi^*$ absorption band in the visible region and UV, respectively. Minor or major changes in spectroscopic character of these materials is possible via adding substituents to the azobenzene rings. One particular change is ortho- or para-substitution with an electron-donating group (usually an amino, $-NH_2$). This change leads to a new class of compounds, Aminoazobenzenes. In aminoazobenzenes, the $n \rightarrow \pi^*$ and $\pi \rightarrow \pi^*$ absorption bands are much closer. In fact, the $n \rightarrow \pi^*$ may be completely buried beneath the intense $\pi \rightarrow \pi^*$. Even though, azobenzenes are fairly indifferent to solvent polarity, aminoazobenzene absorption bands can shift to higher energy in non-polar solvents. As for polar solvents a shift to lower energy was observed. A "push/pull" substitution pattern i.e. substituting azobenzene at the 4- and 4′-positions with an electron-donor and an electron-acceptor (such as an amino and a nitro group, $-NH_2$ $-NO_2$) results in a highly asymmetric distribution of electrons. This change shifts the $\pi \rightarrow \pi^*$ absorption band to lower energy, toward the red and after the $n \rightarrow \pi^*$. The third spectroscopic class, the pseudo-stilbenes (in analogy to stilbene, phenyl–C=C–phenyl) is defined by this reversed ordering of the absorption bands. Because of high sensitivity of the pseudo-stilbenes to local environment, they can be used in some applications [5, 6].

Azobenzenes are strong and multipurpose moieties and have been extensively studied as small molecules, as pendants on other molecular structures or incorporated into a vast variety of amorphous, crystalline, or liquid crystalline polymeric systems. Incorporation of azobenzene moieties in a polymer matrix can be done covalently and noncovalently. The latter can be with or without specific interactions with the polymer matrix [6]. Because of photo responsive properties and potential applications of polymers containing azobenzene in fields such as; optical devices, artificial muscles and drug-delivery, they have attracted significant attention [7]. Covalently bounding of azobenzene to polymer, influence the orientation of the polymer chains. The nonabsorbing fragments of the polymer can be included in the reorientation process by the photo-induced orientation of the azobenzene fragments [4]. Chemical nature of chromophore and polymer and the method of involvement of the chromophore in polymer, determine the efficiency and stability of photoinduced processes in polymers [8].

Most of the works in the research on polymers containing azobenzene are aimed at diblock copolymers. Few reports concern copolymers including three different blocks. Mixture systems based on multi-block copolymers and homopolymer haven't been studied thoroughly for azobenzene-containing polymer systems [7]. We have studied quantitatively the solvent media effects on the photo-physical properties of photochromic copolymathacrylates with two blocks. For materials in this study, one block contains planer cyanoazobenzene groups and other block includes butyl acetate group as side chains (PCN50 and PCN80). Also, the other compound, PCNHOMO, formed only with blocks containing planer cyanoazobenzene group. The structure of these three polymers is illustrated in Table 1. Spectra of these compounds were recorded, which are broad and shift according to the solvent polarity. The study of absorption and fluorescence emission spectra of substances in different solvents, show shifts due to the polarity difference, which is the result of interactions between the solute and solvent molecules. Their absorption bands show a red shift, which indicate rather strong guest-host interactions between the polymeric azo molecules and the host environment. Furthermore, it was observed that the steric hindrance between butyl acetate and cyanoazobenzene block or between neighbor cyanoazobenzene groups in these materials results in an interesting and strange spectroscopic behavior [9].

In a similar study, we have investigated the solvent effect on absorption and fluorescence spectra for some photochromic copolymathacrylates with azo sulfonamide chromophores in side chains [10]. The structure of the polymers are shown in Table 2. In this study, behavior of these polymeric azo compounds

Table 1 Molecular structure of polymeric cyanide azo dyes [9]

Dyes	PCN50	PCN80	PCNHOMO
Weight of Azo block	50%	80%	100%
Molecular structure	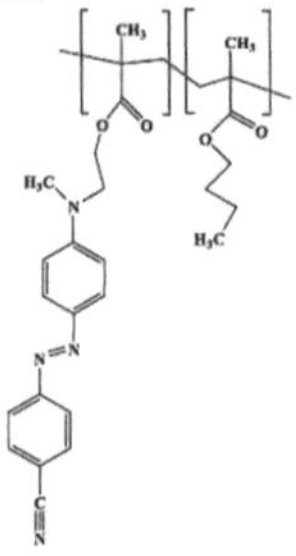	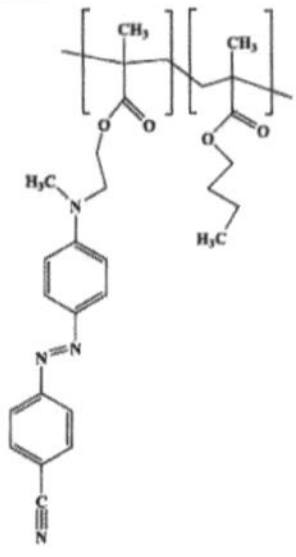	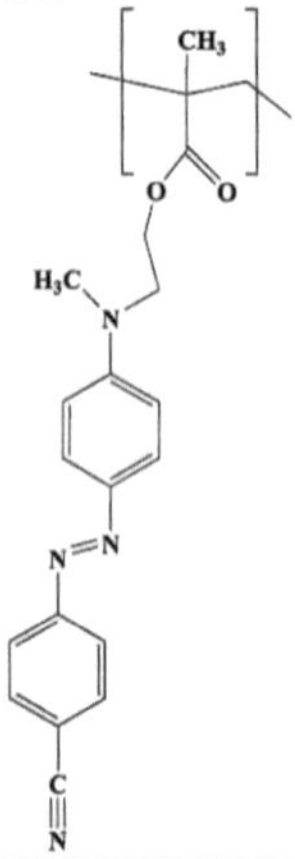

Table 2 Molecular structure of polymeric azo sulfonamide compounds [10]

Dye	IZO-3 (Copolymer)	PIZO-3(Homopolymer)	MB2I
Molecular Structure	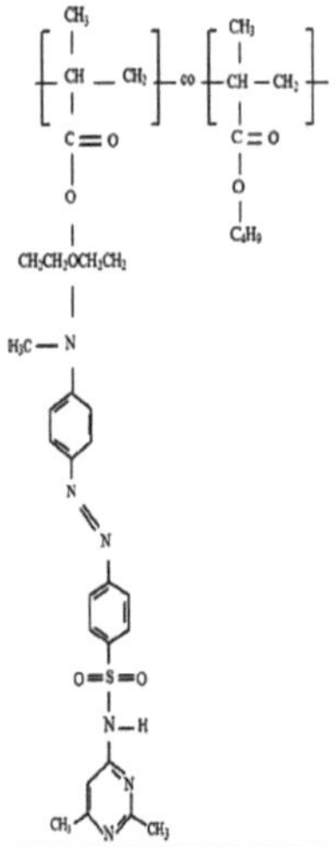	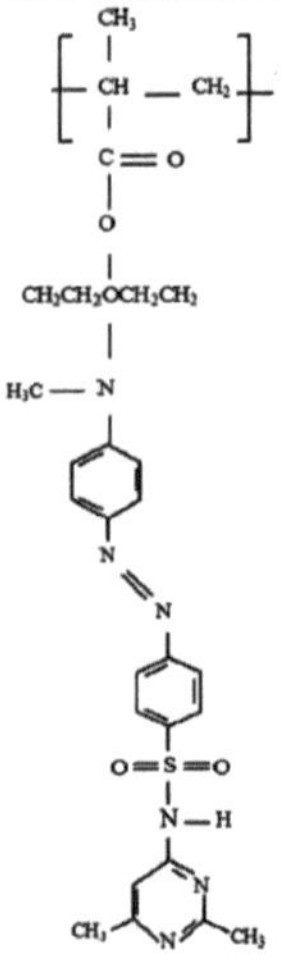	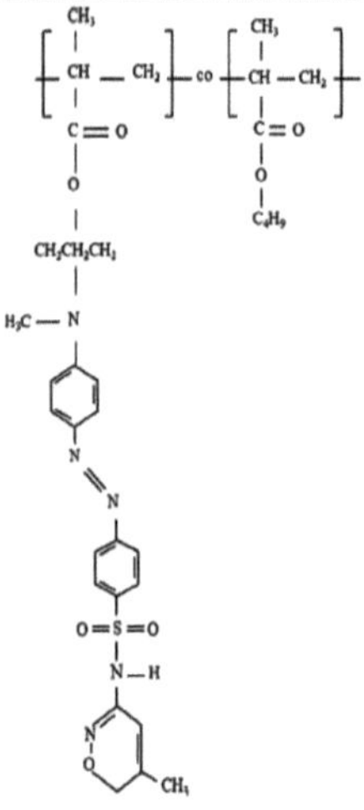

(Izo-3 copolymer, MB2I and PIzo Homopolymer) are very complex and strongly dependent on the included groups in structures and the nature of solvents. Izo-3 (Copolymer) and MB2I are formed with two blocks; one block includes azo sulfonamide groups and the other block contains butyl acetate groups. As for PIzo, there are exactly the same block containing azo sulfonamide groups. The steric hindrance between butyl acetate and azo sulfonamide blocks or between neighbor azo sulfonamide groups leads to a fascinating and strange spectroscopic behavior. Spectral shift in absorption and emission is the result of interactions between the solvent and solute molecules. In PIzo, there is a less intense structural steric resistance effects as a result of specific interactions (hydrogen bond acceptor ability) of solvents, and in return the excited state is stabilized [10].

2.2 *Photoisomerization Process*

A fast and reversible trans–cis isomerization of a double bond in azobenzene and its derivatives make them one of the fascinating and interesting materials. The trans isomer is more stable by approximately 50 kJ/mol. Barrier to isomerization, i.e. energy barrier to the photo-excited state, is on the order of 200 kJ/mol. Therefore, most azobenzene molecules, in the dark, will be found in the trans form. Transazobenzenes are converted to the sphere-shaped cis state upon absorption with photons of a certain wavelength. This conversion reduces the molecular size (the distance between 4 and 4′ carbons decreases from 9 to 5.5 Å as shown in Fig. 4). Even though this process occurs at a temperature well below the glass transition temperature of the polymer, it disrupts the molecular conformation and results in notable deformation of the sample. In this sense, behavior of an azobenzene chromophore resembles a photo-switch. This concept is conceptually related to the fundamental principle of vision in the human eye. Being completely reversible and free from side reactions, prompt Rau to characterize photoisomerizationas "one of the cleanest photoreactions known". When the light source is switched off, cis form thermally revert to the trans form, as the trans is more stable (rate is determined by the molecule's particular substitution pattern). Reconversion of cis to trans form could be done with illumination of visible light. Under illumination of a bulk azo sample or solution with a steady-state trans/cis composition, a photostationary state will be achieved based on the opposing effects of; photoisomerization into the cis state, thermal relaxation back

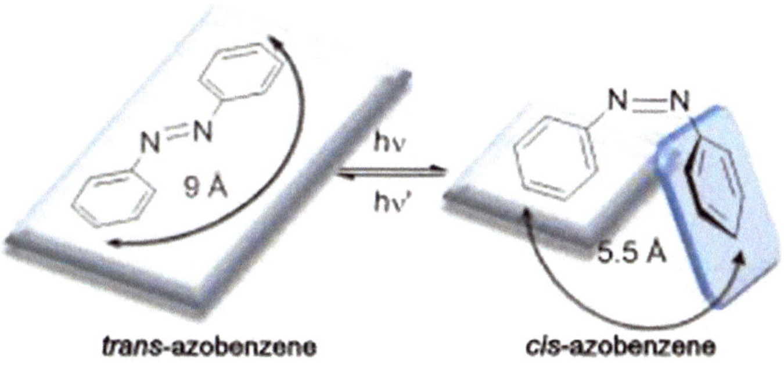

Fig. 4 Photoisomerization process of azobenzene [5]

to the trans state and probable cis reconversion upon light absorption. The steady-state composition relates to the quantum yields for the two processes and the thermal relaxation rate constant and it is unique for each system. Furthermore, this composition depends upon intensity of irradiation, wavelength of light, temperature and the matrix (gas phase, solution, liquid crystal, sol–gel, monolayer, polymer matrix, etc.). Azos are photochromic, meaning their color changes upon illumination, since the effective absorption spectrum (a combination of the trans and cis spectra) changes with light intensity. The timescale of photoisomerization process is usually in the order of pico second. Depending on the substitution pattern and local environment, thermally reconversion of azos from cis to trans state happens with a timescale extending from milliseconds to hours. Under moderate irradiation, the composition of the photostationary state is predominantly cis for azobenzenes, a mix of cis and trans for aminoazobenzenes and mainly trans for pseudo-stilbenes. More specifically, lifetimes for azobenzenes, aminoazobenzenes and pseudo-stilbenes are usually in the order of hours, minutes and seconds, respectively. In some cases, the cis state can persist for days, if the material possesses some bulky substituents, which can inhibit the cis → trans relaxation process [1, 6, 11].

Joo et al. investigated absorption spectrum of poly(malonic ester) containing *p*-cyanoazobenzene group polymer film during illumination of laser beam at 365 and 488 nm. The polymer film has two broad absorption bands. As it is well known, the large absorption band around 362 nm and much smaller one around 450 nm are due to the trans conformer of azobenzene and the cis one, respectively. The study of absorption band of polymer during illumination of pump beam with 365 nm wavelength, showed that the absorption band of trans conformer rapidly reduced with increasing the exposure time of irradiation, while for the cis form, it increases gradually. After irradiating over 30 s, the concentration of trans and cis forms reaches photostationary state. On the other hand, upon illumination of pumping beam of 488 nm, absorption bands of both trans and cis conformers remained unchanged. These results mean that, the initial concentration of unstable cis conformer is very low, and also that the photochemical trans–cis transition rate constant, by the pumping beam of 488 nm, is fairly small [12]. We also investigated the effect of illumination of polymethacrylate polymer film bearing cyanoazobenzene

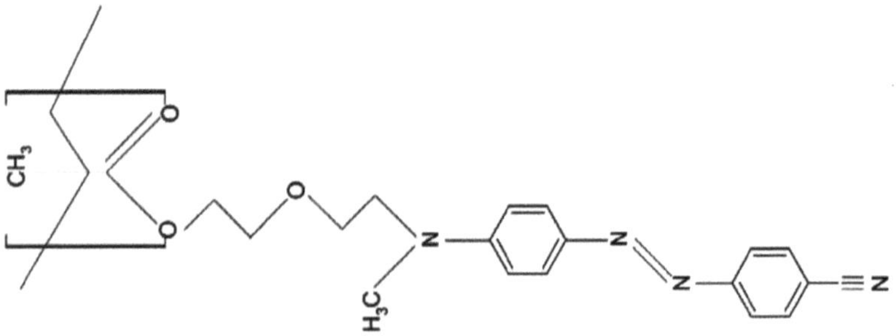

Fig. 5 Chemical structure of pCNOhomo polymer [13]

in side chain (pCNOhomo, belonging to pseudo-stilbene type), with 532 nm pump beam, at various irradiation times. The structure of pCNO homo polymer is given in Fig. 5. Because of $\pi - \pi^*$ transition of the azobenzene groups in the trans form, it's absorption band is located around 430 nm. By increasing the exposure time the intensity of absorption band enhances [13].

2.3 *Isomerization Mechanism*

Since two opposing pathways have been proposed for isomerization process; the reaction may proceed either by rotation around the N=N double bond, or it may happen by flip-flop inversion of one of the nitrogen atoms, the mechanism of cis–trans isomerization has not explicitly established, yet [14]. Upon $S_0 \rightarrow S_1$ excitation, an in-plane bending of either or both of the CNN angle leads to the photoisomerization process and a so-called inversion mechanism occurs. Upon excitation of $S_0 \rightarrow S_2$, a probable out-of-plane twisting around the N=N bond yields to the rotation mechanism. Thus, the inversion mechanism requires a much smaller free volume than the rotation [6, 15]. The mechanisms are illustrated in Fig. 6. The two pathways, rotation and inversion, are competitive, but depending on the electronic nature of the substituents covalently bonded to one or both phenyl rings and the polarity of the reaction medium, one may be favored [14]. It has been seen that, in a solvent of high polarity, the rotation mechanism is dominant. In compounds with matrix or molecular restrictions of the azobenzene isomerization, inversion mechanism is favored [6].

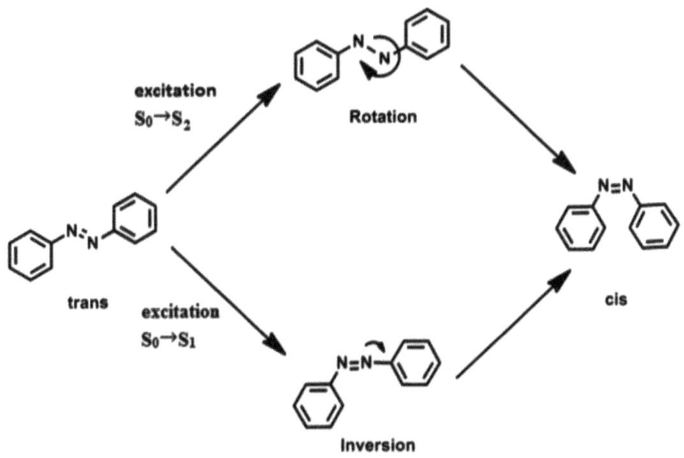

Fig. 6 Mechanisms of azobenzene isomerization

2.4 *Application of Photoisomerization Process in Drug Delivery Systems*

One of the most attractive stimuli to control the spatial and temporal triggering of the release in drug delivery systems is light, beside other stimuli like temperature, ultrasound, electricity, pH, ionic strength and biological signals (enzymes, biomolecules) [16]. Azobenzenes as photo-responsive materials are good candidate to accomplish light-induced conformational change which can be used in drug release from photo-responsive delivery systems [17, 18]. In a recent work, Pearson et al. synthesized a drug delivery system containing glycopolymer micelles with azobenzene-based core and galactose-based shell for melanoma cells as shown in Fig. 7. The incorporation of azobenzenes, accelerate the release of a hydrophobic drug in melanoma cells. In fact, upon illumination of light, polarity change is observed during trans to cis photoisomerization process and the polar cis forms help to expulsion of a hydrophobic therapeutic molecule. The synthesized micelles exhibit slow relaxation rates of cis to trans isomers. This behavior is due to para-substitution of electron-donating groups which minimize the electron density polarization over the azo bond and in this manner reduce the rate of relaxation. Thus, a long lifetime for cis isomer is desired in drug delivery applications [18].

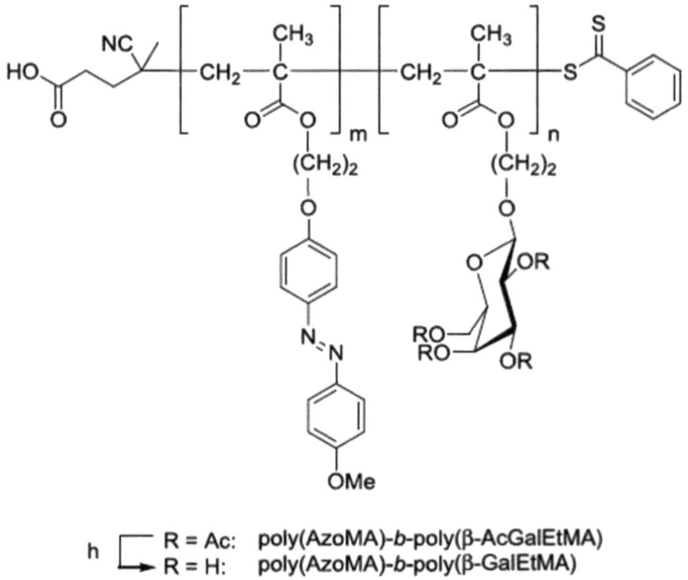

Fig. 7 Block copolymers which are then self-assembled [18]

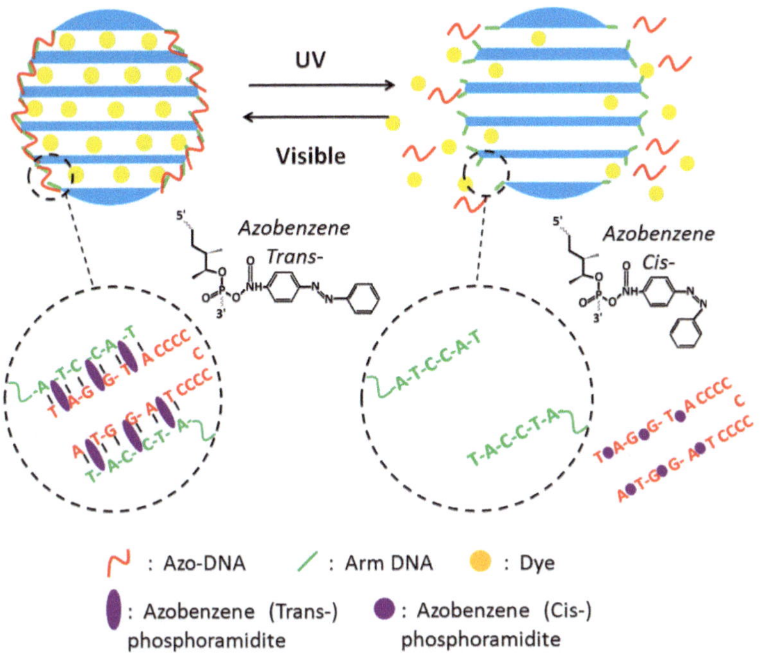

Fig. 8 Schematic of azobezene-modified DNA-controlled reversible release system [19]

In another work, Yuan et al. have studied a new photoresponsive DNA/mesoporous silica hybrid which is formed by introducing azobenzene moieties into the DNA molecule and anticancer drug doxorubicin (DOX) within the interior pores. The schematic of this release system is shown in Fig. 8 which is utilized for chemotherapy of cancer cells. The linker azo-DNA strands comprise azobenzene moieties which are attached to the pore opening at their ends. Rhodamine 6G is used to evaluate the loading and controlled releasing behavior of the composite nanospheres. Cis to trans isomerization cause hybridization of linkers to the arm. In this case, a cap over the pore mouth can be formed. Under illumination of suitable light in absorption band of trans isomers, they will convert to cis form. Thus the linkers dehybridize from the arms and the pore opens to release the drug into the solution. This fast switching between the open and closed states can be used to control drug dose [19].

Because of less damaging effects of near-infrared (NIR) light to biological specimens and also it's much deeper penetration in tissue, compared to UV light, Liu group used NIR light to release anticancer drug which is based on a mesoporous silica-coated UCNPs structure. azobenzene groups and the anticancer drug

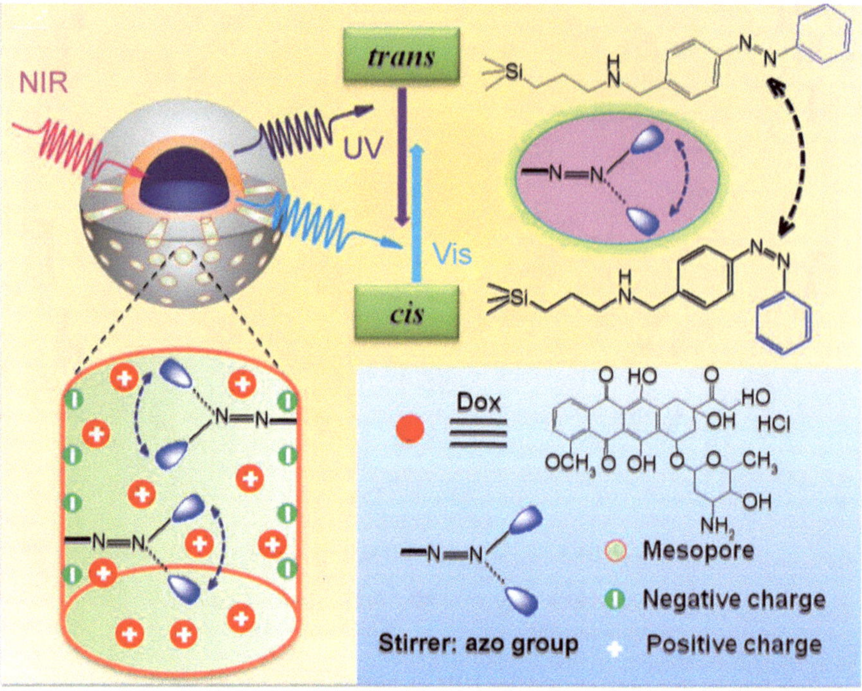

Fig. 9 NIR light-triggered Dox release [20]

doxorubicin (Dox) was loaded into the mesopores (Dox-UCNP@mSiO2-azo). The UCNPs are illuminated by NIR light (980 nm). It is absorbed by nanoparticles and then emit a light in the UV/Vis region. The photo- responsive azo molecules which are located in the pore network of the mesoporous silica layer, would absorb the light with a short wavelength and the photoisomerization process occurs. This process creates a continuous rotation–inversion movement which causes the azo molecules to act as a molecular impeller that drives the release of Dox. In this way, the Dox loaded in the mesopores can be released in a controllable manner [20]. This process in exhibited in Fig. 9.

A solution of Dox-UCNP@mSiO2-azo was illuminated by various intensities of NIR light to evaluate the control of drug release. As shown in Fig. 10, The Dox release amount reached 40 wt% and a maximum of 80 wt% in 16 h, under illumination of NIR light intensity of 2.4 and 8.9 W cm^{-2}, respectively. It indicates that the higher NIR light intensity, the more the release rate. Also, the experiment shows that without NIR irradiation less than 5 wt% of Dox is released into the solution. It shows that only be triggered by NIR exposure, would release the drug which is mostly dependent on the duration and intensity of NIR light [20].

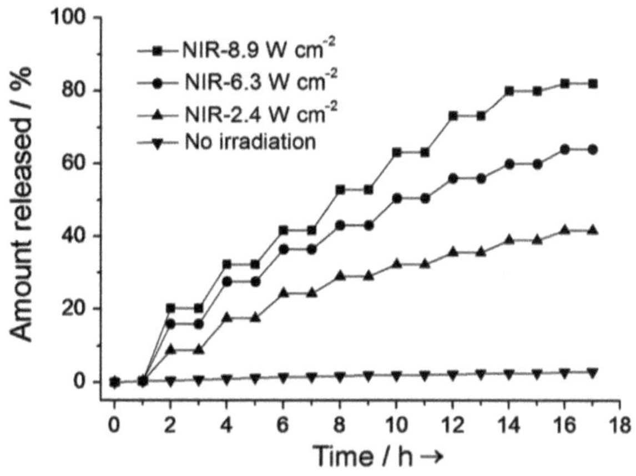

Fig. 10 Drug release in PBS with and without NIR light irradiation [20]

3 Photophysical Behaviors of Azopolymers

When the light polarization is parallel to the transition dipole axis of the azo molecules (often long axis of molecule), they favorably absorb the light. The probability of this absorption varies as $\cos^2 \emptyset$, where $\emptyset$ is the angle between the polarization of light and the dipole axis of azo [6]. Probability of light absorption is related proportionally to the projection of the isomer's transition dipole moment on the electric field vector of the incident light, E. As shown in Fig. 11, the azobenzene molecules oriented along the direction of the light's polarization, have higher possibility of absorbing light than the molecules oriented in all other directions. Absorption of the dye fragments upon light excitation, causes reversible trans–cis isomerizational transitions. The exited molecules eventually relax to the lower energy trans state. However, orientation of molecules after the absorption process is different from the original one. After performing multiple trans–cis transformations, the medium approaches a photo-steady state, where azobenzene fragments are oriented perpendicular to the polarization direction of incoming light. Reaching to this state, the molecules are excluded from further reorientation. Consequently, the vast majority of the azobenzene moieties reorient perpendicular to the vector E. The azobenzene moieties will be in a random state when they are illuminated with circularly polarized light [4, 6].

The concentration of chromophore, interactions between the molecules and anchoring properties at interfaces affect strongly the molecular orientation behaviors [2]. With azobenzenes, photo-reorientation has been observed in various forms of molecular organization. The reorientation mechanism is well understood in amorphous or liquid-crystalline side-chain polymers containing azobenzene derivatives: a change in the orientation of an individual chromophore could generally be the result of photoisomerization from trans to cis isomer with a subsequent isomerization back to the trans form by thermal or light illumination processes. After illumination with light prolonged to the transition moments of an ensemble of

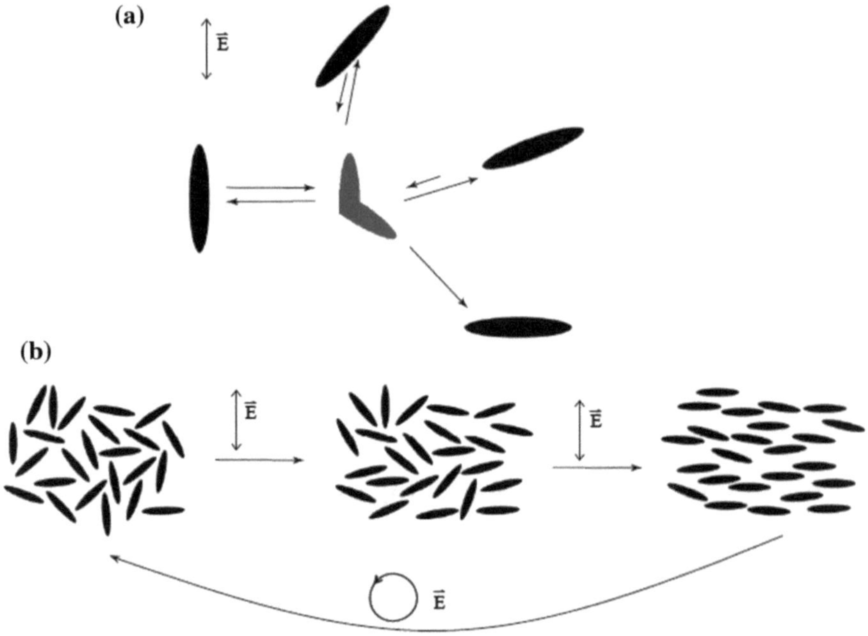

Fig. 11 Photo-orientation of azo molecules. **a** Izomerization and reorientation of azo molecules which are aligned along the polarization direction of the incident light. Those aligned perpendicular cannot absorb and remain fixed. **b** Accumulation of chromophores in the perpendicular direction upon irradiation of an isotropic samples. Isotropy is restored by illumination of circularly polarized light [6]

chromophores, they deplete the direction along the polarization vector of the excitation light, due to the orientation-selective excitation. In azobenzene- containing liquid-crystalline materials, trans–cis isomerization leads to photochemical formation of non-mesogenic cis-azobenzene moieties. Therefore, in light absorbing domains liquid-crystalline phase destabilize and photochemically induced domain rearrangement is induced [21]. Under illumination of nematic liquid crystals containing azobenzene mesogenic groups with ultraviolet (UV) light, the isotropic phase could completely replace the liquid crystalline phase [22]. Langmuir Blodgett (LB) films of side-chain polymers containing azobenzene and aminoazobenzene chromophores, have illustrated photo-reorientation behavior. The primary process of photoreorientation of aminoazobenzenes in LB films was investigated in [23]. They showed that there is a strong aggregation of the trans-form chromophores in the molecular environment. The aggregation could restrict conformation changes, thus photostationary cis concentration isn't detectable that would hint at trans–cis photoisomerization [23].

Different backbone structures of polyethyleneimine (PEI) having azobenzene molecules in the side chain, PEI6M and IP-DAz, were studied by Kim et al. [2] to understand the dependency of molecular orientational properties on backbone

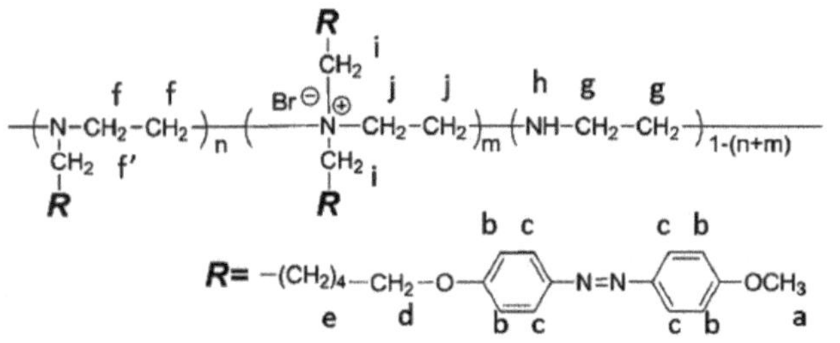

Fig. 12 Chemical structure of PEI6M and IP-DAz [2]

structure. The chemical structure of the polymers is illustrated in Fig. 12. PEI6M exhibited better anisotropy for out-of-plane orientation rather than IP-DAz, because bulky structure of the backbone chains of IP-DAz limit the molecular orientation by the steric hindrance. Under irradiation with linearly polarized light (LPL) at 436 nm, PEI6M showed a higher orientational order in the initial state. PEI6M contains a higher mobility of the azobenzene side groups. Thus, the reorientation tendency was stronger for PEI6M, compared to IP-DAz [2].

Since there are extensive works on in-plane (two-dimensional) reorientation of azobenzenes induced by LPL, few reports have done by nonpolarized light irradiation which induce tilt (three-dimensional) orientation of azobenzene molecules. Haitjema et al. suggested that irradiation of liquid crystalline polymer films, with azobenzene side chains, with nonpolarized visible light normal to the surface of the film change azobenzene orientations, from isotropic to favorably perpendicular directions [24]. So far very few researches have been done on spontaneous out-of-plane molecular orientation of polymeric liquid crystals (PLCs). Kurihara's group reported out-of-plane molecular ordering by annealing in liquid crystalline polyethylene imine having azobenzene side chain group (PEI8M, as its structure shown in Fig. 13). They also studied photoresponsive properties of PEI8M film upon irradiation with UV and visible light. By annealing and non-polarized UV and visible light irradiation, PEI8M film showed orientational changes from random state to out-of-plane and from out-of-plane to random state. This reversible molecular ordering is resulted from the combination of thermal and photo- chemical processes [25].

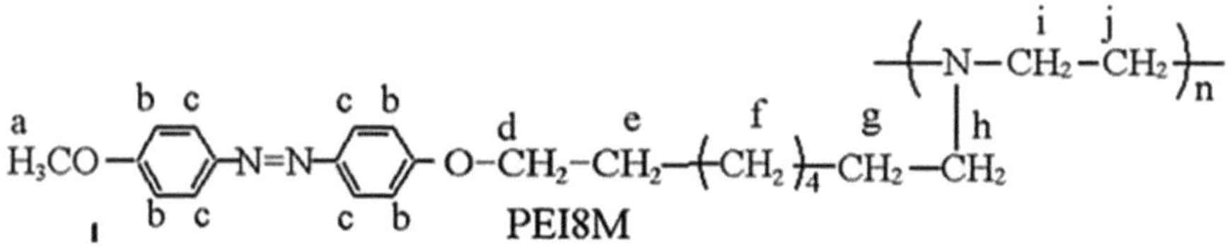

Fig. 13 Structure of polyethylene imine PEI8M [25]

3.1 Anisotropy

The dramatic structural changes induced by azobenzene lead to variations in physical and optical features of the materials. This phenomenon has attracted huge interest in many fields of application such as; photoinduced birefringence, dichroism, optical switching and photoinduced surface relief grating formation [2, 21]. For preparation of anisotropic optical elements such as; anisotropic color filters, polarizers, retarders or anisotropic emitters, this molecular reorientation is a key step. For azobenzene compounds the processes that occur during irradiation were widely studied, for different azobenzene containing media including; viscous solutions of azobenzene molecules, Langmuir–Blodgett films and polymers with azobenzene moieties chemically linked to a backbone azopolymers. The latter materials have extremely stable induced anisotropy. The property of light induced anisotropy can also be applied for alignment of liquid crystals [4]. Generally, it has been observed that in liquid-crystalline (LC) materials, photoinduced anisotropy is higher and more stable, compare to amorphous polymer systems. This is due to strong collective motion and intermolecular interactions of the photochromic units of LC materials [27]. The orientational order induced in a series of azobenzene methacrylate copolymers is studied in [28]. This orientational order is bigger for the LC polymers and increase with azobenzene content. Photoinduced anisotropy disappears in amorphous polymers when they are heated above Tg. However, the LC polymer shows a thermotropic increase of birefringence, which can be related to the aggregation of azobenzene moieties [28].

Parameters of induced anisotropy rely on the illumination conditions; direction of polarization of excitation light determines the induced axis of anisotropy, while the reliance of the induced birefringence and dichroismison the irradiation dose. In fact, for normal incidence of the probe beam, the induced anisotropy is only in the plane of the films (an "in-plane" anisotropy). However, it is natural to consider a three-dimensional distribution of azobenzene fragments. Indeed, there are many directions perpendicular to the polarization direction of the incoming light. These directions form a perpendicular plane to the electric field vector E. So, the perpendicular ordering of the fragment to the film surface (an "out-of-plane" anisotropy) should be considered. Sergan et al. investigated photoresponsive properties of two polymers; P1 and P2, with chemical structure shown in Fig. 14. They showed that upon irradiation of polymer films with nitro groups (P1) by polarized light, the azobenzenes align planar in the direction perpendicular to the UV light polarization. Irradiation of films of polymer with alkyl groups (P2) by polarized light induces fanlike structures, with small in-plane birefringence and desired out-of-plane orientation of the azobenzene fragments. Illumination of this polymer via high UV doses provides highly ordered homeotropic alignment, presenting high out-of-plane birefringence [4].

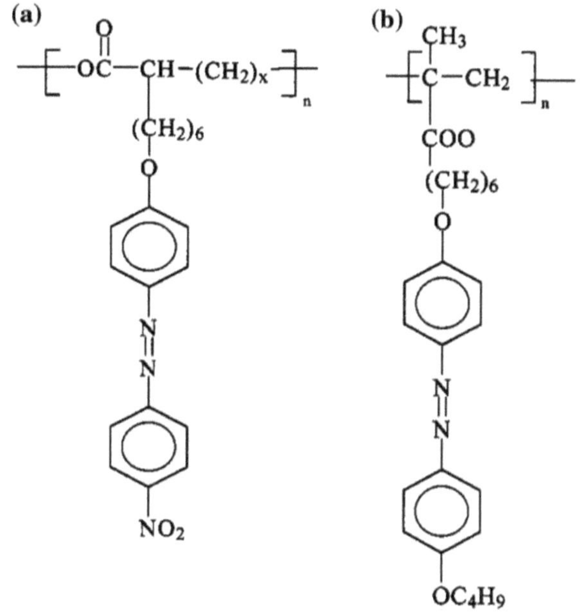

Fig. 14 Structure of polymers *P*1 (**a**) and *P*2 (**b**), respectively [4]

3.2 Birefringence

For azobenzene moieties in polymeric systems, under irradiation with LPL of suitable wavelength, uniaxial molecular orientation, perpendicular to the polarization direction is induced via a photoisomerization process. So, the orientational distribution of azobenzene molecules becomes photoselectively anisotropic (the Weigert effect), which results in photo-induced birefringence. Birefringence is an optical property of a material which shows the dependency of refractive index on the polarization and propagation direction of light. Linearly polarization light, with polarization parallel and perpendicular to the optical axis of material, will experience unequal refractive indices, denoted n_e and n_o where, the subscripts stand for extraordinary and ordinary, respectively. The preferred orientation of the azobenzene fragments was estimated, considering the assumption that the material possesses a higher polarizability and hence, a higher refractive index for the extraordinary wave—the light polarized in the direction of long molecular axis. So, azobenzene moieties possess positive birefringence, similar to low weight rodlike liquid crystals [4]. This can be evaluated for the in-plane orientation, normal to the monitored light [2]. Orientational order of the azobenzene groups, through trans–cis–trans isomerization cycles, results in birefringence, which consequently results in an additional photochromic entities oriented perpendicular to the pump light's polarization direction. A low intensity probe beam is used for measuring the induced birefringence in polymer film. A major part of the photoinduced orientation is conserved when the pump beam is turned off.

To induce birefringence, a pump beam with wavelength inside the absorption band of material and a probe beam outside the absorption band is used. For measuring the birefringence, the polymer film is placed between crossed polarizers and transmission of probe beam through the film is recorded as a function of time. Polarization of pump beam is set at 45° angle, with respect to the probe beam polarization. When the pump beam is switched off, an optical anisotropy in induced in the film. Birefringence modulus Δn is given by:

$$\Delta n = \frac{\lambda_0}{\pi d} \arcsin\left[\sqrt{\frac{I(t)}{I_0}}\right] \tag{1}$$

Where, λ_0, d, I(t) and I_0 are wavelength of the probe beam, thickness of the film, intensity of transmitted and incident probe beam, respectively. Figure 15 demonstrates a typical birefringence excitation—relaxation sequence. After the pump beam is turned on, the birefringence signal increases rapidly and reaches saturation, at point A. When the pump beam is turned off, the birefringence decreases and reaches a constant value that will keep constant for a long period of time in the dark, at point B [29].

Dynamics of birefringence's decay, in the absence of pump beam, have been studied by biexponential function, for fast and slow decays. The biexponential function is given by Eq. (2):

$$\Delta n(t) = A + B\ \exp(-\tau_1 t) + C\ \exp(-\tau_2 t), \tag{2}$$

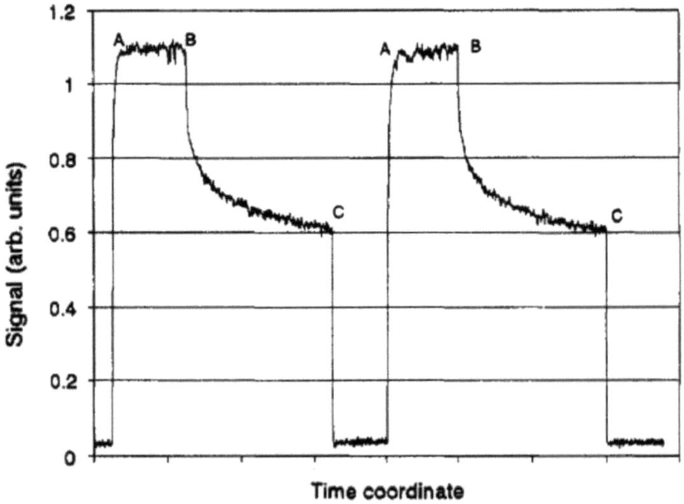

Fig. 15 Writing and erasing sequence on an amorphous azo polymer: **a** Writing with linearly polarized pump beam on; **b** Polarized pump beam off; **c** Erasing with circularly polarized pump beam [29]

where, A is the birefringence conserved for a long times, τ_1 and τ_2 are the relaxation rates with amplitudes of B and C, respectively. We showed that low-Tg azopolymers exhibit stable and saturated value of Δn_s. The photo induced birefringences of polymers decrease with the reduction of Tg. The relaxation rates, τ_1 and τ_2 of the photoinduced birefringence increase with the reduction of Tg [30].

Via illumination with circularly polarized light, induced birefringence can be erased. Recently, we studied the birefringence induced in pCNOhomo polymer with a structure shown in Fig. 15. It showed a sudden increment of transmitted beam followed by relaxation, under illumination of linear polarized pump beam. This behavior is due to reorientation of azo molecules and could be the result of phase transition to a higher state of order. Once the linear polarized pump beam is switched off, light induced birefringence is conserved. We also did the experiment with circularly polarized light and noticed that illumination of circularly polarized pump beam induces birefringence in polymer film [13]. Photoinduced birefringence of polymer film with linearly and circularly polarized light is illustrated in Fig. 16.

In the past decade, molecular architecture effects on orientational characteristics of photochromic polymers was studied by several groups. Sekkat et al. [31] have investigated that illumination of azobenzene-containing polyglutamate film with linearly polarized UV light (360 nm) results in alignment of cis form perpendicular to the initial light polarization and large anisotropy in the cis-isomer orientation. Furthermore, Natansohn et al. [32] have investigated optical properties of polymer with the aminoazobenzene-type side chain and compared with two azo polymers with structure similarities. They showed that as the dipole moment of azo fragments increases, the birefringence induced in polymer films would increase. They found that the increment and relaxation of induced birefringence depends on the dipole moment, glass transition temperature and the molecular weight of the polymer.

It has been reported that, Bisazo-containing polymers yield higher and more stable photoinduced birefringence, in comparison to the corresponding

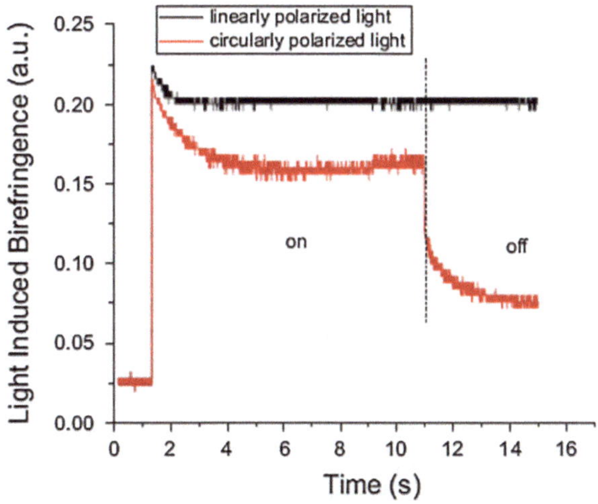

Fig. 16 Photoinduced birefringence of pCNOhomo polymer with linearly and circularly polarized light

monoazo-functionalized polymers. This is due to their large length-to-width ratio and low side-chain mobility. In a recent study by Wang et al., it was shown that the substitution pattern of the bisazochromophore and the excitation wavelength have great influence on the photoinduced birefringence and SRG formation in epoxy based bis azobenzene polymers. In another recent study, Wu et al. showed that the magnitude and temporal stability of the photoinduced birefringence of a covalently functionalized azopolymer were enhanced through hydrogen bonding between guest azobenzene units and azo groups of the host polymer, therefore forming a supramolecular bisazopolymer [27].

3.3 *Types of Motions in Azobenzenes*

Under illumination, azobenzene undergoes the process of photoisomerization. This cycle of photoisomerization results in a series of motions of the chromophores. When the azobenzene is bound to a polymer chain, it can be more. When the azobenzenes are dissolved in a polymer matrix, they are non-bound, but can still affect their environment. These motions can be roughly classified into three levels, as described by Natansohn and Rochon, illustrated in Fig. 17. The first level of these motions is the chromophore motion. It is affected by the polarization of light. Under illumination by linearly polarized light, the photoisomerization is only activated when the chromophore's transition dipole moment axis possesses a component parallel to the light polarization. Perpendicular direction of light polarization is not optically activated and will become enriched in chromophores. The concentration of chromophores, reoriented perpendicular to light polarization, increases steadily under irradiation with polarized light, until it reaches a saturation level. There are two directions that could be considered perpendicular to the light's polarization. One is in the plane of film. Chromophores in this direction can be monitored. The other direction perpendicular to polarization is in the direction of the light propagation, which is perpendicular to the surface of polymer film, i.e., in the direction of the film's thickness. The chromophores aligned specially in the homeotropic direction are usually "invisible" to the normal monitoring of the film. After photoinducing a preferred orientation in the polymer film and termination of irradiation, depending on the nature of the polymer film, this orientation may be conserved or not. The second level of motion is at the "domain" level, roughly nanoscale level. This type of motion, as the next one, requires the chromophore be bound to the polymer matrix or be part of a tightly organized structure (liquid crystalline, Langmuir-Blodgett or monolayer films). Usually, it also needs the matrix to have some degree of intrinsic order, which can be liquid crystalline, or semicrystalline. Another way to define this type of motion would be the motion of a limited chromophore. The limitation can be either a liquid crystalline domain, a Langmuir-Blodgett monolayer or anything else at this domain scale. When the chromophores direct themselves into an ordered structure, photoisomerization and the first type of motion are hindered, since it would destroy the intrinsic order. Nevertheless, quantum yield for

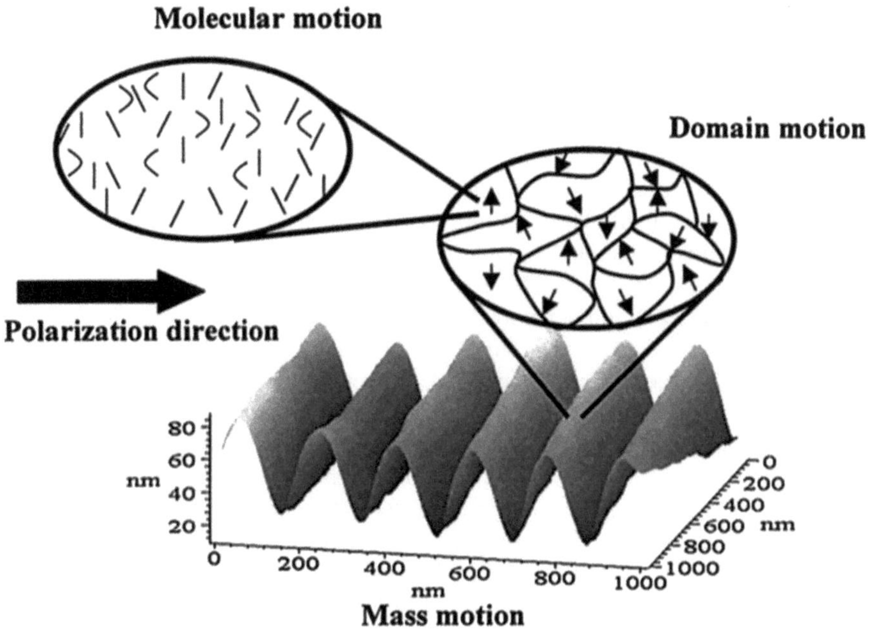

Fig. 17 Illustration of the three levels of polymer motion produced with light. At the molecular level light trans–cis–trans isomerization excludes the chromophores from the direction of polarization, at the domain level the polar chromophore movement reorients polar domains and at the mass level macroscopic movement of the polymer can be induced by light illumination [33]

photoisomerization in azobenzenes is quite high and the driving force for selection along the "blind" perpendicular orientation is very strong. Thus, the whole liquid crystalline or crystalline domains reorient in the direction perpendicular to the light polarization. This is known as cooperative motion, which is very common in ordered materials. Through this motion, the order parameter within the domain does not change, but an overall orientation of whole domains occurs. Therefore, it creates a very strong overall orientation, much stronger than the one created in amorphous polymers. Since these motions happen at the level of liquid crystalline or crystalline domains within the material, the amount of material by this motion is greater, in comparison to the first case—chromophore motion. The domain size varies with the thermal history of the material, but usually it is at the nanoscale level. This second kind of motion has very interesting consequences for helical structure of macro-molecules, in photoinduced phase transitions, for amplification phenomena and even for photoinduced and photoswitched chirality in achiral macromolecules. Finally, the third type of motion is at even a larger scale; it can be called macroscopic motion. This motion too requires that the chromophore be bound to the polymer, and it involves massive motion of the polymer material. This motion produces patterns on the film surface, visible even to the naked eye, with depth and spacing at micrometer scale. This was an extremely unexpected finding and generated a huge literature after year 1995 [33].

3.4 Surface Relief Grating (SRG)

Irradiation of azo polymer films, for a period of time longer than that required for photoinduced orientation, produces substantial movements of the polymer material, from light to the neighboring dark regions and an unexpected modification of the film surface. This modification of film surface can be induced with an interference pattern of coherent light. Thus, gratings with depths, up to one micron, could be obtained. Maximum heights of light induced SRG correspond to light intensity minima, for glassy polymers. This reorganization of polymer's mass occurs well below the glass transition temperature, Tg. SRG can be recorded not only by interference of two light beams, but also by single beam irradiation. Since in this case no interference pattern is applied, some self-organization mechanism is involved in the process of SRG formation. Noisy random fluctuations of the molecular positions start the self-organized process. Local perturbations are incremented through interference of the scattered light waves, thus forming regular patterns. Light initiated spontaneous pattern formation originates from the interplay between self-action (e.g. photoisomerization) and long range interaction (e.g. diffraction). The final state is self-organization as a coherent grating. In the experiments of self-induced SRG formation, a coherent beam excites the photoactive molecules. Molecules exchange position information by the means of scattered light, showing that this is the process which triggers self-organization [34].

The height and pitch of the gratings stored for different pump coherent lengths, were retrieved with a contact-mode atomic force microscope (AFM). The pitch Λ can be obtained by first order diffraction theory in the backward direction which is given by formula (3):

$$\Lambda = \frac{\lambda}{2\sin\theta} \tag{3}$$

In Fig. 18, a typical AFM image self-induced SRGs is shown.

Various mechanisms have been suggested in order to explain the origin of surface relief gratings, in polymers functionalized with azobenzenes. These mechanisms include; thermal gradient mechanisms, asymmetric diffusion based on the creation of a concentration gradient, isomerization pressure and mechanisms based on electromagnetic forces—mean field theory, permittivity gradient theory and gradient electric force [26]. All the above interpretations induce much disagreement and none of the suggested mechanisms seem to provide a suitable explanation for all observations, especially in the case of multistate polarization addressing, using a single beam in an azopolymer film [36]. We have demonstrated multistate addressing by polarization for the first time. A change in polarization direction of an incoming light leads to a grating orientation. Therefore, multistate addressing by polarization could be achieved [30]. We have introduced a mechanism in paper [36]. We described it by means of a phenomenological model, whose simplicity allows the treatment of quite complicated nonlinear optical interactions

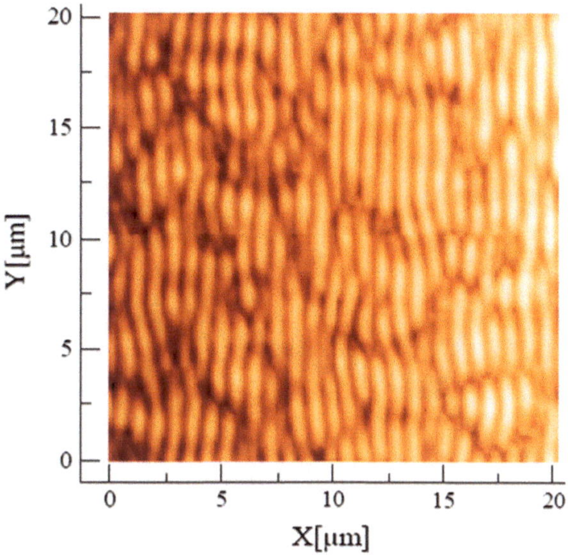

Fig. 18 A typical AFM image self-induced SRGs [35]

involved in the self-organization process. This model is based on Fick's law of diffusion, which scales via light intensity. Furthermore, light is trapped in the sample and couples with counter propagating guided modes, which interact with azopolymer. Using this approach, we could explain the self-patterning of azopolymer film illuminated with a single beam [36].

Even though coherent light is required for self-organization of regular spatial patterns, we demonstrated that atypical light-matter interaction can break this fact. We reported self-organization followed by propagation of a surface relief pattern. A low-power and small-diameter size coherent beam induces self-organization in a polymer film, assisted with a high-power and large-diameter size incoherent and unpolarized beam. The low power beam conveys coherent information about the pitch and orientation of the diffraction grating. While, the high power beam triggers the molecules and allowing the movement [37]. AFM images of SRG in the central part of coherent beam region and within the outer incoherent laser beam region are shown in Fig. 19. As shown in Fig. 19b, random motion of molecules induced by light can't produce any well-defined pattern [37].

Various parameters, regarding gratings, can indeed be accessed by tuning the coherent beam characteristics. To show the importance of pump or signal beam's polarization, we rotated the signal beam's polarization after one hour of interaction. In this case, the direction of diffraction change. It means that when the pump polarization is rotated- also after 1 h of interaction, while keeping the signal beam polarization unchanged- diffraction continues to increase, after about 10 min of time lag. This indicates that, the low power coherent signal beam is leader without considering its polarization direction. In fact, the pump beam permits

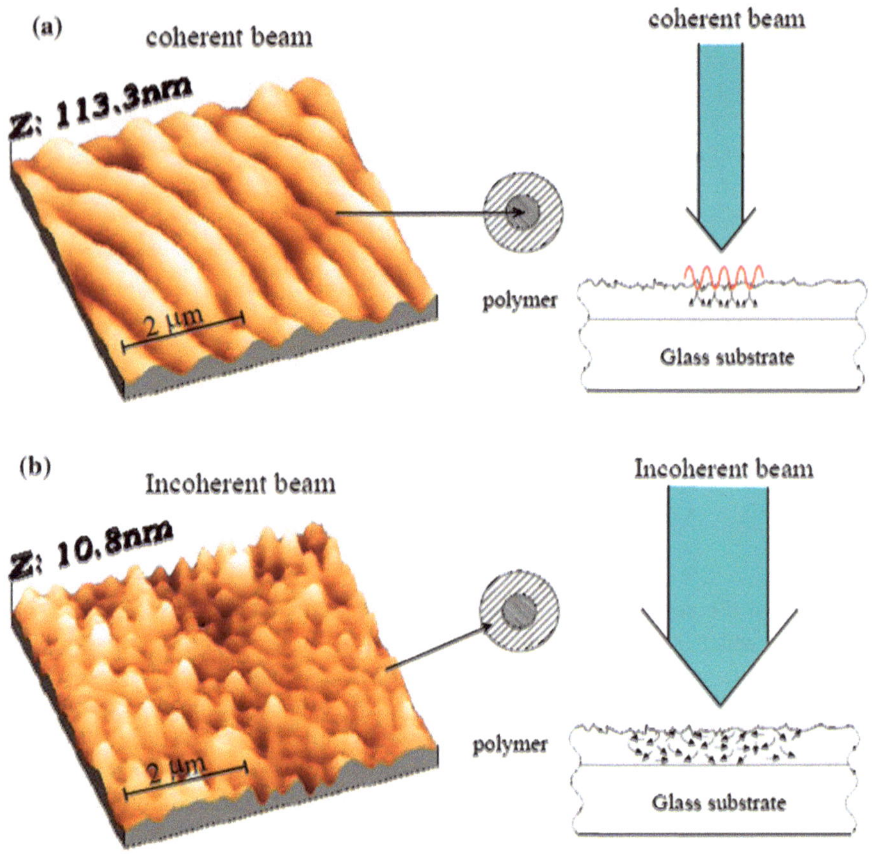

Fig. 19 AFM images of self-patterned SRG structures obtained after 30 min exposure with a low power coherent beam and a large power incoherent beam [37]

collective SRG growth, but does not take part in the decision about the grating's characteristics [38].

There are some parameters that affect strongly the process of SRG such as; molecular weight, the glass transition temperature (Tg) of the host polymer, the concentration and polarity of the azobenzene units. To achieve an efficient surface patterning, sufficiently strong bonding between the chromophores and the polymer backbone is needed. Even though strong intermolecular interactions and liquid crystallinity properties are favorable for photo-orientation behaviors, they can mainly hinder the required photoinduced mass transport for SRG formation. However, there are many reports about LC azopolymer systems which undergo efficient surface mass transport. To fully exploit the potential of the light-induced phenomena of azobenzene-containing polymers, it is important to understand the structure property relationships that control their optical performance [22, 32]. Miniewicz et al. confirmed the possibility of recording holographic gratings in a group of 22 new functionalized polymers with a suitable azobenzene group. Systematic measurements of the grating recording dynamics show the structural units' influence on the grating process. Type of chromophores (i.e. containing single or double N=N bonds), position of chromophores in the polymer chain, length of aliphatic units in the polymer chain and the presence of functional groups (amide and ester groups) are all key parameters. Presence of double N=N bonds or high Tg polymers, make the recording process slow down. Tailoring the structures with azo-groups, in deferent positions in the chain, changes the time constants of hologram recording and durability of recorded gratings [26]. We have studied diffraction pattern of different azopolymers in [35]. The diffracted beam intensity as a function of time is illustrated in Fig. 20. It was shown that both polymer main chain and chromophore structure produce different diffraction efficiencies.

We have studied the diffraction pattern of probe beam after formation of one-beam SRG for polymer with photoresponsive azobenzene derivative bearing heterocyclic sulfonamide moieties. The intensity of diffracted beam with time is illustrated in Fig. 21. As soon as switching off the writing beam, the photoinduced SRG was formed, and the diffracted beam intensity increased. Unlike the case of SRG formed with interference of two beams, after stopping the irradiation of writing beam, the diffracted beam intensity increased rapidly, decreased slowly, reached to a smaller value and remained at that value. When the writing beam is switched on again, after a fast decrease, the intensity of diffracted beam increased again. These phenomena can be related to the some relaxation mechanisms in azo-containing polymer films such as orientational change of azochromophore, cis–trans thermal isomerization and the change of relief height. It is required to clarify

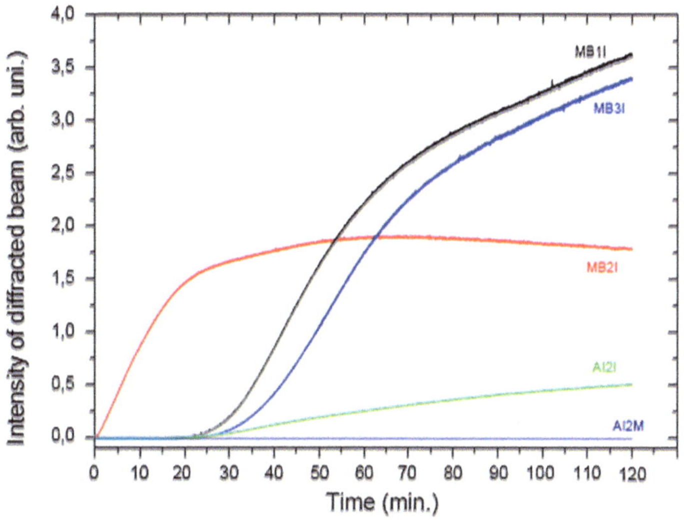

Fig. 20 Diffracted beam intensity with time for various azopolymers [35]

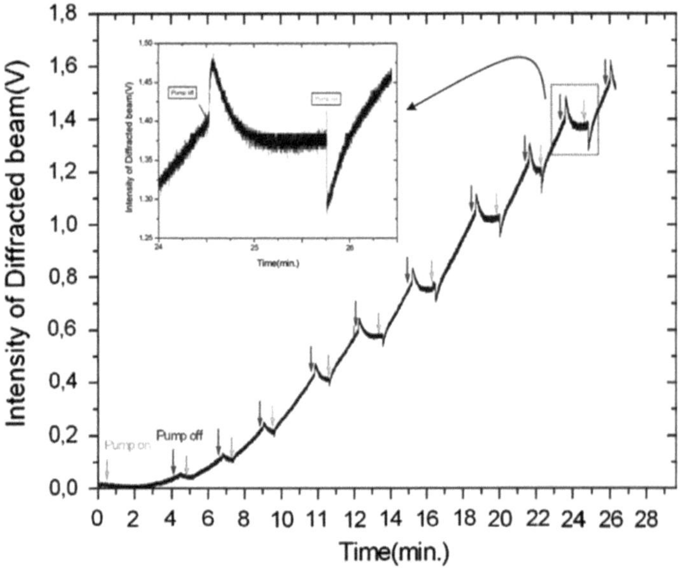

Fig. 21 A diffracted beam intensity as a function of irradiation time. The diffracted beam intensity changes anomalously after switching off the writing beam. The *inset* shows lasts cycle (writing beam off and on) of this relaxation [39]

the role of these mechanisms on the relaxation dynamics of self-photoinduced SRG. To explain the origin of the driving force responsible for SRG formation, we proposed a mechanism. It originates the molecules migration almost parallel to the polarization direction from high intensity to low intensity of light which is the result of photo-isomerization induced translation [39].

Due to potential applications of photosensitive organic molecular thin films in optoelectronic devices, surface relief gratings (SRG) have been extensively studied. In paper [40], we have implemented a very simple neurocomputer, where the individual neuron is the self- induced SRG below a coherent signal beam [40]. With the use of a liquid droplet, we can form surface relief micrograting (SRmG) to improve the optical holographic addressing capability of a polymer. The droplet acts as a truncated sp here interferometer, permitting simultaneous focusing and multiple reflections. This is a new aspect of micro-optics, applied to the imprinting of patterns in photosensitive films [41].

The formation of spontaneous surface relief grating have been done under pulsed exposure with different pulse widths and repetition rates by us. In this study we observed a sharp enhancement of intensity as the pump beam is switched off. Surprisingly, by switching on the writing beam, a sharp decrease followed by an increment, under further illumination can be observed. The sharp enhancement and decrease in diffracted beam intensity have been attributed to cis–trans thermal isomerization and isomeration in whole surface under illumination, respectively. The irradiation time and polarization of the probe beam are parameters which are responsible for anomalies of relaxation processes [42]. Temperature has also an effect on SRG formation; Orientation and motion dynamics of the azochormophores enhance with temperature. As the temperature reaches around Tg point, the mass transport effect, which gives rise to surface relief formation, is hindered [43].

3.5 Nonlinear Behavior of Azo Compounds

Linear and nonlinear optical (NLO) properties of azobenzene-containing polymers are of huge technological interest. To achieve the second order susceptibility, molecular arrangement is required in bulk materials. There are several methods to get acentrosymmetric electronic distribution on macroscopic scale such as; Langmuir-Blodgett technique, corona poling and all optical poling. For materials with ionic structure, we can use all optical poling method [44]. We can maximize the nonlinearity in poled polymers, by synthetic design of molecules with large dipole moment and nonlinear coefficients and through maximizing the applied poling field's strength or by employing steric forces. We can modify the molecular parameters in synthetic approaches, such as; the relative electron affinities of the donor and acceptor groups in the dipole, and the length and nature of the connecting system [45]. The NLO chromophore can be used as, a solute in a polymer matrix (guest-host system) or as a side-chain fragments linked to polymer main chain, covalently [44]. Nonlinear phenomenon of second harmonic generation (SHG) was induced by noncentrosymmetric ordering of Guest–host (GH) systems as the first organic materials, under electrical poling. A wide variety of centrosymmetric NLO chromophores can be used for SHG activity. NLO properties of chromophore-functionalized polymers offer various advantages over the guest–host systems [45]. We have investigated optical nonlinearity of guest-host type system with methacrylate copolymers as a host and three azo amide chromophores containing long aliphatic chain, as a guest. The first order nonlinear optical susceptibility, value of d33, obtained for Amid N (with nitro group) is equal to 93.2 pm/V, which indicates that this compound could be a useful chromophore for NLO applications [44]. SHG measurements for some amorphous and liquid crystalline methacrylic copolymers, with cyano azobenzene chromophores in the side chain, have been done by Villacampa group. The non-linear d33 coefficient decreases with the azocontent [28]. Using single beam Z-scan and optical Kerr effect (OKE) techniques, nonlinear optical properties of an azobenzene-containing ionic liquid crystalline polymer were investigated. Nonlinear refractive index of electronic origin (3.1×10^{-19} m^2/W) and nonlinear absorption coefficient (3.63×10^{-13} m/W) were determined, with 800 nm femtosecond laser pulses at a repetition rate of 1 kHz. The estimated response time of the observed nonlinearities was as fast as 300 fs. These results indicate the azopolymer as a promising candidate for applications in all-optical switching modulators and nonlinear photonic devices [46].

4 Conclusion

Azobenzenes comprise two isomers, more stable trans and meta-stable cis. The photoisomerization process of these materials make them exhibit unique properties. Under illumination of proper light, these materials isomerize from trans to cis. Reconversion of cis–trans can be done thermally or by irradiation of light with suitable wavelength of cis isomers' absorption band. They are classified in three classes according their spectroscopic properties; azobenzenes, aminoazobenzenes and pseudo-stilbenes. They undergo inversion or rotation mechanism during photoisomerization process. Azobenzenes can be studied in guest-host systems or incorporation covalently or noncovalently into polymers. Under illumination of polarized light, they orient in perpendicular direction of light polarization. Thus, the photoisomerization phenomenon is responsible for inducing anisotropy and birefringence in these materials. Also, orientation can be done with nonpolarized light. An application of azobenzenes in light-responsive drug delivery systems is discussed. Upon illumination of light with a wavelength in absorption band of azo dyes, different motions can be observed, molecular, domain and macroscopic levels of motion. Surface relief grating is the result of macroscopic motion which is induced under illumination of light for a long period of time. SRGs can be induced either with two-beam interference pattern or single-beam irradiation. In the latter case, there are some self-organization mechanisms forming SRG. The dependency of SRG formation on polymer structure and the concentration and polarity of the azobenzene units is also discussed. SRG can be formed with illumination of a polarized and a nonpolarized light. In this case, the polarized and nonpolarized light are responsible for orientation of diffraction grating and movement of molecules, respectively. Thus, there is no regular pattern in AFM image of this experiment. Azo compounds have nonlinearities of second and third order. These materials are a good candidate for NLO applications.

References

1. Iqbal, D., Samiullah, M.H.: Photo-responsive shape-memory and shape-changing liquid-crystal polymer networks. Materials (Basel) **6**(1), 116–142 (2013)
2. Kim, S., Nakamura, T., Yagi, R., Kuwahara, Y., Ogata, T., Ujiie, S., Kurihara, S.: Photo-response orientation behaviors of polyethylene imine backbone structures with azobenzene side chains. Polym. Int. **63**(4), 733–740 (2014)
3. Gregory, P.: The Chemistry and Application of Dyes. Plenum Press, Berlin (1990)
4. Yaroschuk, O., Sergan, T., Lindau, J., Lee, S.N., Kelly, J., Chien, L.C.: Light induced structures in liquid crystalline side-chain polymers with azobenzene functional groups. J. Chem. Phys. **114**(12), 5330–5337 (2001)
5. Merino, E., Ribagorda, M.: Control over molecular motion using the cis–trans photoisomerization of the azo group. Beilstein J. Org. Chem. **8**, 1071–1090 (2012)

6. Yager, K., Barrett, C.: Light-induced nanostructure formation using azobenzene polymers. Polym. Nanostruct. Appl. **0**, 1–38 (2006)
7. Lin, S., Wang, Y., Cai, C., Xing, Y., Lin, J., Chen, T., He, X.: Tuning self-assembly and photo-responsive behavior of azobenzene-containing triblock copolymers by combining homopolymers. Nanotechnology **24**(8), 085602 (2013)
8. Tarasenko, V., Nadtoka, O., Syromyatnikov, V.: Synthesis of azobenzene-containing polymers based on polymethacrylic acid and their photooptical properties studies. Fr.–Ukr. J. Chem. **1**(1), 89–93 (2013)
9. Zakerhamidi, M.S., Sorkhabi, S.G., Ahmadi-Kandjani, S., Moghadam, M., Ortyl, E.: Polymeric cyanide azo compounds dipole moments and photo-physical properties in solvents media. Spectrochim. Acta Part A Mol. Biomol. Spectrosc. **114**, 148–153 (2013)
10. Zakerhamidi, M.S., Sorkhabi, S.G., Ortyl, E., Shahabadi, S.: Substituent and solvent effects on the dipole moments and photo-physical properties of a group of polymeric azo sulfonamide substances. J. Mol. Struct. **1048**, 441–447 (2013)
11. He, L.H.: Deformation of amorphous azobenzene-containing polymer films induced by polarized light. Acta Mech. Sin. Xuebao **28**(4), 1203–1208 (2012)
12. Joo, W.J., Oh, C.H., Song, S.H., Kim, P.S., Han, Y.K.: Photoinduced birefringence in poly (malonic ester) containing p-cyanoazobenzene with photoexcitation of cis conformer. J. Phys. Chem. B **105**(35), 8322–8326 (2001)
13. Bagherzadeh-Khajeh Marjan, E., Ahmadi-Kandjani, S., Zakerhamidi, M.S., Nunzi, J.-M.: Cyano azobenzene polymer films: photo-induced reorientation and birefringence behaviors with linear and circular polarized light. Opt. Mater. (Amst) **38**, 228–232 (2014)
14. De Maria, P., Fontana, A., Gasbarri, C., Siani, G., Zanirato, P.: Kinetics of the Z–E isomerization of monosubstituted azobenzenes in polar organic and aqueous micellar solvents. Arkivoc **2009**(8), 16–29 (2009)
15. Shao, J., Lei, Y., Wen, Z., Dou, Y., Wang, Z.: Nonadiabatic simulation study of photoisomerization of azobenzene: Detailed mechanism and load-resisting capacity. J. Chem. Phys. **129**(16), 1443–1447 (2008)
16. Kim, S., Kim, J.H., Jeon, O., Kwon, I.C., Park, K.: Engineered polymers for advanced drug delivery. Eur. J. Pharm. Biopharm. **71**(3), 420–430 (2009)
17. Hrubý, M., Filippov, S.K., Štěpánek, P.: Smart polymers in drug delivery systems on crossroads: Which way deserves following? Eur. Polym. J. **65**, 82–97 (2015)
18. Pearson, S., Vitucci, D., Khine, Y.Y., Dag, A., Lu, H., Save, M., Billon, L., Stenzel, M.H.: Light-responsive azobenzene-based glycopolymer micelles for targeted drug delivery to melanoma cells. Eur. Polym. J. **69**, 1–12 (2015)
19. Yuan, Q., Zhang, Y., Chen, T., Lu, D., Zhao, Z., Zhang, X., Li, Z., Yan, C.H., Tan, W.: Photon-manipulated drug release from a mesoporous nanocontainer controlled by azobenzene-modified nucleic acid. ACS Nano **6**(7), 6337–6344 (2012)
20. Liu, J., Bu, W., Pan, L., Shi, J.: NIR-triggered anticancer drug delivery by upconverting nanoparticles with integrated azobenzene-modified mesoporous silica. Angew. Chemie—Int. Ed. **52**(16), 4375–4379 (2013)
21. Bogdanov, A.V., Vorobiev, A.K.: Photo-orientation of azobenzene-containing liquid-crystalline materials by means of domain structure rearrangement. J. Phys. Chem. B **117**(44), 13936–13945 (2013)
22. Xue, X.-G., Zhao, L., Lu, Z.-Y., Li, M.-H., Li, Z.-S.: Molecular dynamics simulation study on the isomerization and molecular orientation of liquid crystals formed by azobenzene and (1-cyclohexenyl)phenyldiazene. Phys. Chem. Chem. Phys. **13**(25), 11951–11957 (2011)
23. Schönhoff, M., Mertesdorf, M., Lösche, M.: Mechanism of photoreorientation of azobenzene dyes in molecular films. J. Phys. Chem. **100**(18), 7558–7565 (1996)

24. Han, M., Ichimura, K.: Tilt orientation of p-methoxyazobenzene side chains in liquid crystalline polymer films by irradiation with nonpolarized light. Macromolecules **34**(1), 82–89 (2001)
25. Kamruzzaman, M., Kim, S., Kuwahara, Y., Ogata, T., Kurihara, S.: Thermal and photo alignment behavior of polyethylene imine having methoxy substituent azobenzene side chain group. Open J. Polym. Chem. **03**(04), 92–98 (2013)
26. Sobolewska, A., Miniewicz, A., Grabiec, E., Sek, D.: Holographic grating recording in azobenzene functionalized polymers. Cent. Eur. J. Chem. **4**(2), 266–284 (2006)
27. Koskela, J.E., Vapaavuori, J., Hautala, J., Priimagi, A., Faul, C.F.J., Kaivola, M., Ras, R.H.: Surface-relief gratings and stable birefringence inscribed using light of broad spectral range in supramolecular polymer-bisazobenzene complexes. J. Phys. Chem. C **116**(3), 2363–2370 (2012)
28. Rodríguez, F.J., Sánchez, C., Villacampa, B., Alcalá, R., Cases, R., Millaruelo, M., Oriol, L.: Optical anisotropy and non-linear optical properties of azobenzene methacrylic polymers. Polymer (Guildf) **45**(7), 2341–2348 (2004)
29. Xie, S., Natansohn, A., Rochoni, P.: Recent developments in aromatic azo polymers research. Chem. Mater. **5**(4), 403–411 (1993)
30. Ahmadi Kandjani, S., Barille, R., Dabos-Seignon, S., Nunzi, J. M., Ortyl, E., Kucharski, S.: Multistate polarization addressing using one single beam in an azo polymer film. In: Conference on Lasers and Electro-Optics Europe—Technical Digest, pp. 1–11 (2005)
31. Sekkat, Z., Wood, J., Knoll, W.: Reorientation mechanism of azobenzenes within the trans. fwdarw. cis photoisomerization. J. Phys. Chem. **99**(47), 17226–17234 (1995)
32. Ho, M.-S., Natansohn, A., Barrett, C., Rochon, P.: Azo polymers for reversible optical storage. 8. The effect of polarity of the azobenzene groups. Can. J. Chem. **73**(11), 1773–1778 (1995)
33. Natansohn, A., Rochon, P.: Photoinduced motions in azo-containing polymers. Chem. Rev. **102**(11), 4139–4175 (2002)
34. Barille, R., Ahmadi-Kandjani, S., Ortyl, E., Kucharski, S., Nunzi, J.-M.: Cooperative interaction in azopolymers upon irradiation. New J. Chem. **33**(6), 1207 (2009)
35. Ahmadi-Kandjani, S., Tajalli, P., Khoshsima, H., Barille, R., Nunzi, J.M., Kucharski, S.: Birefringence properties and surface relief grating formation on methylacrylate polymers with photochromic side chains. **4**(1) (2010)
36. Leblond, H., Barille, R., Ahamadi-kandjani, S., Nunzi, J.-M., Ortyl, E., Kucharski, S.: Spontaneous formation of optically induced surface relief gratings. J. Phys. B: Atom. Mol. Opt. Phys. **42**(20), 205401 (2009)
37. Kandjani, S.A., Barille, R., Dabos-Seignon, S., Nunzi, J.M., Ortyl, E., Kucharski, S.: Incoherent light-induced self-organization of molecules. Opt. Lett. **30**(23), 3177–3179 (2005)
38. Barille, R., Ahmadi-Kandjani, S., Ortyl, E., Kucharski, S., Nunzi, J.M.: Cognitive ability experiment with photosensitive organic molecular thin films. Phys. Rev. Lett. **97**(4), 1–16 (2006)
39. Ahmadi Kandjani, S.D.-S.S., Barille, R., Nunzi, K.J.-M., Ortyl, E.: The light polarization control of self-organized surface relief gratings in an Azo-polymer. Nonlinear Opt. Quantum Opt. **36**, 217–227 (2007)
40. Barille, R., Dabos-Seignon, S., Nunzi, J.-M., Ahmadi-Kandjani, S., Ortyl, E., Kucharski, S.: Nonlocal communication with photoinduced structures at the surface of a polymer film. Opt. Express **13**(26), 10697–10702 (2005)
41. Barille, R., Nunzi, J.M., Ahmadi-Kandjani, S., Ortyl, E., Kucharski, S.: One step inscription of surface relief microgratings. Opt. Commun. **280**(1), 217–220 (2007)
42. Mazaheri, L., Kandjani, S.A.: Spontaneous nano pattern formation under pulsed exposure: Relaxation kinetics study. Opt. Quantum Electron. **44**(1–2), 17–25 (2012)

43. Mazaheri, L., Ahmadi-Kandjani, S., Nunzi, J.M.: Influence of temperature on the relaxation kinetics of spontaneous pattern formation in an azo-polymer film. Opt. Commun. **298–299**, 150–153 (2013)
44. Marański, K., Kucharski, S., Ortyl, E., Janik, R., Nunzi, J.-M., Ahmadi-Kandjani, S., Barille, R.: Second harmonic generation and photochromic grating in guest-host polymer system containing azo amide chromophores. e-Polymers, **10**(1), 1–8 (2010)
45. Yesodha, S.K., Sadashiva Pillai, C.K., Tsutsumi, N.: Stable polymeric materials for nonlinear optics: a review based on azobenzene systems. Prog. Polym. Sci. **29**(1), 45–74 (2004)
46. Zhao, F., Wang, C., Zeng, Y., Zhang, J.: Femtosecond third-order Optical nonlinearity of an azobenzene-containing ionic liquid crystalline polymer. Opt. Express **20**(24), 26845–26851 (2012)

Photochemical Process of Wood-Thermoplastic Composites

Yao Peng and Jinzhen Cao

1 A Brief Introduction of WPCs

1.1 Definition of WPCs

According to the definition by American Society for Testing and Materials (ASTM), the term WPCs refer to composite materials that contain wood fiber/flour and various thermoplastics or thermosets. Traditionally, there are two definitions of WPCs, wood-plastic composites and wood polymer composites.

Wood polymer composites are produced by impregnating the wood with different monomers, for example, styrene (St), methylmethacrylate (MMA), acrylonitrile (AN), and unsaturated polyester (UPE). These monomers can transfer into homopolymer or grafted on the wood cell wall during polymerization process, which can be initiated by γ-radiation, high temperature, or chemical catalysts.

Wood-plastic composites can be further divided into wood-thermosetting composites and wood-thermoplastic composites. Thermosets or thermoset polymers have the property of becoming permanently hard and rigid once been heated or cured. Such polymers contain phenol resins, epoxy resins, unsaturated polyester resins, and so on. Among them, phenol formaldehyde (PF) and urea formaldehyde (UF) are the most widely used thermosets for wood composites. The birth of wood-thermosetting composites can date back to the early 1900s. In 1907, Dr. Leoh

Y. Peng · J. Cao (✉)
MOE Key Laboratory of Wooden Material Science and Application, College of Materials Science and Technology, Beijing Forestry University, Qinghua East Road 35, Beijing 100083, China
e-mail: caoj@bjfu.edu.cn

Y. Peng
e-mail: pengyao0304@163.com

D. Rosu and Visakh P. M. (eds.), *Photochemical Behavior of Multicomponent Polymeric-based Materials*, Advanced Structured Materials 26,
DOI 10.1007/978-3-319-25196-7_10

Bend made the composite that composed of PF and wood flour (WF). The first commercial application of this composite was used as a gearshift knob for Rolls Royce car in 1916 under the registered trademark of Bakelite [1]. On the contrary, thermoplastics can be repeatedly melted without any change in their inherent properties, including polypropylene (PP), high and low density polyethylene (HDPE and LDPE), polyvinyl chloride (PVC), and so on. In this chapter, the abbreviation of WPCs represents wood-thermoplastic composites, which are usually simply referred to as wood-plastic composites (WPCs). Similarly, the common understanding of plastics always refers to thermoplastics.

1.2 Significance of WPCs

In the past few decades, people have become more and more dependent on plastic products, from daily living sundries to advanced scientific equipment. However, along with the convenient life, the development of plastic industry also brings serious environmental problems, named white pollution.

The world plastic industry has been experiencing a high speed development. It has been reported that more than 288 million tons of plastics were produced in 2013 with an average annual growth rate of approximately 3.9 % in the world [2]. At the same time, waste plastics also contributes 15–20 % to the solid waste in city, most of which come from the disposable plastic packaging. The disposition of these waste plastics is a harsh problem. There are three main disposal methods, including combustion, landfill, and recycling, while the first two methods will contaminate the air and filed, respectively. Considering the diminishing of petroleum resources, the cycling of waste plastics is proven to be a promising method, though accounting only for 15 % of the total plastics consumption because of the higher cost of recycling process [3].

In addition, environmental groups have increasingly realized the decreasing of forest resources. As a result, governments are launching more stringent legislation, which limits or inhibits the trees felled for future generations. Therefore, factories and people are encouraged to maximize the efficacy use of woody materials, as well as to minimize the creation of wastes. Traditionally, wood is used only in its solid form as large timbers or lumber. Meanwhile, the processing of wood usually produces 25–35 % sawdust or wood fiber, which are regarded as wastes. With the decreasing availability of large-diameter trees and the increasing price, composite industries showed an increasing interest in the usage of these wood wastes, such as smaller-diameter trees, saw and pulp mill wastes. It has been widely accepted that WPCs is a very promising and sustainable material, which could benefit both plastic recycling and wood resources conservation.

1.3 Development History of WPCs

Wood-plastic composites was born in Europe and experienced dramatic growth in North America in 1990s. In 1973, a company named SonessonPlast AB (Sweden) promoted a PVC/WF composite product with the trade name "Sonwood". This was probably the first commercial WF/PVC composite in the world. Ten years later, in 1983, "Woodstock" appered in Italy, which consisted of 50 % WF and 50 % PP. WF and PP were extruded into a flat sheet that was then formed into different shapes. It was then applied in interior automotive paneling by Ford Motor Company, becoming the first major applications of WPCs in the United States.

In the early 1990s, Trex (Winchester, VA) promoted WF/PE composites, which were sold as industrial flooring, decking, landscape timbers, and outdoor tables. Some of them were milled into window and door components. In 1993, Andersen Corporation (Bayport, MN) began producing WF/PVC composite structural profiles for windows and doors through extruding and injection molding technology. After that, the market for WPCs window and door profiles continues to grow.

In 1996, a new technology was used in WPCs industry. Several companies in America invited a pelletized feedstock from WF (or natural fibers) and plastic. These companies selled compound pellets to many composite manufacturers who do not want to blend the raw materials. One year later, Doroudiani blended HDPE with WF and kraft pulp to inject WPCs panels. Since then, the activity of WPCs industry has increased significantly. In 2001, CertainTeed Corporation (Vally Forge, PA) and its product line for WF/PVC composites "Boardwalk" decking and railing received the approval of the National Evaluation Service (NES). Today, the decking market is still the largest and fastest growing part of WPC market [4].

With the prosperity of market, more and more manufacturers and scientists realized the importance of innovating WPCs technology. In 1991, in order to share ideas and new technology on WPCs, about 50 researchers and industrial representatives from both plastics and wood products industrials attended the First International Conference on Wood fiber-Plastic Composites, in Madison, Wisconsin, America. In the following year, a similar conference entitled "Progress in Wood fiber-Plastic Composites" was convened in Toroto, Ontario, Canada. These relative conferences experienced a steady growth through the 1990s and have been held not only in North America but also somewhere around the world. In 2011, more than 500 members attended the 11th International Conference on Wood & Biofiber Plastic Composites in Madison.

During the last 15 years, WPCs have gained many markets and experienced a significant growth [5]. The global average annual growth rate was 38 % from 2003 to 2007. In the same period, the rate was as high as 48 % in the Europe market. The worldwide capacity of WPCs is expected to increase from 0.36 million metric tons in 2007 to 3.45 million metric tons in 2020 [6]. In North America, according to PricipiaParlners (an industrial consulting firm), the demand for WPCs products was 700 million pounds in 2001, 875 million pounds in 2002, and 1.3 billion pounds in 2007 [7, 8]. In America, the demand for WPCs increased nearly 12 % annually between 2000 and 2010. Therein to, the market share of PE-based composites was about 83, 9 % for PVC, and 7 % for PP-based composites. It was reported that the demand for WF/PVC composites increased about 200 % from 2000 to 2010, 130 % for WF/PP composites, and 40 % for WF/PE composites [9, 10]. China is the second largest producer of WPCs products in the world with average annual growth rate of 30–50 %. In China, there are currently over 170 WPCs manufacturers on the market and the market shipments are predicted to 8 billion dollars in near future. The driving force comes from the advantages of WPCs products, including environmental-friendly, recyclable, biodegradable, low cost, and low density, and the require less routine maintenance [11].

1.4 Application of WPCs

WPCs have shown significant market expansion in several applications, for example, construction and automotive industries. Although, the current prices of WPCs are relatively higher than those of equivalent solid wood or treated wood lumber, many widow and door manufacturers are considering replacing solid wood by WPCs due to their lower maintenance, lack of cracking or splintering, and high durability. The service life of WPCs is currently being debated, while most manufacturers offer a 10-year warranty.

The majority of WPCs products are used in constructions, including window and door frames, decking, roofing, fencing, and flooring, which makes up about 75 % of the market for the total number of WPCs applications. The composite decking segment has realized compound annual growth rates in excess of 20 % over the last 10 years. According to another set of numbers, WPCs decking reached $ 956 million in market value in 2005, and $ 1.2 billion in 2006. A PE matrix is used most often for decking, accounting for 50 % of the market.

PVC-based composites are most often applied to windows and doors, representing 22 % of the market. Fiber contents vary from 30 to 70 % in WF/PVC composites. Compared to unfilled PVC, WF/PVC composites are more expansive. However, they are gaining favor because that they can offer acceptable mechanical properties, rot-roof ability, chemical, and moisture resistance, as well as a long lifetime [10].

Automotive applications have 14 % of the market share, which is dominated by PP-based composites. WPCs are being increasing used in automotive industries as the replacements of fiberglass and steels in interior parts. They are usually used in dashboard, parcel shelves, cabin linings, door panels, backrests, and set cushions. The main advantages of WPCs in the auto interior parts include enhanced mechanical strength, lower cost, improved passenger safety, and improved biodegradability [12]. The most important advantage is the weight savings of between 10 and 30 % and the corresponding cost savings. For instance, the Mercedes-Benz E-Class has achieved a remarkable weight reduction of about 20 %, while the mechanical properties, which is extremely important for passenger protection in an accident, were improved [13].

In Germany, almost all the major car manufacturers (Daimler Chrysler, Mercedes, Volkswagen, Audi Group, BMW, Ford and Opel) use WPCs in various applications. From 1996 till 2003, the use of WPCs in the German automotive industry increased from 4000 tons to 18,000 tons per year [14]. In 1996, Mercedes began using jute-based door panels in its E-class vehicles in 1996. In 2000, Audi used flax/sisal/polyurethane composites as door trim panels in its A2 midrange car. Now, besides interior materials, Daimler Chrysler has increased its research on flax reinforced polyester composites for exterior applications. From the point of view of environmental protection, it is more important for automakers to improve recyclability of newly produced vehicles. According to the requirements by the European Commission, the recyclable percentage of a car should reach 95 % by 2015. Vehicles must contain 95 % recyclable materials, with 85 % recoverable through reuse or mechanical recycling and 10 % through energy recovery or thermal recycling. This will definitely lead to an increased use of WPCs [9].

2 Composition and Manufacturing of WPCs

2.1 Wood Materials

Wood-thermoplastic composites usually consist of wood flour, thermoplastic, and various additives. The wood used in WPCs mostly refers to wood flour or wood fibers. Apart from woody materials, the agricultural plant residues, such as stems, stalks, leave, and seed hairs could also be used as a filler to manufacture WPCs. Generally, wood fiber is the most abundantly used plant fiber due to their extensive use in pulp and paper industries. Commercial composite products typically contain approximately 50 % wood. In some cases, some products only contain little wood, while others contain as much as 70 %. It has been proved that material compositions is highly related to the weathering of WPCs. Therefore, it is imperative to understand the chemical constituents and physical properties of wood.

2.1.1 Physical Characterization

Wood is composed of numerous cells in different sizes, morphology, and arrangements. Cell is the fundamental structural unit of wood, which contains several layers, namely the middle lamella, primary wall (P), and layers of the outer (S1), middle (S2), and inner (S3) secondary walls (Fig. 1). In the cell wall, each layer has their own particular arrangement of cellulose microfibrils. For example, in P, cellulose microfibrils are randomly oriented at the cell surface. However, in the secondary layers, the cellulose microfibrils have different parallel orientations in respect to the

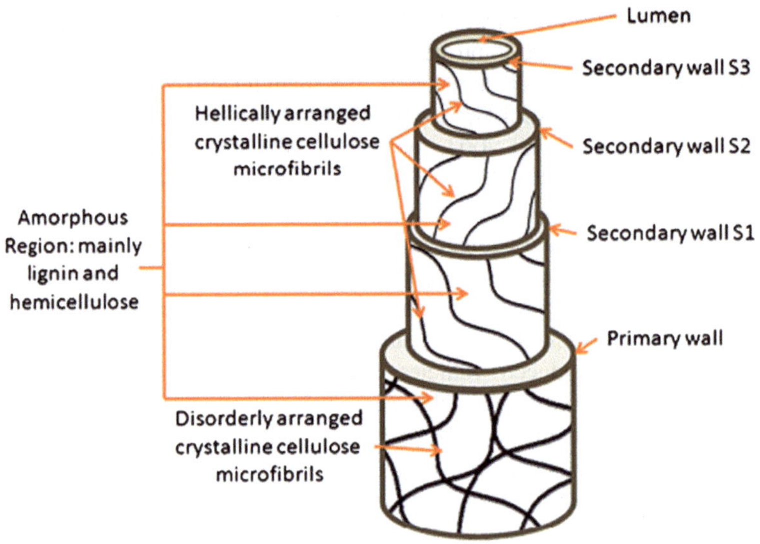

Fig. 1 Three-dimensional structure of the biofibre [17]

axis of the cell. In S1, the microfibrils are arranged in a slight slope perpendicular to the fiber direction, while the micorfibrils in S2 appear to be at an acute angle to the long axis of the cell. The microfibrils in S3 layer are nearest the lumen of the cell, which also show the almost perpendicular orientation to the long axis of the cell [15, 16].

The microfibrils are composed of approximately 40 individual cellulose molecules, which are arrayed in bundles. The cellulose molecules are highly ordered within these bundles and thus contain crystalline region. In wood cell wall, cellulose serves as a structural framework and is largely responsible for the tensile properties of wood. Hemicellulose acts as the amorphous matrix substance, which is present between the microfibrils and contribute significantly to the stiffness of wood. Lignin is usually regarded as the encrusting substance, which help bind the wood cells together. Thereby the lignin provides the rigidity to the cell wall [18].

2.1.2 Chemical Composition

The major constituents in wood cell are cellulose (60–75 %), lignin (20–30 %), hemicellulose (15–35 %), and little ash and extractives.

Cellulose: Cellulose is the main and essential structural component of wood. It is a kind of linear and high molecular mass polymer consisting of β-D-glucose units. These units joint together by β-1,4-glycosidic bonds, making cellulose a 1,4-β-D-glucan (Fig. 2). The molecular weight of cellulose ranges from 50,000 to 2.5 million [19].

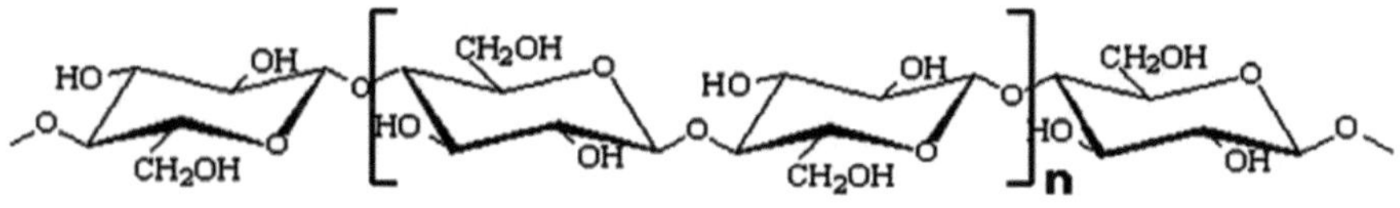

Fig. 2 Structure of cellulose [18]

The formation of molecular cellulose is largely related to the surface hydrogen groups. In pulp wood, the degree of crystallinity can reach nearly 60–70 %. There are two main forms of hydroxyl groups in cellulose. One is located between glucose units in the same molecule (intramolecular linkages), and the other appears between two adjacent molecules (intermolecular linkages). These hydroxyl groups play an important role in the reactive nature of cellulose. For example, cellulose is a polar and hygroscopic molecule and it can combine water through hydrogen bonds. However, only those free hydroxyl (not linked with each other) can be available for water to react with. Therefore, water can only access to the amorphous regions of cellulose, where are also available for chemical reactions [20].

Hemicellulose: Hemicellulose is predominantly found in the primary and second cell walls in wood. Compared to cellulose, hemicellulose usually exhibits lower molecular weight and lower degree of polymerization, which may be tens of hundreds of repeating units. Hemicellulose comprises of polysaccharides, e.g., arabino-xylans, gluco-mannans and galactans, of which xylans are predominant [21]. Unlike cellulose, the composition of hemicellulose varies between trees and species. Besides, hemicelluloses are more hydrophilic than cellulose and are extremely susceptible to alkali and acid hydrolysis because of their branched and amorphous nature [22]. The combination of cellulose and hemicellulose is known as holocellulose.

Lignin: Lignin is widely distributed throughout the secondary cell wall and highly deposited in the middle lamella. The content of lignin in softwoods (25–35 %) is usually higher than that in hardwoods (18–25 %). Lignin is responsible for part of the stiffness in wood, as well as for the exclusion of water from the moisture-sensitive carbohydrates. It is reported to be less reactive than cellulose due to its hydrophobic nature.

Similarly, lignin is also comprised of carbon, hydrogen, and oxygen. However, it consists of a complex arrangement of substituted phenolic units. Lignin is a kind of amorphous, highly complex, three-dimensional aromatic polymer that consist of phenylpropane units [18] through C–O–C and C–C linkages. There are three precursors of lignin biosynthesis (Fig. 3), including *p*-coumaryl alcohol (structure 1), coniferyl alcohol (structure 2), and sinapyl alcohol (structure 3). Coniferyl alcohol is the major precursor in softwood lignins. According to different structural elements, phenylpropane units could be divided into three groups: guaiacyl (G), syringyl (S), and p-hydroxyphenyl (H) moieties. There is a wide variation of structures within different wood species. Generally, softwoods have a much higher percentage

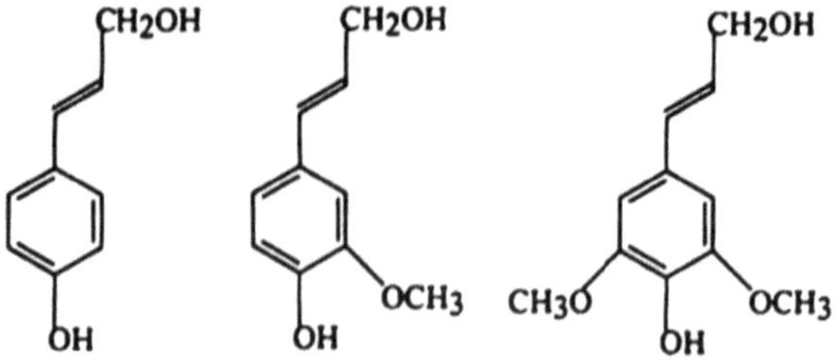

Fig. 3 Chemical structures of lignin precursors: (1) *p*-coumaryl alcohol, (2) coniferyl alcohol, and (3) sinapyl alcohol [18]

of G lignin. For example, the ratio of G:S:H in loblolly pine (*Pinustaeda*) is 86:2:13.

Extractives: Extractives are organic chemicals that can be extracted from wood with solvents. They are generally low molecular weight compounds with the content accounting for 2–5 % of the wood. Comparatively, softwoods have a higher extractives content than hardwoods and most of them are located in the heartwood. The component of extractives is responsible for some wood properties such as the color, smell, density, hygroscopicity, flammability, and durability.

The extractives are a group of chemicals mainly consisting of organic waxes, oils, tannins, resins, fats, gums, carbohydrates, coloring matter and phenols. These chemicals exist as monomers, dimers, and polymers. In general, there are three types of wood extractives, namely, aliphatic compounds, terpenes and terpenoids, and phenolic compounds.

2.2 *Thermoplastic Matrix Materials*

The WPCs industry presently consume over 700,000 tons of thermoplastics annually. Only those thermoplastics, which can be processed at temperatures below 200 °C, are applicable in WPCs. Polyethylene resin (PE) is the dominant resin in market that accounts for 83 % of the total volume. Polypropylene (PP) and polyvinyl chloride (PVC) account for 9 and 7 % of the total, respectively. Only about 1 % of other resins are used in the WPCs industry.

2.2.1 Polyethylene (PE)

Polyethylene (PE) is the largest volume plastic produced in the world. PE is a linear molecule and has relatively lower melting temperature (106–130 °C), depending on its density or branching. According to its different densities, PE could be divided into low density polyethylene (LDHE, 0.91–0.925 g/cm^3), medium density polyethylene (MDHE, 0.926–0.94 g/cm^3), and high density polyethylene (HDHE, 0.941–0.97 g/cm^3) [23].

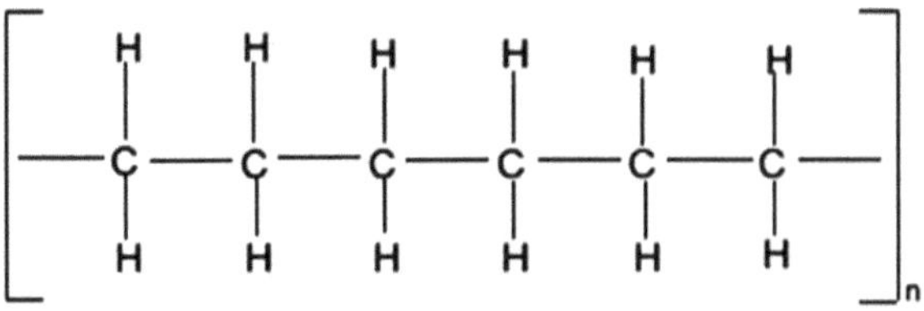

Fig. 4 Structure of PE

Polyethylene is soft, making PE-based composite easier to nail, screw, cut, and saw. PE is also a kind of hydrophobic materials and shows near-zero moisture absorption. PE is highly durable and shows resistance to chemicals corrosion, including strong acids, such as sulfuric, hydrochloric, and nitric.

PE is a semicrystalline polymer consisting of a carbon backbone with hydrogen as repeating unit (Fig. 4). At ambient temperatures, PE exhibits two rather distinct phrases, namely crystalline and amorphous. In general, the glass transition temperature (T_g) of PE is −20 °C, imparting the plastic with excellent ductile and reasonable good low-temperature resistance. Bellow T_g the polymer is brittle, while above T_g the polymer becomes flexible. The brittle temperature of PE ranges from −20 to −70 °C, depending on its M_w. The branching degree can influence the melt temperature (T_m) of PE. For LDPE, T_m ranges from 108 to 126 °C, MDPE ranges from 126 to 130 °C, while HDPE ranges from 126 to 137 °C.

LDPE typically has long side-chain branching off the main molecular chain and therefore is a more amorphous polymer. It shows lower shrinkage compared to a more crystalline HDPE. MDPE is regarded as a mixture of LDPE and HDPE. However, this polymer is not favored by WPCs manufacturers. HDPE is stronger and stiffer than LDPE, but is more prone to warpage. The tensile strength of HDPE is usually two to three times that of LDPE, and it has an excellent compressive strength (4600 Pa), whereas LDPE typically does not break on compression.

2.2.2 Polypropylene (PP)

The share of PP-based WPC decking in the market is relatively small, accounting for about 10 % of the WPCs market.

Similar to PE, PP is also a linear molecule and exhibits a lower density (0.9–0.91 g/cm^3). PP is polymerized from the propylene ($CH_3(CH)=CH_2$) and has a methyl (CH_3) side group at every other carbon along the molecular backbone [23]. According to the different arrangements of the methyl groups, the structure of PP can be either isotactic or atactic. PP can be called isotactic when all the methyl groups appear on one side of the carbon backbone (Fig. 5), while PP with random arrangements of methyl side groups is called atactic. Commercial PP usually contain about 90–95 % isotactic, with other structures of atactic and syndiotactic.

PP is superior to PE in a number of properties. PP is lighter, stronger, and stiffer and it shows improved creep resistance, less wear, and less slippery. The mechanical properties of PP is better than PE, with the tensile strength over

Fig. 5 Structure of isotactic PP

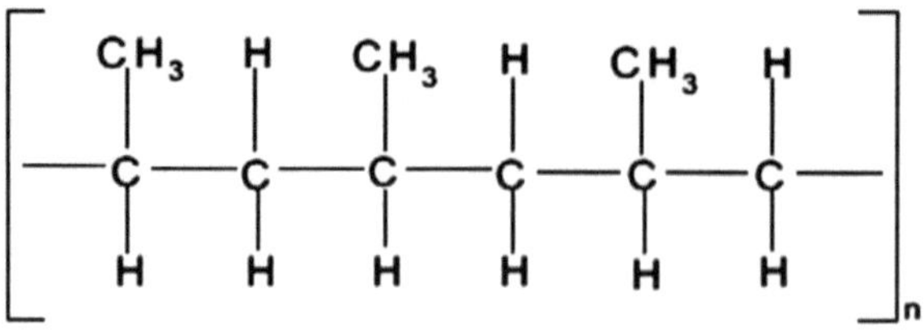

30 MPa. The flexural strength of PP ranges from 6000 to 7000 Pa, which is four times higher than PE. However, it is more brittle than PE, particularly at low temperatures. For a comparison, PP is difficult to fasten, cut or saw because of its high stiffness. That is why PP-based composites are recommended to be installed using special fastening systems. PP is also hydrophobic and shows negligible water absorption (less than 0.01 % after 24 h underwater submersion). PP is durable and shows resistance to the majority of chemicals, except for concentrated sulfuric acid, concentrated nitric acid, and strong oxidations [24].

2.2.3 Polyvinyl Chloride (PVC)

Compared to PP-based composites, the total market share of PVC-based materials is even smaller. There are only three commercially available brands in the WPC decking market [24].

PVC resin is white or light yellow powder and consists of the repeating unit (CH_2–CHCl) (Fig. 6). The density of PVC (1.32–1.44 g/cm^3) is higher than both PP and PE. Similar to PP, PVC also has three chemical structures, syndiotactic (chlorine atoms in the polymer chains on the same side of plane), isotactic (on an alternate sides), and atactic (in a random arrangement). Commercial PVC usually contains 55–65 % syndiotactic polymers, while the rest are atactic polymers. The average molecular weight of PVC is often between 100,000 and 200,000, with the number-average molecular weight of 45,000–64,000.

Compared to PE and PP, PVC has a poor thermal stability. The glass transition temperature and melting temperature of PVC are very low, with the number of 80 and 140 °C, respectively. When the temperature goes up 170 °C, PVC becomes unstable and release toxic hydrogen chloride (HCl) during thermal degradation. Therefore. PVC is commonly considered as the most environmentally damaging polymer among thermoplastics. However, PVC also has some principal beneficial

Fig. 6 Structure of isotactic PVC

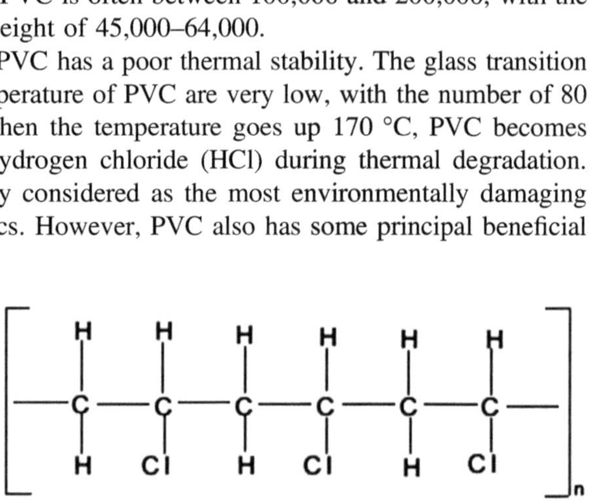

properties, one of which is its inherently flame resistant. PVC contains about 57 % chlorine, contributing to its flame retardancy. The limiting oxygen index (LOI) value of PVC is about 47 (17.4 for PE). Flame spread index for a PVC-based composites is 25, whereas it is in the range of 50–170 for HDPE-based composites [24].

In addition, PVC exhibits much better mechanical properties than PP and PE. The flexural strength for PVC (6000–16,000 Pa) is higher than that for PP (6000–7000 Pa) and PE (around 1400 Pa). The flexural modulus for PVC (350,000–600,000 Pa) is also higher compared to PP (165,000–250,000 Pa) and PE (125,000–240,000 Pa). The water absorption of PVC is about 0.1 % after submersion for 24 h, which is close to PP and PE.

2.3 Additives

Apart from wood materials and thermoplastic, the manufacture of WPCs also involves some additional materials such as coupling agents, light stabilizers, pigments, lubricants, fungicides, and foaming agents. All the additives are added in small amounts and aimed to affect WPCs' processing and performance. Some additive suppliers are specifically targeting the WPCs industry.

2.3.1 Coupling Agents

Among the various kinds of additives in WPCs, coupling agents is the most widely used.WPCs have one principal problem related to their integrity. The wood component of WPCs is hydrophilic and the thermoplastic is hydrophobic. Hence, the interface and adhesion between plastic and wood filler is typically weak, and it will not benefit to the stress transfer between the two phases.

In order to improve the interfacial bonding of the two different phases, the terms coupling agents, or compatibilizers are in use. A proper coupling agent usually contain two functional groups: one group capable to form entanglements, or segmental crystallization with the polymer matrix, and the other one can strongly interact with wood component through covalent bonds, ionic interactions, hydrogen bonds, and so on [25]. Coupling agents can greatly improve mechanical properties of WPCs, for example, they can double the flexural and tensile strength of composites by increasing flexural modulus (stiffness) by as much as 40 %, double and triple the impact strength, decrease water absorption by two–four times, increase the density and durability of the composites, including oxidation, microbial degradation, and so on [26].

There are dozens kinds of coupling agents, while only few of them have been introduced to WPCs industry. These coupling agents can be subdivided into (a) maleatedpolyolefins, which bind with wood fiber via chemical bonds; (b) oligomers, which can interact with inorganic fillers via an ion-pair bond;

(c) silanes, which could covalentlyinteract with hydroxyl groups of wood fiber, forming a Si–O–C linkage; (d) acrylic-modified polytetrafluoroethylene (PTEF); (e) chloroparafins; and (f) other coupling agents, which benefit to a better dispersion of wood fillers in the polymer matrix.

Maleatedpolyolefins Maleatedpolyolefins are the most widely used coupling agents, which are usually made by grafting maleic anhydride onto the polymer backbone (via radical reactions). Maleic anhydride grafted polypropylene (MAPP) is one of the most often used maleatedpolyolefins in WPCs. They contain two functional groups: the maleic anhydride, which is able to from covalent bonds with wood fiber at extrusion temperatures, via hydrogen, or ionic bonds (Fig. 7), and a polyolefin (HDPE or PP), which is able to form entanglements with the polymer matrix (Fig. 8). The addition level of maleatedpolyolefins is usually maintained at 1–5 % by weight in a WPCs formulation.

Organosilanes Silanes have been proven to be a widely used coupling agent in plastics filled with minerals [27]. The general formula of the silanes can be represented by $Y(CH_2)_nSi(OR)_3$, in which Y could be amino, epoxy, vinyl, or alkyl group and possess an affinity for polymer matrix. Alkoxy (-R) could react with water vapor in atmosphere to form silanols: $Y–(CH_2)_n–Si(OR)_3 \cdot H_2O \rightarrow Y–(CH_2)_n–Si(OH)_3$ [28]. The hydroxyl group could link with wood fiber via hydrogen bond. The addition level of silanes is 3–10 % by weight of a WPC formulation.

There are many different kinds of organofunctional groups available on organosilanes. In order to get optimum performance of the composites, it is important to choose the proper silane, which should have an organofunctional group that will react with the polymer matrix. For instance, an epoxy resin has the

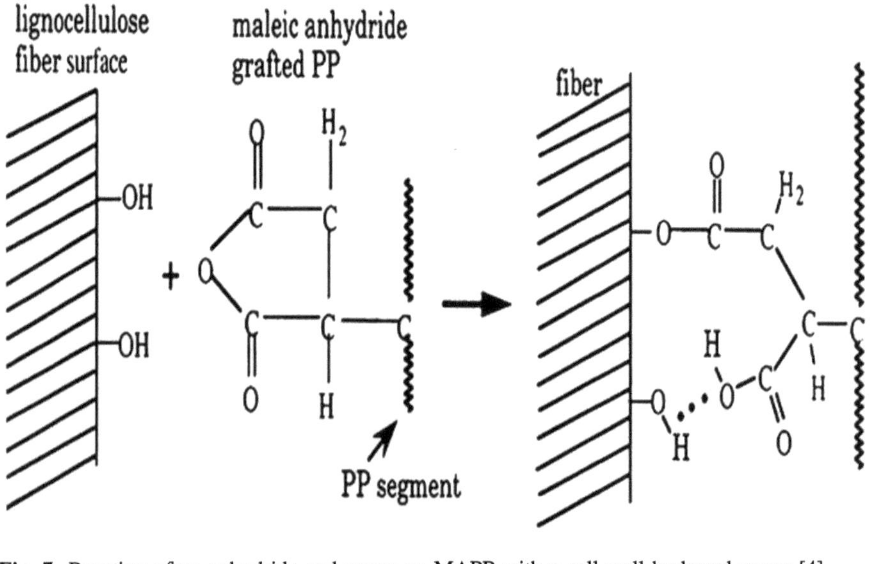

Fig. 7 Reaction of an anhydride end group on MAPP with a cell wall hydroxyl group [4]

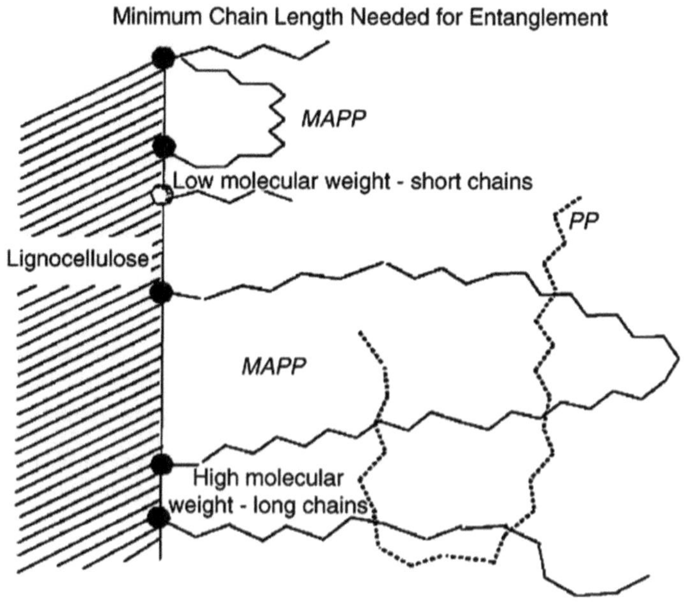

Fig. 8 Entanglement of the PP on MAPP with the matrix PP [4]

proper reactive groups to form covalent bonds with especially epoxy and amino silanes. The proper silane can improve compatibility with the polymer, resulting in an improvement of physical and mechanical properties.

Other coupling agents Long-chain chlorinated paraffins (LCPP) with chlorine content around 70 %, could be used as a non-reaction coupling agent. LCPP is reported to show an effect of a coupling agent and increases the flexural properties of the products, as well as the UV and moisture resistance. The amount of LCPP is 10 %, along with 3 % of a lubricant in the WPCs formulation.

2.3.2 Light Stabilizer

The solar radiation that reaches the earth's surface is limited to the range between 290 and 3000 nm. Visible light with the wavelengths from 400 to 800 nm accounts for 40 % of the solar radiation, and infrared radiation (800–3000 nm) accounts for 55 %. Wavelengths less than 400 nm are ultraviolet (UV) light, which only account 6 % of the total solar radiation. However, UV light plays a negative effect on the deterioration of materials, because the covalent bonds in materials can be ruptured by UV light [29]. In order to improve the photostability of WPCs, light stabilizers are often used. The light stabilizers can be classed according to the degradation mechanism they act. The main three types of light stabilizers are light screens, UV absorbers (UVAs), UV quenchers, and free radical scavengers.

Light screens Light screens could reflect the majority of UV light. The common light screens are usually inorganic, including TiO_2, ZnO, carbon black, and so on. Their protection action depends on concentration and particle size.

UV absorbers UVAs can strongly absorb UV light and convert the absorbed energy into low-frequency bond vibration or heat, and thereby prevent photodegradation by physical means. The most studied UVAs are hydroxybenzophenones and hydroxyphenylbenzotriazoles.

UV quenchers UV quenchers are usually metal complexes, which can decrease the activity of UV light. UV quenchers can take over the energy absorbed by molecules and dissipate it as heat or as fluorescent or phosphorescent radiation.

Free radical scavengers Free radical scavenger cannot reflect or absorb UV light directly, but it can eliminate free radicals, slowing down the process of photodegradation.

2.3.3 Antioxidants

In general, antioxidants can either scavenge free radicals or prevent free radicals formation during photodegradation process. There are many types of antioxidants, such as hindered phenolic compounds, hindered amines, secondary aromatic amines, and phosphite compounds [30]. They can convert free radicals into stable compounds by chemical means, or prevent free radicals formation by converting their predecessors into stable compounds [31].

2.3.4 Other Additives

Lubricants are often used as purpose to decrease the friction force during WPCs manufacturing process. The main kinds of lubricants are fatty acids, fatty acid esters, hydrocarbon (wax), methyl silicone oil, and metallic soaps. The addition level of such additives are usually between 1 and 2 % by weight of WPCs. Moreover, in order to enrich the colors of WPCs and then expand their applications, some colorants, both organic and inorganic are also used. As an example, Accel Company has been supplying the colorants for WPCs for almost 15 years.

2.4 Manufacturing Process

The manufacture of WPCs is comprises two steps: blending raw materials and making composites blends into a product [32]. Raw materials can be mixed in either batch or continuous mixers. The forming techniques include extrusion, injection molding, and compression molding. However, some factors, such as fiber type, fiber content, fiber orientation, moisture content of fiber, should be taken into consideration when choosing the suitable manufacturing process [33].

2.4.1 Extrusion

Extrusion is a popular and efficient way to produce WPCs, whose process is similar to the squeezing of toothpaste. There are two common types of extruders, single screw and twin screw extruder. Single screw extruders are not positive displacement, so they are not effective mixing device. Therefore, the majority of WPCs are manufactured in twin screw extruders, in which the composite materials are conveyed down in a heated barrel and forced through a die to make a continuous profile of the desired shape [34]. For example, window sashes, door jambs, and decking can be fabricated by attaching a die at the end of the extruder and directly forming the cross-section desired. In addition, the extruder can produce the form of spaghetti-like strands, which are further cut into pellets to be used in other processing methods.

2.4.2 Injection Molding

Injection molding is the most widely used polymeric fabrication method and can produce WPCs products with complex three-dimensional shapes. In this process, WPCs pellets are heated and then the molten composites are forced into mold, often with a plunger action. The mold is cool, and after the part cools the mold opens and ejects it. Injection molded WPCs panels usually appear in a skin-core morphology. In such composites, fibers in the core layer are oriented perpendicularly to flow, while those in the skin layer are oriented parallel to flow [35, 36]. Temperature, pressure and flow are the three main variables that can influence the properties of composites. For example, a low mold temperature leads to a large skin thickness, while smaller thickness can appear when using higher barrel temperature, higher screw and injection speed.

2.4.3 Compression Molding

Compression molding involves two steps, including mixing and compressing. Firstly, the wood flour, thermoplastic and additives are mixed in a high-speed mixture. Then, the mixture is dried and matted. A hot press is used to produce the composites by compressing the mat between two mold halves at a high temperature and high pressure. After hot pressing, the formed mat is usually cooled down at room temperature in a cold press, also under a high pressure [37]. This alternative processing method has advantages when processing of a continuous piece is not desired.

It is believed that different manufacturing processes could result in the different mechanical properties of the produced composites. For instance, injection molding process is reported to lead to higher tensile and flexural strength values compared to the other processes. Compression molding process can lower odor concentrations of the products, which is favorable for the automotive application [38]. It is because that the decomposition of materials during injection molding process is more severely, leading to higher odor concentrations.

3 Photodegradation and Mechanism of WPCs

3.1 Testing Methods

3.1.1 Weathering Parameters

There are several environmental elements that affecting the photodegradation process of wood-thermoplastic composites (WPCs), namely sunlight, wind, rain, snow, hail, atmospheric pollution, and temperature variations. After long-time exposition, the materials will consequently degrade.

Sunlight The wavelengths of sunlight that reaches the earth's surface is limited to the range between 290 and 3000 nm. Visible light (400–800 nm) and infrared radiation (800–3000 nm) account for 40 and 55 % of the solar radiation, respectively. The so-called ultraviolet (UV) light with wavelengths less than 400 nm only accounts 6 % of the total solar radiation. Sunlight is composed of photons, varying in energy from about 0.5–4 eV. In addition, as the wavelength of a photo decreases, its energy increases. Therefore, the energy of an UV photon is higher than 3 eV, which could break many chemical bonds, especially in organic materials.

Wind Wind can exert forces on WPCs composites, resulting in the formation of material fatigue and cracks. Moreover, wind also carries dust and debris, causing some physical damages to the materials' surface.

Moisture Moisture variations are usually caused by rain, snow, hail, frost, and dew. Changes in humidity lead to changes in equilibrium water content. When moisture increases, water vapor is absorbed by WPCs, particularly wood. When moisture decreases, wood shrinks. The changes in moisture accelerate the deterioration of wood flour/plastic interface bonding.

Atmospheric pollution The atmosphere is made up of three main compositions, nitrogen (78 %), oxygen (21 %), and water vapor (0.3–3 %). Among them, nitrogen is not chemical reactive, while oxygen and water vapor can take part in some chemical reactions and cause the formation of various oxides and hydroxides. Some other atmospheric constituents including CO_2, SO_2, and NO_2, originated from combustion, can dissolve in water and then form acids. These acids can promote chemical corrosive of the materials. Meanwhile, the soot from combustion, which is a strong absorber of sunlight, deposits on materials' surface, promoting the weathering process [29].

3.1.2 Weathering Methods

The current used weathering methods are either natural or accelerated weathering regimes. The main difference between the two weathering methods is that the factors such as UV radiation, moisture, and temperature can be controlled in accelerated weathering.

Natural weathering Natural weathering is generally carried outdoors and it is reliable to accurately predict the service-life of a material. The long-term behavior of materials can be evaluated by real time observations in such process [39]. However, natural weathering is time consuming and it usually takes several years. Due to the changing environmental conditions, natural weathering is lack of repeatability and reproducibility [40]. Therefore, natural weathering is proved to be impractical due to the amount of time required, although it is still needed to obtain a reliable prediction of service life.

Accelerated weathering In order to imitate natural conditions and shorten testing time, accelerated weathering has been adopted. Accelerated weathering simulates the factors of UV radiation, moisture, and heat in a controlled manner in ageing chambers. The main advantage includes faster testing than natural exposure and is repeatable [41]. Moreover, accelerated weathering could provide basic understanding on the degradation mechanism of the materials [42].

Generally, accelerated weathering equipment usually contains UV light exposure and water spray. There are four commercial available light sources for UV radiation, including xenon-arc lamps, fluorescent UV lamps, carbon arc lamps, and mercury lamps [43]. The first two kinds of lamps are the most popular and widely used today. Fluorescent UV lamps contains UVA lamps and UVB lamps. UVA lamps simulate natural radiation well in a certain range (320–400 nm), while UVB lamps (275–320 nm) are not comparable with natural light. Comparatively, xenon-arc lamps are most close to natural UV radiation when properly filtered. Carbon arc lamps have strong radiation below 300 nm and are different from natural light, therefore, causing unrealistic damage to materials. Mercury lamps can well simulate natural conditions and have main emissions at 254, 290, and 365 nm [43].

Weathering standards There are several testing standards about natural or accelerated weathering. The most commonly used standards are made by American Society for Testing and Materials (ASTM). ASTM has launched a wide range of weathering procedures for both wood and plastic, while no standards for testing WPCs. The standard commonly used to weather WPCs is ASTM G154-06, *Standard Practice for Operating Fluorescent Light Apparatus for UV Exposure of Nonmetallic Materials*. The procedures requires the use of a UVA-340 lamp, which gives a simulation of sunlight in the critical short wavelength region from 365 nm down to 295 nm. The irradiance should be 0.89 W/m^2 at 340 nm wavelength. The weathering cycle consists of 8 h UV radiation at 60 °C and 4 h condensation at 50 °C. In addition, some other weathering procedures are shown in Table 1 [44].

3.1.3 Physical and Mechanical Properties Tests

Surface color A significant feature of WPCs is its aesthetic value, which undergoes severe discoloration during weathering. Color measurement of composites is analyzed according to ASTM D 2244 (*Standard Practice for Calculation of Color Tolerances and Color Differences from Instrumentally Measured Color Coordinates*) and is conducted in a chroma meter.

The most widely used color system for measuring the color of a material is the CIE $L^*a^*b^*$ color system (Fig. 9). In this color system, the L^* represents the lightness coordinate and varies from 100 (white) to 0 (dark). The a^* represents red (+100) to green (−100) coordinate, whereas b^* represents the yellow (+100) to blue (−100) coordinate [45, 46]. The difference in color between two samples (ΔE) is calculated with the following equation:

$$\Delta E = \sqrt{\Delta L^{*2} + \Delta a^{*2} + \Delta b^{*2}}$$

where, ΔL^*, Δa^*, and Δb^* are the difference of initial and final values of L^*, a^*, and b^*, respectively.

Flexural testing Flexural properties are evaluated by the three-point bending tests according to the procedures outlined in ASTM D 790-10 (*Standard Test Methods for Flexural Properties of Unreinforced and Reinforced Plastics and Electrical Insulating Materials*). Before testing, all the specimens should be dried in an oven at 50 °C for 24 h. As shown in Fig. 10 the sample is put on two stationary supports with load applied in the middle of the sample. At least five specimens for each sample are tested for calculation. The modulus of elasticity (MOE) and modulus of rupture (MOR) were calculated to evaluate the flexural properties. To clearly show the influence of weathering, sometimes, the MOR and MOE retention ratios after weathering were defined as follows.

Table 1 Summary of common ASTM International for weathering conditions

ASTM and light source	Irradiance ($W/m^2/nm$)	Wavelength (nm)	Weathering cycle
G 155 D2565 Xenon-arc	0.35	340	102 min irradiation at 63 (±2.5) °C BPT 18 min irradiation and H_2O spray (air tem. Not controlled)
	0.35	340	102 min irradiation at 63 (±2.5) °C Uninsulated BPT 18 min irradiation and H_2O spray (air tem. Not controlled) 6 h dark at 95 (± 4.0)% RH, at 24 (±2.5) °C Uninsulated BPT
	0.35	340	1.5 h irradiation, 70 (±5) % RH, at 77 (±3) °C BPT 0.5 h irradiation and H_2O spray (air temp. Not controlled)
G 154 Fluorescent	0.77	340	8 h UV at 60 (±3) °C BPT 4 h condensation at 50 (± 3) °C BPT
	1.35	340	8 h UV at 60 (±3) °C BPT 4 h condensation at 50 (± 3) °C BPT
	1.35	340	8 h UV at 60 (±3) °C BPT 0.25 h H_2O spray (no irradiation), temp. controlled 3.75 h condensation at 50 (±3) °C BPT
D 4329 Fluorescent	0.35	340	8 h UV with uninsulated BPT controlled at 60 ± 3 °C 4 h condensation with uninsulated BPT controlled at 50 ± 3 °C
	0.35	340	8 h UV with uninsulated BPT controlled at 70 ± 3 °C 4 h condensation with uninsulated BPT controlled at 50 ± 3 °C
	0.35	340	8 h UV with uninsulated BPT controlled at 50 ± 3 °C 4 h condensation with uninsulated BPT controlled at 50 ± 3 °C
D 6662 Xenon-arc	0.70 ± 0.02	340	2 h irradiation only with uninsulated BPT controlled at 70 ± 2.2 °C 2 h irradiation with H_2O spray (on the exposed surface) RH and uninsulated BPT controlled are not applicable

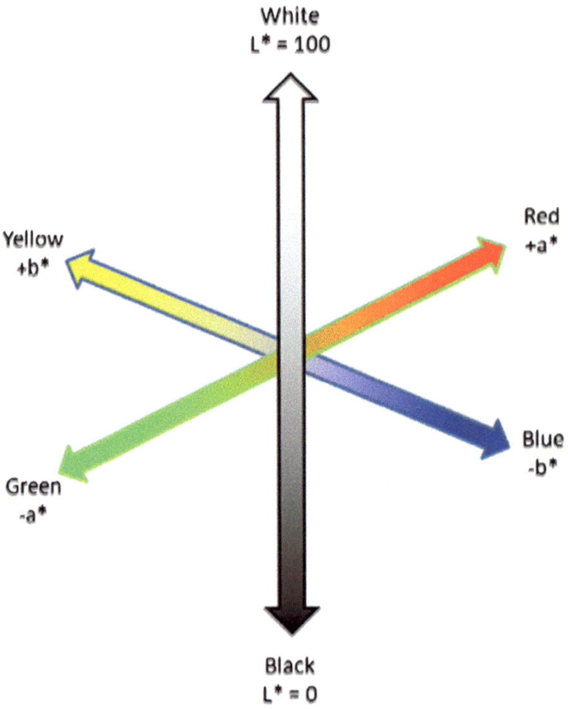

Fig. 9 *L*a*b** color space [47]

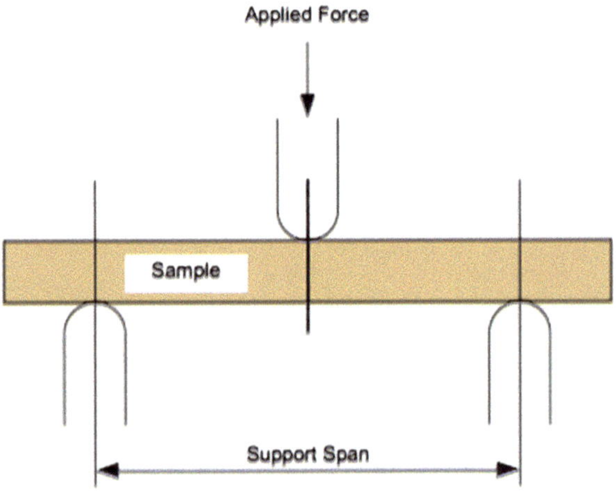

Fig. 10 Three-point set-up [48]

$$\text{MOR retention ratio} = \frac{MOR_t}{MOR_0} \times 100$$

$$\text{MOE retention ratio} = \frac{MOE_t}{MOE_0} \times 100$$

where, MOR_t and MOE_t are the moduli after weathering time t, MOR_0 and MOE_0 are the moduli before exposure.

Impact testing Charpy impact testing is carried out in accordance with the method ISO 179-10 (*Plastics-Determination of Charpy impact properties*). Before testing, the specimens should be conditioned at 23 ± 1 °C and 50 % relative humidity for at least 24 h. The dimensions of the specimens are 80 mm × 8 mm × 3.2 mm with a 0.25 mm single notch (type A).

Tensile testing Tensile testing was carried out according to the ASTM 638-03 (*Standard Test Method for Tensile Properties of Plastics*). Test specimens should be placed in a conditioning chamber at 23 ± 1 °C and 50 % relative humidity for at least 24 h. At least six specimens are tested for each sample.

Hardness Hardness measurements are performed according to ASTM D 2240-05 (*Standard Test Method for Rubber Property-Durometer Hardness*), using two Durometers (Shore D scales) for the composites. The dimensions of the specimens are 16 mm × 16 mm × 6.5 mm. The measurements were performed at room temperature (25 °C).

Surface gloss The measurement of surface gloss for samples is carried out according to the procedures listed in ASTM D 2457-13 (*Standard Test Method for Specular Gloss of Plastic Films and Solid Plastics*), using a glossmeter with the test angle of 60°.

3.1.4 Characterizations

X-ray photoelectron spectroscopy (XPS) XPS is a surface analysis technique that can be used to ascertain surface elemental composition and types of bonds present on the surface. It is also a straightforward, easy, and non-destructive characterization technique. In XPS, different scan resolutions lead to different information. On the one hand, a low-resolution scan can provide the percent of each element, as well as their atomic concentrations. On the other hand, a high-resolution scan can give the types of bonds and concentrations on the surface. Take carbon (C_{1s}) as an example, its high resolution scans can be divided into four components peaks around 285.0, 286.5, 228.0, and 289.5 eV. These subpeaks are C1 (C–C or C–H), C2 (C–OH), C3 (O–C–O or C=O), and C4 (O–C=O), respectively. XPS has frequently been applied to confirm the chemical changes of polymer or fiber after treatment. In addition, XPS can be used to verify the occurrence of surface oxidation by studying the concentration of carbon to oxygen atoms with the following equation [49]:

$$C_{ox/unox} = \frac{C_{oxygenated}}{C_{unoxygenated}} = \frac{C2 + C3 + C4}{C1}$$

Fourier transform infrared (FTIR) spectroscopy FTIR is a useful method for detecting functional groups on materials surfaces. Attenuated total reflectance (ATR) is a method that does not need to change the original morphology of the sample. It can be a solid plate instead of powder. In WPCs photodegradation, ATR-FTIR is often used to monitor the generation of carbonyl groups and the changes in polymer crystallinity, to verify surface oxygenation.

Gel Permeation Chromatography (GPC) This technique is often used for the analysis of polymers. GPC allows the determination of polydispersity index (PDI) and the molecular weight, including the number average molecular weight (M_n), the weight average molecular weight (M_w), the size average molecular weight (M_z), or the viscosity molecular weight (M_v).

Differential Scanning Calorimetric (DSC) DSC is used widely for examining polymeric materials to determine their thermal transitions. The changes in crystallinity contents, the melting temperatures, and the glass transition temperatures of plastics during weathering can be estimated by DSC. The percentage crystallinity X_c of plastics is calculated by the following equation [50]:

$$X_c(\%) = (100 \times \Delta H_m)/(\Delta H_m^0 \times w)$$

where X_c is crystallinity (%); ΔH_m^0 is enthalpy of melting 100 % crystallized polymer; ΔH_m is enthalpy required for melting each sample, and w is weight fraction of polymer in composites.

Scanning electron microscope (SEM) SEM is generally used to observe the surface morphology of materials during weathering. The magnification ranges from 20× to approximately 30,000×, spatial resolution from 50 to 100 nm. Before observation, the sample is stuck on carbon tape adhered to the stage, and then is sputter coated with a thin layer of gold.

3.2 *Surface and Mechanical Properties Influenced by Photodegradation*

Degradation processes during weathering involve photo-radiation, thermal degradation, photo-oxidation and hydrolysis. These irreversible processes can cause some changes in the chemical, physical and mechanical properties of WPCs products, for example, causing yellowing, discoloration, weight loss, surface roughening, and loss of mechanical properties.

3.2.1 Surface Color Change

Surface color is a significant feature of WPCs products, which becomes unaesthetic during weathering thorough severe discoloration. The color measurement of composites is usually evaluated by two parameters, the total color change (ΔE) and lightness (L^*).

Studies showed that for both natural and accelerated weathering, longer exposure time increased the degree of color change and lightness. For most commercial WPCs boards, the initial L^* value is between 40 and 55. In some cases, the value of the most light-colored boards is close to or around 60 [24]. However, after a relatively short exposure, the value of L^* goes up to 90 units and higher. Meanwhile, the ΔE value can reach 40–60 units after weathering. When subjected to sunlight, the high-energy UV light breaks polymer chains and other organic compounds on the surface [42, 49].

During weathering, wood initially becomes yellow due to the breakdown of lignin to water soluble products, and then grey due to leaching of the degraded products [51]. Comparatively, WPCs show a much better color stability than wood. Study by Chetanachan et al. [52] reported that the ΔE value of WF/PVC composites was just one third of the solid wood after 1 year natural exposure. However, the surface of WPCs will eventually become white after a long period of weathering.

Study by Butylina et al. [53] proved that higher wood fiber content in WPCs led to more significant discoloration. When fiber content exceeds 40 %, the encapsulation of wood fiber by polymer matrix becomes impossible. Degradation becomes easier and promotes the color fading of materials. The depth of color change is generally limited to 0.5 mm, while below this depth, the condition is similar to those of the non-weathered composites.

Except for wood fiber content, the discoloration of composites also depends on the types of thermoplastics. Falk et al. compared the weathering behaviors between two WPCs, PP-based composites and HDPE-based composites. Results showed that PP-based composites faded at a more rapid rate than did the HDPE-based composites, especially during the first 400 h of exposure [54].

Fabiyi et al. [55] studied the effects of wood species on the accelerated weathering behaviors of HDPE-based WPCs. The selected wood species are poplar, Douglas-fir, black locust, white oak, and ponderosa pine. Results revealed that hybrid poplar and ponderosa pine had good color stability among the wood species examined for WPCs.

There is not much of a difference between discoloration of the extruded and injection-molded composites made of the same materials. For the extruded samples, the ΔL^* valueis around 30 after 1000 h of weathering, while it is 28 for the injected-molded ones. After 3000 h, L^* values for both the composites are close to each other [56].

Weathering in the presence of water, enhances the rate of degradation of WPC. Wood cell walls swell when penetrated by water, facilitating light penetration further into the wood providing sites for further degradation. Moreover, periodic water spraying during weathering resulted in a cyclical and continuous erosion of the surface. The lightness of the injection molded composites increased 87 % when exposed to UV light and water, but only increased 28 % when exposed to UV light only. Similar trends were found for the extruded composites; the lightness increased 46 % (UV light and water) and 13 % (UV light), respectively [57]. That is because that washing can also remove water soluble extractives that impart color to wood.

3.2.2 Surface Morphology

Research on the photodegradation of WPCs has also focused on the changes in surface morphology, which usually exhibits a chalky appearance after weathering. As a result of wetting and drying cycle, WPCs were differentially contracted between the surface and interior sections. The absorption and desorption of wood fiber create microcracks in the matrix. The degradation of the polymer matrix leads to the formation of highly crystallised polymer zones, which increases the embrittlement of polymer [58]. Both the degradation of wood fiber and polymer matrix contribute to the cracking on composite surface.

The surface degradation process of WPCs consist of four main cyclic stages: (1) surface layer is eroded, creating several cavities on the surface of composites; (2) the number and size of the cavities increase with increasing exposure time; (3) small cracks accumulated on the weathered surface; (4) the upper degraded layer are washed out and expose the new surface to degradation [59].

Studies has shown that the degradation rate of the polymer matrix is faster in the presence of wood fiber, owing to the larger number of chromophores in wood.

3.2.3 Mechanical Properties

Mechanical properties of WPCs also undergo changes after weathering, which can be attributed to composite surface oxidation, matrix crystallinity changes, and interfacial degradation.

Flexural properties Weathering has been proven to decrease the flexural properties of both pure polymer and composites significantly. The flexural properties (both MOR and MOE) of WPCs are significantly affected only after 240 h of accelerated UV exposure [60]. Moreover, composites with higher wood loading showed higher loss of flexural properties after weathering than composites having a lower wood content [61, 62]. Take PP-based composites as an example, the strength for unfilled PP decreases approximately 10 % after 8 weeks of aging, while the data for 50 wt% wood fiber decreases approximately 20 %. Stark et al. compared the effects of different processing methods on flexural properties of wood flour/HDPE composites during weathering [57]. For injection-molded samples, the decrease in MOE is significant for each 1000 h of weathering, while the flexural strength is not significant until 2000 h of weathering. For extruded composites, they lost a larger percentage of their total mechanical properties only after the first 1000 h of exposure. After 3000 h of weathering, injection-molded composites show much better flexural properties compared to those of extruded ones [57].

Impact strength Compared to flexural properties, weathering poses a more obvious effect on the impact strength of WPCs. The impact strength for unfilled PP drops 50 % after 8 weeks of natural weathering, which is only 10 % for flexural strength. This reduction is likely to be due to the embrittlement of polymer matrix, along with formation of surface cracks [63]. Study by Beg et al. also reported that the impact strength of PP reduced dramatically from 10.5 to 3.3 kJ/m^2 after 1000 accelerated weathering [64]. However, composites with higher wood fiber content (50 wt%) display no significant embrittlement after 8 weeks' aging.

Tensile properties Similar to impact strength, the tensile strength decreases significantly in the unfilled PP, while the changes are less pronounced in the filled samples. The elastic modulus changes are limited during the first 150 h of weathering, while it starts to decrease significantly for longer exposure times. The decrease of the deformability is dramatic in the unfilled PP, while is much less significant in the filled composites, which stays almost constant upon the increasing UV exposure [65].

Mechanisms The deterioration of flexural properties could be attributed to the high moisture absorption and UV penetration during the weathering. Weathering results in the degradation of the main chemical compositions of wood fiber, i.e. cellulose, hemicellulose, and lignin [66]. In addition, the exposure of composites to UV radiation leads to the polymer matrix undergo chain scissions, and the shorter chains are easier to rearrange into a crystalline phase by chemicrystallization, decreasing the flexural properties of the composites. The microcracks on weathered surfaces restricted the efficiency of stress transfer from polymer matrix to wood fiber, contributing to poor flexural properties [67, 68].

The reduction in tensile properties is mainly due to the degradation of the wood fibers and the fiber/matrix interfacial bonding. Cellulose determines the tensile properties of the fiber, as well as the composite, because it is the stiffness and the strongest organic constituent in wood [69]. Long exposure time leads to the breaking down of cellulose, as well as the poor interfacial bonding between the polymer matrix and the wood fiber. The degradation of lignin also contributes to the interface degradation, since lignin acts mainly as bonding agents in fibers [70, 71]. Therefore, the poor interfacial bonding in turn results in a lower tensile properties of the composites after weathering.

3.3 Photodegradation Mechanism of Thermoplastics

As mentioned above, long exposure usually leads to the photodegradation of polymers, causing color changes, embrittlement, and loss of mechanical properties. Industrially produced polymers contain various light-absorbing impurities that are introduced during polymer manufacturing, processing, or storage [72]. These impurities contain chromophores, such as hydroperoxide groups, carbonyl groups, and double bonds, which come from pigments, dyes, thermal, and photostabilizers, and metal compounds.

The presence of chromophoric groups is considered to induce the photodegradation of polymers. Photodegradation involves two main reactions, namely, photo-oxidation and thermo-oxidation. UV light and oxygen are the two most important influential factors. The energies of photos in the UV region are much higher than the bond energies in polymers (e.g., C–C and C–H bonds), which is capable of initiating bond scission within the polymer backbone via rupturing and chain reactions through the formation of radical species [73]. In the presence of oxygen, almost all polymers degrade faster than in an inert environment. This type of degradation is called photo-oxidative degradation, which is a radical-based auto-oxidative process [72].

Oxidation can be divided into three stages, as illustrated in Fig. 3.3, namely: initiation step (free radicals are generated); propagation step (reactions are thermal reactions); and termination step (reactions are bimolecular). During the initiation step, RH transfers to the radical precursor R· (1-1) under UV radiation and heat. This initiation step tends to progress to the propagation step due to the reaction of R· with oxygen (O_2) to generate a peroxy radical (ROO·) intermediate (1-2). Then, ROO· attack the polymer by reacting with another polymer molecule RH and give birth to hydroperoxide and R· (1-3). The number of these free radicals tend to multiply during the propagation step. Meanwhile, the decomposition of hydroperoxide leads to the formation of an alkoxyl radical (RO·) (1-4). Except for that, hydroperoxide could react with another polymer molecule and generates RO· and R· (1-5), the former one further abstracts a liable hydrogen from another polymer molecule and forms ROH (1-6). Steps from 1-3 to 1-6 increase the number of the free radicals that are generated during weathering. Finally, termination occurs when two free radicals combine with one another (1-7 to 1-9) [72] (Fig. 11).

These irreversible processes affect the overall physical and mechanical properties of the polymer, such as melt flow, molecular weight, viscosity, and mechanical strength, making it unsuitable for further applications.

3.3.1 Photodegradation Mechanism of Polyethylene (PE)

The photodegradation of PE has been extensively examined [58, 74, 75]. It has been postulated that carbonyl groups are the main light-absorbing species responsible for the photochemical-induced degradation reactions for UV-exposed PE [76].

Fig. 11 Sequence of oxidation reaction in polymers

Initiation:

$$RH \xrightarrow[\text{Heat}]{\text{UV radiation}} R\cdot + H\cdot \quad (1\text{-}1)$$

$$R\cdot + O_2 \longrightarrow ROO\cdot \quad (1\text{-}2)$$

Propagation:

$$ROO\cdot + RH \longrightarrow ROOH + R\cdot \quad (1\text{-}3)$$

$$ROOH \longrightarrow RO\cdot + OH\cdot \quad (1\text{-}4)$$

$$ROOH + RH \longrightarrow RO\cdot + R\cdot + HOH \quad (1\text{-}5)$$

$$RH + RO\cdot \longrightarrow ROH + R\cdot \quad (1\text{-}6)$$

Termination:

$$2\,ROO\cdot \longrightarrow \text{Inert products} \quad (1\text{-}7)$$

$$ROO\cdot + R\cdot \longrightarrow ROOR \quad (1\text{-}8)$$

$$R\cdot + R\cdot \longrightarrow R\text{-}R \quad (1\text{-}9)$$

The degradation reactions proceed from carbonyl group precursors according to Norrish type I and II reactions [77]. If degradation of the proceeds according to a Norrish I reaction, the resulting free radicals can attack the polyolefin Fig. 12, which then leads to termination by chain scission or crosslinking. As for the Norrish II reaction, degradation results into the formation of carbonyl groups and terminal vinyl groups (Fig. 12, and chain scission occurs. Ketones, carboxylic acids, and vinyl groups are the three major functional groups that accumulate with the photodegradation of PE. The two reactions, chain scission and crosslinking are competing during photodegradation. In natural weathering, chain scission is more predominant [78, 79].

PE is a kind of semi-crystalline polymer, in which the crystalline phase is much tighter than the amorphous phase. Therefore, the diffusion of oxygen in crystalline phase is limited and the degradation process mainly happens in the amorphous region [77]. In general, chain scission occurs in amorphous regions, whereas crosslinking occurs in imperfect crystalline phase. Chain scission leads to the generation of shorter chains, which are more mobile and are able to crystallize more readily. This recrystallization process increases the degree of crystallinity of PE [80]. However, crosslinking does not affect polymer crystallinity. The crystalline component of PE has a higher modulus of elasticity (MOE) than the amorphous component [81]. Therefore, the MOE of PE usually increases after weathering. Fabiyi et al. [82] reported the crystallinity of HDPE in WPCs increased from 40 to 53 and 48 after accelerated and natural wreathing, respectively. Chain scission lowers the molecular weight, while crosslinking raises the molecular weight by

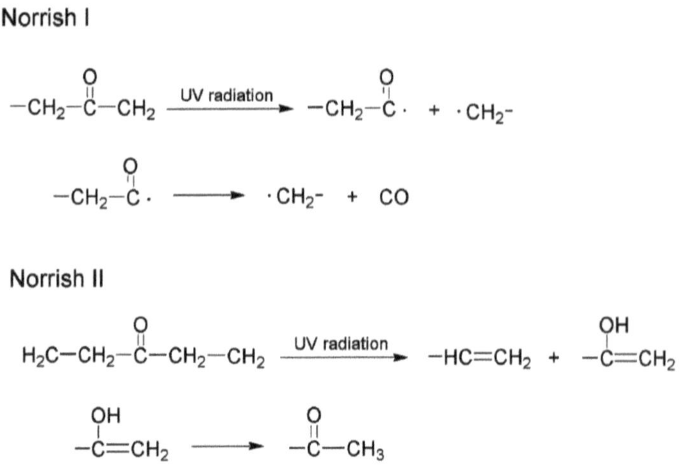

Fig. 12 Norrish degradation mechanisms from carbonyl groups

increasing the bonding between polymer chains. Except for these microstructure changes, photodegradation also leads to undesirable surface cracks, which origins from a development of surface stresses during weathering. The accumulation of cracks causes embrittlement and thus negatively influences the mechanical properties of the polyolefins.

3.3.2 Photodegradation Mechanism of Polypropylene (PP)

The photodegradation process of PP is similar to that of PE. The main degraded products of PP are carbonyl compounds and hydroperoxides [72].

Under UV radiation without oxygen, the degradation of PP involves chain scission and crosslinking:

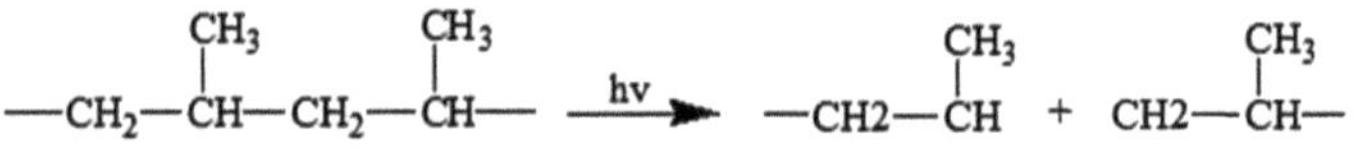

(2-1)

Another possible reaction is the dissociation of C–H on tertiary carbon atom:

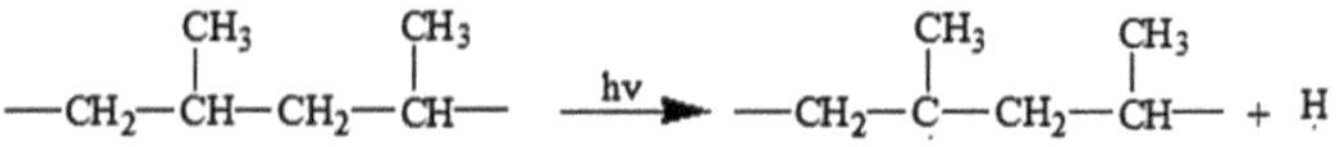

(2-2)

The disproportionation of formed free radicals lead to chain scission:

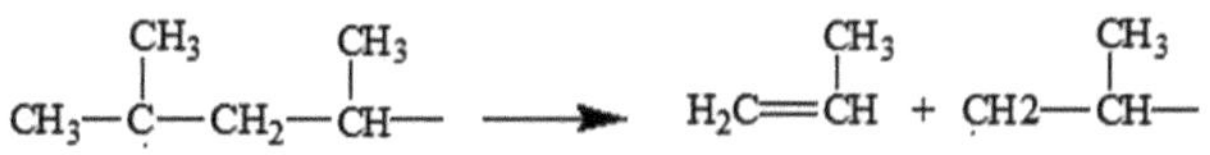

(2-3)

Two free radicals combine and form crossking:

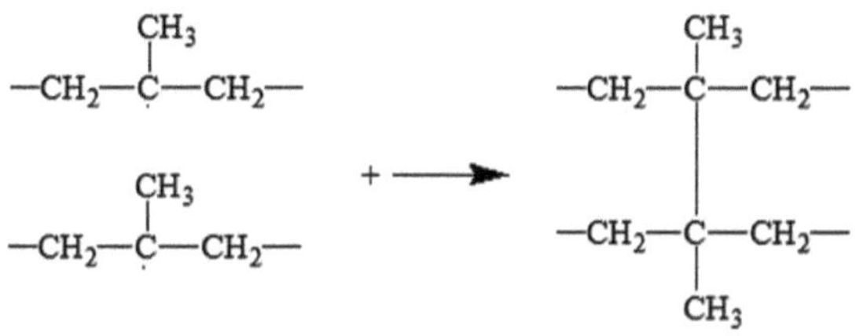

(2-4)

In the presence of oxygen, PP undergoes rapid photo-oxidation, which is also a radical-based chain reaction. Firstly, PP is oxidized and produces hydroperoxides.

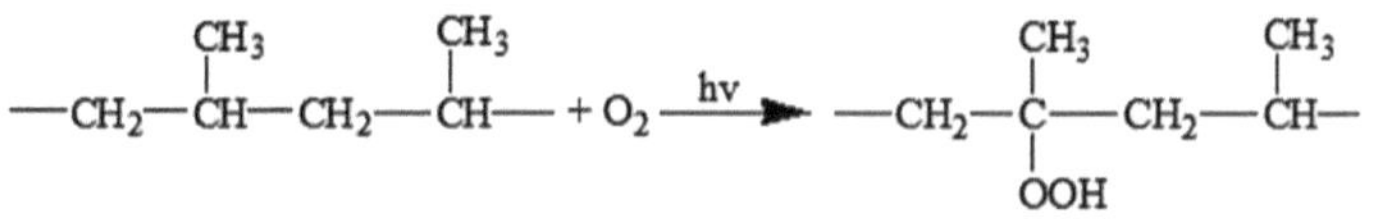

(2-5)

Under UV radiation, the formed hydroperoxides decompose into alkoxyl radicals and hydroxyl radicals:

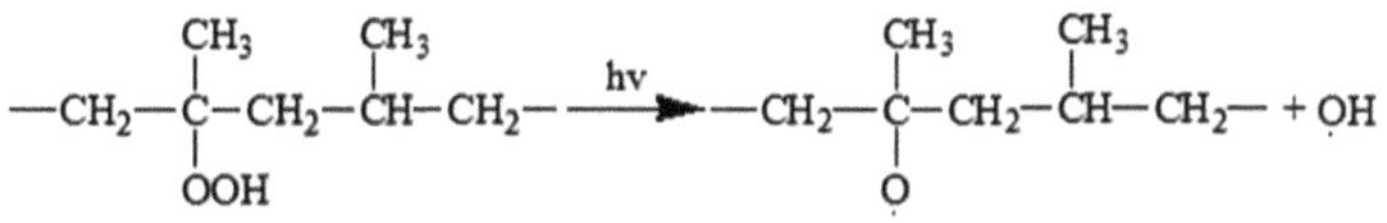

(2-6)

Hydroperoxides undergo β cleavage reaction (2-7) or produce ketones (2-8):

(2-7)

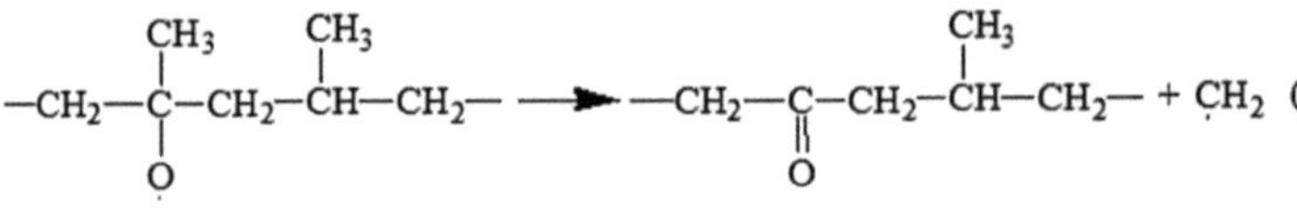

(2-8)

During the photodegradation of PP, β cleavage reaction is considered as the main reason for the backbone fracture. Lone exposure of PP causes embrittlement, stiffening, and cracking.

3.3.3 Photodegradation Mechanism of Polyvinyl Chloride (PVC)

The structure of PVC consists of three kinds of chemical boned, C–H, C–C, and C–Cl. Different from PE and PP, the photodegradation of PVC contains three main mechanisms: chain scission, crosslinking, and dehydrochlorination [83]. The chain scission and crosslinking reactions of PVC are also proceed according to radical-based chain reaction, which are expected to have deleterious effects on the mechanical properties. In addition, photodegradation of PVC results in rapid color change due to the formation of alkene or polyene linkages, as well as the dehydrochlorination of the allylic chlorides in the polymeric matrix. Simultaneously, large amounts of HCl gas evolve by elimination [84]. This behavior may be represented as follows (Fig. 13).

PVC exposed outdoors for a lone time may present either whitening, or yellowing, loss of gloss, as well as a decrease in the mechanical properties.

3.4 *Photodegradation Mechanism of Wood*

The photodegradation in wood is a more complicated process. This process leads to changes on wood surface such as discoloration, roughening, cracking, loss of gloss, and deterioration of mechanical and physical properties. The research on the photodegradation of wood began some decades ago. There are a lot of environmental parameters that contribute significantly to that process [85], including humidity (rainfall, snow and dew), temperature, solar radiation, atmospheric ozone content, and pollution. Among them, UV radiation is the most damaging component. In order to have a precise understand of wood weathering, UV radiation and temperature are usually controlled during investigations, which typically consist of

Fig. 13 The dehydrochlorination process of PVC under UV radiation [83]

$$[\text{—CH}_2\text{—CHCl—}]_n \xrightarrow{h\nu} [\text{—CH=CH—}]_n + \text{HCl} \quad \textbf{(dehydrochlorination)}$$

two steps: first exposure of wood surface to UV light and then detecting the changes.

All wood chemical components (lignin, cellulose, hemicellulose, and extractives), which contain internal chemical groups such as carboxyl, carbonyl, aldehyde, unsaturated double bonds, phenolic hydroxyl, and external entities such as fats, waxes, and metal ions, are capable of absorbing sunlight or UV light. The absorbed energy can cause the dissociation of bonds in the molecules of the wood constituents. Similar to thermoplastics, the photodegradation of wood also involves a serials of radical-based reactions.

3.4.1 Depth of Photodegradation

As for the depth of penetration, researchers have drawn different conclusions. In 1978, Hon and Ifju reported the UV radiation penetrated only 75 μm from wood surface, whereas visible light penetrated 200 μm. Other researchers believe that the limit to the depth of degradation was 900 μm [86]. Park et al. [87] studied the weathering of hinoki(*Chamecyparis obtuse* Endl.) and found degradation at depths of 750–850 μm. Kataoka et al. [88] examined cross-sections of sugi (*Cryptomeria japonica*) using a micro FTIR technique after xenon arc exposure. They reported that the changes in the carbonyl and aromatic absorptions could be measured at depths of 600–700 μm.

The differences in the reported depth of degradation are related with some factors that affecting the penetration of UV radiation into wood, for example, the wood density and the wavelength distribution of the used light. Wood with higher density is penetrated less by UV radiation or visible light, while longer wavelengths can penetrate more. It is also possible that the depth of photodegradation might occur deeper into the wood than the depth at which the free radicals were generated. That is because that these free radicals could be carried deeper into the wood by the action of water [89, 90].

3.4.2 Photodegradation Mechanism of Lignin

Lignin contains numerous chromophores that are very sensitive to UV light [91] (Fig. 14). Lignin accounts for 80–95 % of the total amount of UV light absorbed by wood. The potential chromophoric groups in lignin include phenolic hydroxyl groups, double bonds, carbonyl groups, quinones, quinonemethides, biphenyls and so on.

The partial structure model and main linkages of lignin is given in Fig. 15. Lignin contains many kinds of reactive groups including ethers of various types, primary and secondary hydroxyl groups, carbonyl groups, and carboxyl groups. There also exist number of aromatic and phenolic sites, as well as activated locations, which are capable of initiating free-radical chain reactions under radiation. The two chromophoric groups labelled as A and B in Fig. 15 are very sensitive to

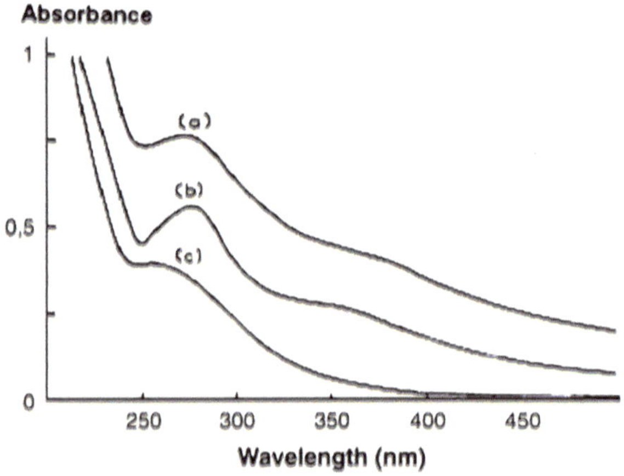

Fig. 14 Reflection spectra of wood (**a**), lignin (**b**) and cellulose (**c**) [93]

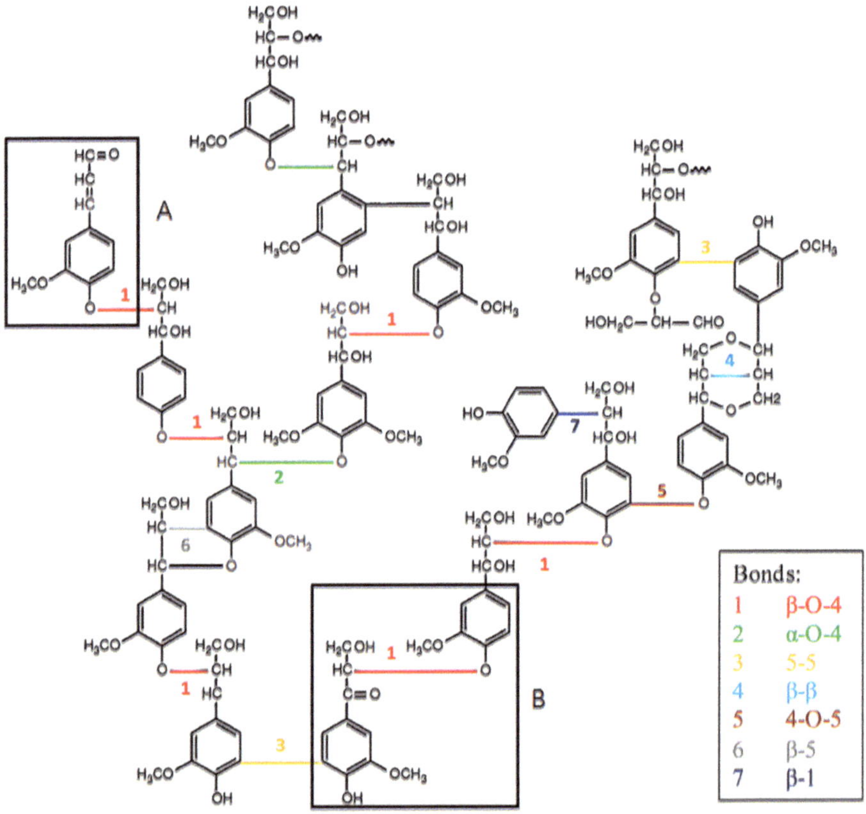

Fig. 15 Model and main linkages of softwood lignin [94]

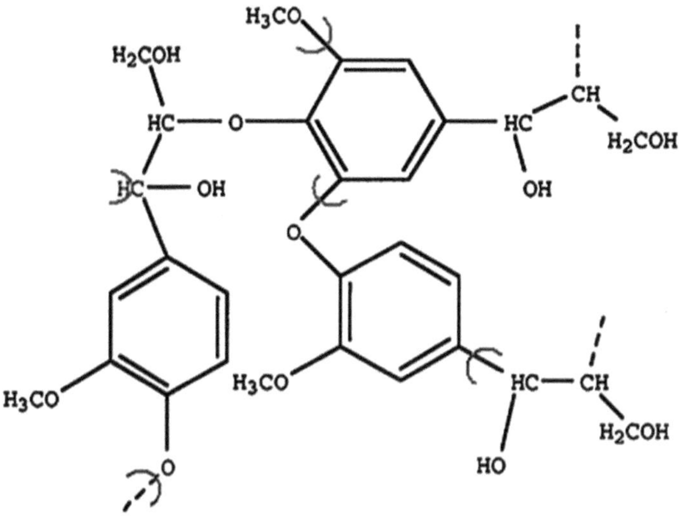

Fig. 16 Possible cleavage sites for lignin under UV radiation [95]

UV light. The possible cleavage sites in lignin under UV radiation are shown in Fig. 16, especially β–O–4 linkage [92].

The UV degradation process is initiated by the formation of free radicals and presumably begins with oxidation of phenolic hydroxyl. Figure 17 shows the three different mechanisms of free radical formation from lignin degradation [96, 97]. The degradation process results in a decrease in methoxyl and lignin contents. It also results in the formation free radicals, such as cetyl radicals (A), phenoxy radicals (B), peroxy radicals (C), and phenacyl radicals (D). Among them, phenoxy radicals are the major intermediate formed radicals, due to larger amount of the phenolic hydroxyl groups and ether bonds in lignin. These free radicals are not stable and further take part in the reactions with oxygen to produce *ortho*- and *para*-quinone structures by inducing demethylation of the guaiacyl unit in lignin [98, 99].

The formation pathways of quinone structures are illustrated in Fig. 18. The degradation starts with the oxidation of phenolic hydroxyls [91, 100]. Under UV radiation, the hydrogen abstraction on the phenolic hydroxyl generates a phenoxy radical, which can be further oxidized and triggers the phenoxylquinone redox cycle. Both the structures are capable of being excited by light or reacting with oxygen to induce demethylation and ultimately yields *ortho*- and *para*-quinoid-chromophores [101].

The degradation process is accompanied by the decrease in the methoxyl and lignin contents, and the formation of chromophoric groups such as carboxyls, carbonyls, quinones, peroxides, hedroperoxides, and conjugated double bonds, which are related to the discoloration of wood. During weathering, water mechanically abrades the wood surface and removes degradation products. As a result, the loss of lignin also makes the surface more hydrophilic. Study by Kalinins

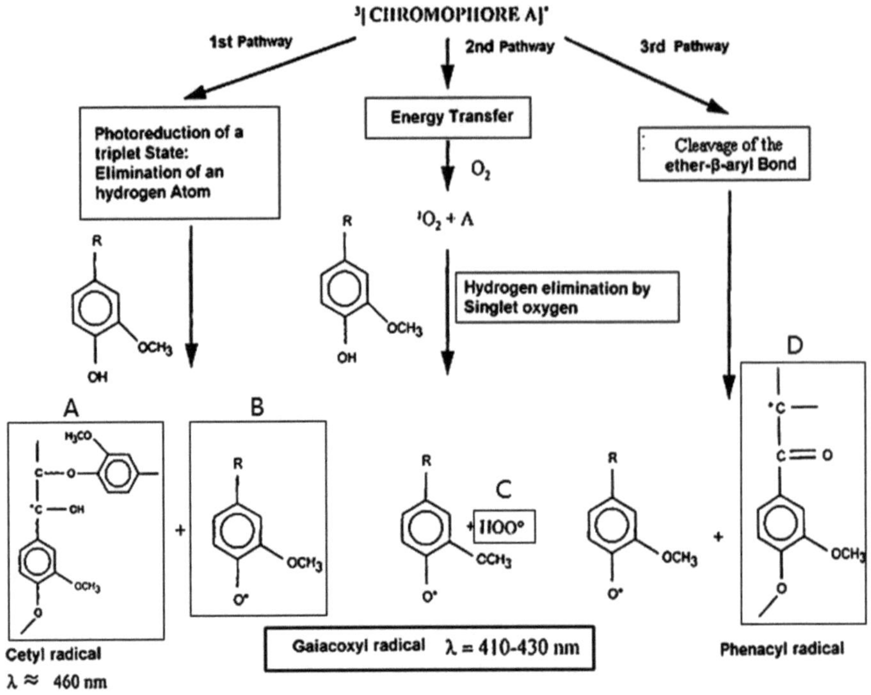

Fig. 17 Mechanism of free radical formation from lignin degradation [93]

et al. [102] reported that the contact angle of weathered western red cedar dropped dramatically from 77° to 55° after 4 weeks of natural weathering [103].

3.4.3 Photodegradation Mechanism of Holocellulose

Contrary to lignin, pure cellulose in not a good light absorber. Cellulose absorbs light strongly below 200 nm, with some absorption between 200 and 300 nm and some absorption extending to 400 nm [104]. The possible chain scission and dehydrogenation sites of the cellulose are shown in Fig. 19, marked by grey lines [95].

When subjected to sunlight, the glycosidic linkages of cellulose are cleaved, causing a loss of strength and degree of polymerization. When exposed to light with wavelengths longer than 340 nm, alkoxy and carbon radicals form due to the cleavage of glycosidic bonds. Dehydrogenation takes place (preferentially at the C-l and C-5 positions) when the radiation wavelengths longer than 280 nm. Photodegradation of cellulose results in the formation of both aldehyde and keto group on C-2 and C-3 positions, accompanied by the volatile degradation products, including acetaldehyde, propionaldehyde, methyl formate, acetone, methanol, ethanol, methane and ethane [95, 105].

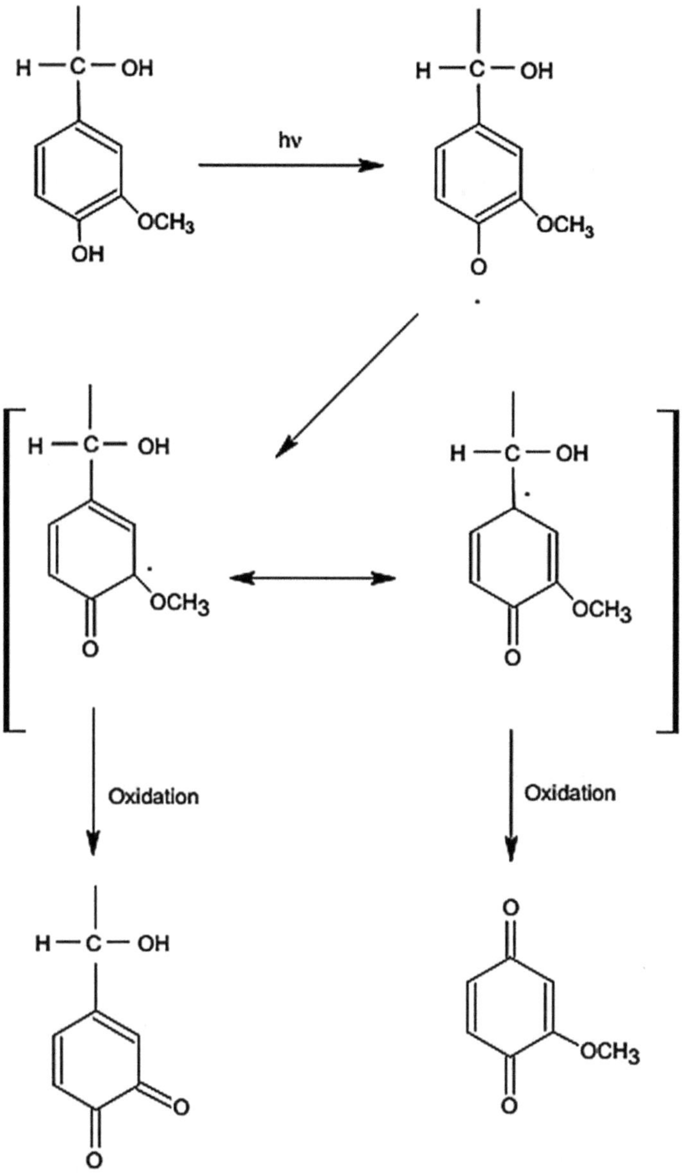

Fig. 18 Formation of *ortho-* and *para-*quinoidchromophores [100]

Cellulose is more durable than lignin and becomes the main composition on wood surface after weathering, which contributes to the hydroxyl concentration on the surface. When water contacts the weathered wood, a larger amount of interactions between the hydroxyl groups and water would undoubtedly occur. Therefore, weathering definitely increase the wettability of the wood.

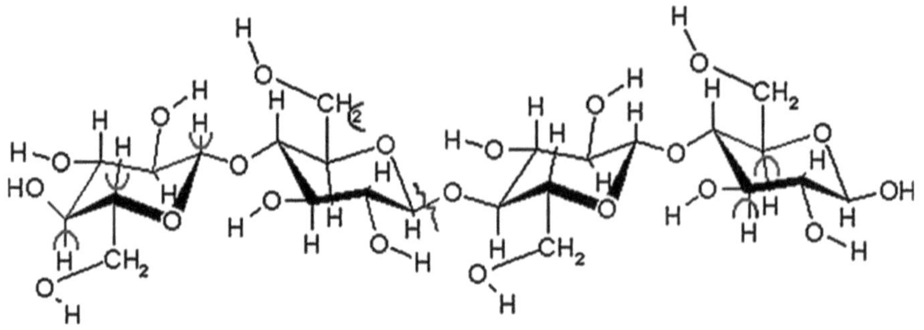

Fig. 19 Possible chain scission by glicosidic linkages and dehydrogenation sites for cellulose under UV radiation [95]

Because of structural similarity, the photodegradation mechanisms of hemicellulose resemble those of cellulose. However, oxidation of hemicellulose seems to occur more readily. The oxidation of hemicellulose creates aldehyde groups on the C-2 and C-3 sites in the anhydroglucose units.

3.4.4 Photodegradation Mechanism of Extractives

Extractives, the minor components of wood, are also susceptible to light. They can be classified into three types according to their chemical structure: polyphenols, terpenoids, and others. Pietta [106] indicated that the phenolics in plant extractives have effective antioxidant activity by catching the reactive oxygen species and stabilizing the free radicals. Wu et al. [107] pointed out that the phenolic antioxidants from the heartwood of *A. confusa* benefited excellent antioxidant and free radical scavenging activity. Chang et al. [108] examined the role of extractives in the photodegradation of wood and reported that the protective ability of extractives is due to the presence of unsaturated phenolics. These phenolic compounds are oxidized primarily after absorbing light energy. Therefore, the lignin decreased less when the wood containing extractives was exposed to the ultraviolet light.

Chang et al. [109] discussed the stabilizing effects of extractives in *A. confuse* wood during weathering and proposed the possible mechanisms, as shown in Fig. 20. These extractives, such as melanoxetin, transilitin, and okanin, could quench the new-formed radicals by hydrogen exchange. Ultimately, these extractives transfer into *ortho*-quinone, *para*-quinoid, dimer or other derivatives.

3.5 Photodegradation Mechanism of WPCs

The photodegradation of WPCs is not the simple combination of wood fiber and polymer degradation, whose individual photodegradation mechanisms are

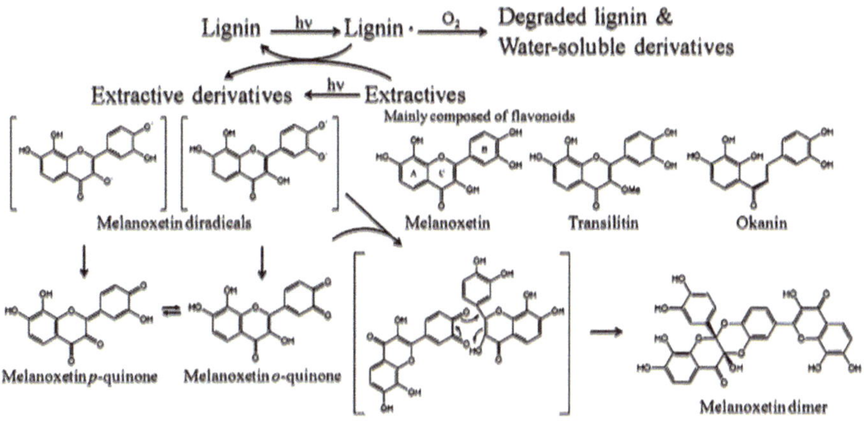

Fig. 20 Mechanism for stabilizing effect of extractives on the photo-oxidation of *A. confusa* wood under UV radiation [109]

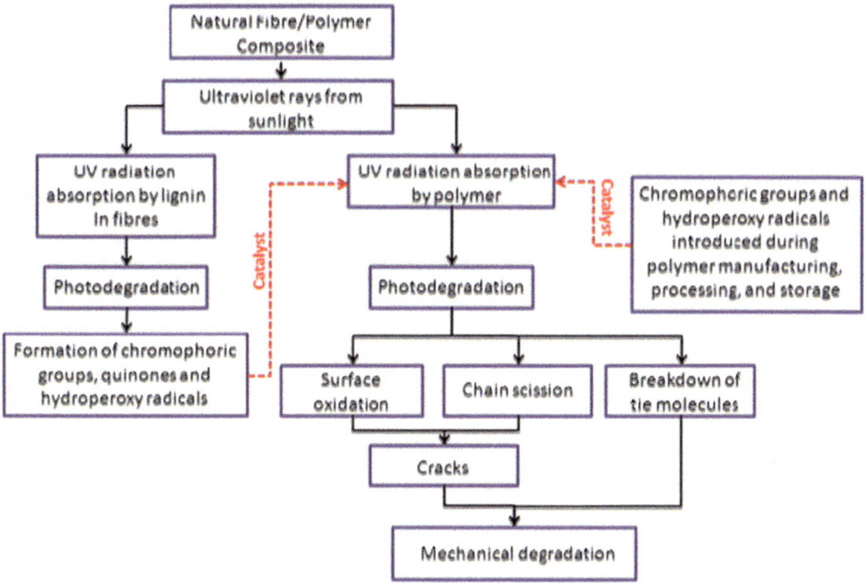

Fig. 21 UV degradation of WPCs and it components [42]

introduced above. The two components also accelerate each other. Figure 21 presents the schematic diagram of the degradation of WPCs due to UV weathering. For wood fibres, degradation begins with UV radiation absorption by lignin, and then reactions of photodegradation happen. The products of degradation are chromophoric groups, quinones, and hydroperoxy radicals, which catalyze the degradation process of polymer in WPCs.

The introduced catalyst residues and chromophoric groups contribute to the UV radiation absorption of polymer. Degradation results in chain scission, crosslinking, and cracks on polymer matrix. Cracks benefit the diffusion of both oxygen and light into the deeper area of WPCs, which play a negative effect on the interface bonding between wood fiber and polymer [66]. Thus, the degradation process leads to a deterioration of the physical and mechanical properties of WPCs.

4 Methods to Retard Photodegradation of WPCs

4.1 Additives

To retard the phtotodegradation process of WPCs, some additives, such as colorants, antioxidants, UV absorbers (UVAs) and hindered amine light stabilizers (HALS), are usually incorporated during composite fabrication. These additives either maintain the color stability or mechanical properties of WPCs against UV weathering.

4.1.1 Colorants

As discoloration is the most obvious change of WPCs during weathering, adding pigments is a common way to improve their color stability. Characteristics of pigments is their extremely low solubility in water or organic solvent and they remain essentially in the solid state during processing. The majority of the pigments added in WPCs are inorganic colored metallic oxide. Stark and Matuana reported that zinc ferrite pigment is more effective in preventing lightening than any other kinds of photostabilizers [110]. Du et al. [111] compared the effects of three different pigments in the WF/HDPE composites, namely ferric oxide, carbon black, and titanium dioxide. The results suggested that the addition of pigments to composites results in less weather-related damage. Four kinds of iron oxide pigments (red, yellow, blue, and black) were chosen to make WF/HDPE composites. The investigation revealed that iron oxide pigments exhibit a very good performance in protecting the composites from discoloration, as well as the mechanical properties, especially for red and black pigments. The color change was found to decrease with the increasing of pigment dosage [112]. Darker color pigments are proved to have a more positive effect on color stability than the other pigments. Butylina et al. investigated three pigments with different color (green, grey, and brown) and concluded the similar results [113]. However, the incorporation of all these inorganic pigments may exert a negative effect on the mechanical properties of the composites [114].

Apart from pigments, dyes are also used to color WPCs by dying wood fiber first. In contrast with pigments, dyes are water soluble and have attraction for fibers [115].

Stark et al. [114] colored the wood flour with either a water-based dye or an oil-based dye and then made composites by injection. Coloring wood flour with an oil-based was more effective for improving composite color stability during 3000 h exposure in a xenon-arc weathering apparatus, which might be due to its hydrophobicity. There are two main staining methods of WPCs, adding pigments and coloring wood flour. Peng et al. [116] compared their effects on composite color stability during accelerated UV weathering. They found that composites containing pigments displayed less color change and less loss of flexural properties after wreathing. Pigments act as photo-blockers at material surface and prevent the penetration of UV radiation. However, their efficacy is short because of the migration and leaching of pigments on weathered surface.

4.1.2 Photostabilizers

Photostabilizers can be classified into two types according to their different action modes: UVAs and HALS.

UVAs UVAs can absorb some UV radiation. Generally, UVAs in current use contain organic and inorganic materials. Inorganic UVAs including TiO_2, ZnO, SiO_2, and Al_2O_3, have been widely used due to their optical and thermal stability as well as nontoxicity. Literature have shown that TiO_2 is more effective on reducing color change and weather-related damage inboth PE and PP-based WPCs [83, 117]. However, as a side-effect, their photocatalytic activity would accelerate the photodegradation process of polymers simultaneously during weathering. The protective action of inorganic UVAs depends on their concentration and particle size.

The most studied organic UVAs are hydroxybenzophenones and hydroxyphenylbenzotriazoles. These compounds are self-regenerating and can return to their original state after UV absorption, thus avoiding photosensitization [73, 77]. The absorption mechanism of hydroxyphenylbenzotriazoles are illustrated in Fig. 22. After absorbing energy, a rapid transformation of internal hydrogen occurs. The absorbed energy is then dissipated as heat, while the UVAs returns to its ground state. Study by Du et al. [118] showed that UV-326 (bumetrizole) provided a more effective protection for rice-hull/HDPE composites during weathering, compared to UV-531 (octabenzone).

HALS HALS is kind of free radical scavengers and offers protection from two mechanisms. In the presence of oxygen, HALS turns into a nitroxyl radical

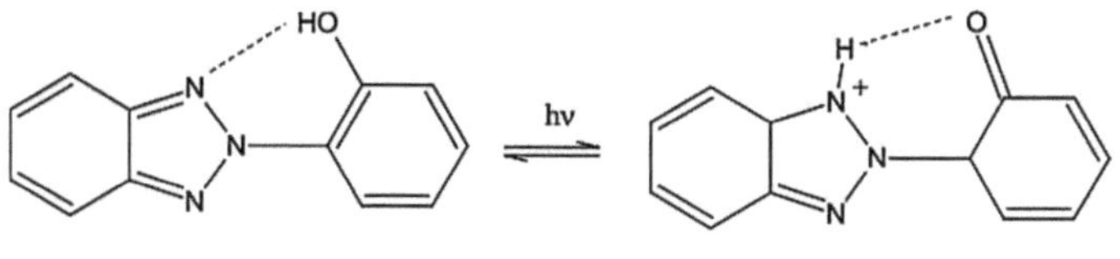

Fig. 22 A kind of mechanisms of UVA photostabilization [48]

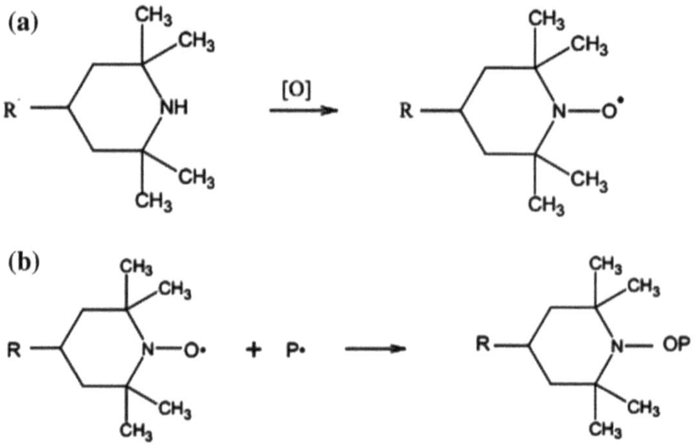

Fig. 23 Protection mechanism of HALS [48]

(Fig. 23a), which is not stable and then react with another polymer radical though a termination reaction (Fig. 23b). In this way, HALS decreases the amount of oxygen available to degrade the polymer chain and stops the chain reaction process [48].

According to Muasher's research, HALS with high molecular weight exhibits greater long term stabilizing ability over that of low and medium molecular weight. That is because that the lower molecular weight HALS is easy to loss by surface evaporation, leading to their short efficacy after longer exposure time [51]. Stark et al. [119] examined the effect of a HALS, a benzotriazole UVA, and a pigment on the changes in lightness and mechanical properties of WPCs during weathering. Results showed that only the UVA and pigment significantly reduced composite lightening. However, the addition of a benzotriazole UVA to HALS shows synergistic effect on decreasing color fading. García et al. [46] mixed HALS with UVAs and antioxidants, and the blend exhibited better than using HALS alone, with no cracking or fracture on the aged surface. Study by Bouza et al. [120] also proved that the combination of UVA and HALS successfully prevented the deterioration of UV radiation on composite surface.

4.1.3 Antioxidants

Antioxidants have been widely used in the manufactory processing of polymers. According to different action types, antioxidants can be divided into primary and auxiliary antioxidants. They usually include hindered phenols antioxidants, phosphorous antioxidants, thioester antioxidants, and compound antioxidants.

Combinations of hindered phenols and phosphorous antioxidants have been widely used as stabilisers, which are reported to have a synergistic effect on improving the durability, as well as physical and mechanical properties in polyolefins [121]. Hindered phenols can terminate oxidation by trapping alkyl peroxy,

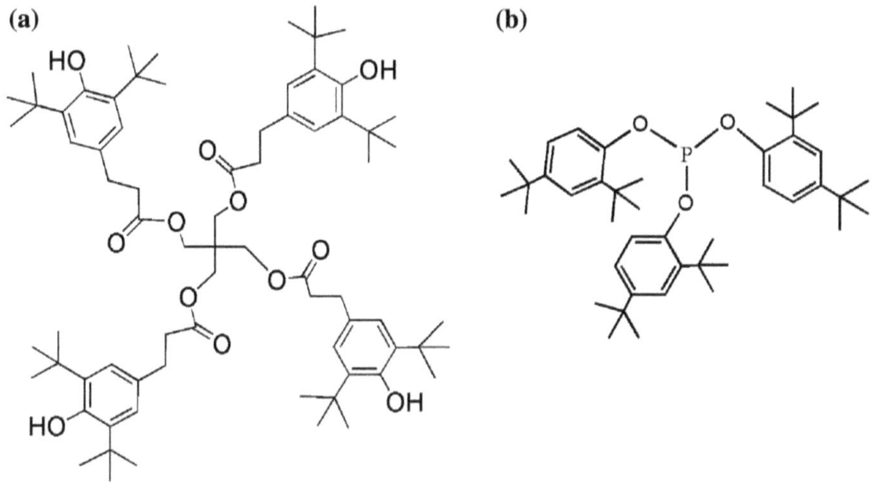

Fig. 24 Chemical structures of **a** AO-1010 and **b** AO-168 [125]

alkoxy, and alkyl radicals [122]. Phosphorous antioxidants inhibit degradation reactions by reacting with hydroperoxide groups, carbon-centred, peroxy, and oxy radicals [123]. Ojeda et al. [124] reported that sterically hindered phenol and phosphite antioxidant additives added to PP slowed its photo-oxidative degradation under natural weathering. The combination of Irganox 1010 and Irganox HP 136 showed a significant effect on the UV stabilisation of PP [121].

The addition of antioxidants can alleviate the photodegradation of both wood fiber and thermoplastics. Peng et al. [125] combined two kinds of antioxidants, namely antioxidant 1010 (AO-1010, hindered phenols) and antioxidant 168 (AO-168, phosphite), and then made WF/PP composites at different compound proportions. The chemical structures of the two antioxidants are given in Fig. 24. The compound systems showed a synergistic effect on maintaining the surface color and flexural properties of the composites, especially at the ratio of 1:5 (AO-1010: AO-168).

The possible antioxidant mechanisms of AO-1010 and AO-168 in WPCs are given in Fig. 25. Under UV radiation, PP and WF undergo rapid degradation (Reactions 7–9). The antioxidant effect of AO-1010 is achieved by donating a hydrogen atom to the free radicals originating from the photodegradation of PP and lignin (Reaction 1 and 2). Therefore, AO-1010 inhibits the chain scission of PP and AO-1010 itself is oxidized to quinones (Reaction 3 and 4), which is responsible for the discoloration of WPCs. Fortunately, as an auxiliary antioxidant, AO-168 can react with quinones and hydroperoxide, and then change into achromous aryl phosphate (Reaction 5 and 6). In this way, the reaction products of AO-1010 are decomposed by AO-168 and the compound system successfully alleviates the photodegradation of the composites.

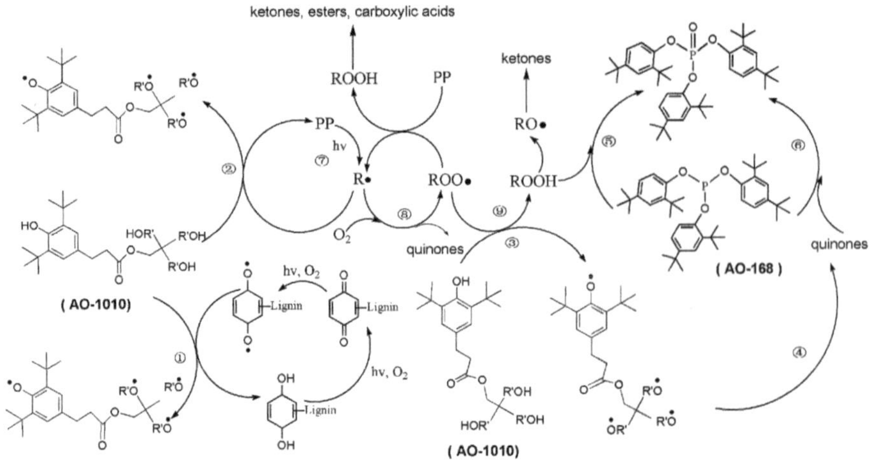

Fig. 25 Possible antioxidant mechanisms of AO-1010 and AO-168 in WPCs [125]

However, apart from composites properties, health safety has vital importance in many application areas. Therefore, some natural antioxidants have been studied and incorporated in WPCs. For example, vitamin E is a kind of natural nontoxic antioxidants. The stabilizing effect of Vitamin E (Ve, α-tocopherol) against oxidative degradation of polyolefins (LDPE, HDPE and PP) has been extensively studied in the past 20 years. Peng et al. studied the anti-weathering effect of Ve on WF/PP composites at different loading levels. The addition of Ve demonstrated excellent efficacy and successfully hindered the chain scission process of PP. Composites containing Ve showed less color fading, less loss of flexur properties, less cracks and disclosure of wood fibres on surface after weathering, especially at the content of 0.4 wt%. It is widely accepted that Vitamin E can stop the degradation process by donating a hydrogen atom to free radicals, including peroxy radicals and alkyl radicals caused by irradiation [126, 127]. Then, the chemical structure of Ve changes from phenol into paraquinone (Fig. 26), which contributes to the discoloration of WPCs as a side effect [128].

In order to eliminate paraquinone, as well as to lower the costs, AO-168 was tested in combination with Ve to increase the weathering resistance of WPCs [129]. AO-168 showed a synergistic effect with Ve to further improve the anti-weathering ability of the composites. The compound system exhibited the best performance at the ratio of 1:5 (Ve:AO-168).

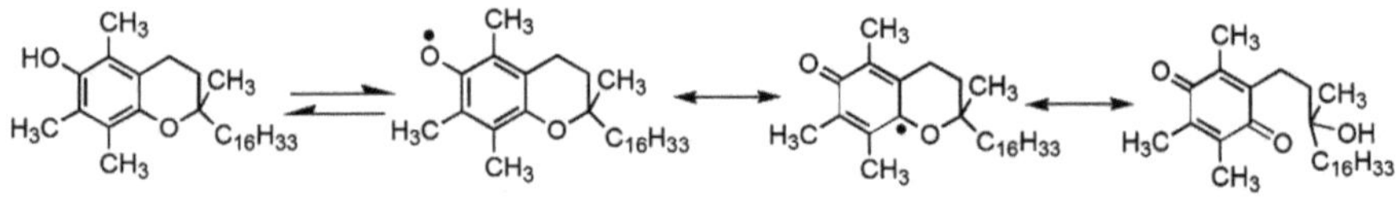

Fig. 26 Chemical structure change of Ve during UV accelerated UV weathering

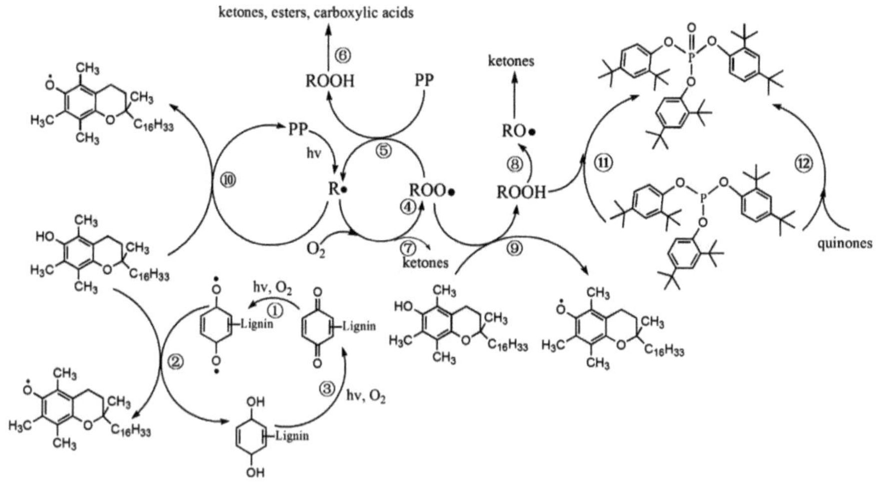

Fig. 27 Reaction pathways and stabilizing mechanisms of VE and AO [126, 129]

Fig. 28 Chemical structure of curcumin [130]

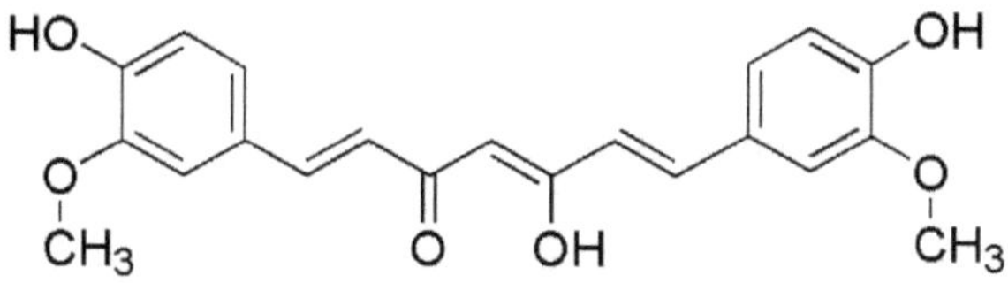

The mechanisms of the compound system are shown in Fig. 27. During UV radiation, free radicals were produced in PP because of radiolytic bond scission, and triggered the oxidation cascade (Fig. 26, Reactions 4–8). The stabilization mechanism of Ve in PP is illustrated in Reactions 9 and 10 by its capacity of stabilizing peroxy radicals and reacting with alkyl radicals. During this process, the chemical structure of Ve changed from phenol into paraquinone and led to the formation of carbonyl groups. Meanwhile, AO as a kind of auxiliary antioxidant could combine with hydroperoxide and quinones to form stable phosphoester (Fig. 26, Reactions 11 and 12), terminating the oxidation chain reaction.

Another kind of natural oxidant, curcumin Fig. 28 is also proved to have stabilizing efficiency. Results from Kirschweng et al. [130] showed that curcumin hinders the oxidation of PE and the formation of long chain branches during processing. Curcumin is not only a hydrogen donor but its unsaturated linear moiety can also scavenge alkyl and oxygen centred macroradicals.

4.1.4 Nanoclays

In recent years, some nanoclays have been widely used to improve the physical and mechanical properties of polymers or WPCs. Meanwhile, a number of studies investigated their effects on materials during weathering.

Layered double hydroxides (LDH) are a well-known type of ionic clays made up of positively charged brucite-like layers with an interlayer containing charge compensating anions and solvation molecules. LDH are available as both naturally occurring minerals and synthetic materials, whose structure is shown in Fig. 29. The general formula of LDH can be represented by $[M^{2+}_{1-x}M^{3+}_{x}(OH)_2]^{x+}A^{n-}_{x/m} \cdot nH_2O$, in which M^{2+} and M^{3+} are divalent and trivalent metallic cations, respectively, x is normally between 0.2 and 0.4, and A^{n-} symbolizes various organic or inorganic anions, n is the number of water molecules located in the interlayer region [131, 132].

LDH-based UV-shielding materials have shown significantly enhanced performance compared with organic UV absorbents. The multi nestification layered structure of LDH imparts the inorganic layer sheets with the physical shield effect against UV light, while the metal elements of layer sheets and negative ions between layer sheets provide LDH with chemical absorbability of UV light [133, 134]. This combined two effects result in excellent UV resistance of LDH in the organic materials. Previous studies have reported that LDH can inhibit the photo-oxidation of asphalt during weathering, especially at high loading levels [135]. Study by Peng et al. [136] reported that $Mg_3Al–NO_3$–LDH not only enhanced the flexural and thermal stability of WF/PP composites, but also showed a dramatic influence on slowing down their photo-oxidation process. Compared to other photostabilizers, the effect of LDH is much longer and more effective. That is could be attributed to the better anti-leaching properties of LDH. Moreover, because of the exchangeability of their interlayer anions, LDH can be modified to impart new functions by intercalation of appropriate guest anions into the galleries between the host layers. For

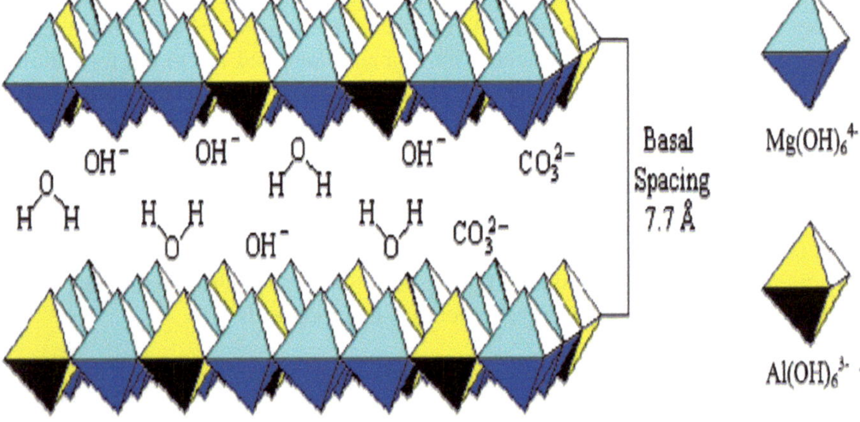

Fig. 29 Structure of LDH

example, some researchers have developed several organic UVAs-intercalated LDH systems, which significantly enhance the photo-stability of polymers during UV weathering. The UV blocking properties of the UVAs-intercalated LDH materials resulted from both shielding by the LDH layers and UV light absorption by the interlayer organic species.

Similar to LDH, montmorillonite (MMT) is also a typical and most used natural clay, which belongs to the genera family of 2:1 layered silicates [137]. According to the different types of cations, MMT could be divided into inorganic and organic MMT, such as sodium MMT (Na-MMT) and organically modified MMT (OMMT), respectively. At the early stage of weathering, both the MMT can alleviate the discoloration and surface cracking of materials during weathering, because of its physical shield effect. However, with the increasing of exposure time, MMT migrates to the surface through micro cracks [138]. The decomposition of the interlayer ions in MMT may create acidic sites on the layered silicates, which then result in some active sites. The formation of these active sites accelerate the radical-based reactions [139]. Therefore, MMT poses a catalytic effect in the degradation of the polymer matrix at the late stage of weathering. Compared to OMMT, Na-MMT is much effective in maintaining the color stability of materials. However, the loading level of Na-MMT in materials should be maintained at an appropriate content, considering its hydrophilicity [140].

4.2 Coating and Chemical Modification

4.2.1 Surface Painting

In order to obtain a long-term protection of wood, the most simple and economic way is surface painting by using various decorative and protective finishes. As we discussed above, moisture and UV radiation are the two main parameters during wood weathering. As a primary function, coating can prevent the penetration of moisture from surface. Moreover, the blocking of UV radiation can be achieved by the addition of UVAs or pigments to a coating, such as TiO_2, SiO_2, CeO_2, and HALS. These additives in coating prohibit UV radiation from reaching the materials surface by absorbing, reflecting, or scattering effects. The efficacy of different coatings on wood substrates has been widely studied, while little such work has been done on WPCs.

Pattamasattayasonthi et al. [141] choose a kind of acrylic coating which contained CeO_2, and used it as coating layer for WF/PVC composite samples. After 504 h UV exposure, the mechanical (tensile and flexural) properties and color stability of composites demonstrated a significant decrease, accompanied by an increase of the hydrophilicity on composites surface. However, composites with a coating layer containing CeO_2 showed less changes on hydrophilicity and mechanical properties, due to the strong UV absorption of CeO_2. The absorbed energy is dissipated as light and heat, which also have a bactericidal effect. The efficiency of the added particles to a coating depends much on their particle size and nano-scaled particles usually perform better.

4.2.2 Chemical Modification

Wood constituents (lignin, cellulose, and hemicellulose), especially some groups on cell wall take an active part in the photodegradation process during weathering. Thus, the chemical modification of such groups, mainly OH-substitution, seems to the best way to improve some properties of wood, such as water-repellency, dimensional stability, and UV stability. This reaction usually corresponds to the acetylation of lignin, in which the hydroxyl groups in the aliphatic and the aromatic part are substituted by acetyl groups (Fig. 30). In this way, the reactivity of lignin under UV radiation is largely weakened and the formation of phenoxy radicals can also be prevented. Study by Rosu et al. [142] showed that the succinic anhydride-modified (SA) wood is more stable to the artificial light than the non-modified wood. The photochemical stability of the modified wood is found to be highly related with the concentration of SA solution used in modification.

Huang et al. [143] investigated the natural weathering properties of bamboo plastic composites (BPCs) reinforced with bamboo fibers, which was first acetylated by acetic anhydride. The results indicated that the acetylation of bamboo fibers enhanced the durability and decay resistance of BPCs. The acetylated BPCs also exhibited higher MOR and MOE retention ratios than the unmodified one during the 1080 days exposure.

To the best of our knowledge, the researches about coating and chemical modification of WPCs are much fewer than those involving in additives. There are two main reasons for this: (1) the process of coating or chemical modification is relatively complicated; (2) the performance of such methods is usually inferior to additives.

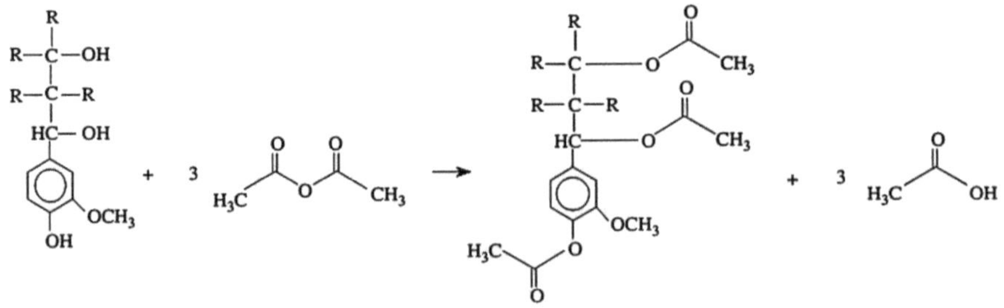

Fig. 30 Mechanism of lignin acetylation [93]

4.3 Changing Wood Fiber Composition

Due to the different photodegradation mechanisms of lignin, cellulose, hemicellulose, and extractives in wood, their behaviors and effects on WPCs during weathering are also various. Fabiyi et al. [55] prepared extractive-free fibers and holocellulose from pine by extraction and delignification, and made HDPE-based composites. It was found that composites containing holocellulose fibers showed the best color stability during weathering. However, composites filled with control and extractive-free fibers showed more obvious total color change, as well as higher oxidation degree, due to the sensitivity of lignin to UV light. Therefore, holocellulose-based composites may be suitable for applications where color stability is of high priority.

Peng et al. [144, 145] produced several polypropylene composites reinforced with WF, lignin, and cellulose at different loading levels, according to the lignin content in different plants. The formulations are listed in Table 2.

The results indicated that the discoloration of composites was accelerated by the presence of lignin, especially at higher lignin content. Interestedly, after 960 h UV exposure, composites containing lignin showed less loss of flexural strength and modulus (Fig. 31), less cracks (Fig. 33), and better hydrophobicity on weathered surface than other groups, which confirms its functions of stabilization and antioxidation. Similar to previous studies, cellulose-based composites exhibited better color stability during weathering (Fig. 33). However, they underwent significant deterioration in flexural properties after weathering, with the final MOR and MOE retention ratios of only 36.0 and 27.0 %, respectively. Therefore, cellulose-based composites could not be used as load-bearing structure in outdoor applications (Fig. 32).

Based on the above results, it may be unfair to blame all the weather-related damages of WPCs on lignin. The above evidence has proved that lignin could act as antioxidants due to its structure is similar to hindered phenols. It is beneficial to

Table 2 Compositions and labeling of the studied formulations [144]

Labels	WF (wt%)	Cellulose (wt%)	Lignin (wt%)	PP (wt%)
C (control)	40	0	0	60
PP	0	0	0	100
CL	0	40	0	60
L1	0	30	10	60
L2	0	20	20	60
L3	0	10	30	60

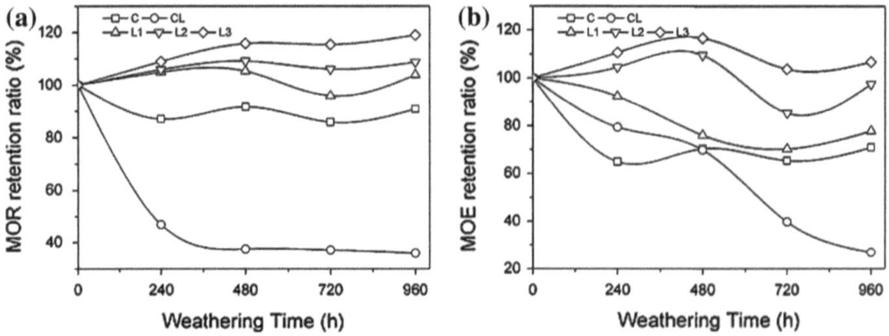

Fig. 31 Change in **a** MOR retention ratio and **b** MOE retention ratio of samples as a function of weathering time [145]

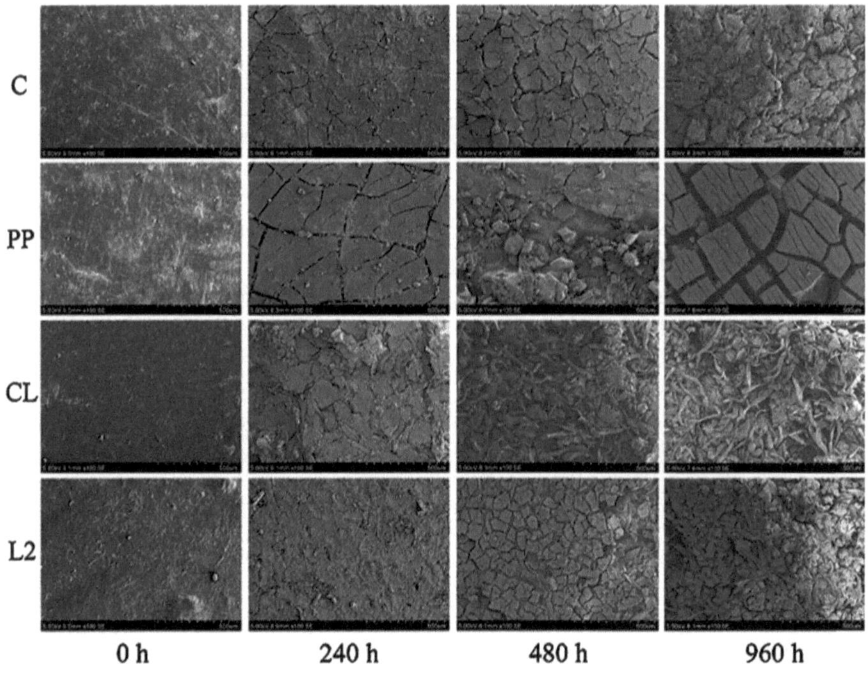

Fig. 32 SEM image of the surface layer of different types of composites as a function of weathering time [144]

maintain the mechanical properties of WPCs during weathering. In addition, studies by Gregorová et al. [146] also reported that the antioxidant efficiency of lignin in the PP matrix increased with its loading level, especially when combined with commercial antioxidants (Fig. 33).

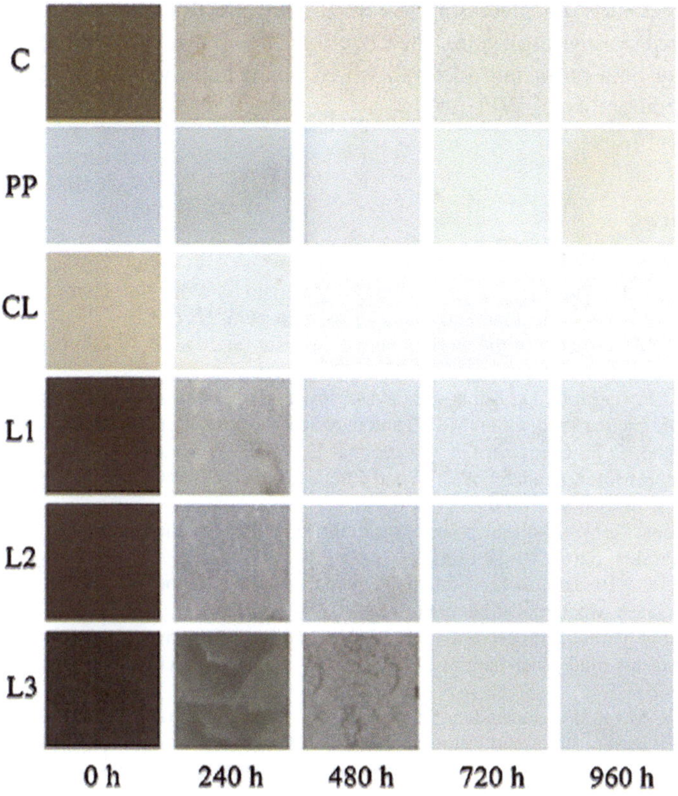

Fig. 33 Visual photos of all groups of composites as a function of weathering time [144]

5 Summary

Wood-plastic composites (WPCs) are a kind of environmental-friendly and promising materials, which are widely used in both indoor and outdoor applications. The photodegradation of WPCs used outdoors is a natural phenomenon. Weathering tests are necessary to estimatethe service life of composites and to find solutions to avoid or slow down this deterioration. The photodegradation process of WPCs involves a series of complex chemical reactions, based on the degradation of both wood fiber/flour and plastics. After long-term artificial UV light irradiation or solar irradiation, photodegradation of WPCs leads ultimately to changes in their aesthetic, physical, chemical, and mechanical properties.

In order to extend the service life of WPCs, various additives have been added to stabilize them against the damaging effects of UV light. However, the majority of these additives show a negative effect on the original mechanical properties of

WPCs. In addition, it is necessary to compare the effect of these additives on a more standardized weathering cycle. The development of new organic-inorganic hybrid UVAs may be a promising method, which could improve both mechanical properties and durability of WPCs.

References

1. Gordan, J.E.: The New Science of Strong Materials (or Why You Don't Fall Through the Floor), 2nd edn. Princeton University Press, Princeton, NJ (1988)
2. Zhu, Y.M.: Progress of the world's plastics industry in 2013–2014. China Plast. Ind. **43**(3), 1–40 (2015)
3. Gug, J.I., Cacciola, D., Sobkowicz, M.J.: Processing and properties of a solid energy fuel from municipal solid waste (MSW) and recycled plastics. Waste Manag. **35**, 283–292 (2015)
4. Caulfield, D.F., Clemons, C., Jacobson, R.E., et al.: Wood thermoplastic composites. In: Roger, R.M. (ed.) Handbook of Wood Chemistry and Wood Composites, p. 365. Taylor & Francis, Boca Raton (2005)
5. Clemons, C.: Wood-plastic composites in the United States: the interfacing of two industries. For. Prod. J. **52**(6), 10–18 (2002)
6. Faruk, O., Bledzki, A.K., Fink, H.P., et al.: Progress report on natural fiber reinforced composites. Macromol. Mater. Eng. **299**(1), 9–26 (2014)
7. Principia Partner, Proceedings of 7th International Conference on Wood Fiber-Plastic Composites (and other natural fibers), USDA Forest Products Laboratory, Madison, Wis, USA (2003)
8. Reddy, M.M., Vivekanandhan, S., Misra, M., et al.: Biobased plastics and bionanocomposites: current status and future opportunities. Prog. Polym. Sci. **38**(10), 1653–1689 (2013)
9. Ashori, A.: Wood-plastic composites as promising green-composites for automotive industries! Bioresour. Technol. **99**(11), 4661–4667 (2008)
10. Jiang, H., Kamdem, D.P.: Development of poly (vinyl chloride)/wood composites. A literature review. J. Vinyl Additive Technol. **10**(2), 59–69 (2004)
11. Clemons, C.: Proceedings of 3rd International Wood and Natural Fiber Composites Symposium, Kassel, Germany (2000)
12. Bismarck, A., Baltazar-Y-Jimenez, A., Sarlkakis, K.: Green composites as Panacea? Socio-economic aspects of green materials. Environ. Dev. Sustain. **8**(3), 445–463 (2006)
13. Chen, Y., Sun, L., Chiparus, O., Negulescu, I., Yachmenev, V., Warnock, M.: Kenaf/ramie composite for automotive headliner. J. Polym. Environ. **13**(2), 107–114 (2006)
14. Karus, M.S., Ortmann, O.D., Vogt, G.: Use of natural fibres in composites in the German automotive production 1996 till 2003. Nova-Institut GmbH (2004). www.nova-institut.de/NR
15. Rowell, R.M., Pettersen, R., Han, J.S., et al. Cell wall chemistry. In: Roger, R.M. (ed.) Handbook of Wood Chemistry and Wood Composites, pp. 35–74. Taylor & Francis, Boca Raton (2005)
16. Dittenber, D.B., GangaRao, H.V.S.: Critical review of recent publications on use of natural composites in infrastructure. Compos. A Appl. Sci. Manuf. **43**, 1419–1429 (2012)
17. John, M., Thomas, S.: Biofibres and biocomposites. CarbohydrPolym **71**, 343–364 (2008)
18. Saka, S.: Chemical composition and distribution. In: Hon, D.N.S., Shiraishi, N. (eds.) Wood and Cellulosic Chemistry. 2nd edn, pp. 51–82. Marcel Dekker, Inc., New York (1991)
19. Fujita, M., Harada, H.: Ultrastructure and formation of wood cell wall. In: Hon D.N.S., Shiraishi, N. (eds.) Wood and Cellulosic Chemistry, 2nd edn, pp 1–50. Marcel Dekker, Inc., New York (2001)

20. Fengel, D., Wegener, G.: Wood: Chemistry, Ultrastructure. Reactions, p. 613. Walter de Gruyter Berlin, New York (1989)
21. Thakur, V.K., Thakur, M.K., Gupta, R.K.: Review: raw natural fiber-based polymer composites. Int. J. Polym. Anal. Charact. **19**(3), 256–271 (2014)
22. Baeza, J., Freer, J.: Chemical characterization of wood and its components. In: Hon, D.N.S., Shiraishi, N. (eds.) Wood and Cellulosic Chemistry, 2nd edn, pp. 275–384. Marcel Dekker, Inc., New York (2001)
23. Chanda, M., Roy, S.K.: Plastics Technology Handbook, 3rd edn, p. 1195. Published by Marcel Dekker Inc., New York (1998)
24. Klyosov, A.A.: Wood-plastic composites. Wiley, Hoboken (2007)
25. Lu, J.Z., Qinglin, W., McNabb Jr., H.S.: Chemical coupling in wood fiber and polymer composites: a review of coupling agents and treatments. Wood Fiber Sci. **32**(1), 88–104 (2000)
26. Simonsen, J., Jacobsen, R., Rowell, R.: Wood-fiber reinforcement of styrene-maleic anhydride copolymers. J. Appl. Polym. Sci. **68**(10), 1567–1573 (1998)
27. Gassan, J., Bledzki, A.K.: The influence of fiber-surface treatment on the mechanical properties of jute-polypropylene composites. Compos. A Appl. Sci. Manuf. **28**(12), 1001–1005 (1997)
28. Gassan, J., Bledzki, A.K.: Dynamic-mechanical properties of natural fiber-reinforced plastics: the effect of coupling agents. In: Proceedings of the 4th International Conference on Wood Fiber-Plastic Composites, Madison, WI, 12–14 May, pp. 76–80 (1997)
29. Berdahl, P., Akbari, H., Levinson, R., et al.: Weathering of roofing materials—an overview. Constr. Build. Mater. **22**(4), 423–433 (2008)
30. Tocháček, J., Sedlář, J.: Effect of hydrolysability and structural features of phosphites on processing stability of isotactic polypropylene. Polym. Degrad. Stab. **41**(2), 177–184 (1993)
31. Zheng, K., Tang, H., Chen, Q., et al.: Enzymatic synthesis of a polymeric antioxidant for efficient stabilization of polypropylene. Polym. Degrad. Stab. **112**, 27–34 (2015)
32. Fu, S.Y., Lauke, B.: Effects of fiber length and fiber orientation distributions on the tensile strength of short-fiber-reinforced polymers. Compos. Sci. Technol. **56**(10), 1179–1190 (1996)
33. Tucker, C.L., Advani, S.G.: Processing of short-fiber systems. In: Advani, S.G. (Ed.) Flow and rheology in polymer composites manufacturing, pp. 147–202. Elsevier, Amsterdam (1994)
34. Todd, D.B.: Improving incorporation of fillers in plastics, a special report. Adv. Polym. Technol. **19**(1), 54–64 (2000)
35. Clemons, C.M., Caulfield, D.F., Giacomin, A.J.: Dynamic fracture toughness of cellulose-fiber-reinforced polypropylene: preliminary investigation of microstructural effects. J. Elastomers Plast. **31**(4), 367–378 (1999)
36. Migneault, S., Koubaa, A., Erchiqui, F., et al.: Effects of processing method and fiber size on the structure and properties of wood–plastic composites. Compos. A Appl. Sci. Manuf. **40** (1), 80–85 (2009)
37. Graupner, N., Müssig, J.: A comparison of the mechanical characteristics of kenaf and lyocellfibre reinforced poly (lactic acid)(PLA) and poly (3-hydroxybutyrate)(PHB) composites. Compos. A Appl. Sci. Manuf. **42**(12), 2010–2019 (2011)
38. Zampaloni, M., Pourboghrat, F., Yankovich, S.A., et al.: Kenaf natural fiber reinforced polypropylene composites: A discussion on manufacturing problems and solutions. Compos. A Appl. Sci. Manuf. **38**(6), 1569–1580 (2007)
39. Beninia, K., Voorwald, H.J.C., Cioffi, M.O.H.: Mechanical properties of HIPS/sugarcane bagasse fiber composites after accelerated weathering. Procedia Eng. **10**, 3246–3251 (2011)
40. Evans, P.D., Thay, P.D., Schmalzl, K.J.: Degradation of wood surfaces during natural weathering. Effects of lignin and cellulose and on the adhesion of acrylic latex primers. Wood Sci. Technol. **30**(6), 411–422 (1996)

41. Abu-Sharkh, B.F., Hamid, H.: Degradation study of date palm fibre/polypropylene composites in natural and artificial weathering: mechanical and thermal analysis. Polym. Degrad. Stab. **85**(3), 967–973 (2004)
42. Azwa, Z.N., Yousif, B.F., Manalo, A.C., et al.: A review on the degradability of polymeric composites based on natural fibres. Mater. Des. **47**, 424–442 (2013)
43. Adler, E.: Lignin chemistry-past, present and future. Wood Sci. Technol. **11**(3), 169–218 (1977)
44. Fabiyi, J.S.: Chemistry of wood plastic composite weathering. University of Idaho (2007)
45. Matuana, L.M., Jin, S., Stark, N.M.: Ultraviolet weathering of HDPE/wood-flour composites coextruded with a clear HDPE cap layer. Polym. Degrad. Stab. **96**, 97–106 (2011)
46. García, M., Hidalgo, J., Garmendia, I., García-Jaca, J.: Wood-plastics composites with better fire retardancy and durability performance. Compos. A Appl. Sci. Manuf. **40**, 1772–1776 (2009)
47. CIE L*a*b* Color Scale. In: Laboratory, H.A. (ed.) www.hunterlab.com (2008)
48. Stark, N.M.: Photodegradation and photostabilization of weathered wood flour filled polyethylene composites. Michigan Technological University (2003)
49. Stark, N.M., Matuana, L.M.: Surface chemistry changes of weathered HDPE/wood-flour composites studied by XPS and FTIR spectroscopy. Polym. Degrad. Stab. **86**, 1–9 (2004)
50. Mathew, A.P., Oksman, K., Sain, M.: The effect of morphology and chemical characteristics of cellulose reinforcements on the crystallinity of polylactic acid. J. Appl. Polym. Sci. **101** (1), 300–310 (2006)
51. Muasher, M., Sain, M.: The efficacy of photostabilizers on the color change of wood filled plastic composites. Polym. Degrad. Stab. **91**(5), 1156–1165 (2006)
52. Chetanachan, W., Sookkho, D., Sutthitavil, W., et al.: PVC wood: a new look in construction. J. Vinyl Add. Tech. **7**(3), 134 (2001)
53. Butylina, S., Hyvärinen, M., Kärki, T.: A study of surface changes of wood-polypropylene composites as the result of exterior weathering. Polym. Degrad. Stab. **97**(3), 337–345 (2012)
54. Falk, R.H., Lundin, T., Felton, C.: The effects of weathering on wood-thermoplastic composites intended for outdoor applications. In: Proceedings of the 2nd Annual Conference on Durability and Disaster Mitigation in Wood-Frame Housing. Madison, WI, USA, pp. 175–179 (2000)
55. Fabiyi, J.S., McDonald, A.G.: Effect of wood species on property and weathering performance of wood plastic composites. Compos. A Appl. Sci. Manuf. **41**(10), 1434–1440 (2010)
56. Stark, N.M.: Factors influencing the weatherability of wood-polyethylene composites. In: Proceedings of the Global Outlook for Natural Fiber & Wood Composites, Portland, ME, USA. 3–5 Dec 2003
57. Stark, N.M., Matuana, L.M., Clemons, C.M.: Effect of processing method on surface and weathering characteristics of wood-flour/HDPE composites. J. Appl. Polym. Sci. **93**(3), 1021–1030 (2004)
58. Gardette, M., Perthue, A., Gardette, J.L., et al.: Photo-and thermal-oxidation of polyethylene: Comparison of mechanisms and influence of unsaturation content. Polym. Degrad. Stab. **98**(11), 2383–2390 (2013)
59. Fabiyi, J.S., McDonald, A.G., Wolcott, M.P., et al.: Wood plastic composites weathering: Visual appearance and chemical changes. Polym. Degrad. Stab. **93**(8), 1405–1414 (2008)
60. Peng, Y., Liu, R., Cao, J., et al.: Antiweathering effects of vitamin E on wood flour/polypropylene composites. Polym. Compos. **35**(11), 2085–2093 (2014)
61. Lundin, T., Cramer, S.M., Falk, R.H., et al.: Accelerated weathering of natural fiber-filled polyethylene composites. J. Mater. Civ. Eng. **16**(6), 547–555 (2004)
62. Butylina, S., Hyvärinen, M., Kärki, T.: Accelerated weathering of wood–polypropylene composites containing minerals. Compos. A Appl. Sci. Manuf. **43**(11), 2087–2094 (2012)
63. Selden, R., Nyström, B., Långström, R.: UV aging of poly (propylene)/wood-fiber composites. Polym. Compos. **25**(5), 543–553 (2004)
64. Beg, M.D.H., Pickering, K.L.: Accelerated weathering of unbleached and bleached Kraft wood fibre reinforced polypropylene composites. Polym. Degrad. Stab. **93**(10), 1939–1946 (2008)

65. La Mantia, F.P., Morreale, M.: Accelerated weathering of polypropylene/wood flour composites. Polym. Degrad. Stab. **93**(7), 1252–1258 (2008)
66. Yan, L., Chouw, N., Jayaraman, K.: Effect of UV and water spraying on the mechanical properties of flax fabric reinforced polymer composites used for civil engineering applications. Mater. Des. **71**, 17–25 (2015)
67. Colom, X., Carrillo, F., Nogués, F., et al.: Structural analysis of photodegraded wood by means of FTIR spectroscopy. Polym. Degrad. Stab. **80**(3), 543–549 (2003)
68. Homkhiew, C., Ratanawilai, T., Thongruang, W.: Effects of natural weathering on the properties of recycled polypropylene composites reinforced with rubberwood flour. Ind. Crops Prod. **56**, 52–59 (2014)
69. Stokke, D.D., Gardner, D.J.: Fundamental aspects of wood as a component of thermoplastic composites. J. Vinyl Add. Tech. **9**(2), 96–104 (2003)
70. Mikulášová, M., Košíková, B.: Biodegradability of lignin-Polypropylene composite films. Folia Microbiol. **44**(6), 669–672 (1999)
71. Morandim-Giannetti, A.A., Agnelli, J.A.M., Lanças, B.Z., et al.: Lignin as additive in polypropylene/coir composites: thermal, mechanical and morphological properties. Carbohydr. Polym. **87**(4), 2563–2568 (2012)
72. Rabek, J.F.: Photochemical aspects of degradation of polymers. In: Rabek, J.F. (ed.) Polymer Photodegradation: Mechanisms and Experimental Methods. Springer, New York, pp. 24–66 (1995)
73. Gugumus, F.: Light stabilizers. In: Gächter, R., Müller, H. (eds.) Plastics Additives Handbook. Hanser Publisher, New York, pp. 129–262 (1990)
74. Grabmayer, K., Wallner, G.M., Beißmann, S., et al.: Characterization of the aging behavior of polyethylene by photoluminescence spectroscopy. Polym. Degrad. Stab. **107**, 28–36 (2014)
75. Grabmayer, K., Wallner, G.M., Beißmann, S., et al.: Accelerated aging of polyethylene materials at high oxygen pressure characterized by photoluminescence spectroscopy and established aging characterization methods. Polym. Degrad. Stab. **109**, 40–49 (2014)
76. Jabarin, S.A., Lofgren, E.A.: Photooxidative effects on properties and structure of high-density polyethylene. J. Appl. Polym. Sci. **53**(4), 411–423 (1994)
77. Wypych, G.: Artificial weathering versus natural exposure. In: Wypych, G. (ed.) Handbook of Material Weathering. ChemTec Publishing, Toronto, Ontario, Canada, pp. 231–244 (2013)
78. Zhao, B.Y., Yi, X.W., Li, R.Y., et al.: Characterization to the weathering extent of LLDPE/LDPE thin film. J. Appl. Polym. Sci. **88**(1), 12–16 (2003)
79. Sebaa, M., Servens, C., Pouyet, J.: Natural and artificial weathering of low-density polyethylene (LDPE): Calorimetric analysis. J. Appl. Polym. Sci. **45**(6), 1049–1053 (1992)
80. Tavares, A.C., Gulmine, J.V., Lepienski, C.M., et al.: The effect of accelerated aging on the surface mechanical properties of polyethylene. Polym. Degrad. Stab. **81**(2), 367–373 (2003)
81. Gulmine, J.V., Janissek, P.R., Heise, H.M., et al.: Degradation profile of polyethylene after artificial accelerated weathering. Polym. Degrad. Stab. **79**(3), 385–397 (2003)
82. Fabiyi, J.S., McDonald, A.G., McIlroy, D.: Wood modification effects on weathering of HDPE-based wood plastic composites. J. Polym. Environ. **17**(1), 34–48 (2009)
83. Matuana, L.M., Kamdem, D.P., Zhang, J.: Photoaging and stabilization of rigid PVC/wood-fiber composites. J. Appl. Polym. Sci. **80**(11), 1943–1950 (2001)
84. Chaochanchaikul, K., Rosarpitak, V., Sombatsompop, N.: Photodegradation profiles of PVC compound and wood/PVC composites under UV weathering. Exp. Polym. Lett. **7**(2), 146–160 (2013)
85. Colom, X., Carrillo, F., Nogués, F., et al.: Structural analysis of photodegraded wood by means of FTIR spectroscopy. Polym. Degrad. Stab. **80**(3), 543–549 (2003)
86. Wang, S.Y., Lin, S.J.: The effect of outdoor environmental exposure on the main components of wood. MokuzaiGakkaishi **37**(10), 954–963 (1991)
87. Park, B.S., Furuno, T., Uehara, T.: Histochemical changes of wood surfaces irradiated with ultraviolet light. MokuzaiGakkaishi **42**(1), 1–9 (1996)

88. Kataoka, Y., Kiguchi, M.: Depth profiling of photo-induced degradation in wood by FT-IR microspectroscopy. J. Wood Sci. **47**(4), 325–327 (2001)
89. Hon, D.N.S., Ifju, G.: Measuring penetration of light into wood by detection of photo-induced free radicals. Wood Sci. **11**(2), 118–127 (1978)
90. Hon, D.N.S., Ifju, G., Feist, W.C.: Characteristics of free radicals in wood. Wood Fiber **12** (2), 121–130 (1980)
91. Hon, D.N.S.: Weathering and photochemistry of wood. In: Hon, D.N.S., Shiraishi, N. (eds) Wood and Cellulosic Chemistry, 2nd edn. Marcel Dekker, Inc., New York, pp. 512–546 (2001)
92. Popescu, C.M., Popescu, M.C., Vasile, C.: Structural analysis of photodegraded lime wood by means of FT-IR and 2D IR correlation spectroscopy. Int. J. Biol. Macromol. **48**(4), 667–675 (2011)
93. George, B., Suttie, E., Merlin, A., et al.: Photodegradation and photostabilisation of wood-the state of the art. Polym. Degrad. Stab. **88**(2), 268–274 (2005)
94. Windeisen, E., Wegener, G.: Lignin as building unit for polymers. In: Matyjaszewski, K., Martin, M. (eds) Polymer Science: A Comprehensive Reference, pp 255–265. Elsevier, Amsterdam (2012)
95. Popescu, C.M., Spiridon, I., Tibirna, C.M., et al.: A thermogravimetric study of structural changes of lime wood (Tiliacordata Mill.) induced by exposure to simulated accelerated UV/Vis-light. J. Photochem. Photobiol. A **217**(1), 207–212 (2011)
96. Garcia, R., Triboulot, M.C., Merlin, A., et al.: Variation of the viscoelastic properties of wood as a surface finishes substrate. Wood Sci. Technol. **34**(2), 99–107 (2000)
97. Lanzalunga, O., Bietti, M.: Photo-and radiation chemical induced degradation of lignin model compounds. J. Photochem. Photobiol. B **56**(2), 85–108 (2000)
98. Kringstad, K.P., Lin, S.Y.: Mechanisms in the yellowing of high yield pulps by light: structure and reactivity of free radical intermediates in the photodegradation of lignin. Tappi **53**, 2296–2301 (1970)
99. Hon, D.N.S.: Thermomechanical Pulp and Light-Photoactivity of α-Carbonyl Group in Solid Lignin]. J. Wood Chem. Technol. **12**(2), 179–196 (1992)
100. Pandey, K.K., Vuorinen, T.: Comparative study of photodegradation of wood by a UV laser and a xenon light source. Polym. Degrad. Stab. **93**(12), 2138–2146 (2008)
101. Müller, U., Rätzsch, M., Schwanninger, M., et al.: Yellowing and IR-changes of spruce wood as result of UV-irradiation. J. Photochem. Photobiol. B **69**(2), 97–105 (2003)
102. Kalnins, M.A., Feist, W.C.: Increase in wettability of wood with weathering. For. Prod. J. **43** (2), 55 (1993)
103. Teacă, C.A., Roşu, D., Bodîrlău, R., et al.: Structural changes in wood under artificial UV light irradiation determined by FTIR spectroscopy and color measurements—a brief review. BioResources **8**(1), 1478–1507 (2013)
104. Browne, F.L.: The penetration of light into wood. For. Prod. J. **7**, 308–314 (1957)
105. Hon, D.N.S.: Photooxidative degradation of cellulose: reactions of the cellulosic free radicals with oxygen. J. Polym. Sci. Polym. Chem. Ed. **17**(2), 441–454 (1979)
106. Pietta, P.G.: Flavonoids as antioxidants. J. Nat. Prod. **63**(7), 1035–1042 (2000)
107. Wu, J.H., Tung, Y.T., Wang, S.Y., et al.: Phenolic antioxidants from the heartwood of Acacia confuse. J. Agric. Food Chem. **53**(15), 5917–5921 (2005)
108. Chang, T.C., Chang, H.T., Wu, C.L., et al.: Influences of extractives on the photodegradation of wood. Polym. Degrad. Stab. **95**(4), 516–521 (2010)
109. Chang, T.C., Chang, H.T., Wu, C.L., et al.: Stabilizing effect of extractives on the photo-oxidation of Acacia confusa wood. Polym. Degrad. Stab. **95**(9), 1518–1522 (2010)
110. Stark, N.M., Matuana, L.M.: Influence of photostabilizers on wood flour-HDPE composites exposed to xenon-arc radiation with and without water spray. Polym. Degrad. Stab. **91**(12), 3048–3056 (2006)
111. Du, H., Wang, W., Wang, Q., et al.: Effects of pigments on the UV degradation of wood-flour/HDPE composites. J. Appl. Polym. Sci. **118**(2), 1068–1076 (2010)

112. Zhang, Z., Du, H., Wang, W., et al.: Property changes of wood-fiber/HDPE composites colored by iron oxide pigments after accelerated UV weathering. J. For. Res. **21**(1), 59–62 (2010)
113. Butylina, S., Hyvärinen, M., Kärki, T.: Weathering of wood-polypropylene composites containing pigments. Eur. J. Wood Wood Prod. **70**(5), 719–726 (2012)
114. Stark, N.M., Mueller, S.A.: Improving the color stability of wood-plastic composites through fiber pre-treatment. Wood Fiber Sci. **40**(2), 271–278 (2008)
115. Devi, R.R., Gogoi, K., Konwar, B.K., et al.: Synergistic effect of nanoTiO_2 and nanoclay on mechanical, flame retardancy, UV stability, and antibacterial properties of wood polymer composites. Polym. Bull. **70**(4), 1397–1413 (2013)
116. Peng, Y., Guo, X., Cao, J., et al.: Effects of two staining methods on color stability of wood flour/polypropylene composites during accelerated UV weathering. Polym. Compos. DOI: 10.1002/pc.23683 (2015)
117. Butylina, S., Martikka, O., Kärki, T.: Weathering properties of coextruded polypropylene-based composites containing inorganic pigments. Polym. Degrad. Stab. **120**, 10–16 (2015)
118. Du, H., Wang, W., Wang, Q., et al.: Effects of ultraviolet absorbers on the ultraviolet degradation of rice-hull/high-density polyethylene composites. J. Appl. Polym. Sci. **126**(3), 906–915 (2012)
119. Stark, N.M., Matuana, L.M.: Ultraviolet weathering of photostabilized wood-flour-filled high-density polyethylene composites. J. Appl. Polym. Sci. **90**(10), 2609–2617 (2003)
120. Bouza, R., Abad, M.J., Barral, L., et al.: Efficacy of hindered amines in woodflour-polypropylene composites compatibilized with vinyltrimethoxysilane after accelerated weathering and moisture absorption. J. Appl. Polym. Sci. **120**(4), 2017–2026 (2011)
121. Rychlý, J., Mosnáčková, K., Rychlá, L., et al.: Comparison of the UV stabilisation effect of commercially available processing stabilizers Irganox HP 136 and Irganox 1010. Polym. Degrad. Stab. **118**, 10–16 (2015)
122. Scott, G.: Mechanisms of polymer stabilization. Pure Appl. Chem. **30**(1–2), 267–290 (1972)
123. Pénzes, G., Domján, A., Tátraaljai, D., Staniek, P., Földes, E., Pukánszky, B.: High temperature reactions of an aryl–alkyl phosphine, an exceptionally efficient melt stabilizer for polyethylene. Polym. Degrad. Stab. **95**(9), 1627–1635 (2010)
124. Ojeda, T., Freitas, A., Birck, K., Dalmolin, E., Jacques, R., Bento, F., Camargo, F.: Degradability of linear polyolefins under natural weathering. Polym. Degrad. Stab. **96**(4), 703–707 (2011)
125. Peng, Y., Liu, R., Cao, J.: Effects of antioxidants on photodegradation of wood flour/polypropylene composites during artificial weathering. BioResources **9**(4), 5817–5830 (2014)
126. Bracco, P., Brunella, V., Zanetti, M., et al.: Stabilisation of ultra-high molecular weight polyethylene with vitamin E. Polym. Degrad. Stab. **92**(12), 2155–2162 (2007)
127. Al-Malaika, S., Issenhuth, S.: The antioxidant role of vitamin E in polymers. IV. Reaction products of dl-α-tocopherol with lead dioxide and with polyolefins. Polymer **42**(7), 2915–2939 (2001)
128. Peng, Y., Liu, R., Cao, J., et al.: Antiweathering effects of vitamin E on wood flour/polypropylene composites. Polym. Compos. **35**(11), 2085–2093 (2014)
129. Peng, Y., Liu, R., Cao, J., et al.: Effects of vitamin E combined with antioxidants on wood flour/polypropylene composites during accelerated weathering. Holzforschung **69**(1), 113–120 (2015)
130. Kirschweng, B., Tátraaljai, D., Földes, E., et al.: Efficiency of curcumin, a natural antioxidant, in the processing stabilization of PE: concentration effects. Polym. Degrad. Stab. **118**, 17–23 (2015)
131. Wang, Q., Wu, J., Gao, Y., et al.: Polypropylene/Mg 3 Al–tartrazine LDH nanocomposites with enhanced thermal stability, UV absorption, and rheological properties. RSC Adv. **3**(48), 26017–26024 (2013)

132. Chai, H., Lin, Y., Evans, D.G., et al.: Synthesis and UV absorption properties of 2-naphthylamine-1, 5-disulfonic acid intercalated Zn-Al layered double hydroxides. Ind. Eng. Chem. Res. **47**(9), 2855–2860 (2008)
133. Zhang, X., Zhou, L., Pi, H., et al.: Performance of layered double hydroxides intercalated by a UV stabilizer in accelerated weathering and thermal stabilization of PVC. Polym. Degrad. Stab. **102**, 204–211 (2014)
134. Zhang, L., Lin, Y., Tuo, Z., et al.: Synthesis and UV absorption properties of 5-sulfosalicylate-intercalated Zn–Al layered double hydroxides. J. Solid State Chem. **180** (4), 1230–1235 (2007)
135. Pang, L., Liu, K., Wu, S., et al.: Effect of LDHs on the aging resistance of crumb rubber modified asphalt. Constr. Build. Mater. **67**, 239–243 (2014)
136. Peng, Y., Wang, W., Cao, J., et al.: Effects of a layered double hydroxide (LDH) on the photostability of wood flour/polypropylene composites during UV weathering. RSC Adv. **5** (51), 41230–41237 (2015)
137. Qin, H., Zhao, C., Zhang, S., et al.: Photo-oxidative degradation of polyethylene/montmorillonitenanocomposite. Polym. Degrad. Stab. **81**(3), 497–500 (2003)
138. Pagacz, J., Chrzanowski, M., Krucińska, I., et al.: Thermal aging and accelerated weathering of PVC/MMT nanocomposites: structural and morphological studies. J. Appl. Polym. Sci. **132**(24), 42090–42101 (2015)
139. Redhwi, H.H., Siddiqui, M.N., Andrady, A.L., et al.: Durability of LDPE nanocomposites with clay, silica, and zinc oxide II. weatherability of the nanocomposites. Polym. Compos. **34**(11), 1878–1883 (2013)
140. Morlat-Therias, S., Mailhot, B., Gonzalez, D., et al.: Photooxidation of polypropylene/montmorillonitenanocomposites. 2. Interactions with antioxidants. Chem. Mater. **17**(5), 1072–1078 (2005)
141. Pattamasattayasonthi, N., Chaochanchaikul, K., Rosarpitak, V., et al.: Effects of UV weathering and a CeO_2-based coating layer on the mechanical and structural changes of wood/PVC composites. J. Vinyl Add. Tech. **17**(1), 9–16 (2011)
142. Rosu, D., Teaca, C.A., Bodirlau, R., et al.: FTIR and color change of the modified wood as a result of artificial light irradiation. J. Photochem. Photobiol. B **99**(3), 144–149 (2010)
143. Hung, K.C., Chen, Y.L., Wu, J.H.: Natural weathering properties of acetylated bamboo plastic composites. Polym. Degrad. Stab. **97**(9), 1680–1685 (2012)
144. Peng, Y., Liu, R., Cao, J., et al.: Effects of UV weathering on surface properties of polypropylene composites reinforced with wood flour, lignin, and cellulose. Appl. Surf. Sci. **317**, 385–392 (2014)
145. Peng, Y., Liu, R., Cao, J.: Characterization of surface chemistry and crystallization behavior of polypropylene composites reinforced with wood flour, cellulose, and lignin during accelerated weathering. Appl. Surf. Sci. **332**, 253–259 (2015)
146. Gregorová, A., Cibulková, Z., Košíková, B., et al.: Stabilization effect of lignin in polypropylene and recycled polypropylene. Polym. Degrad. Stab. **89**(3), 553–558 (2005)

Photochemical Behavior and Optoelectronic Applications of Some Conjugated Polymers

Ufana Riaz and S.M. Ashraf

Abstract The chemical modification of π-conjugated materials offers a wide range of possibilities to design and tailor the structural properties such as conjugation length, molecule-to-molecule interaction, molecular weight, and band gap. This ability to design and integrate materials showing controlled optical and electronic properties is significant for fabricating novel optoelectronic devices such as light emitting diode (LEDs) and solar cells. Conjugated polymer based solar cells (CPSCs), reflect high potential for commercialization as they can be designed to emit desirable electroluminescence, along with ease of fabrication, low production costs as well as low environmental impact. The present chapter focuses on some conjugated polymers used for the preparation of light emitting diodes (CPLEDs) and Organic Solar Cells with special emphasis on poly(phenylenevinylene)s, polyfluorenes, polycarbazoles and poly(thiophene)s.

1 Introduction

Optoelectronic properties of π-conjugated polymers such as polyaniline (PANI), polypyrrole (PPy), poly(thiophene) (PTh) and their numerous derivatives have been extensively explored as the band gap of these polymers can be easily tuned to design a wide range of solar cells/light emitting diodes (LEDs) [1–5]. Moreover, these amazing organic molecules exhibit electroluminescence and photoluminescence when stimulated by light of suitable wavelength. Due to economic advantages over many other device fabrication techniques, conjugated polymers find potential application in designing/fabricating solar cells and LEDs [6–10].

S.M. Ashraf: Now retired.

U. Riaz (✉) · S.M. Ashraf
Materials Research Laboratory, Department of Chemistry, Jamia Millia Islamia (A Central University), New Delhi 110025, India
e-mail: ufana2002@yahoo.co.in

D. Rosu and Visakh P. M. (eds.), *Photochemical Behavior of Multicomponent Polymeric-based Materials*, Advanced Structured Materials 26,
DOI 10.1007/978-3-319-25196-7_11

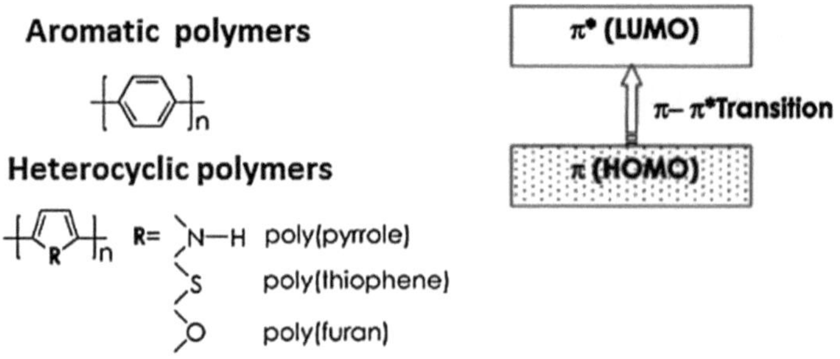

Fig. 1 Formation of HOMO and LUMO in conjugated polymers

Conjugated polymers are generally prepared by chemical/electrochemical polymerization techniques which are initiated by oxidation of the monomer followed by propagation through coupling of two radical cations leading to the formation of polymers in oxidized state [11–15]. The electrical conductivity in these polymers can be increased to several orders of magnitude by altering their band gap and doping is one such technique [16–20]. Conducting polymers exhibit absorption in the visible spectrum which is generally associated with π-π* transitions [21, 22]. The π-conjugated systems are generally formed by the overlap of p_z orbitals of carbon. In nitrogen containing conjugated polymers, the p_z orbitals of nitrogen also participate in conjugation. The combination of the electron donating/withdrawing substituents are therefore the deciding factors for the semiconducting properties of these polymers and by changing the bond alteration via the introduction of donor/acceptor functional groups, the band gap can be narrowed/widened [23, 24].

Light in these polymers is generated by the fast decay of excited molecular states, producing color which depends on the energy difference between the excited states and the molecular ground level. As most of the organic conjugated polymers possess low electron affinity, they exhibit hole-transporting properties [25, 26]. The electron affinity of such polymers corresponds to the lowest unoccupied molecular orbital (LUMO) while the ionization potential corresponds to the highest occupied molecular orbital (HOMO), Fig. 1. The ionization potential in these polymers generally lies between 2–3 eV and can be suitably altered by using various functional groups. Most of the conjugated polymer form films that are amorphous and the impurities formed during the synthesis lead to the creation of inner gap states.

2 Fabrication of Polymer Light Emitting Diode (PLEDs) Based on Conjugated Polymers

PLED device can be fabricated by two ways: single layer and multi-layer structures. The former depends on highly emitting material for effective injection and transport of electrons and holes (the majority charge carrier) and also cause the radiative decay of excited molecular states to ground state [27, 28]. The simple technique of fabrication

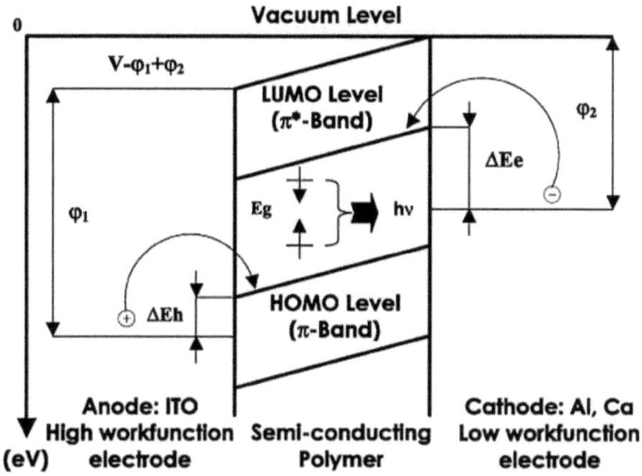

Fig. 2 Mechanism of charge generation by conjugated polymers in PLED. Reproduced with permission from Elsevier, A. Pron, P. Rannou, Processible conjugated polymers: from organic semiconductors to organic metals and superconductors, Prog. Polym. Sci., 27(1), 2002, 135–190

of a polymer based light emitting diode consists of an electroluminescent conjugated polymer and two electrodes. One of the electrodes must be transparent to allow transmission of light during electroluminescence effect and is usually referred to the hole injecting electrode (ITO). Doped conjugated polymers having a high work function (ϕ) are also used as hole injecting electrodes while the electron injecting electrode is made up of a low work function material, Fig. 2. Opposite charge carriers are injected into the conjugated polymer from anode to cathode. Holes are ejected from the valence band (π) (HOMO level) whereas electrons are introduced in the conduction band (π^*) (LUMO level) of the conjugated polymer. There exist barriers on both electrodes for the injection of electrons as well as holes, i.e., ΔE_h and ΔE_e and the barrier for electron injection is higher than the barrier for hole injection. The singlet or triplet states are created by the injected opposite charge carriers and the singlet state radiatively decays to give light. conjugated polymers have either holes or electrons as majority charge carriers [29, 30]. In single layer PLED devices, the electroluminescence takes place in the vicinity of the cathode due to higher mobility of holes in the conjugated polymers which lowers the luminescence efficiency due to non-radiative recombination process. However, it is difficult to balance the injection as well as transport of two charges in a single-material matrix over a specified voltage range and the imbalance between the charges leads to an increase in the voltage causing loss of luminance efficiency. This can be prevented by the addition of an electron transporting layer via the formation of a two-layer device which separates the cathode from the electroluminescent polymer layer [31–35].

3 Synthesis of Conjugated Polymers

Conjugated polymers are synthesized using various organic polymeric reactions to obtain desired chemical structures. Some of these reactions are discussed in the proceeding section.

3.1 Dehydrohalogenation Reactions

These reactions are carried out in presence of strong bases such as such as potassium *tert*-butoxide, sodium hydride [36]. Dehydrohalogenation has been used to synthesize unsubstituted polyphenylenevinylene (PPV) using dichloroxylene/sodium hydride and DMF solvent, Fig. 3.

Glich polymerization is a dehydrohalogenation reaction which has been widely adopted to synthesize alkyl/alkoxy substituted PPV derivatives [37]. The reaction is a two-step process and is reported to proceed through the formation of quinodimethane intermediate which is generated by radical/living chain anionic polymerization, Fig. 4.

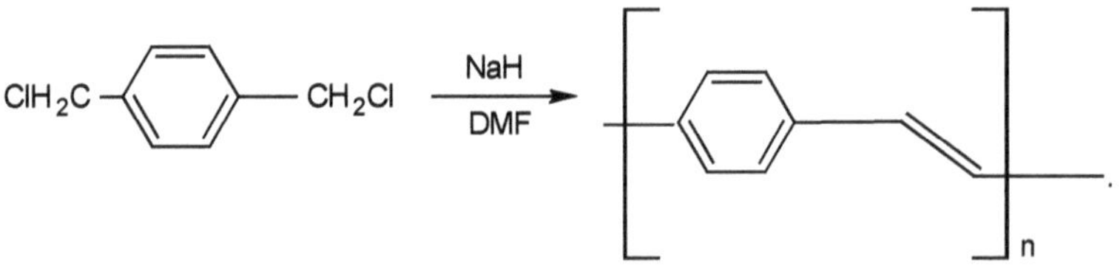

Fig. 3 Dehydrohalogenation polymerization of PPV

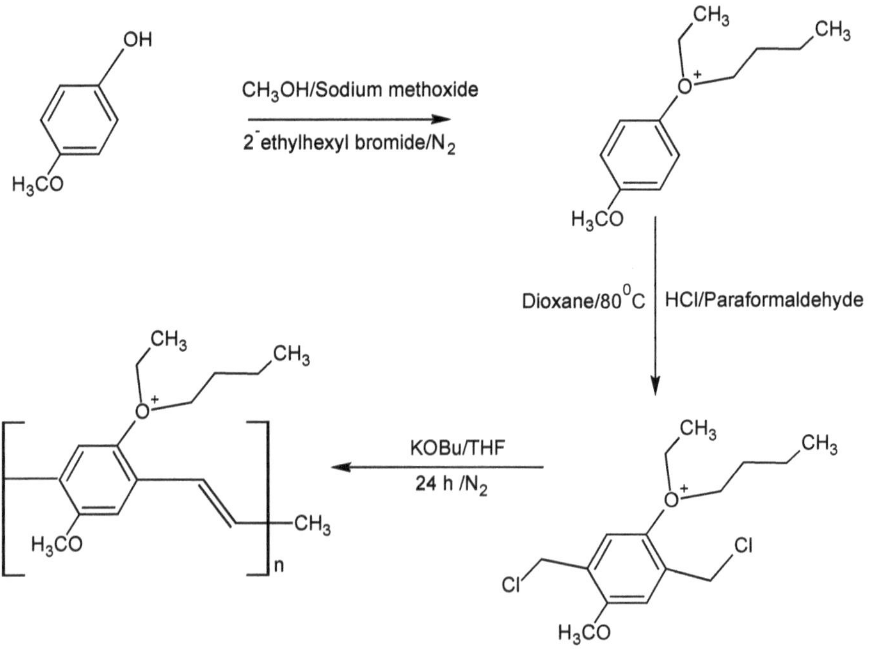

Fig. 4 Glich polymerization of alkoxy substituted PPV derivative

The molecular weight of the polymer can be controlled by optimizing the reaction parameters such as temperature, time, solvent, concentration of the monomer, and amount of base. It has been widely used for the synthesis of homopolymers and copolymers of PPV.

3.2 *Precursor Route Polymerization*

This polymerization strategy is extensively used for the synthesis of PPV and its substituted derivatives. Wessling and Zimmerman [38] developed a method for the synthesis of PPV via thermo-conversion of sulfonium intermediate (prepolymer) into PPV to yield its film. In the Wessling precursor route 1,4-xylylene-bis-(dialkyl sulfonium)-dichloride is used which upon elimination of tetrahydrothiophene and HCl yields PPV polymer, Fig. 5. The thermo-conversion mechanism yields pin-hole free thin films of conjugated polymers applicable for PLED fabrication. The thermo-conversion temperatures can be lowered to as low as 100 °C using bromide derivatives which can help in the fabrication of flexible PLED devices.

In the halo-precursor route, p-xylylene dichloride is used to obtain a halo precursor which upon elimination of HCl at 250 °C produces either PPV or its derivative. Tetrahydrothiophene salts are generally preferred due to low temperature stability of the prepolymer and also due to ease of conversion to PPV.

3.3 *Transition Metal Catalyzed Coupling Reactions*

This technique is generally adopted for the synthesis of heterocycles possessing strongly electron-withdrawing groups which generally tend to increase the reaction rate and product yield.

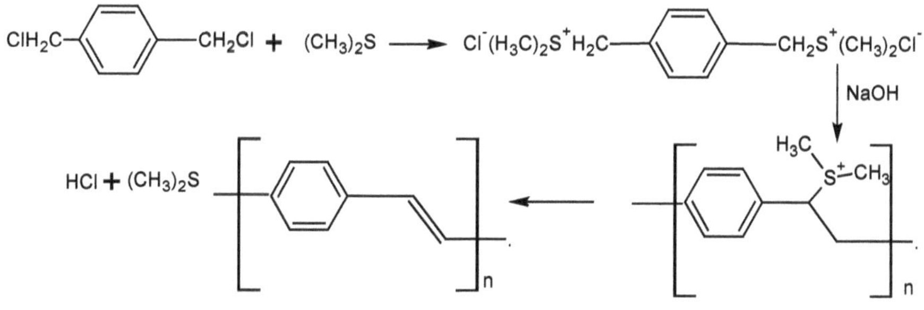

Fig. 5 Wessling Zimmerman Reaction for the synthesis of PPV

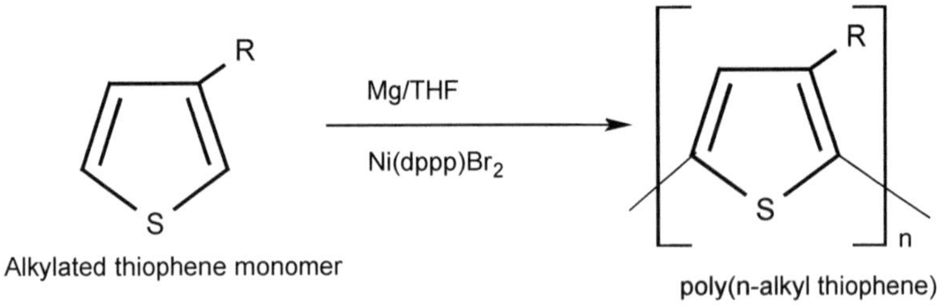

Fig. 6 Synthesis of alkylated thiophenes via Kumada coupling reaction

3.3.1 Kumada Coupling Reaction

This is a cross-coupling reaction, that is utilized for producing C=C bonds via the reaction of Grignard reagents with functional groups such as with alkyl, vinyl or aryl in presence of tetrahydrofuran (THF). Poly(3-alkyl thiophenes) have been synthesized by this reaction, Fig. 6 [39].

3.3.2 McCullough Coupling Reaction

McCullough method has been widely adopted to synthesize poly(3-alkylthiophene). In this method, 2-bromo-3-alkylthiophene is treated with lithium diisopropylamide (LDA) at −40 °C to form 2-bromo-3-alkyl-5-lithiothiophene which is converted into a Grignard reagent by reacting with $MgBr_2{\cdot}(OEt_2)$ to yield 2-bromo-5-(magnesiobromo)-3-alkylthiophene, Fig. 7. Ni(dppp)Cl_2 catalyzed cross coupling of 2-bromo-5-(magnesiobromo)-3-alkylthiophene is then done to produce poly (3-alkylthiophene) [39, 40]. In the Grignard metathesis polymerization, 2,5-dibromo-3-alkythiophene monomer is used and is reacted with Grignard reagent to give 2-bromo-5-(magnesiobromo)-3-alkylthiophene followed by cross-coupling reaction in the presence of nickel catalyst to produce poly(3-alkylthiophene).

3.3.3 Suzuki Coupling Reaction

This is a Pd catalyzed coupling reaction in which organic halides are reacted with boronic acids to form aryl derivatives of poly(para-phenylenes), polyfluorenes, Fig. 8 [40].

3.3.4 Heck Coupling

This coupling reaction of Pd mediated olefin arylation is the reaction between an organic halide with a vinylbenzene derivative that produces carbon-carbon double

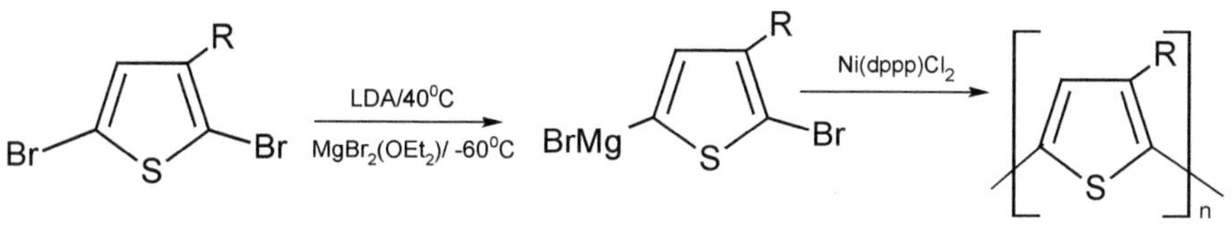

Fig. 7 Synthesis of alkylated thiophenes via *McCullough coupling* reaction

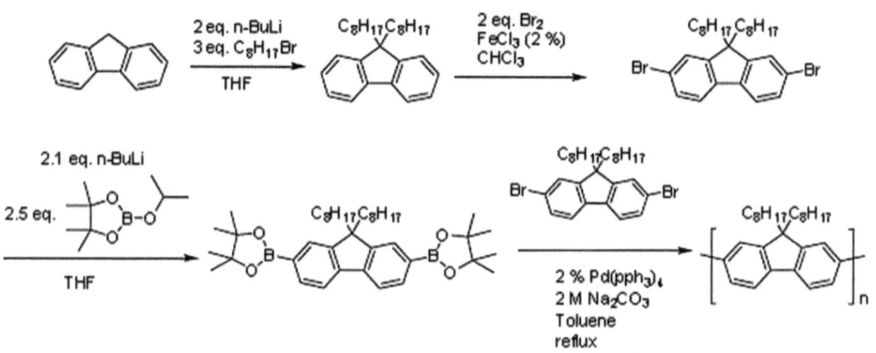

Fig. 8 Synthesis of PF via Suzuki coupling reaction

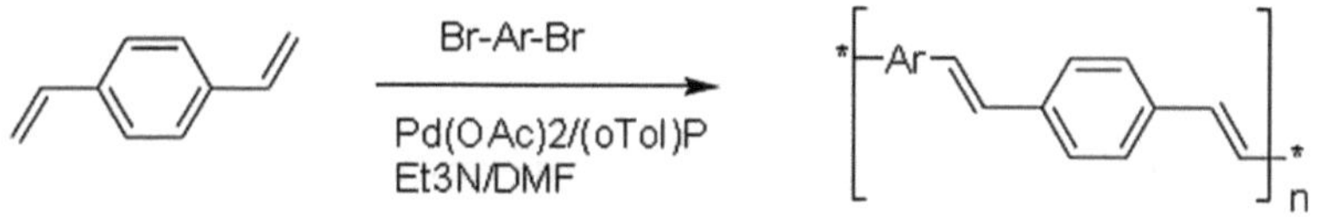

Fig. 9 Synthesis of PPV derivative via Heck reaction

bond, with *trans*-selectivity. It is generally preferred for the preparation of PPV related block copolymers, Fig. 9 [41].

3.4 Condensation Polymerizations

Condensation polymerization reactions are also utilized to develop light emitting conjugated polymers which are classified as given below.

3.4.1 Wittig Reaction

Wittig polycondensation route has been widely used for the preparation of alternating copolymers in which carbonyl compounds such as aldehyde and ketone are

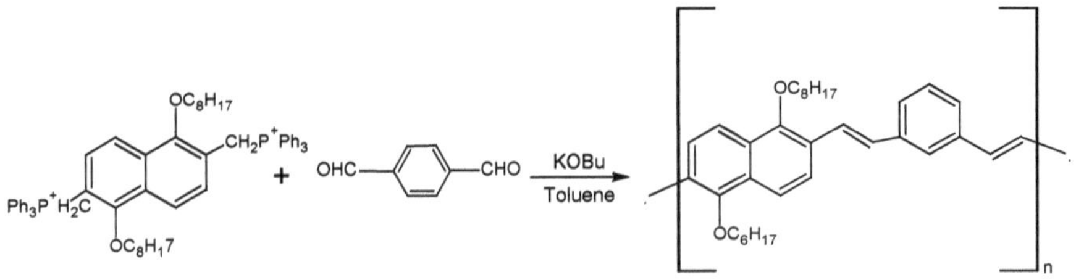

Fig. 10 Synthesis of poly(m-phenylenevinylene) derivative via Wittig reaction

reacted with phosphonium salts of conjugated monomers in presence of base such as KOBu, Fig. 10 [42, 43].

3.4.2 Knoevenagel Condensation Polymerization

This reaction takes place between aldehyde groups with active methylene species and requires strong electron withdrawing substituent groups such as cyano group [43, 44]. Employing Knoevenagel condensation, numerous PPV related homo and copolymers with CN containing vinylene units have been reported which are prepared using monomers such as 1,5-bis(hexyloxy)-2,6-naphthalenediacetonitrile and 1,5-bis(hexyloxy)-2,6-naphthalenedicarbaldehyde in the presence of a strong base is shown in Fig. 11.

3.4.3 Horner-Emmons Condensation Polymerization

Horner-Emmons condensation consists of substituted phosphonate ester reacted with terephthaldehyde in presence of potassium *tert*-butoxide to produce alternating copolymer, Fig. 12 [45].

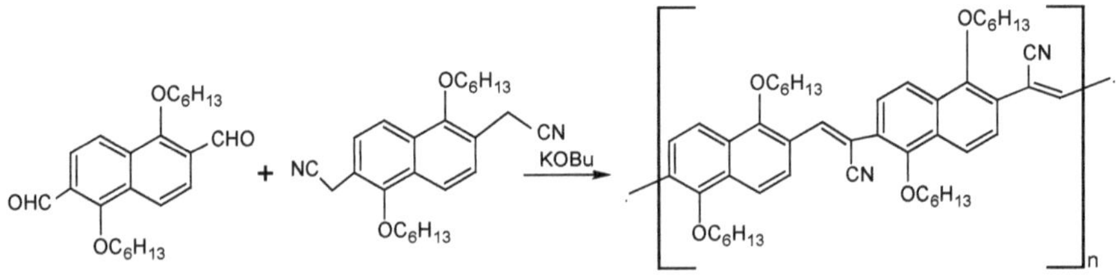

Fig. 11 Synthesis of poly(2,6,naphthylenevinylene) via Knoevenagel method

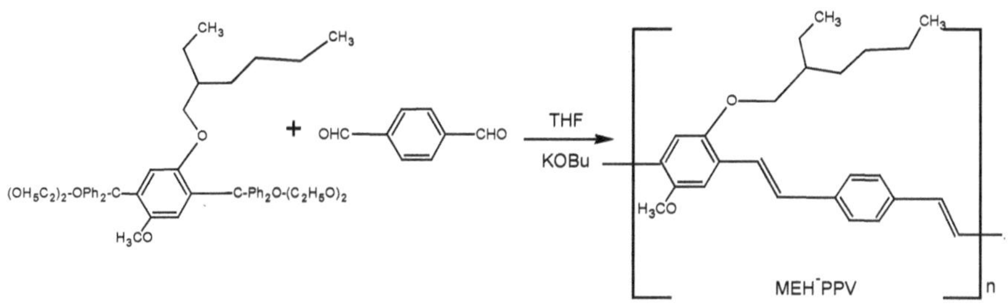

Fig. 12 Synthesis of MEH-PPV via Horner-Emmons condensation polymerization

4 Common Conjugated Polymers Used in PLEDs

Many conjugated polymers are used in PLEDs such as poly(phenylenevinylene) (PPV) [46] and its derivatives [47], poly(carbazole) and its modified derivatives [48], polyfluorenes [49], polyphenylenes [50], poly(phenylene ethynylenes) [51], and polyalkylthiophenes [52]. For most electroluminescent polymers, Calcium provides an effective polymer-metal junction and effectively injects electrons into the conjugated polymer. As internal electric field in a conjugated polymer is lower than that of a semiconductor, it slows down the migration of ions from the ITO to the hole transporting layer. Hence by varying the choice of metal (showing high work function), enhancement of drive voltage and device efficiency are observed. This observation substantiates the tunneling model for cathodic charge injection through the Schottky barrier [53–56].

4.1 Polyfluorene and Its Derivatives

Polyfluorenes are polymers with C6 or higher substituents at C9 position which are soluble in conventional organic solvents and are strong blue emitters when excited with UV irradiation [57]. Copolymers consisting of 9,9-dialkylfluorene and various aromatic amines have been investigated [58, 59]. The hole mobility of amine copolymers are quite high and therefore the preparation and properties of alternating copolymers of fluorine with various monomers such as thiophene, bithiophene, triarylamine, etc. have been extensively explored [60–65]. It has been reported that cyano-stilbene based copolymers emit green light, while bithiophene based fluorene copolymers emitted yellow light. Thus, the choice of comonomer can be used as an effective tool to design and optimize polymers with well-balanced hole-electron-transport characteristics, Fig. 13 [66].

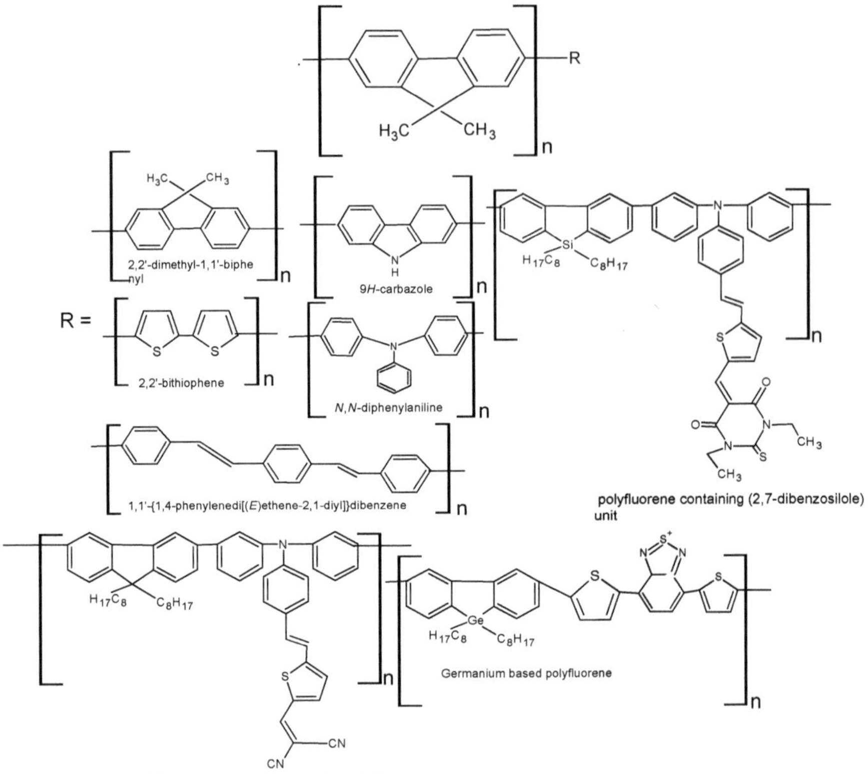

Fig. 13 Various derivatives of polyfluorene

4.2 Polyphenylene (PPP) and Its Derivatives

Poly(p-phenylene) (PPP) is a widely explored conjugated polymer in the field of PLED materials, Fig. 14 [67]. The presence of large band gap in the polymer reveals its blue emission characteristics. PPPs are insoluble and intractable in nature and therefore researchers have explored routes to synthesize soluble PPP films and a variety of PPP precursor routes have been discussed in literature [68]. PPP precursor routes involve the thermal elimination such as the elimination of two equivalents of acetic acid (per monomer unit) from poly-1,4-(5,6-diaceto-2,3-cyclohexene). This

Fig. 14 Strucutre of poly (phenylene)

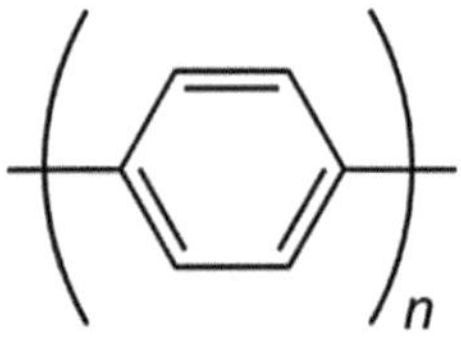

enables aromatization of the polymer chain rings and formation of PPP structure [68, 69]. As the reactions require relatively high temperatures, reduction of molecular weight takes place. Therefore, limited success has been achieved in the precursor approach to synthesize PPP and more interest has been focused on the preparation and characterization of PPPs bearing alkyl, alkoxy, and aryl units [70–75]. Polymerization reaction in these derivatives has been carried out via aryl/Suzuki coupling approaches. High molecular weights have been achieved which are good enough to cast films, and blue-emitting diodes having reasonable efficiencies have been formulated [74, 75].

Schlüter and Wegner [76] have formulated PPP derivatives with alkyl side chains through transition-metal-catalyzed polycondensations, Suzuki coupling of arylboronic acids and bromoarenes while Yamammoto [77] reported nickel-catalyzed coupling of substituted dihalobenzenes, hydroquinone bistriflates, and bis(mesylates) [78–80]. The inclusion of solubilizing side groups on the PPP backbone suffers from a major drawback that the additional substituents considerably twist the substituted phenylene rings out of plane causing a decrease in the interaction of the aromatic π-electron system accompanied by a drop in fluorescence quantum yield. The twist cannot occur if the PPP backbone is planarized by transforming it into a ladder polymer. Scherf and Müllen [81] synthesized regular ladder-type polymers derived through a two-step route in which a soluble poly (p-phenylene) was prepared by Suzuki aryl-aryl coupling of monomers. The ladder formation occurred with high regularity, with no defects such as cross-linking or incomplete cyclization. The molar masses of soluble, ladder-type poly (p-phenylene)s (LPPPs) were reported to be high. The phenylene rings in the ladder polymers are highly planar and the homopolymer shows a broad emission between 650 and 450 nm in the solid state, in contrast to the polymers sharp fluorescence emission around 450 nm in solution. The large Stokes shift (yellow emission) is especially enhanced after annealing, and has been attributed to the formation of excimers owing to interchain interactions and π-stacking between the planarized conjugated segments. The mechanism of generation of an intra-chain polaron-exciton involves coupling between electron and vibrational excitations as each inserted charge causes strong geometric distortion of the lattice, which displays a quinonoid-based structure. When radiant combination takes place, photons with an energy higher than that which separates the energy levels of two polarons (or bipolarons) are formed, which generates lattice coupling. The excited electron–hole pair can be localized on the same molecule (Frenkel exciton) or over different molecules (Wannier exciton) [82, 83].

4.3 Poly(Phenylene Vinylene) (PPV) and Its Derivatives

Poly(phenylene vinylene) (PPV) exhibits linear rigid-rod-like structure, with a good tendency to form crystalline morphology, Fig. 15 [84–88]. However, due to lack of functional groups to enhance solubility, this system is intractable and not

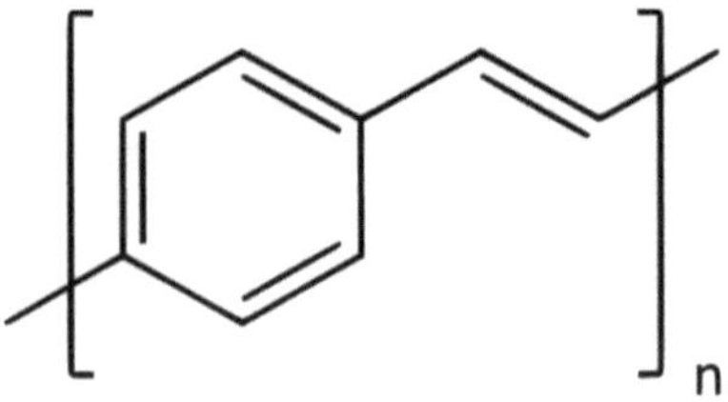

Fig. 15 Chemcial structure of poly(phenylene vinylene)

processible directly from solution but shows the tendency to emit yellow light under electrical stimulation. A variety of solvent-processable versions of PPV, and PPV copolymers have been synthesized [89, 90]. Since solvent processability is a desirable property for polymeric emitting systems, PPVs that can be cast as films have been developed [90]. Solution processable PPV precursors have been synthesized via thermal elimination of tetrahydrothiophene and hydrogen chloride at temperatures between 150–300 °C via Wessling condensation route [38].

2-methoxy-5-(2¢-ethylhexyloxy) derivative (MEH-PPV) is the most investigated derivative of PPV. The presence of alkoxy substituent have shown to impart red emission and hence, MEH-PPV exhibits an orange electroluminescence as compared to yellow-green emitted by PPV. The Wessling sulfonium reaction is generally adopted for the synthesis of the dialkoxy PPVs which involves base-initiated dehydrohalogenation polymerization of 1,4-bis(halomethyl)-2,5-dialkoxybenzene monomers [83, 84]. This reaction produces highly soluble and highly pure PPVs. LED devices based on soluble PPVs have shown very good performance and have led to considerable commercial interest in developing materials of this type. The choice of the monomer is significant because those monomer units which cause weak π interactions between phenyl rings, result in the production of a large band gap. The introduction of non-conjugated sequences, which diminish the degree of conjugation, have been used to induce a blue shift in the emission of PPV.

PLEDs based on poly(phenylene ethynylene) (PPE) have been developed via Sonogashira coupling of diethynylbenzenes and dihalobenzenes [91–93]. Poly (phenylene ethynylene) exhibits yellow emission (λ_{max} = 600 nm) [94] whereas its alkoxy derivative reveals red-orange emission [94]. A blue green emission can be attained via the introduction of 2,5-pyridinediyl unit (λ_{max} = 480 nm). Poly (3,4-dialkyl-1,6-phenylene ethynylene)s reveal blue emission (λ_{max} = 410 nm) which can be prepared by palladium cross-coupling reaction of a 3,4-dialkyl-1,6-diiodobenzene with 3,4-dialkyl-1,6-diethynylbenzene [94]. Apart from several other factors, steric interactions in PPEs cause twisting of the polymer chains thereby reducing the effective conjugation length, which limits their effective application as stable PLEDs.

4.4 Polythiophene and Its Derivatives

Thiophene derivatives represent a good example of how color of emission can be varied in PLEDs by modifying the polymer structure, Fig. 15 [95–99]. Soluble poly (3-alkylthiophene)s have been prepared by polymerization of 3-alkylthiophene monomers in the presence of excess ferric chloride, or by nickel-catalyzed coupling of dihalothiophenes with thiophene-bis(magnesium halide)s [100, 101]. 3-alkylthiophenes homopolymers and copolymers exhibit red electroluminescence (λ_{max} = 640 nm) having alkyl chain lengths in the range of 6–22 carbon atoms [102–105].

Doping these alkyl thiophenes with fluorescent dyes is reported to enhance the device efficiency and the frequency of emission is found to be dependent on regularity of the side chains on the polymers, Fig. 16. Alkylated polythiophenes were reported by Hadziioannou et al. [105] who found a correlation between emission maxima and the inverse number of alkylated thiophenes between two consecutive head-to-head coupled dyads.

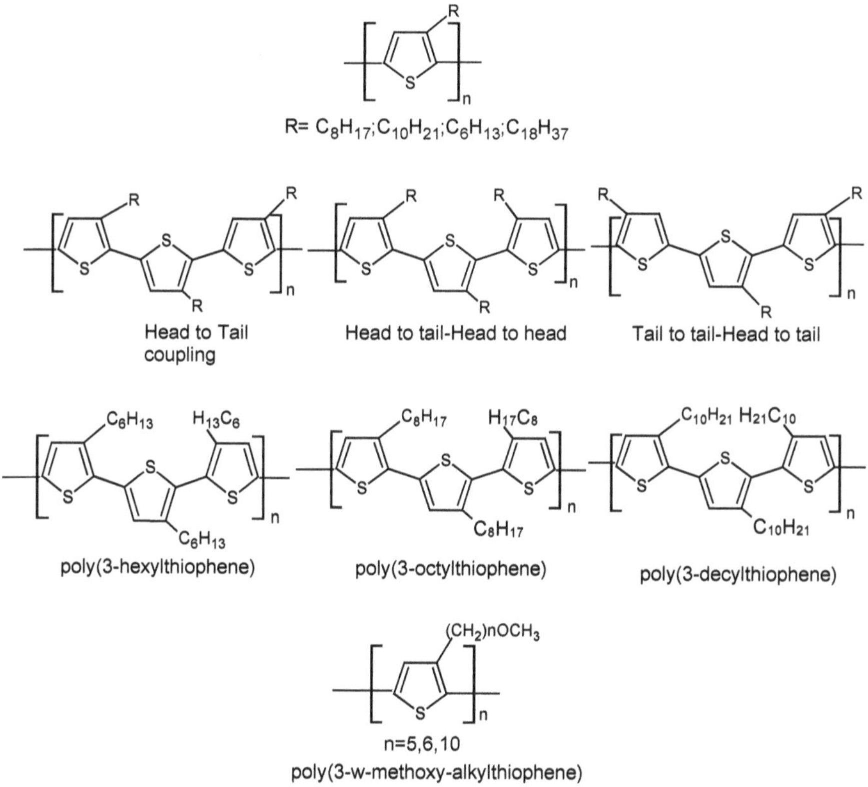

Fig. 16 Derivatives of thiophenes

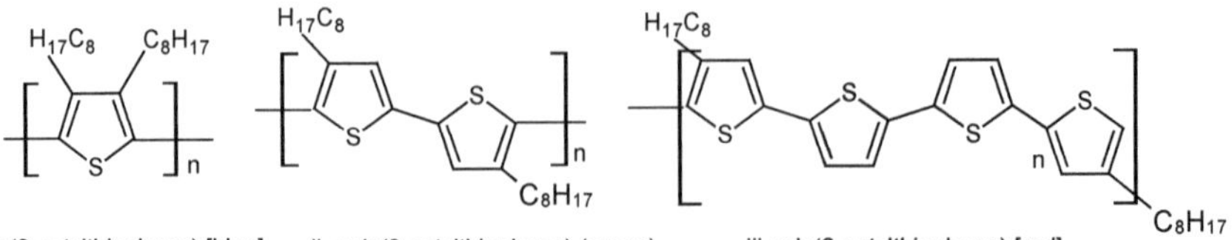

i. poly(3-octylthiophene) [blue] ii. poly(3-octylthiophene) (green) iii.poly(3-octylthiophene) [red]

Fig. 17 Chemical structures of poly(3-octyl thiophene)

Thus, (i) showed an emission maximum at 460 nm (2.7 eV), while (ii) and (iii) had emission maxima at 530 (2.35 eV) and 550 nm (2.25 eV), respectively. Similarly, poly(3-hexylthiophene) revealed a blue-shift with increasing amounts of head-to-head dyads as reported by Holdcroft and Xus, Fig. 17 [98, 99]. The blue shifts were obtained by introducing bulky substituents and also by substitution at both the 3rd and 4th positions of poly(3-hexylthiophene) [104–106]. The blue shifts can be attributed to twisting of the conjugated backbone due to the presence of bulky substituents. Greater control of emission colors requires the ability to control polymer regioregularity. When thiophenes were polymerized using ferric chloride as initiator, the resulting polymers showed region-randomness with head-to-tail (2,5′-) coupling, varying from 52 to 80 % [105, 106], while coupling of oligomeric alkylthiophenes resulted in the formation of regioregular copolymers [107, 108]. More general routes for the synthesis of regioregular (head-to-tail) poly (3-alkylthiophene)s have been developed by McCullough et al. [108, 109] and Rieke et al. [100–111]. They carried out nickel-catalyzed coupling of 2-halo-5-metallothiophenes. Poly(3-alkylthiophene) copolymers with side chains containing terminal azide groups have also been synthesized revealing polarized emission from stretch-oriented polythiophenes [111]. Oriented films of poly[3-(10-methoxydecyl) thiophene] deposited by Langmuir ± Blodgett techniques were reported to emit polarized electroluminescence [105–111].

5 Fabrication of Solar Cells Using Conjugated Polymers

Organic semiconductor films are generally photo-sensitive and photo-induced electron transfer from a donor-acceptor-type moeity/functional group introduces free charge carriers. The positive charge carriers are created on the donor layer, i.e., p-type, whereas negative charge carriers are created on the acceptor layer, i.e., n-type. These donor-acceptor-type bilayer devices thus work like p-n junctions, Fig. 18. The most common representatives of hole conducting donor-type semiconducting polymers are: derivatives of phenylene vinylene such as poly[2-methoxy-5-(3,7-dimethyloctyloxy)]-1,4-phenylenevinylene) (MDMOPPV), poly(3-hexylthiophene) (P3HT), fluorine derivatives such as (poly(9,9′-dioctylfluorene-*co*-bis-*N*,*N*′-(4-butylphenyl-1,4-phenylenediamine) (PFB), Fig. 19. These polymer-based photovoltaic materials are commonly solution processed at low temperatures [112–117].

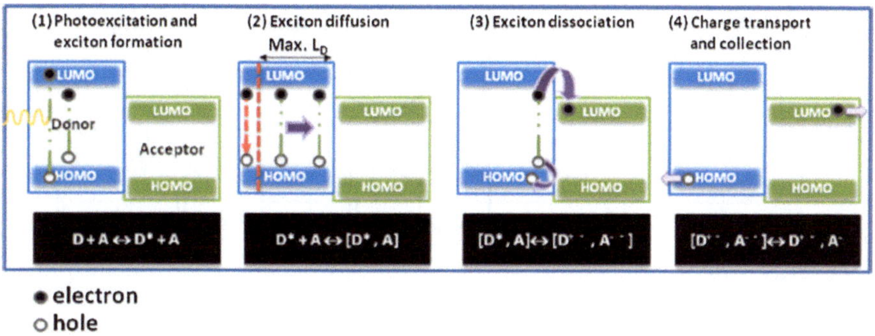

Fig. 18 Mechanism for donor-acceptor heterojunction solar cells. (*1*) Photoexitation of the donor to generate a Coulomb correlated electron-hole pair, an exciton. (*2*) Exciton diffusion to the D-A interface. A distance longer than the maximum diffusion length (max *LD*) will lead to relaxation of the exciton. (*3*) Bound exciton dissociation at the D-A interface to form a geminate pair. (4) Free charge transportation and collection at electrodes. Reprinted with permission from ACS, Y.-J. Cheng, S.-H. Yang, and C.-S. Hsu, Synthesis of Conjugated Polymers for Organic Solar Cell Applications Chem. Rev. 2009, 109, 5868–5923

Soluble precursor polymers can also be converted into the final semiconducting form with a post-deposition conversion reaction [118]. For the formulation of organic solar cells spin-coating, doctor blading, as well as screen-printing methods are generally applied as they lead to upscaling of the production with low-energy consumption. For the fabrication of a bilayer heterojunction device, p-type and n-type organic semiconductors are sequentially stacked on top of each other in various combinations [119]. The excitons created within a distance of 10–20 nm are only able to reach the interface which results in loss of absorbed photons away from the interface leading to low quantum efficiencies [120]. Hence, the efficiency of bilayer solar cells is limited by the generation of charges within the range of 10–20 nm from donor-acceptor interface. The thickness of the film also plays a crucial role and thick film affects the absorbing material, resulting in a minimum photocurrent at the maximum of optical absorption spectrum. In bulk hetero junction devices, the donor-acceptor moieties are taken in bulk volume which creates donor-acceptor phase separation within 10–20 nm length scale and each interface is within a distance less than the exciton diffusion length from the absorbing site. This increases the interfacial area between the donor and acceptor phases and results in improved solar cell efficiency [121, 122].

In the bilayer heterojunction devices, the donor-acceptor phases are separated from each other and can selectively contact the anode and cathode, whereas in the bulk heterojunction both phases are intimately mixed and there is no preferred direction for the internal fields of separated charges, Fig. 20. The electrons and holes are thus created within the volume having concentration gradient (diffusion) as driving force. The separated charges require percolated pathways and the donor-acceptor phases form bicontinuous interpenetrating network [123]. Bulk hetero-junction devices are sensitive to the morphology in the blend [124]. Majority of

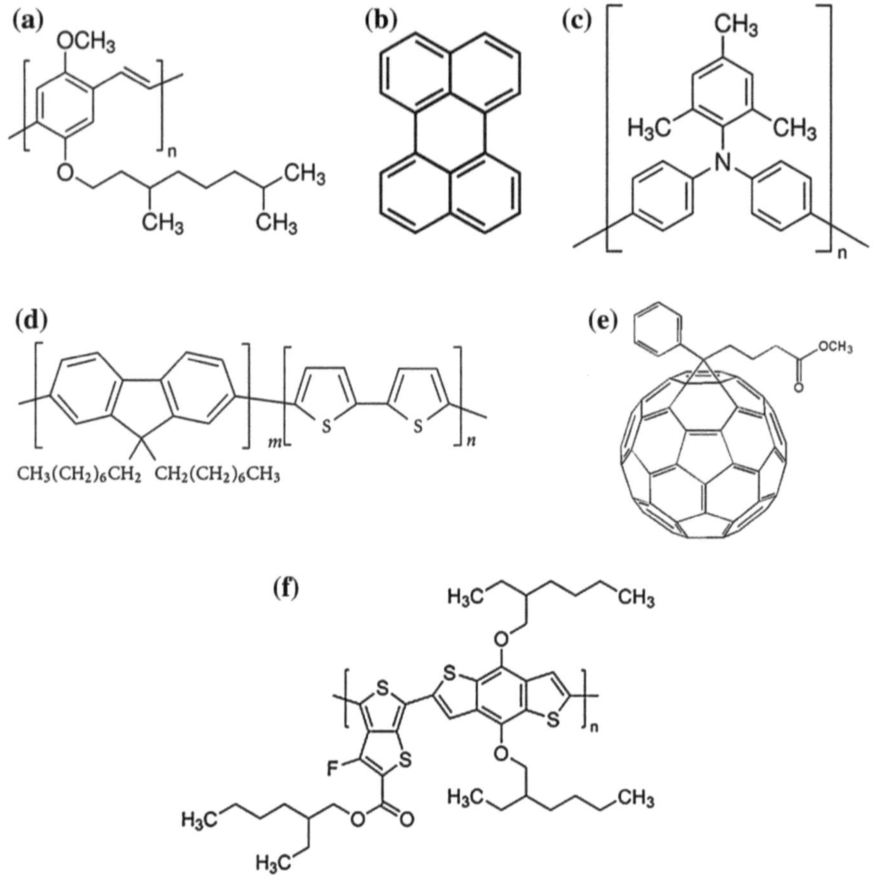

Fig. 19 Common organic semiconductors used in organic solar cells **a** poly[2-methoxy-5-(3′,7′-dimethyloctyloxy)-1,4-phenylenevinylene] (MDMO-PPV), **b** perylene, **c** poly(triaryl amine), **d** poly[(9,9-di-n-octylfluorenyl-2,7-diyl) ((F8T2)), **e** Phenyl-C61-butyric acid methyl ester-alt-2,2′-bithiophene-5,5′-diyl)] (PCBM), **f** poly({4,8-bis[(2-ethylhexyl)oxy]benzo[1,2-*b*:4,5-b′] dithiophene-2,6-diyl}{3-fluoro-2-[(2-ethylhexyl)carbonyl]thieno[3,4-*b*]thiophenediyl}) (PTB7)

conjugated polymers have band gaps higher than 2 eV, which limits the possibility of harvesting solar photons to about 30 %. This is because organic materials have a large band gap (2–3 eV), which can absorb only a small portion of the incident solar light. A band gap of 1.1 eV can absorb 77 % of the solar irradiation and because the absorption coefficients of organic materials are as high as 105 cm^{-1}, a layer of 100 nm thickness is enough to absorb most of the photons [124, 125]. The primary photo-excitations in organic materials do not directly produce free charge carriers but to coulombically bound electron-hole pairs, called excitons. Hence for optimum dissociation of excitons, strong electric fields are necessary which can be supplied via externally applied electrical fields or via interfaces. The exciton diffusion length should be of the same order of magnitude as the donor acceptor phase

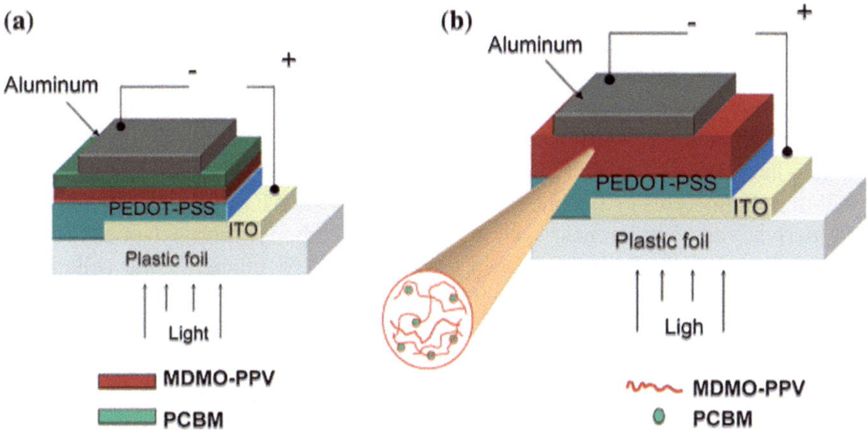

Fig. 20 **a** Bilayer Device fabrication and **b** Bulk heterojunction device fabrication. Reprinted with permission from ACS, S. Gunes, H. Neugebauer, and N. S. Sariciftci, Chem. Rev. 2007, 107, 1324–1338

separation length otherwise, their energy is lost for the power conversion before reaching the interface due to exciton decay via radiative/nonradiative pathways. The exciton diffusion lengths in conjugated polymers are usually reported to be in the range of 10–20 nm.

Blending of conjugated polymers with electron acceptors is a very efficient way to break photo-excited excitons into free charge carriers. Ultrafast photophysical studies reveal that photo-induced charge transfer in such blends is much faster than other competing relaxation processes. For producing highly efficient photovoltaic devices, the charges generated need to be transported to the appropriate electrodes within their lifetime and the charge carriers need a driving force to reach the electrodes. A gradient in the chemical potentials of electrons and holes (quasi Fermi levels of the doped phases) is built up in a donor-acceptor junction which is determined by the difference between the highest occupied molecular (HOMO) level of the donor and the lowest unoccupied molecular orbital (LUMO) level of the acceptor. This internal electrical field determines the maximum open circuit voltage (*V*oc) and contributes to a field-induced drift of charge carriers.

6 Conjugated Polymers Used in Organic Solar Cells

6.1 Phenylene-Vinylenes and Its Derivatives

Poly(p-phenylenevinylene) is reported to be insoluble as discussed earlier and therefore difficult to process in solid state. Wittig reaction has been used to develop

PPV but the major drawbacks of precursor approach include the generation of toxic side products during the solid-state, incomplete thermal conversion or oxidation causing structural defects as well as undefined molecular weight. The incorporation of long alkyl or alkoxy chains into the phenylene monomer is seen to produce solubility, Soluble derivatives of phenylene-vinylenes, such as poly(2-methoxy-5-((2′-ethylhexyl)oxy)-1,4-phenylenevinylene) (MEH-PPV), poly[2-methoxy-5-(3′,7′-dimethyloctyloxy)-1,4-phenylene vinylene) (MDMO-PPV) mixed with soluble derivatives of fullerenes, such 1-(3-methoxycarbonyl)propyl-1-phenyl-[6, 6]-methanofullerene (PCBM) are known to produce 2.5 % efficient solar cells [126]. Shaheen et al. [123] prepared MDMO-PPV: PCBM in the weight ratio of 1:4 for obtaining a power conversion efficiency of 2.5 % using chlorobenzene as a solvent. The electron mobility of pure PCBM has been reported to be higher than the hole mobility of pure MDMO-PPV [124]. Hoppe et al. [75] demonstrated that increasing PCBM content increases the size of these PCBM nanoclusters. To study the relationship between morphology and performance in bulk heterojunction solar cells, MDMO-PPV/ PCBM blends have been investigated in detail [126–128]. Mihailetchi et al. [128] developed a model for quantitative description of the behavior of PPV:PCBM bulk heterojunction solar cells.

The introduction of cyano linkages in place of vinylene linkages is seen to lower both LUMO and HOMO levels by ~0.5 eV, with little effect on the magnitude of the band gap. With the introduction of cyano groups, the conjugated backbone undergoes twisting due to steric hindrance. Cyano functionalized PPVs exhibit high electron affinities and electron-transport properties as a result of the electron-withdrawing effect of the cyano side group. It therefore acts as a suitable electron acceptor in photovoltaic devices in either bilayer or bulk heterojunction device configuration [129–133].

To further lower the band gap of cyan based PPVs below 2 eV, thiophene units with lower aromaticities have been incorporated into the main chain to form a D-A arrangement. Vanderzande et al. [132] developed copolymers based on the bis (1-cyano-2-thienylvinylene) phenylene prepared by Knoevenagel condensation to construct cyanovinylene linkages, Fig. 21.

Reynold et al. [133] developed CN-PPV derivatives containing dioxythiophene moieties in the main chain which revealed narrow band gaps ranging between 1.5–1.8 eV and good solubilities in common organic solvents, Fig. 22.

6.2 Poly(3-Alkylthiophenes) and Its Derivatives

Poly(3-alkylthiophenes) (P3ATs) and regioregular poly(3-alkylthiophenes) (RRP3AT) (P3HT:poly(3-hexylthiophene), P3OT:poly(3-octylthiophene), and P3DDT:poly(3 dodecylthiophene) are used as electron donors in heterojunction solar cells showing conversion efficiencies up to 5 % [134–137]. The influence of the alkyl side chain length of regioregular P3HT, P3OT, and P3DDT on the electrochemical and optical properties was studied by Al Ibrahim et al. [134]. It was

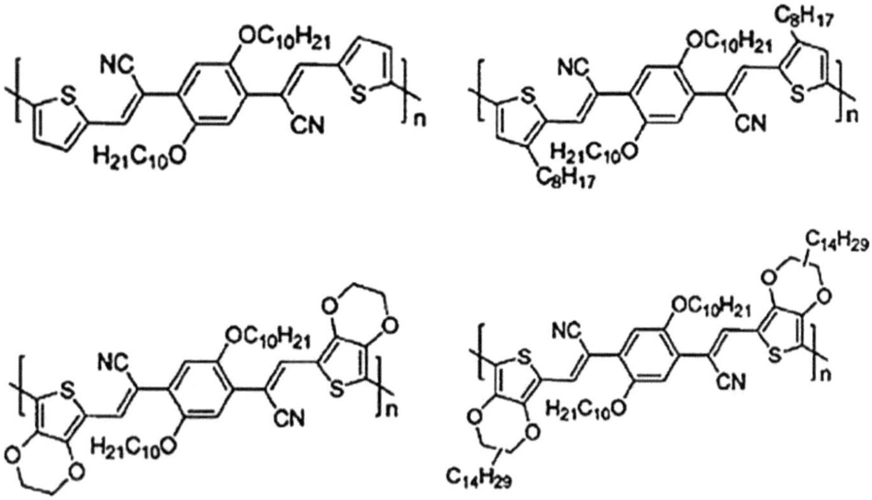

Fig. 21 Structures of CN-PPV derivatives

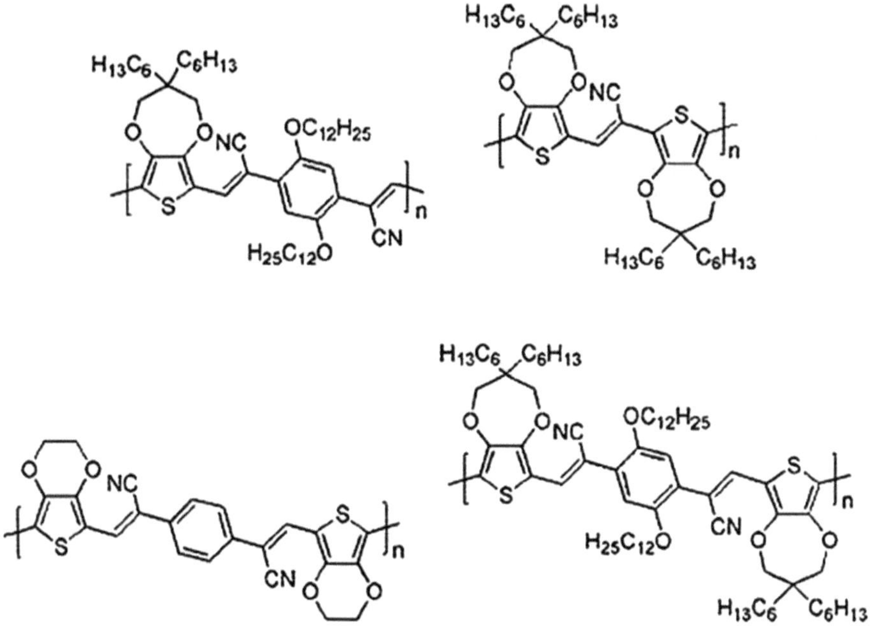

Fig. 22 Structures of CN-PPV derivatives containing dioxythiophene

found that with longer side chain length, the band gaps were slightly increased while absorption coefficient revealed a decrease upon increasing the side chain length containing polythiophenes due to chromophore dilution. Using regioregular

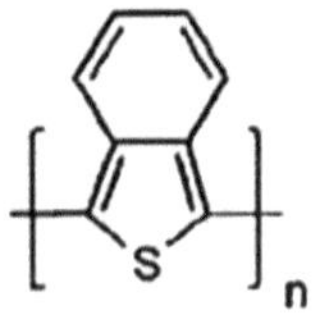

Fig. 23 Structure of Polyisothianaphthene (PITN)

poly(3-hexylthiophene) (RR-P3HT) as donor and PCBM as acceptor, bulk heterojunction solar cells have been developed revealing 75 % quantum efficiencies and power conversion efficiencies up to 5 % [135, 136]. The high efficiency of these devices has been reported to be due to the presence of microcrystalline lamellar stacking in the solid state packing which reduces recombination of electrons and holes. Regioregular P3HT shows interchain interactions that cause a red shift of the optical absorption due to stacking. Padinger et al. [138] investigated the effect of annealing on P3HT based solar cells by applying an external voltage which was shown to cause recrystallization along with reduction in the free volume as well as density defects at the interface. Chiravaze et al. [139] investigated that P3HT showed pronounced red shift upon thermal annealing at a temperature above 110 ° C for 4 min. The hole mobility in the P3HT component was found to increase by more than 3 orders of magnitude.

Polyisothianaphthene (PITN), Fig. 23, is also a polythiophene derivative where all the thiophene rings are fused to a benzene ring at the 3,4-positions of thiophene. It possesses a band gap of 1 eV, which is smaller than that of polythiophene. The PITN main chain favors the quinoid form to preserve aromaticity causing a change in the thiophene aromaticity [140].

Hillmyer and co-workers [141] developed Isothianaphthalene (ITN)—thiophene copolymer using distannyl-isothianaphthene (obtained via lithiation of isothianaphthalene in presence of n-butyl lithium in tetramethylethylenediamine (TMEDA) and THF followed by reaction with trimethyltin chloride). It was found that this molecule could be used as a precursor to prepare the ITN-fluorene copolymer/ITN–thiophene through Stille coupling with 9,9-dihexyl-2,7-dibromofluorene/ bis(bromothienyl)-isothianaphthene, Fig. 24.

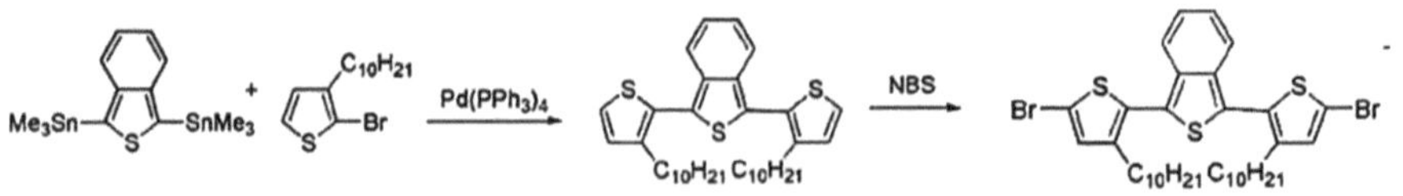

Fig. 24 Synthesis of ITN—thiophene copolymer using distannyl-isothianaphthene and bis (bromothienyl)-isothianaphthene

6.3 Polypyrrole and its Derivatives

Pyrrole ring containing polymers have been synthesized via Stille coupling of 2,5-bis(5-(trimethylstannyl)-2-thienyl)-*N*-dodecylpyrrole and 4,7-dibromo-2,1,3-benzothiadiazole) using $Pd(PPh_3)_2Cl_2$ as the catalyst and higher molecular weights are attained using distannyl monomer, Fig. 25 [142–144]. This polymer reveals a low band gap of 1.6 eV and efficient photoinduced electron transfer, was reported by photoinduced absorption and photoluminescence quenching of the former. The solar cell device based on the pyrrole ring derivative with /PCBM exhibited 1 % efficiency when used in the weight ratios P131:PCBM ratio of 1:3. The addition of diketopyrrolopyrrole unit has shown to impart high electron and hole mobility both close to 0.1 $cm^2/(V\ s)$ in an FET device. Janssen and co-workers [144] reported a conjugated polymer synthesized using quaterthiophene and diketopyrrolopyrrole units, Fig. 26.

6.4 Porphyrin Based Polymers

Porphyrin derivatives reveal good photochemical and thermal stabilities due to their large π-conjugation systems which have been found suitable for organic photonic and electronic applications [145–147]. Porphyrin molecules have been well established to serve as potential photosensitizers in dye-sensitized solar cells. Soluble porphyrin-dithienothiophene copolymers have been synthesized using Sonogashira

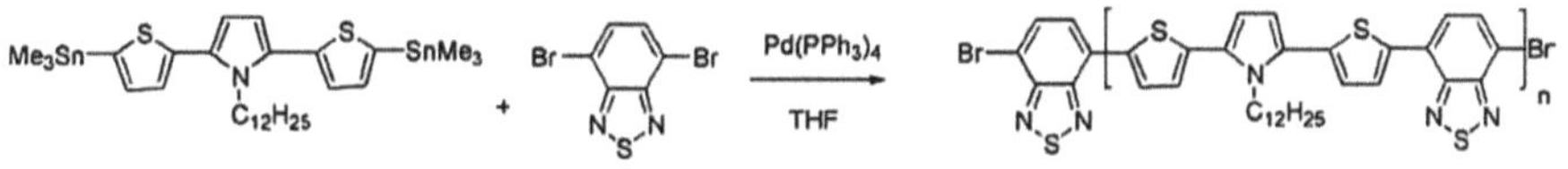

Fig. 25 Synthesis of pyrrole containing copolymer via Stille coupling of 2,5-bis(5-(trimethylstannyl)-2-thienyl)-*N*-dodecylpyrrole and 4,7-dibromo-2,1,3-benzothiadiazole

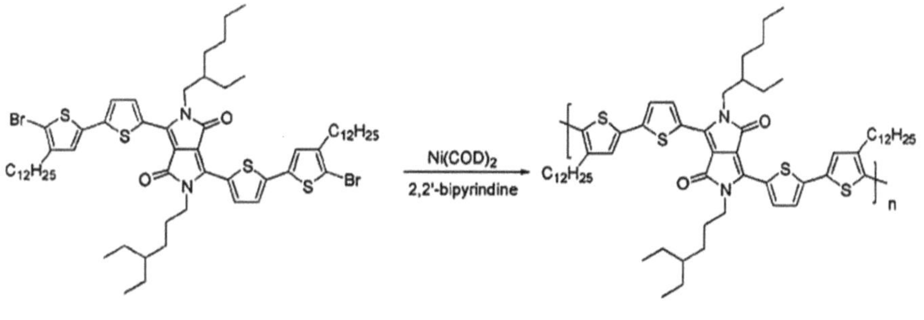

Fig. 26 Synthesis of diketopyrrolopyrrole containing polymer via Yamamoto coupling

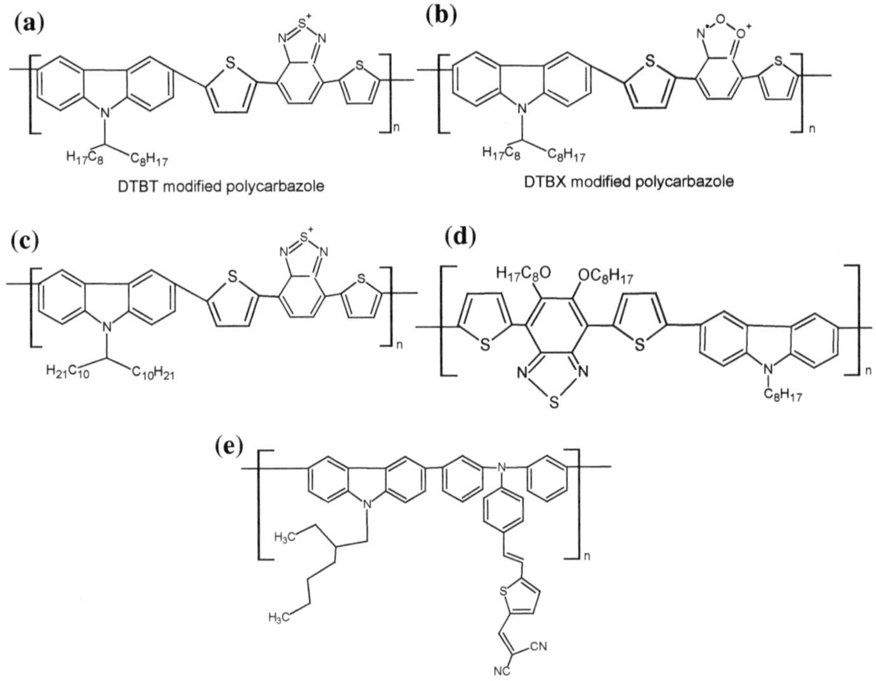

Fig. 27 Various derivatives of carbazole containing conjugated polymers

coupling reaction [147]. The introduction of diethynyldithienothiophene into the porphyrin main chain is found to reduce steric hindrance, and enhances conjugation, which helps in improving the charge-transport property. However, to ensure sufficient solubility of the porphyrin-containing polymer for solution processing, a large aliphatic chain insulating portion must be incorporated into the porphyrin unit, and this is known to have a negative effect on the charge transport.

6.5 *Poly(Carbazole) and Its Derivatives*

The carbazole monomer has a fused pyrrole ring which makes tricyclic carbazole fully aromatic and electron-rich due to the presence of nitrogen group. The introduction of alkyl chains showing good solubility when alkylation is carried out at 9-position, Fig. 27. Leclerc et al. [148, 149] developed 4,7-dithien-2-yl-2,1,3-benzothiadiazole (DTBT) based poly(N-alkyl-2,7-carbazole) (PC) exhibiting enhanced thermal stability, relatively high molecular weight and good solubility. He also developed 4,7-dithien-2-yl-2,1,3-benzothiadiazoxaline (DTBX) based polycarbazole as accepting unit which showed good hole mobility due to its symmetrical backbone and well organized structure.

In order to improve the solubility and close packing of polymer, Hashimoto developed a 10-carbon chain alkyl derivative of polycarbazole which also exhibited similar photovoltaic parameters as those of DTBT based polycarbazole. Qin et. al. [150] designed a carbazole derivative showing planar polymer conformation by introducing two octyloxy chains onto benzothiazole (BT) ring and an octyl chain onto carbazole ring. Cao [151] developed a series of carbazole-based two-dimensional polymers having HOMO energy levels as the conjugated backbone but their LUMO levels were governed by narrow band gap side chains. The 2-ethyl-hexyl chains had little influence on the photovoltaic performance but heptadecanyl chains on carbazole led to a significant decrease in the photovoltaic performance. Hence, alkyl chains on carbazole unit had important influence on the hole transporting ability of the resultant polymers. The presence of short alkyl chains resulted in high mobility than bulkier alkyl side-chains. The bulky alkyl chains possibly cause imbalanced electron transport and consequently decreased fill factor of polymer [152].

7 Conclusion and Future Prospects

Several strategies have been adopted to tailor the band gap of conjugated polymers and significant progress has been made in synthesis, as well as structural modification of conjugated polymers. These materials show impressive applicational potential in energy harvesting, as well as designing of opto-electronic devices However, large-scale synthesis of these materials with optical tunability is highly desirable. Fabrication of highly crystalline conjugated polymers is still a challenge. The reproducibility and stability of doping level is also required. Conjugated polymer based organic-inorganic hybrid structures is a highly promising field of research, to enhance the above-mentioned properties and to address the corresponding challenges multifunctional conjugated polymers need to be designed and developed to open up new avenues of research in the field of energy harvesting and opto-electronic device fabrication.

Acknowledgment The corresponding author Dr Ufana Riaz wishes to acknowledge the Department of Science and Technology (DST)—Science and Engineering Research Board (SERB), India, vide sanction no. SB/S-1/PC-070-2013 for financial assistance.

References

1. Nordén, B., Krutmeijer, E.: The Nobel Prize in Chemistry: Conductive Polymers. The Royal Swedish Academy of Sciences (2000)
2. Wynne, K.J., Street, G.B.: Conducting polymers. A short review. Ind. Eng. Chem. Prod. Res. Dev. **21**, 23–28 (1982)
3. Brédas, J.L., Street, G.B.: Polarons, bipolarons, and solitons in conducting polymers. Acc. Chem. Res. **18**, 309–315 (1985)

4. Moliton, A., Hiorns, R.C.: Review of electronic and optical properties of semiconducting π-conjugated polymers: applications in optoelectronics. Polym. Int. **53**, 1397–1412 (2004)
5. Choo, D.J., Talaie, A., Lee, Y.K., Jang, J., Park, S.H., Huh, G., Yoo, K.H., Lee, J.Y.: Examination of EL and PL properties of MCHM-PPV and MEH-PPV: a study towards introduction of a new series of thin film EL devices. Thin Solid Films **363**, 37–41 (2000)
6. Waltman, R.J., Bargon, J.: Electrically conducting polymers: a review of the electropolymerization reaction, of the effects of chemical structure on polymer film properties, and of applications towards technology. Can. J. Chem. **64**, 76–95 (1986)
7. Bakhshi, A.K., Bhalla, G.: Electrically conducting polymers: materials of the twentyfirst century. J. Sci. Ind. Res. India **63**, 715–728 (2004)
8. Blouin, N., Michaud, A., Leclerc, M.: A low-band gap poly(2,7-Carbazole) derivativefor use in high-performance solar cells. Adv. Mater. **19**, 2295–2300 (2007)
9. Blouin, N., Michaud, A., Gendron, D., Wakim, S., Blair, E., Neagu-Plesu, R., Belletete, M., Durocher, G., Tao, Y., Leclerc, M.: Toward a rational design of poly(2,7- carbazole) derivatives for solar cells. J. Am. Chem. Soc. **130**(2), 732–742 (2008)
10. Boudreault, P.L.T., Michaud, A., Leclerc, M.: A new poly(2,7-dibenzosilole) derivative in polymer solar cells. Macromol. Rapid Commun. **28**(22), 2176–2179 (2007)
11. Boudreault, P.L.T., Wakim, S., Blouin, N., Simard, M., Tessier, C., Tao, Y., Leclerc, M.: Synthesis, characterization, and application of indolo[3,2-b]carbazole semiconductors. J. Am. Chem. Soc. **129**(29), 9125–9136 (2007)
12. Chan, S.-H., Chen, C., Chao, T., Ting, C., Ko, B.-T.: Synthesis, characterization, and photovoltaic properties of novel semiconducting polymers with thiophenephenylene-thiophene (TPT) as coplanar units. Macromolecules **41**(15), 5519–5526 (2008)
13. Burroughes, J.H., Bradley, D.D.C., Brown, A.R., Marks, R.N., Mackay, K., Friend, R.H., Burns, P.L., Holmes, A.B.: Light-emitting diodes based on conjugated polymers. Nature **347**, 539–542 (1990)
14. MacDiarmid, A.G.: "Synthetic metals": a novel role for organic polymers. Curr. Appl. Phys. **1**(4–5), 269–279 (2001)
15. Hou, J., Chen, H.-Y., Zhang, S., Chen, R.I., Yang, Y., Wu, Y., Li, G.: Synthesis of a low band gap polymer and its application in highly efficient polymer solar cells. J. Am. Chem. Soc. **131**(43), 15586–15587 (2009)
16. Liang, Y., Wu, Y., Feng, D., Tsai, S.-T., Son, H.-J., Li, G., Yu, L.: Development of new semiconducting polymers for high performance solar cells. J. Am. Chem. Soc. **131**(1), 56–59 (2009)
17. Kirchmeyer, S., Reuter, K.: Scientific importance, properties and growing applications of poly(3,4-ethylenedioxythiophene). J. Mater. Chem. **15**, 2077–2088 (2005)
18. Hollinger, J., Jahnke, A.A., Coombs, N., Seferos, D.S.: Controlling phase separation and optical properties in conjugated polymers through selenophene–thiophene copolymerization. J. Am. Chem. Soc. **132**(25), 8546–8547 (2010)
19. Berggren, M., Inganas, O., Gustafsson, G., Rasmusson, J., Andersson, M.R., Hjertberg, T., Wennerstrom, O.: Light-emitting diodes with variable colours from polymer blends. Nature **372**, 444–446 (1994)
20. Faraggi, E.Z., Chayet, H., Cohen, G., Neumann, R., Avny, Y., Davidov, D.: Tunable electroluminescence and photoluminescence in phenylenevinylene-naphthylenevinylene copolymers. Adv. Mater. **7**, 742–745 (1995)
21. Tasch, S., List, E.J.W., Hochfilzer, C., Leising, G., Schlichting, P., Rohr, U., Geerts, Y., Scherf, U., Mullen, K.: Efficient red-and orange-light-emitting diodes realized by excitation energy transfer from blue-light-emitting conjugated polymers. Phys. Rev. B **56**(N8), 4479–4483 (1997)
22. Gruner, J., Hamer, P.J., Friend, R.H., Huber, H.-J., Scherf, U., Holmes, A.: A high efficiency blue-light-emitting diode based on novel ladder poly (p-phenylene)s. Adv. Mater. **6**, 748–752 (1994)
23. Gin, D., Conticello, V.: Poly (p-phenylene): new directions in synthesis and application. Trends Polym. Sci. **4**, 217–223 (1996)

24. Yang, Y., Pei, Q., Heeger, A.: Efficient blue polymer light-emitting diodes from a series of soluble poly (paraphenylene)s. J. Appl. Phys. **79**, 934–942 (1996)
25. CIE Report 86Ð1990: Commission Internationale de l'Eclairage. Vienna, Austria (1990)
26. Tessler, N., Cleave, V., Pinner, D.J.: Polymer LEDs as a physics tool. Isr. J. Chem. **40**(2), 147–152 (2000)
27. Steuber, F., Staudigel, J., Stossel, M., Simmerer, J., Winnacker, A., Spreitzer, H., Weissortel, F., Salbeck, J.: White Light emission from organic LEDs utilizing spiro compounds with high-temperature stability. Adv. Mater. **12**, 130–133 (2000)
28. Scharber, M.C., Wuhlbacher, D., Koppe, M., Denk, P., Waldauf, C., Heeger, A., Brabec, C. J.: Design rules for donors in bulk-heterojunction solar cells—towards 10 % energy conversion efficiency. Adv. Mater. **18**(6), 789–794 (2006)
29. Scully, S.R., McGehee, M.D.: Effects of optical interference and energy transfer on exciton diffusion length measurements in organic semiconductors. J. Appl. Phys. **100**, 34–907 (2006)
30. Shaw, P.E., Ruseckas, A., Samuel, I.D.W.: Exciton diffusion measurements in poly(3-hexyl thiophene). Adv. Mater. **20**(18), 3516–3520 (2008)
31. Slooff, L.H., Veenstra, S.C., Kroon, J.M., Moet, D.J.D., Sweelssen, J., Koetse, M.M.: Determining the internal quantum efficiency of highly efficient polymer solar cells through optical modeling. Appl. Phys. Lett. **90**(14), 143506–143508 (2007)
32. Sotzing, G.A., Lee, K.H.: Poly(thieno[3,4-b]thiophene): A p- and n-dopable polythiophene exhibiting high optical transparency in the semiconducting state. Macromolecules **35**(19), 7281–7286 (2002)
33. Tang, W., Hai, J., Dai, Y., Huang, Z., Lu, B., Yuan, F., Tang, J., Zhang, F.: Recent development of conjugated oligomers for high-efficiency bulk-heterojunction solar cells. Sol Energy Mater. Sol C **94**(12), 1963–1979 (2010)
34. Zhou, E., Yamakawa, S., Zhang, Y., Tajima, K., Yang, C., Hashimoto, K.: Indolo[3,2-b] carbazole -based alternating donor–acceptor copolymers: synthesis, properties and photovoltaic application. J. Mater. Chem. **19**(41), 7730–7737 (2009)
35. Hoeg, D.F., Lusk, D.I., Goldberg, E.P.: Poly-p-xylylidene. J. Polym. Sci. Part B Polym. Lett. **2**, 697–701 (1964)
36. Neef, C.J., Ferraris, J.P.: MEH-PPV: improved synthetic procedure and molecular weight control. Macromolecules **33**, 2311–2314 (2000)
37. Jen, K.Y., Oboodi, R., Elsenbaumer, R.T.: Processible and environmentally stable conducting polymers. Polym. Mater. Sci. Eng. **53**, 79–85 (1985)
38. Wessling, R.A., Zimmerman, R.G.: Polyelectrolytes from bissulfonium salts. U.S. Patent No. 3401152 (1968)
39. Elsenbaumer, R.L., Jen, K.Y., Oboodi, R.: Processible and environmentally stable conducting polymers. Synth. Met. **15**, 169–172 (1986)
40. Nehls, B.S., Preis, E., Füldner, S., Farrell, T., Sherf, U.: Microwave-assisted synthesis of 1,5-and 2,6-linked naphthylene-based ladder polymers. Macromolecules **38**, 687–694 (2005)
41. Pasco, S.T., Lahti, P.M., Karasz, F.K.: Synthesis of substituted poly (p-phenylenevinylene) copolymers by the Heck method for luminescence studies. Macromolecules **32**, 6933–6937 (1999)
42. Edmonds, M., Abell, A.: Modern Carbonyl Olefination. Wiley-VCH, Verlag GmbH & Co. KGaA, Weinheim (2004)
43. Liao, J., Wang, Q.: Ruthenium-catalyzed knoevenagel condensation: a new route toward cyano-substituted poly(p-phenylenevinylene)s. Macromolecules **37**, 7061–7063 (2004)
44. Park, L.S., Han, Y.S., Kima, S.D., Kim, D.U.: Synthesis of poly (alkoxyphenylene vinylene-co-phenylene vinylene) type copolymers and electro-optic properties. Synth. Met. **117**, 237–239 (2001)
45. Zheng, Q., Jung, B.J., Sun, J., Katz, H.E.: Ladder-type oligo-p-phenylene-containing copolymers with high open-circuit voltages and ambient photovoltaic activity. J. Am. Chem. Soc. **132**(15), 5394–5404 (2010)

46. Song, S., Park, S.H., Jung, J., Kim, I., Lee, K., Jin, Y., Suh, H.: Increasing of stability depended on the position of alkoxy group in PPV. Synth. Met. **161**, 1186–1193 (2011)
47. Jin, Y., Kim, J.Y., Song, S., Xia, Y., Kim, J., Woo, H.Y., Lee, K., Suh, H.: The influence of tetrakis-ethylhexyloxy groups substituted in PPV derivative for PLEDs. Polymer **49**, 467–473 (2008)
48. Jin, Y., Kim, J.Y., Park, S.H., Kim, J., Lee, S., Lee, K., Suh, H.: Syntheses and properties of electroluminescent polyfluorene-based conjugated polymers, containing oxadiazole and carbazole units as pendants, for LEDs. Polymer **46**, 12158–12165 (2005)
49. Park, J.-W., Park, S.J., Kim, Y.-H., Shin, D.-C., You, H., Kwon, S.-K.: Pure color and stable blue-light emission-alternating copolymer based on fluorene and dialkoxynaphthalene. Polymer **50**, 102–106 (2009)
50. Lee, T.H., Lai, K.M., Leung, L.M.: Hole-limiting conductive vinyl copolymers for AlQ 3-based OLED applications. Polymer **50**, 4602–4611 (2009)
51. Park, M.H., Yun, C., Huh, J.O., Do, Y., Yoo, S., Lee, M.H.: Synthesis and hole-transporting properties of vinyl-type polynorbornenes with ethyl ester linked triarylamine side groups. Synth. Met. **160**, 2000–2007 (2010)
52. Xia, C., Wang, X., Lin, J., Jiang, W., Ni, Y., Huang, W.: Organic light-emitting devices (OLED) based on new triphenylamine derivatives. Synth. Met. **159**, 194–200 (2009)
53. Kim, S.-K., Yang, B., Park, Y., Ma, Y., Lee, J.-Y., Kim, H.-J., Park, J.: Synthesis and electroluminescent properties of highly efficient anthracene derivatives with bulky side groups. Org. Electron. **10**, 822–833 (2009)
54. Wang, J., Leung, L.M.: Synthesis and characterization of blue light emitting *N,N*-dimethylpyren-1-amine and its vinyl polymer. Adv. Mater. Res. **557–559**, 722–726 (2012)
55. Leung, L.M., Wang, J., Wong, M.Y., Law, Y.C., Lai, K.M., Lee, T.-H.: Charge balance conductive vinyl polymers for homojuncton organic light emitting diodes (OLEDs). In: IEEE Proceedings 978-1-4577-1964-6/12, 2012, SCET2012: 594–597
56. Grell, M., Bradley, D., Inbasekaran, M., Woo, E.: A glass-forming conjugated main-chain liquid crystal polymer for polarized electroluminescence applications. Adv. Mater. **9**, 798–802 (1997)
57. Bernius, M., Inbasekaran, M., Woo, E., Wu, W., Wujkowski, L.: Fluorene-based polymers-preparation and applications. J. Mater. Sci. Mater. Electron **11**, 111–116 (2000)
58. Redecker, M., Bradley, D., Inbasekaran, M., Wu, W., Woo, E.: High Mobility hole transport fluorene-triarylamine copolymers. Adv. Mater. **11**, 241–246 (1999)
59. Ohmori, Y., Tada, N., Fujii, A., Ueta, H., Sawatani, T., Yoshino, K.: Organic electroluminescent device with RGB emission. Thin Solid Films **331**, 89–95 (1998)
60. Redecker, M., Bradley, D., Inbasekaran, M., Woo, E.: Nondispersive hole transport in an electroluminescent polyfluorene. Appl. Phys. Lett. **74**, 1400–1403 (1999)
61. Redecker, M., Bradley, D., Baldwin, K., Smith, D., Inbasekaran, M., Wu, W., Woo, E.: An investigation of the emission solvatochromism of a fluorene-triarylamine copolymer studied by time resolved spectroscopy. J. Mater. Chem. **9**, 2151–2153 (1999)
62. Zeng, G., Yu, W.-L., Chua, S.-J., Huang, W.: Spectral and thermal spectral stability study for fluorene-based conjugated polymers. Macromolecules **35**(18), 6907–6914 (2002)
63. Park, J.-W., Park, S.J., Kim, Y.-H., Shin, D.-C., You, H., Kwon, S.-K.: Pure color and stable blue-light emission-alternating copolymer based on fluorene and dialkoxynaphthalene. Polymer **50**, 102–106 (2009)
64. Veldman, D., Ipek, O., Meskers, S.C.J., Sweelssen, J., Koetse, M.M., Veenstra, S.C., Kroon, J.M., Bavel, S.S., Loos, J., Janssen, R.A.J.: Compositional and electric field dependence of the dissociation of charge transfer excitons in alternating polyfluorene copolymer/fullerene blends. J. Am. Chem. Soc. **130**(24), 7721–7735 (2008)
65. Gadisa, A., Mammo, W., Andersson, L.M., Admassie, S., Zhang, F., Andersson, M.R., Inganäs, O.: A new donor–acceptor–donor polyfluorene copolymer with balanced electron and hole mobility. Adv. Funct. Mater. **17**(18), 3836–3842 (2007)

66. Herguth, P., Jiang, X., Liu, M.S., Jen, A.K.-Y.: Highly efficient fluorene- and benzothiadiazole-based conjugated copolymers for polymer light-emitting diodes. Macromolecules **35**(16), 6094–6100 (2002)
67. Vilbrandt, N., Gassmann, A., Seggern, H., Rehahn, M.: Blue-greenish electroluminescent poly(p-phenylenevinylene) developed for organic light-emitting diode applications. Macromolecules **49**(5), 1674–1680 (2016)
68. Liu, L., Han, T., Wu, X., Qiu, S., Wang, B., Hanif, M., Xie, Z., Ma, Y.: Aggregation Behaviors of ladder-type poly(p-phenylene) in dilute solutions and spin-coated films. J. Phys. Chem. C **119**(21), 11833–11838 (2015)
69. Grem, G., Leditzky, G., Ullrich, B., Leising, G.: Realization of a blue-light-emitting device using poly(p-phenylene). Adv. Mater. **4**(1), 36–37 (1992)
70. Stille, J.K., Gilliams, Y.: Poly(p-phenylene). Macromolecules **4**(4), 515–517 (1971)
71. Kovacic, P., Jones, M.B.: Dehydro coupling of aromatic nuclei by catalyst-oxidant systems: poly(p-phenylene). Chem. Rev. **87**(2), 357–379 (1987)
72. Baker, K.N., Fratini, A.V., Resch, T., Knachel, H.C.: Crystal structures, phase transitions and energy calculations of poly(p-phenylene) oligomers. Polymer **34**(8), 1571–1587 (1993)
73. Yamamoto, T., Morita, A., Miyazaki, Y., Maruyama, T., Wakayama, H., Zhou, Z.H., Nakamura, Y., Kanbara, T., Sasaki, S., Kubota, K.: Preparation of π-conjugated poly (thiophene-2,5-diyl), poly(p-phenylene), and related polymers using zerovalent nickel complexes. Linear structure and properties of the π-conjugated polymers. Macromolecules **25**(4), 1214–1223 (1992)
74. Kang, J.T.-S., Harrison, B.S., Bouguettaya, M., Foley, T.J., Boncella, J.M., Schanze, K.S., Reynolds, J.R.: Near-infrared light-emitting diodes (LEDs) based on poly(phenylene)/Yb-tris(β-diketonate) complexes. Adv. Funct. Mater. **13**(3), 205–210 (2003)
75. Bredas, J.L., Silbey, R., Boudreaux, D.S., Chance, R.R.: Chain-length dependence of electronic and electrochemical properties of conjugated systems: polyacetylene, polyphenylene, polythiophene, and polypyrrole. J. Am. Chem. Soc. **105**(22), 6555–6559 (1983)
76. Rehahn, M., Schlüter, A.-D., Wegner, G.: Soluble poly(para-phenylene)s. 1. Extension of the Yamamoto synthesis to dibromobenzenes substituted with flexible side chains. Polymer **30**(6), 1054–1059 (1989)
77. Yasuda, T., Yamamoto, T.: Synthesis and Characterization of new luminescent 1,10-phenanthroline- and pyridine-containing π-conjugated polymers. Their optical response to protic acid, M^{n+}, and solvents. Macromolecules **36**, 7513–7519 (2003)
78. Yamamoto, T., Fang, Q., Morikita, T.: New soluble poly(aryleneethynylene)s consisting of electron-accepting benzothiadiazole units and electron-donating dialkoxybenzene units. Synthesis, molecular assembly, orientation on substrates, and electrochemical and optical properties. Macromolecules **36**, 4262–4267 (2003)
79. Fahlman, M., Lögdlund, M., Stafstrtöm, S., Salaneck, W.R., Friend, R.H., Burn, P.L., Holmes, A.B., Kaeriyama, K., Sonoda, Y., Lhost, O., Meyers, F., Brédas, J.L.: Experimental and Theoretical Studies of the electronic structure of poly(p-phenylenevinylene) and some ring-substituted derivatives. Macromolecules **28**, 1959–1965 (1995)
80. Yamamoto, T., Muramatsu, Y., Lee, B.L., Kokubo, H., Sasaki, S., Hasegawa, M., Yagi, T., Kubota, K.: Preparation of new main-chain type polyanthraquinones. chemical reactivity, packing structure, piezochromism, conductivity, and liquid crystalline and photonic properties of the polymers. Chem. Mater. **15**, 4384–4393 (2003)
81. Scherf, U., Müllen, K.: The synthesis of ladder polymers. In: Synthesis and Photosynthesis Volume 123 of the series Advances in Polymer Science, pp. 1–40 (2005)
82. Garay, R.O., Baier, U., Bubeck, C., Müllen, K.: Low-temperature synthesis of poly (p-phenylenevinylene) by the sulfonium salt route. Adv. Mater. **5**, 561–564 (1993)
83. Heller, C.M., Campbell, I.H., Laurich, B.K., Smith, D.L., Bradley, D.D.C., Burn, P.L., Ferraris, J.P., Müllen, K.: Solid-state-concentration effects on the optical absorption and emission of poly (p-phenylene vinylene)-related materials. Phys. Rev. B **54**, 5516–5520 (1996)
84. Wudl, F., Barbara, S., Srdanov, G.: U.S. Patient No. 5189136 (1993)

85. Choo, D.J., Talaie, A., Lee, Y.K., Jang, J., Park, S.H., Huh, G., Yoo, K.H., Lee, J.Y.: Examination of EL and PL properties of MCHM-PPV and MEH-PPV: a study towards introduction of a new series of thin film EL devices. Thin Solid Films **363**, 37–41 (2000)
86. Jeong, H.Y., Lee, Y.K., Talaie, A., Kim, K.M., Kwon, Y.D., Jang, Y.R., Yoo, K.H., Choo, D.J., Jang, J.: Synthesis and characterization of the first adamantane-based poly (p-phenylenevinylene) derivative: an intelligent plastic for smart electronic displays. Thin Solid Films **417**, 171–174 (2002)
87. Wudl, F., Sigurd, H.: United States Patent. 5679757 (1997)
88. Ahn, T., Ko, S.W., Lee, J., Shim, H.K.: Novel Cyclohexylsilyl- or phenylsilyl-substituted poly(p-phenylene vinylene)s via the halogen precursor route and Gilch polymerization. Macromolecules **35**, 3495–3505 (2002)
89. Swanson, L.S., Lu, F., Shinar, J., Ding, Y.W., Barton, T.J.: Proc. SPIE Int. Soc. Opt. Eng. **1993**, 101–110 (1910)
90. Weder, C., Wrighton, M.S.: Efficient solid-state photoluminescence in new poly (2,5-dialkoxy-p-phenyleneethynylene)s. Macromolecules **29**, 5157–5165 (1996)
91. Schmitt, G., Eiselt, P., Giesa, R., Schmidt, H.-W.: Synthesis and properties of para-linked aromatic polycondensates containing 3,3,4,4 -biphenyltetracarboxylic-N, N-bis (4-aminobiphenyl) diimide units. Macromol. Chem. Phys. **200**, 1879–1888 (1999)
92. Swanson, L.S., Shinar, J., Ding, Y.W., Barton, T.J.: Photoluminescence, electroluminescence, and optically detected magnetic resonance study of 2,5-dialkoxy derivatives of poly (p-phenyleneacetylene) (PPA) and PPA-based light-emitting diodes. Synth. Met. **55**, 1–6 (1993)
93. Weder, C., Wagner, M.J., Wrighton, M.S.: Solid state structure and photoluminescence properties of poly(2,5-dialkoxy-p-phenylene ethynylene)s. In: L.R. Dalton (ed.) Organic Solid State Materials III. Mat. Res. Soc. Symp. Proc. **413**, 77–84 (1996)
94. Tada, K., Onoda, M., Hirohata, M., Kawai, T., Yoshino, K.: Blue-green electroluminescence in copolymer based on poly (1, 4-phenylene ethynylene). Jpn. J. Appl. Phys. **35**, L251–L253 (1996)
95. Ohmori, Y., Uchida, M., Muro, K., Yoshino, K.: Visible-light electroluminescent diodes utilizing poly(3-alkylthiophene). Jpn. J. Appl. Phys. **30**(2)3, L1938–L1944 (1991)
96. Dyreklev, P., Berggren, M., Inganäs, O., Andersson, M.R., Wennerstrom, O., Hjertberg, T.: Polarized electroluminescence from an oriented substituted polythiophene in a light emitting diode. Adv. Mater. **7**, 43–45 (1995)
97. Bolognesi, A., Bajo, G., Paloheimo, J., Östergärd, T., Stubb, H.: Polarized electroluminescence from an oriented poly (3-alkylthiophene) langmuir–blodgett structure. Adv. Mater. **9**, 121–124 (1997)
98. Mao, H., Holdcroft, S.: Grignard synthesis of π-conjugated poly (3-alkylthiophenes): controlling molecular weights and the nature of terminal units. Macromolecules **25**, 554–558 (1992)
99. Mao, H., Xu, B., Holdcroft, S.: Synthesis and structure-property relationships of regioirregular poly(3-hexylthiophenes). Macromolecules **26**, 1163–1169 (1993)
100. Marsella, M.J., Swager, T.M.: Designing conducting polymer-based sensors: selective ionochromic response in crown ether-containing polythiophenes. Polym. Prepr. **33**, 1196–1197 (1992)
101. Tian, J., Wu, C.-C., Thompson, M.E., Sturm, J.C., Register, R.A., Marsella, M.J., Swager, T. M.: Electroluminescent properties of self-assembled polymer thin films. Adv. Mater. **7**, 395–398 (1995)
102. Gill, R.E., Malliaras, G.G., Wildeman, J., Hadziioannou, G.: Tuning of photo-and electroluminescence in alkylated polythiophenes with well-defined regioregularity. Adv. Mater. **6**, 132–135 (1994)
103. Tian, J., Wu, C.-C., Thompson, M.E., Sturm, J.C., Register, R.A.: Photophysical properties, self-assembled thin films, and light-emitting diodes of poly (P-pyridylvinylene) s and poly (P-pyridinium vinylene)s. Chem. Mater. **7**, 2190–2198 (1995)

104. Marsella, M.J., Fu, D.-K., Swager, T.M.: Synthesis of regioregular poly (methyl pyridinium vinylene): an isoelectronic analogue to poly (phenylene vinylene). Adv. Mater. **7**, 145–147 (1995)
105. Gill, R.E., Malliaras, G.G., Wildeman, J., Hadziioannou, G.: Tuning of photo-and electroluminescence in alkylated polythiophenes with well-defined regioregularity. Adv. Mater. **6**, 132–135 (1994)
106. McCullough, R.D., Lowe, R.D.: Enhanced electrical conductivity in regioselectively synthesized poly (3-alkylthiophenes). J. Chem. Soc. Chem. Commun. 70–72 (1992)
107. McCullough, R.D., Lowe, R.D., Jayaraman, M., Anderson, D.L.: Design, synthesis, and control of conducting polymer architectures: structurally homogeneous poly (3-alkylthiophenes). J. Org. Chem. **58**, 904–912 (1993)
108. Chen, T.-A., Rieke, R.D.: The first regioregular head-to-tail poly(3-hexylthiophene-2,5-diyl) and a regiorandom isopolymer: nickel versus palladium catalysis of 2(5)-bromo-5(2)-(bromozincio)-3-hexylthiophene polymerization. J. Am. Chem. Soc. **114**(10), 087–10088 (1992)
109. Chen, T.-A., Wu, X., Rieke, R.D.: Regiocontrolled synthesis of poly(3-alkylthiophenes) mediated by rieke zinc: their characterization and solid-state properties. J. Am. Chem. Soc. **117**, 233–244 (1995)
110. Sandstedt, C.A., Rieke, R.D., Eckhardt, C.J.: Solid-state and solvatochromic spectra of a highly regular polythiophene. Chem. Mater. **7**, 1057–1059 (1995)
111. Wang, H.L., Marsella, M.J., Fu, D.-K., Swager, T.M., MacDiarmid, A.G., Epstein, A.J.: Photophysics of poly(p-pyridine): Blue electroluminescent devices from a soluble conjugated polymer. Polym. Mater. Sci. Eng. **73**, 473–474 (1995)
112. Nalwa, H.S.: Handbook of Conductive Molecules and Polymers, vol. 1–4. Wiley, Chichester (1997)
113. Lane, P.A., Rostalski, J., Giebeler, C., Martin, S.I., Bradley, D.D.C., Meissner, D.: Electroabsorption studies of phthalocyanine/perylene solar cells. Sol. Energy Mater. Sol. Cells **63**, 3–13 (2000)
114. Rostalski, J., Meissner, D.: Photocurrent spectroscopy for the investigation of charge carrier generation and transport mechanisms in organic p/n-junction solar cells. Sol. Energy Mater. Sol. Cells **63**(1), 37–47 (2000)
115. Hoppe, H., Sariciftci, N.S.: Organic solar cells: an overview. J. Mater. Chem. **19**, 1924–1945 (2004)
116. Singh, B., Sariciftci, N.S.: Progress in plastic electronics devices. Ann. Rev. Mater. Res. **36**, 199–230 (2006)
117. Winder, C., Sariciftci, N.S.: Low bandgap polymers for photon harvesting in bulk heterojunction solar cells. J. Mater. Chem. **14**, 1077–1086 (2004)
118. Bundgaard, E., Krebs, F.C.: Low band gap polymers for organic photovoltaics. Sol. Energy Mater. Sol. Cells **91**, 954–985 (2007)
119. Kroon, R., Lenes, M., Hummelen, J.C., Blom, P.W.M., de Boer, B.: Small bandgap polymers for organic solar cells (polymer material development in the last 5 years). Polym. Rev. **48**, 531–582 (2008)
120. Thompson, B.C., Fre´chet, J.M.J.: Polymer–fullerene composite solar cells. Angew. Chem. Int. Ed. **47**, 58–75 (2008)
121. Mayer, A.C., Scully, S.R., Hardin, B.E., Rowell, M.W., McGehee, M.D.: Polymer-based solar cells. Mater. Today **10**, 28–33 (2007)
122. Roncali, J.: Synthetic principles for bandgap control in linear π-conjugated systems. Chem. Rev. **97**, 173–206 (1997)
123. Shaheen, S., Brabec, C.J., Sariciftci, N.S., Padinger, F., Fromherz, T., Hummelen, J.C.: 2.5 % efficient organic plastic solar cells. Appl. Phys. Lett. **78**, 841–844 (2001)
124. Yokoyama, H., Kramer, E.J., Rafailovich, M.H., Sokolov, J., Schwarz, S.A.: Structure and diffusion of asymmetric diblock copolymers in thin films: a dynamic secondary ion mass spectrometry study. Macromolecules **31**, 8826–8830 (1998)

125. Hoppe, H., Niggemann, M., Winder, C., Kraut, J., Hiesgh, R., Hinsch, A., Meissner, D., Sariciftci, N.S.: Nanoscale morphology of conjugated polymer/fullerene-based bulk-heterojunction solar cells. Adv. Funct. Mater. **14**, 1005–1011 (2004)
126. Gebeyehu, D., Brabec, C.J., Padinger, F., Fromherz, T., Hummelen, J.C., Badt, D., Schindler, H., Sariciftci, N.S.: The interplay of efficiency and morphology in photovoltaic devices based on interpenetrating networks of conjugated polymers with fullerenes. Synth. Met. **118**, 1–9 (2001)
127. Martens, T., Hoen, J.D., Munters, T., Beelen, Z., Goris, L., Monca, J., Oliesloeger, M.D., Vanderzende, D., De Schopper, L., Andriessen, R.: Disclosure of the nanostructure of MDMO-PPV: PCBM bulk hetero-junction organic solar cells by a combination of SPM and TEM. Synth. Met. **138**, 243–247 (2003)
128. Mihailetchi, V., Van Duren, K.K., Blom, P., Hummelen, J.C., Janssen, R., Kroon, J., Rispens, M., Verhees, W., Wienk, M.: Charge transport and photocurrent generation in poly (3-hexylthiophene): methanofullerene bulk-heterojunction solar cells. Adv. Funct. Mater. **16**, 699–708 (2006)
129. Granstro¨m, M., Petritsch, K., Arias, A.C., Lux, A., Andersson, M.R., Friend, R.H.: Laminated fabrication of polymeric photovoltaic diodes. Nature **395**, 257–260 (1998)
130. Halls, J.J.M., Walsh, C.A., Greenham, N.C., Marseglia, E.A., Friend, R.H., Moratti, S.C., Homles, A.B.: Efficient photodiodes from interpenetrating polymer networks. Nature **376**, 498–500 (1995)
131. Gupta, D., Kabra, D., Kolishetti, N., Ramakrishnan, S., Narayan, K.S.: An efficient bulk-heterojunction photovoltaic cell based on energy transfer in graded-bandgap polymers. Adv. Funct. Mater. **17**, 226–232 (2007)
132. Colladet, K., Fourier, S., Cleij, T.J., Lusten, L., Gelan, J., Vanderzande, D., Nguyen, L.H., Neugebauer, H., Sariciftci, S., Aguirre, A., Janssen, G., Goovaerts, E.: Low band gap donor-acceptor conjugated polymers toward organic solar cells applications. Macromolecules **40**, 65–72 (2007)
133. Thompson, B.C., Kim, Y.G., Reynolds, J.R.: Spectral broadening in MEH-PPV: PCBM-based photovoltaic devices via blending with a narrow band gap cyanovinylene-dioxythiophene polymer. Macromolecules **38**, 5359–5362 (2005)
134. Al Ibrahim, M., Roth, H.K., Schroedner, M., Kalvin, A., Zhokhavets, U., Gobsch, G., Scharff, P., Sensfuss, S.: The influence of the optoelectronic properties of poly (3-alkylthiophenes) on the device parameters in flexible polymer solar cells. Org. Electron. **6**, 65–77 (2005)
135. Ahn, T., Choi, B., Ahn, S.N., Han, S.N., Lee, H.: Electronic properties and optical studies of luminescent polythiophene derivatives. Synth. Met. **117**, 219–221 (2001)
136. Chirvaze, D., Chiguvare, Z., Knipper, M., Parisi, J., Dyakonov, V., Hummelen, J.C.: Electrical and optical design and characterisation of regioregular poly (3-hexylthiophene-2, 5diyl)/fullerene-based heterojunction polymer solar cells. Synth. Met. **138**, 299–304 (2003)
137. Takahashi, K., Tsuji, K., Inote, K., Yamaguchi, T., Kanura, T., Murata, K.: Enhanced photocurrent by Schottky-barrier solar cell composed of regioregular polythiophene with merocyanine dye. Synth. Met. **130**, 177–183 (2004)
138. Padinger, F., Rittberger, R., Sariciftci, N.S.: Effects of postproduction treatment on plastic solar cells. Adv. Funct. Mater. **13**, 85–88 (2003)
139. Chirvaze, D., Parisi, J., Hummelen, J.C., Dyakonov, V.: Influence of nanomorphology on the photovoltaic action of polymer? Fullerene composites. Nanotechnology **15**, 1317–1326 (2004)
140. Wudl, F., Kobayashi, M., Heeger, A.J.: Poly (isothianaphthene). J. Org. Chem. **49**, 3382–3384 (1984)
141. Qin, Y., Kim, J.Y., Frisbie, C.D., Hillmyer, M.A.: Distannylated isothianaphthene: a versatile building block for low bandgap conjugated polymers. Macromolecules **41**, 5563–5570 (2008)

142. Dhanabalan, A., van Duren, J.K.J., van Hal, P.A., van Dongen, J.L.J., Janssen, R.A.J.: Synthesis and characterization of a low bandgap conjugated polymer for bulk heterojunction photovoltaic cells. Adv. Funct. Mater. **11**, 255–262 (2001)
143. van Duren, J.K.J., Dhanabalan, A., van Hall, P.A., Janssen, R.A.J.: Synth. Met. **121**, 1587 (2001)
144. Wienk, M.M., Turbiez, M., Gilot, J., Janseen, R.A.J.: Narrow-bandgap diketo-pyrrolo-pyrrole polymer solar cells: the effect of processing on the performance. Adv. Mater. **20**, 2556–2560 (2008)
145. Rochford, J., Chu, D., Hagfeldt, A., Galoppini, E.J.: Tetrachelate porphyrin chromophores for metal oxide semiconductor sensitization: effect of the spacer length and anchoring group position. Am. Chem. Soc. **129**, 4655–4665 (2007)
146. Feng, J., Zhang, Q., Li, W., Li, Y., Yang, M., Cao, Y.J.: Novel porphyrin-grafted poly (phenylene vinylene) derivatives: Synthesis and photovoltaic properties. Appl. Polym. Sci. **109**, 2283–2290 (2008)
147. Huang, X., Zhu, C., Zhang, S., Li, W., Guo, Y., Zhan, X., Liu, Y., Bo, Z.: Porphyrin−Dithienothiophene π-conjugated copolymers: synthesis and their applications in field-effect transistors and solar cells. Macromolecules **41**, 6895–6902 (2008)
148. Blouin, N., Michaud, A., Leclerc, M.: A low-bandgap poly(2,7-carbazole) derivative for use in high-performance solar cells. Adv. Mater. **19**(17), 2295–2300 (2007)
149. Peng, B., Najari, A., Liu, B., Berrouard, P., Gendron, D., He, Y., Zhou, K., Zou, Y., Leclerc, M.: A new dithienylbenzotriazole-based poly(2,7-carbazole) for efficient photovoltaics. Adv. Mater. **211**, 2026–2033 (2010)
150. Qin, R., Li, W., Li, C., Du, C., Veit, C., Schleiermacher, H.-F., Andersson, M., Bo, Z., Liu, Z., Inganas, O., Wuerfel, U., Zhang, F.: A planar copolymer for high efficiency polymer solar cells. J. Am. Chem. Soc. **131**(41), 14612–14613 (2009)
151. Duan, C., Chen, K.-S., Huang, F., Yip, H.-L., Liu, S., Zhang, J., Jen, A.K.Y., Cao, Y.: Synthesis, characterization, and photovoltaic properties of carbazole-based two-dimensional conjugated polymers with donor-π-bridge-acceptor side chains. Chem. Mater. **22**(23), 6444–6452 (2010)
152. Wang, M., Hu, X., Liu, P., Li, W., Gong, X., Huang, F., Cao, Y.: Donor-acceptor conjugated polymer based on naphtho[1,2-c:5,6-c]bis[1, 2, 5]thiadiazole for high-performance polymer solar cells. J. Am. Chem. Soc. **133**(25), 9638–9641 (2011)

Optical Performance of Organic Distributed Feedback Lasers Based on Holographic Polymer Dispersed Liquid Crystals

Lijuan Liu, Li Xuan and Ji Ma

1 Introduction

Liquid crystals (LCs) have become ubiquitous in our daily life from flat-panel displays to optical devices as LCs are optical and electrical anisotropic. The properties along the direction parallel and perpendicular to the LC molecule are different and can be changed by external fields such as electric field, temperature, mechanical force, optical field and magnetic field, which make LC devices are sensitive to external changes [1–9]. In this research area, LC and polymer composites have drawn more attention for many years since LC/polymer composites further expand LC practical applications [10–13]. Depending on different concentrations in composites, LC/polymer composites can be regarded as polymer-stabilized liquid crystals (PSLCs) with polymer network in the system [14–18] or polymer-dispersed liquid crystals (PDLCs) with polymer matrix in the system [19, 20]. Holographic polymer dispersed liquid crystals (HPDLCs) have been developed from 1993 [21, 22] and show potential in optical communications [23–27], reflective displays [28–33], optical switches [34–37], organic lasers and other applications.

HPDLCs are fabricated by photo-polymerization induced phase separation (PIPS) method by exposing a mixture of LC(s) and photo-sensitive monomer(s) to a laser optical interference field [38–41]. The laser optical field is created by two or multiple coherent laser beams. The periodic structure in HPDLCs is composed of alternating lamellas of polymer-rich layer and phase-separated LC-rich layer, in

L. Liu
College of Physics and Engineering, Qufu Normal University, Qufu 273165, China

L. Xuan · J. Ma
State Key Laboratory of Applied Optics, Changchun Institute of Optics,
Fine Mechanics and Physics, Chinese Academy of Sciences, Changchun 130033, China

J. Ma (✉)
Liquid Crystal Institute, Kent State University, Kent, OH 44240, USA
e-mail: jma2@kent.edu

D. Rosu and Visakh P. M. (eds.), *Photochemical Behavior of Multicomponent Polymeric-based Materials*, Advanced Structured Materials 26,
DOI 10.1007/978-3-319-25196-7_12

which the morphology is corresponding to the interference patterns of the laser optical filed. Organic distributed feedback (DFB) lasers based on HPDLCs became well-known novel HPDLC application, recently. To form a laser, the capability of amplifying light in a cavity or resonator is needed while the periodic structure in HPDLCs with different refractive indices by polymer-rich layer and LC-rich layer is ideal for light feedback. Therefore HPDLC gratings have been developed as cavities or external feedback layers for organic DFB lasers. Compared with other fabrication techniques, such as hot embossing [42], electron beam lithography [43] or nanoimprint lithography [44], DFB lasers based on HPDLCs exhibit merits like easy fabrication, simple optical construction and tunability capability by external fields.

In this chapter, we aim to present a snapshot of the research carried out so far on DFB lasers based on HPDLCs. First we briefly introduce HPDLC materials, fabrication setups and morphologies. Then we focus on the progress of HPDLCs for DFB lasers as laser cavities and external light feedbacks. The emission characteristics of HPDLC DFB lasers and the ways to improve optical performance are reviewed. The novel applications of multiple wavelength HPDLC lasers and tunable laser properties by external fields are also discussed.

2 Holographic Polymer Dispersed Liquid Crystals

Fabrication methods and morphologies of the HPDLCs are important since they will influence the output lasing performance. Different configurations or structures will give different device properties. Therefore, we briefly introduce materials, writing setups and morphologies of HPDLCs in this section.

2.1 Materials for HPDLCs

One of fundamental research on HPDLCs is to study material systems and material formulations such as components and concentration of each component in the material system in order to improve HPDLC performance. The material systems used for HPDLCs mainly contain two underlying components: LC(s) and monomer (s). Acrylate-based and thiol-ene-based monomers are usually used in HPDLCs. Acrylate-based monomers with different functionalities have been studied including dipentaerythritol hydroxy pentaacrylate (DPHPA) [21, 22, 45, 46], phthalic diglycol diacrylate (PDDA) [47, 48], 2-ethylhexyl acrylate (EHA) [49, 50], trimethylolpropane triacrylate (TMPTA) [50–53], polyurethane acrylate oligomers (PUA) [46, 54–56], pentaerythritol tetraacrylate (PETA-3) [57, 58] and dipentaerythritol pentaacrylate (SR399) [59]. Thiol-ene monomers with multifunctional thiol and vinyl monomers have been used, such as Norland optical adhesives (NOA 61, NOA 65, and NOA 68) [60–65]. The molecular structures of typical polymers are shown in Table 1. For LCs, nematic LCs are usually used. Ideally, the ordinary

Table 1 Chemical structures of typical prepolymers/monomers used in HPDLCs

Acrylate-based monomers			Thiol-ene-based monomers		
Name	Abbreviation	Chemical structure	Name	Abbreviation	Chemical structure
Dipentaeryth-ritol hydroxyl pentaaciy late	DPHPA	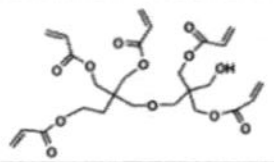	Pentaerythri-tol tetra-3-mercaptopropionate	PETMA	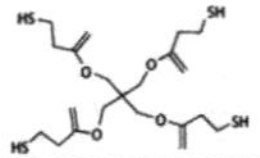
Phthalic dialycol diacrylate	PDDA	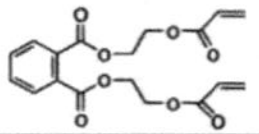	Trimethylol-propane diallyl ether	TMPDE	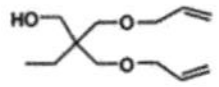
Trimethylolp-ropane triacrylate	TMPTA	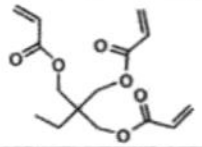	Isophorone diisocyanate	NOA65(I)	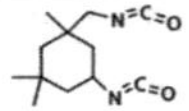
Pentaerythrit-ol triacrylate	PETA	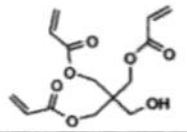	Benzophen-one	NOA65(II)	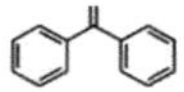

refractive index of the LC approximately matches the refractive index of the surrounding polymer matrix, and the extraordinary refractive index of the LC is different with the refractive index of the polymer. Once an electric field is applied, the HPDLC will become optically homogeneous and the grating is in its "off" state. Upon removal of the field, the LCs will return to their original orientation and the grating reverts back to its "on" state. In addition, in order to induce photo-polymerization, photo-initiators and co-initiators are necessary. Table 2 shows some examples of the photo-initiators for different writing laser beam wavelengths. Sometimes, chain extenders are added to optimize grating morphologies, such as *N*-vinylpyrrolidinone (NVP) [66]. All materials will be mixed together to form a homogeneous mixture for HPDLC fabrications.

2.2 Writing Setups for HPDLCs

The homogeneous mixture will be injected into an empty glass cell, which is made by two glass substrates, and the cell gap is controlled in the range of several micrometers. Transmission and reflection HPDLCs are two common types in HPDLCs, according to the holographic recording setups. As shown in Fig. 1a, for transmission gratings, the two writing coherent beams are in the same side of the sample. The recorded fringes by LC-rich layers and polymer-rich layers are perpendicular to the cell surface. On the other hand, as shown in Fig. 1b, for reflection gratings, the writing light coherent beams are in the different sides of the sample. The recorded fringes are parallel to the cell surface. The spatial period of the gratings can be controlled by the angle (θ) between the two writing beams by

$$\Lambda = \frac{\lambda_w}{2n\,\sin(2/\theta)} \tag{1}$$

where Λ is the grating period, λ_w is the writing wavelength, and n is the average refractive index of the material mixture [38].

2.3 HPDLC Morphologies

Different materials, writing setups and curing conditions will result in different HPDLC morphologies in the gratings. We can categorize HPDLC grating morphologies to three types. The first type is LC droplet-like morphologic transmission [50] or reflection gratings [67–69], which can be made from acrylate-based monomers or thiol-ene-based monomers [65]. The LC molecules diffuse and form LC droplets in the grating in PIPS process, as shown in Fig. 2a. High curing light intensity (~ 100 mW/cm^2) and high effective functionality of monomers (>3.5) will produce LC droplet-like morphology in the grating. In this type of HPDLCs, the

Table 2 Chemical structures of typical photo-initiators used in HPDLCs

Name	Chemical structure	Writing wavelength (nm)	Material system
2,4,6-Trimethylbenzoyl-diphenylphosphine oxide 2-Hydroxy-2-methyl-1-phenylpropan-1-one	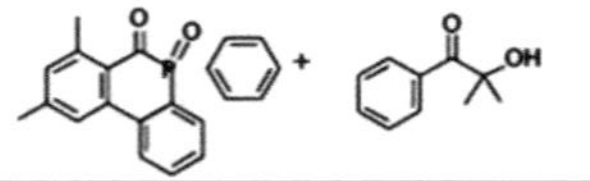	351	Acrylate-based
Irgacure 784	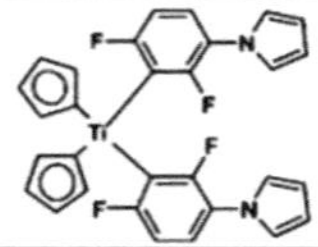	532	Acrylate-based
Rose Bengal (RB) *N*-phenylglycine (NPG)	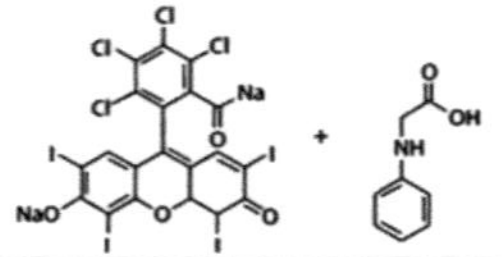	488, 514, 532	Acrylate-based
Irgacure 1173	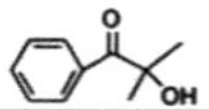	351, 363	Thiol-ene-based
Rhodamine 6G (Rh6G) Benzoyl peroxide (BzO2)	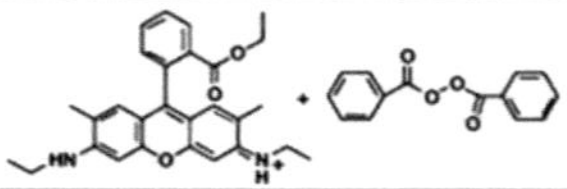	488, 514, 532	Thiol-ene-based
Methylene blue (MB) p-Touenesulfonic acid	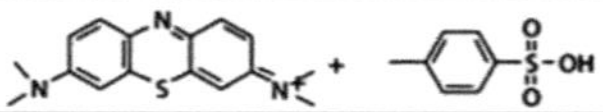	633	Thiol-ene-based
Oxazine 170 perchlorate (Qxzl70)	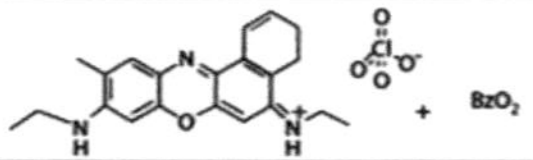	647	Thiol-ene-based

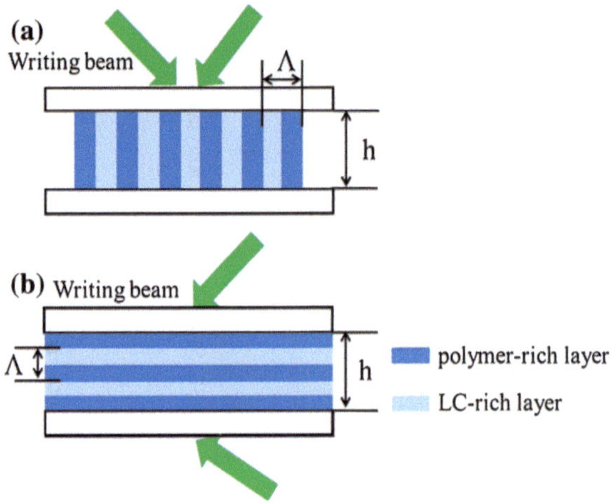

Fig. 1 Typical optical writing setups for HPDLCs: **a** transmission grating and **b** reflection grating. h, cell gap

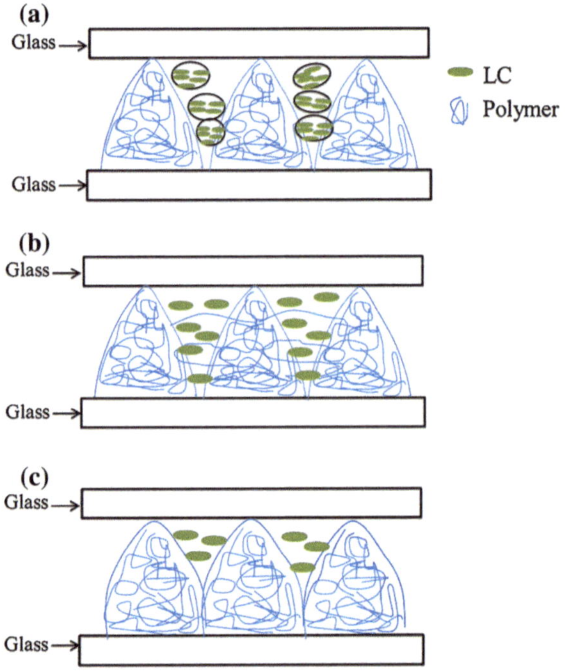

Fig. 2 Schematic configuration of HPDLC transmission grating: **a** LC droplet-like morphology, **b** polymer scaffolding morphology and **c** sliced polymer morphology

light scattering is high since the diameter of LC domains in LC droplets is comparable with the visible-light wavelengths.

The second type is polymer scaffolding morphology [49, 51, 70–72], in which the transmission gratings are made from acrylate-based monomers with a slow curing process. The LC domains or LC layers instead of small-sized LC droplets exist in the LC-rich layer while polymer filaments are transverse between the neighboring polymer-rich planes to form scaffolding structures [49], as shown in Fig. 2b. Low polymerization rate is fundamental factor to form scaffolding morphology grating, which means either relatively low curing light intensity or low effective functionality of monomers can be utilized to obtain such morphology. The scattering loss is lower than that in droplet-like morphologic gratings in general. The LC alignment is along the grating vector, i.e., orthogonal to the holographic planes.

The third type is sliced polymer morphology (also called POLICRYPS) due to unique curing condition [36, 37, 62, 73–77], in which pure and well-aligned nematic LC "slice" is separated by polymer slice, as shown in Fig. 2c. The gratings made by thiol-ene-based material are fabricated at a temperature above the LC clearing point with a slow curing process. Because the LC concentration is low (15–30 wt%), the LC layers are narrower than the polymer layers. LCs are aligned along the grating vector since LCs are anchored by a hybrid layer with some polymer networks at the LC/polymer interface.

3 HPDLCs for Organic DFB Lasers

3.1 Optical Pumping for Opto-Optical HPDLC DFB Lasers

For laser applications, gain media such as laser dyes or semiconducting materials are doped in HPDLC systems. Laser emission from HPDLC structures is achieved through optical pumping, using either picosecond or nanosecond pulses from solid-state lasers such as Nd:YAG. The wavelength dependence on the gain media absorption defines the excitation wavelength of pump laser. The schematic setup of optical pumping of one example is shown in Fig. 3. The pump laser beam is focused through a cylindrical lens onto the surface of the HPDLC to form a gain area along the direction perpendicular to the grating layers for light feedback to form a laser.

Fig. 3 Schematic diagram of optical pumping for HPDLC DFB laser

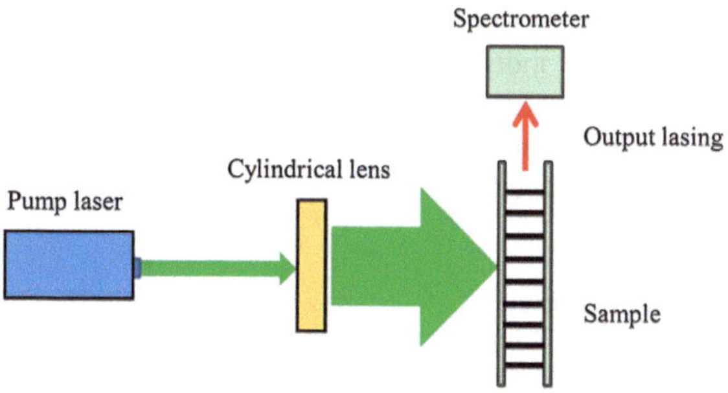

3.2 HPDLC as Light Feedbacks

3.2.1 HPDLC as DFB Laser Cavities

HPDLCs have periodic refractive index distribution by polymer-rich layer and LC-rich layer, so they can be used as DFB laser cavities. In the DFB laser cavity, the light feedback for lasing modes relies on *Bragg* scattering by the periodic microstructures [78, 79]. The DFB lasing wavelength from the device should satisfy *Bragg* condition [80]

$$m\lambda_{las} = 2n_{eff}\Lambda \tag{2}$$

where λ_{las} is the wavelength of output lasing, n_{eff} is the effective refractive index of the laser mode and m is the Bragg order. That is to say, for a given periodic configuration, there is a particular set of wavelengths that will be diffracted. For optical pumping, the lower the Bragg feedback order is, the lower the output lasing threshold [81–84]. For this reason, the first and second order DFB lasers are most widely investigated in this topic.

To form a waveguide structure, the HPDLC is usually sandwiched by two glass substrates in which laser dyes or laser gain material is needed. The typical configuration of HPDLC as a laser cavity is shown in Fig. 4. The laser dye such as pyrromethene 597 [85], coumarin 485 (C485) [68], 4-(dicyanomethylene)-2-methyl-6-(*p*-dimethylaminostyryl)-4H-pyran (DCM) [86] or rhodamine 6G [87, 88] was doped in the prepolymer mixture. Light is stimulated from laser dyes by optical pumping and is amplified in HPDLC grating layer by light feedback. The mode which satisfies both *Bragg* condition and waveguide condition [89] would form a laser.

The reflection HPDLC grating was first demonstrated as a laser cavity for obtaining surface-emitting laser emission [68, 85, 87, 88]. Later on, the transmission HPDLC gratings were used more often for DFB laser cavities [90]. Compared with the reflection grating, the transmission grating has a much longer feedback length, which facilitates better performance laser. The representative lasing spectrum and the dependence of lasing intensity on the pumping energy from a droplet-like morphologic transmission HPDLC grating [86] are shown in Fig. 5. The laser emission was at 609 nm. The threshold was about 120 μJ and the full width at half maximum (FWHM) of this lasing was about 1.8 nm at the pumping laser wavelength of 532 nm by a Q-switched frequency doubled Nd:yttrium–

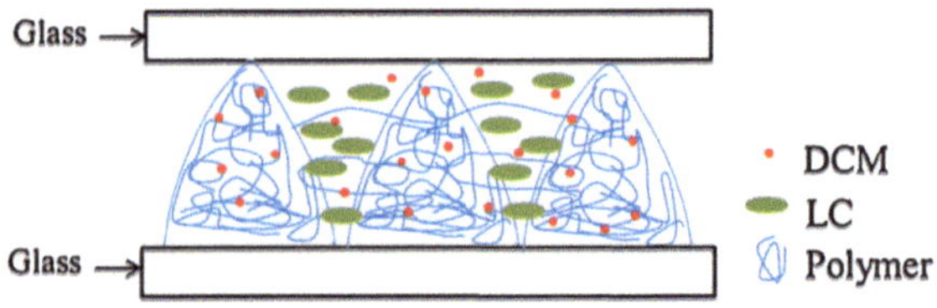

Fig. 4 Schematic configuration of laser dye doped polymer scaffolding morphologic HPDLC transmission grating as laser cavity for DFB laser

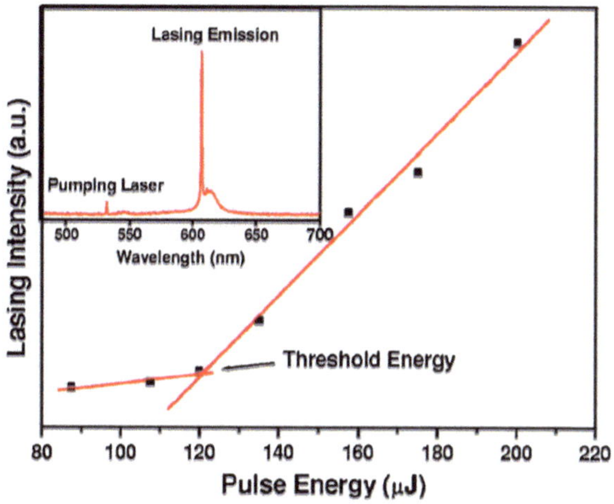

Fig. 5 Lasing output intensity as a function of pump intensity for the DCM-doped HPDLC transmission grating laser. The *inset* shows the corresponding lasing spectra. Reprinted with permission from [86], Copyright 2006, American Institute of Physics

aluminum–garnet pulsed laser. The threshold was considerably lower than previous reports due to the uniform grating structure, which provides more uniform coherent distributed feedback [86].

3.2.2 Methods to Improve Optical Performance (HPDLC as DFB Laser Cavities)

To improve optical performance of DFB lasers, such as to decrease laser threshold, many ways have been proposed to improve HPDLC itself by investigation of material system (concentrations or polymer types), fabrication conditions and HPDLC structures. For example, by optimizing LC concentration in the material system, a better performance of lasing output from DCM-doped LC droplet-like transmission grating can be obtained. At a lower LC concentration, the volume fraction of LC-rich lamellae was decreased and the LC droplets became smaller, resulting less scattering loss. The full width at half maximum (FWHM) of lasing was reduced to 0.8 nm when the LC concentration was 14.6 wt% and the lowest threshold pumping energy was found to be about 5 μJ/pulse when the LC concentration was 19.7 wt% [86]. By using thio-ene monomers, uniform size and spherical shape LC droplets in gratings could be formed and lowered scattering loss in LC droplet-like reflection HPDLC grating. The threshold was reduced to 0.07 mJ/cm^2 in the grating with thio-ene monomer from 0.17 mJ/cm^2 in the grating with multi-functional acrylate monomer [67]. Another way is to increase fabrication temperature to the nematic–isotropic transition point of the LC, so the LC molecules would align randomly in the pure LC layer in droplet-like HPDLC transmission grating and resulted in a much higher refractive index contrast between the polymer-rich layer and the phase-separated LC-rich layer [48]. In this case, more effective light feedback was obtained. The FWHM of the lasing output was about only 0.6 nm and the threshold of the lasing output was 6 μJ/pulse. Two dimensional (2D) transmission HPDLC grating had also been demonstrated to improve laser performance [91–95]. 2D structure was regarded to show an enhanced coupling coefficient [96, 97] and the lasing output beam gave small divergence, which was closed to the diffraction limit [98]. The threshold values of the 2D HPDLC DFB lasers are currently in the range of 0.2–2 mJ/cm^2 [91–95], which is lower than 1D HPDLC DFB laser.

Polymer scaffolding morphologic HPDLC transmission gratings were studied recently for DFB laser applications, since this type grating exhibits high diffraction efficiency, low scattering loss and high degree of phase separation [99–105]. The scattering loss is <6 %, meaning that more efficient light pumping and light feedback can be achieved and higher conversion efficiency of the pumping input to the lasing output can be achieved. The threshold of the lasing output from the DCM-doped scaffolding morphologic transmission grating was 2.75 μJ/pulse with a FWHM of only 0.5 nm [106]. Different laser dyes or dye blends were also studied on optical performance in this type HPDLC. Amplified spontaneous emission (ASE) measurement is usually used to study laser gain materials [107] and ASE is also a useful tool to explore effective lasing conditions [108, 109], such as lasing output wavelength region and threshold distribution at different wavelengths. HPDLC DFB lasers using dye blends as gain media has been studied to form an efficient Forster energy transfer system. The Forster energy transfer from host emitter to guest emitter has been demonstrated an effective way to enhance the device performance [110, 111]. Figure 6a shows the ASE for PM567, DCM and their blends of 2, 10 and 20 % of DCM in PM567 and Fig. 6b shows the lasing threshold as a function of output wavelength. The lowest threshold obtained for each dye system was corresponding to ASE maximum where the net gain is the highest [106]. The lowest threshold energy (1.35 μJ/pulse) in the 10 % DCM in PM567 doped system by Forster energy transfer reduced by factor 2 compared to that without energy transfer process (2.75 μJ/pulse). We believe that more efforts on HPDLCs and gain systems will further improve optical performance of DFB lasers.

3.2.3 HPDLCs as External Feedbacks

Dyes need substantial dilution to avoid quenching, which reduces conversion efficiency of pumping light. To avoid quenching, the concentration of dyes is usually lower than 1 wt% [112]. This concentration of dyes limits optical performance of lasing output from dye–doped HPDLC gratings. Semiconducting materials, as gain media, have been demonstrated to combine with HPDLC gratings for DFB laser applications. Different from dyes, semiconducting polymer can have higher photoluminescence quantum yield in the solid state. Poly[2-methoxy-5-(2-ethylhexyloxy)-1,4-phenylenevinylene] (MEH-PPV) as a gain medium has been adopted in polymer scaffolding morphologic HPDLC DFB lasers [101, 102, 113, 114]. The device configuration is schematically shown in Fig. 7. In this structure, HPDLC

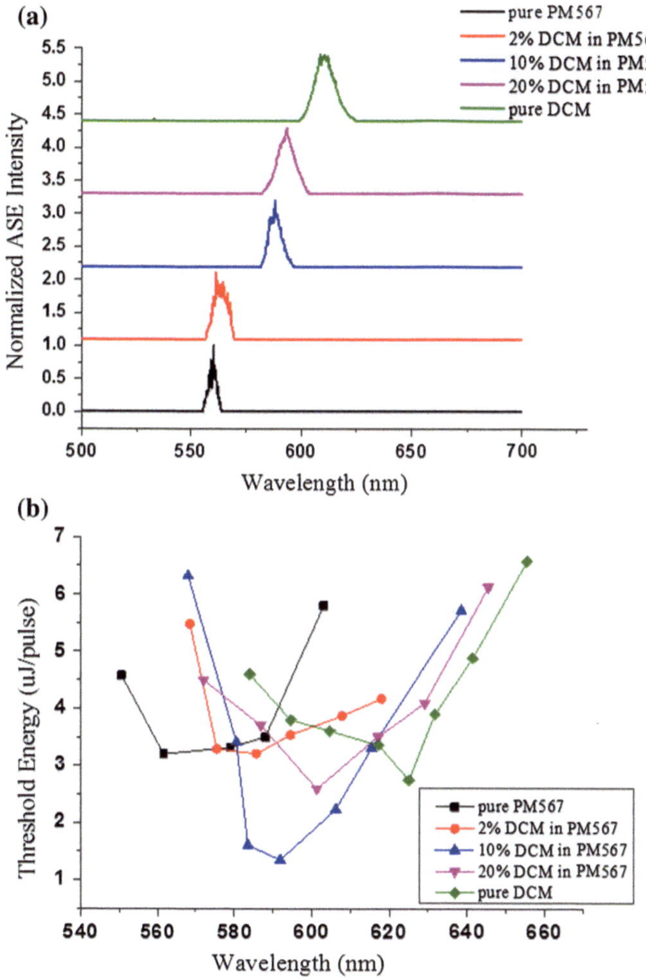

Fig. 6 **a** Amplified spontaneous emission (ASE) of PM567, DCM and their blends and **b** dependence of threshold energy on the lasing wavelength for PM567, DCM and their blends. Reprinted with permission from [106], Copyright 2013, Taylor & Francis

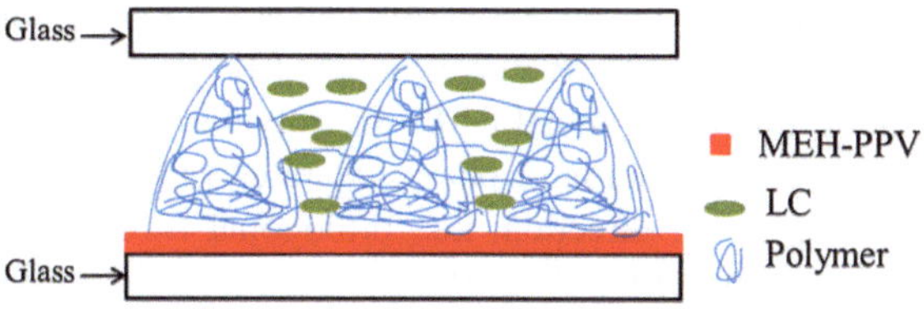

Fig. 7 Schematic configuration of the device containing conductive polymer as a gain medium and HPDLC transmission grating as an external feedback

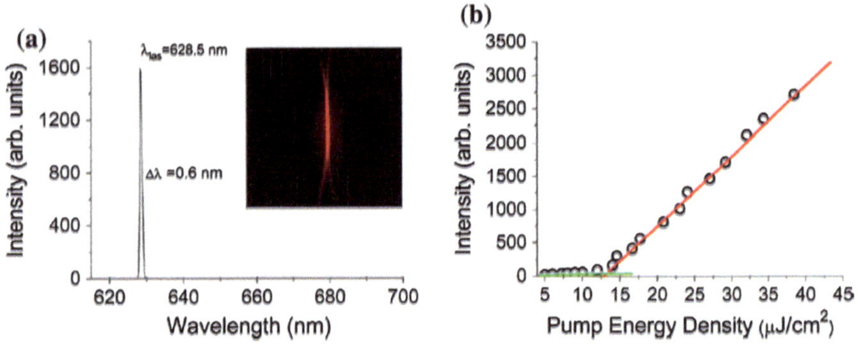

Fig. 8 **a** Emission spectra from the sample (the *inset* shows the vertical laser emission on a paper screen) and **b** emission laser intensity versus pump energy density. Reprinted with permission from [113]. Copyright 2013, Elsevier

grating/MEH-PPV/glass substrate defined a typical asymmetric slab waveguide. The MEH-PPV layer acted as both gain medium and waveguide core layer while the HPDLC grating acted as an external feedback structure in the cladding layer. The evanescent wave spread from the MEH-PPV layer into the HPDLC layer to obtain the feedback. The emitted lasing was totally TE polarized. The MEH-PPV layer thickness was around 75 nm. When the MEH-PPV layer is too thin (lower than 60 nm) or too thick (higher than 140 nm), the emission spectra exhibited broad band rather than lasers. The neat 75 nm MEH-PPV film could absorb about 70 % pump light, which would benefit for better performance of the lasing output. In contrast, 1 wt% dye in HPDLC grating could absorb <20 % pump light. Figure 8 shows laser performance from the second-order MEH-PPV HPDLC DFB laser. The threshold was 13.3 μJ/cm^2 [113], which was much less than that from dye-doped HPDLC DFB laser.

3.2.4 Method to Improve Optical Performance (HPDLC as External Feedbacks)

To further improve laser performance of DFB lasers with HPDLC as external feedback layers, increasing the feedback efficiency was proposed [103]. By controlling the orientation of phase separated LCs in the HPDLC, the refractive index difference between polymer-rich layer and LC-rich layer in the light feedback direction can be increased for better light feedback. Alignment layer such as a rubbed polyimide (PI) layer adopted on the substrate can be used to increase the index difference in HPDLC. In polymer scaffolding morphologic HPDLCs, the averaged orientation of phase separated LC molecules is usually aligned along the grating vector direction, i.e., orthogonal to the holographic planes. For feedback lights propagating along the grating vector, the refractive index difference comes from the polymer (index n_p) and the phase separated LC (ordinary index n_o). These two values of index were very close and the typical difference can be as low as 10^{-5} [90]. Thus the effective light feedback for lasing output is not high. Adopting low functional monomers, the surface anchoring strength of the polymer was much smaller since weak polymer network would be formed. Thus the rubbed PI alignment layer could suppress the anchoring effect of the polymer network, then altered the orientation of phase separated LCs along the direction of the holographic planes. So the index difference was changed to n_p and extraordinary index n_e, which was higher than before. The refractive index difference Δn can be improved from 0.0036 to 0.0236. Figure 9a, b shows the real-time diffraction efficiencies of non-PI sample and the sample with rubbed PI layer and low functional monomer in different polarization states. The p light diffraction efficiency was greater than the s light diffraction efficiency of non-PI sample, indicating the LCs aligned along the grating vector direction. However, the s light diffraction efficiency was greater than the p light diffraction efficiency of the sample with rubbed PI, indicating the LCs

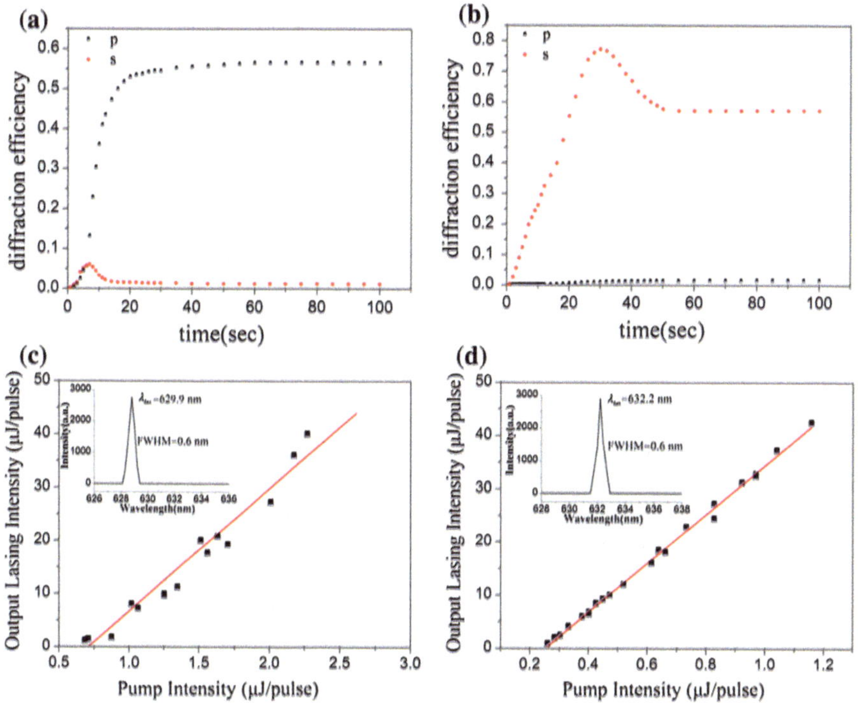

Fig. 9 Real time diffraction efficiency for p polarization (*square*) and s polarization (*sphere*): **a** normal sample without rubbed PI layer and with high functional monomer and **b** sample with rubbed PI layer and low functional monomer. Lasing output intensity as a function of pump intensity for the DFB laser: **c** sample without rubbed PI layer and with high functional monomer and **d** sample with rubbed PI layer and low functional monomer. The *insets* show the corresponding lasing. Reprinted with permission from [103], Copyright 2015, The Royal Society of Chemistry

aligned along the direction of the holographic planes. Figure 9c, d are the lasing output intensity as a function of pump intensity, respectively. For three-order MEH-PPV DFB laser, the threshold of the lasing was reduced from 0.71 to 0.25 μJ/pulse, and the conversion efficiency was increased from 2.3 to 4.6 % [103].

4 HPDLCs for Novel DFB Laser Applications

4.1 HPDLCs for Multi-Wavelength DFB Lasers

Multi-wavelength lasers show potential applications in fiber sensor systems, atmospheric monitoring, THz frequency generator and medical equipment [115–118]. The laser configurations are usually very complex and the controls or demands of the optical apparatus are rigorous. HPDLCs can be used for multiple or dual-wavelength DFB lasers and it is easy to achieve [99, 102]. An dual-wavelength DFB laser empowered by a dye-doped HPDLC grating was reported [99]. According to the Bragg equation (Eq. 2), the workable DCM-doped HPDLC grating period ranges can be calculated for different Bragg orders, as shown in Fig. 10a. In this figure, a grating with a period between 1487 and 1553 nm could be selected to generate a dual-wavelength lasing action via the 7th and 8th Bragg orders simultaneously. The output spectrum of the dual-wavelength, at 586.6 and 670.2 nm simultaneously, was shown in Fig. 10b. The two wavelength lasing behaviors were investigated, as obtained experimentally, as shown in Fig. 10c. The thresholds for the 670.2 and 586.6 nm lasers are 53.2 μJ/pulse with a FWHM of 0.7 nm and 95.44 μJ/pulse with a FWHM of 1.2 nm, respectively [99].

To further enhance the quality of output lasing (lower diverging and improve inhomogeneous beam) and lower the threshold, the configuration of DCM-doped HPDLC grating formed on the top of MEH-PPV layer was proposed [102]. The schematic configuration of the dual-wavelength laser device was shown in Fig. 11a. HPDLC grating provided the feedback for MEH-PPV and DCM via second Bragg order simultaneously, resulting in a coupled dual-wavelength lasing output at one direction on the sample surface. Dual-wavelength emission at 627.8 and 606.6 nm was generated by MEH-PPV and DCM and the FWHM were 0.6 and 0.3 nm respectively, as shown in Fig. 11b. The lasing mechanism of the 627.8 nm MEH-PPV laser can be explained by waveguide theory. The light undergoes stimulated amplification when propagating in the MEH-PPV layer and obtains feedback for specific wavelength by spreading into the grating layer near the interface as the evanescent wave form. While for the 606.6 nm DCM laser, the light propagates in the grating layer as discrete modes and undergoes stimulated amplification by obtaining feedback from the periodic grating structure. The threshold was 0.16 μJ for the 627.8 nm MEH-PPV laser and 0.36 μJ for the 606.6 nm DCM laser, as shown in Fig. 11c. Besides, the wavelength positions of

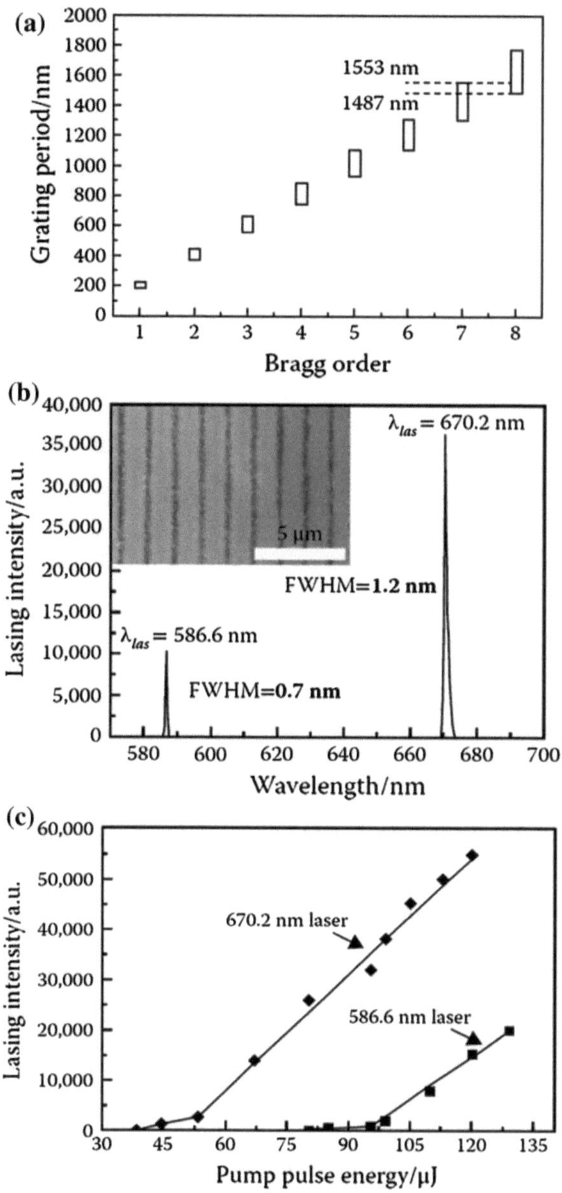

Fig. 10 **a** Prediction of grating period range for different Bragg-order lasing, **b** spectra of the dual wavelength laser based on DCM-doped HPDLC (grating period 1520 nm; *inset* SEM of the grating) and **c** emission laser intensity versus pump energy density. Reprinted with permission from [99], Copyright 2012, The Royal Society of Chemistry

the two lasers could be modulated simultaneously or individually since the dual wavelength laser comes from two different laser media [102].

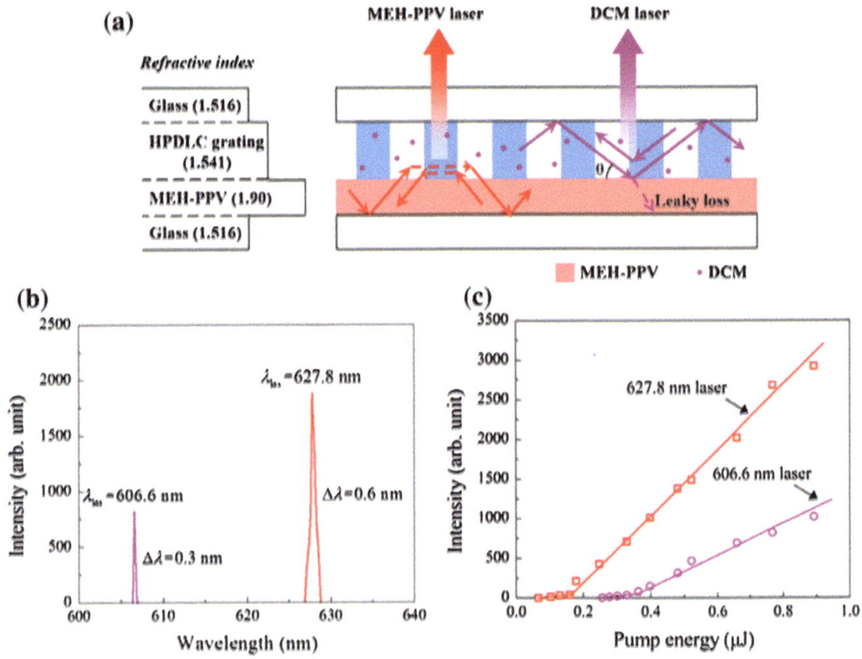

Fig. 11 **a** Schematic device configuration, **b** lasing spectra of the dual-wavelength laser and **c** laser intensity versus pump energy. Reprinted with permission from [102], Copyright 2014, The Royal Society of Chemistry

4.2 Tunable HPDLC Lasing Behaviors

4.2.1 Thermo-Tuning

With the increase of temperature, a redshift in the lasing peaks from 2D HPDLC photonic crystal has been reported [119]. The output laser was measured in TE polarization, i.e., polarization along y direction, as is shown in Fig. 12a. Figure 12b–d shows the lasing peaks are 610/615, 620/626 and 624/631 nm at temperatures of 25, 35, 45 °C, respectively. The LC droplets in 2D HPDLC photonic crystal tends to align the LCs parallel to the z direction at 25 °C. When the temperature was increased, the TE polarization would experience a higher index of LC droplets mainly due to the decreasing of optical anisotropy and it caused the red-shift of the lasing wavelength eventually.

When increasing of temperature, the refractive indices of both LC and polymer would decrease, which will change the property of the HPDLC and its laser applications. Figure 13 shows the wavelength tuning when temperature was changed from 23 to 100 °C [102]. The MEH-PPV HPDLC laser system had a change in laser wavelength from 628 to 625 nm. The DCM-doped system had a larger shift ($\sim$10 nm) in laser wavelength from 607 to 597 nm.

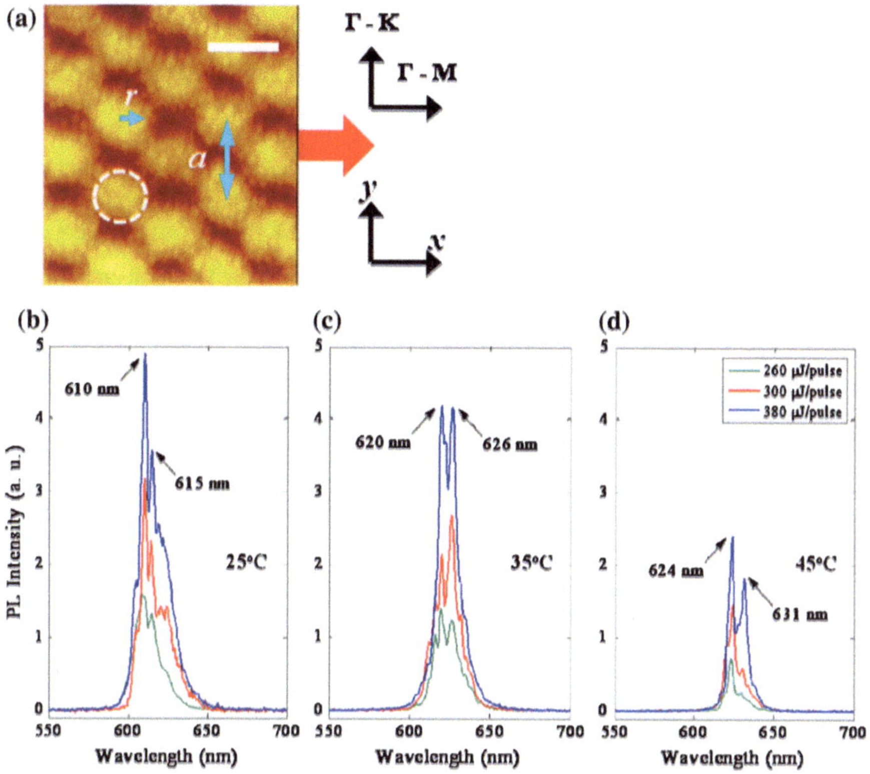

Fig. 12 **a** AFM image of the surface morphology for a 2D hexagonal H-PDLC photonic crystal. *Scale bar* 500 nm and output lasing spectra measured along ΓM direction at temperatures of **b** 25 °C, **c** 35 °C and **d** 45 °C, under different pumping intensity of 260, 300, and 380 μJ/pulse. Reprinted with permission from [119], Copyright 2010, American Institute of Physics

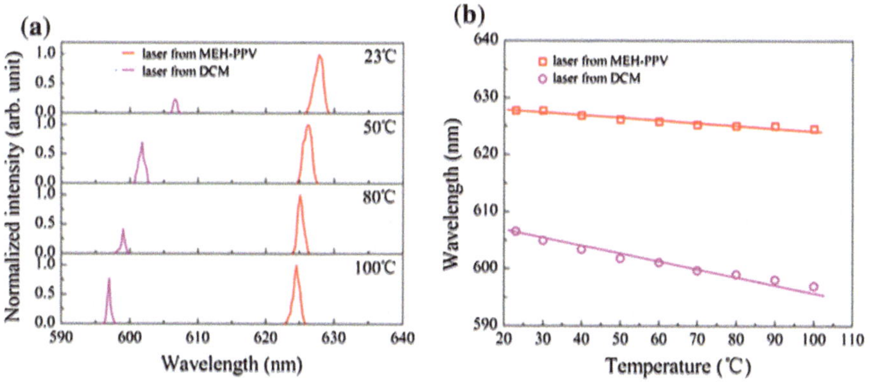

Fig. 13 **a** Lasing spectra and **b** lasing wavelengths at different temperatures for DCM-doped laser and polymer laser. Reprinted with permission from [102], Copyright 2014, The Royal Society of Chemistry

4.2.2 Electro-Tuning

The lasing wavelengths can be tuned by applying an external electric field across the cell to realign the LCs or change the effective refractive index of the laser cavity in the transmission, reflection and 2D HPDLCs. For TE mode laser in polymer scaffolding morphologic HPDLCs, the wavelength can not be tuned since the effective refractive index (RI) of the LC always is n_o regardless of applied electric fields. However, the effective RI of the LC for TM mode (n_{LC}) is equal to $n_o n_e / \sqrt{n_e^2 \cos^2\theta + n_o^2 \sin^2\theta}$, where θ is the angle between the LC molecule and the x-direction that can be changed from 0° to 90° by the electric field, as shown in Fig. 14a. Figure 14b shows the emission spectra from a DCM-doped polymer scaffolding morphologic HPDLC transmission grating under different electric fields [104]. When the applied electric field was changed from 0 to 16.7 V/μm, the peak of the TM_0 mode lasing was tuned from 625.1 to 633.1 nm, showing a total 8 nm redshift. An anisotropic waveguide theory has also been developed for HPDLC lasers to explain the shifting of lasing wavelength and the theoretical simulations and experimental date are good agreement [100].

4.2.3 Photo-Tuning

The HPDLC laser can be photo-tuned due to isomerization of azo-dyes [120, 121]. The intensity of the lasing emission from azo-dye (D2) doped HPDLC grating can

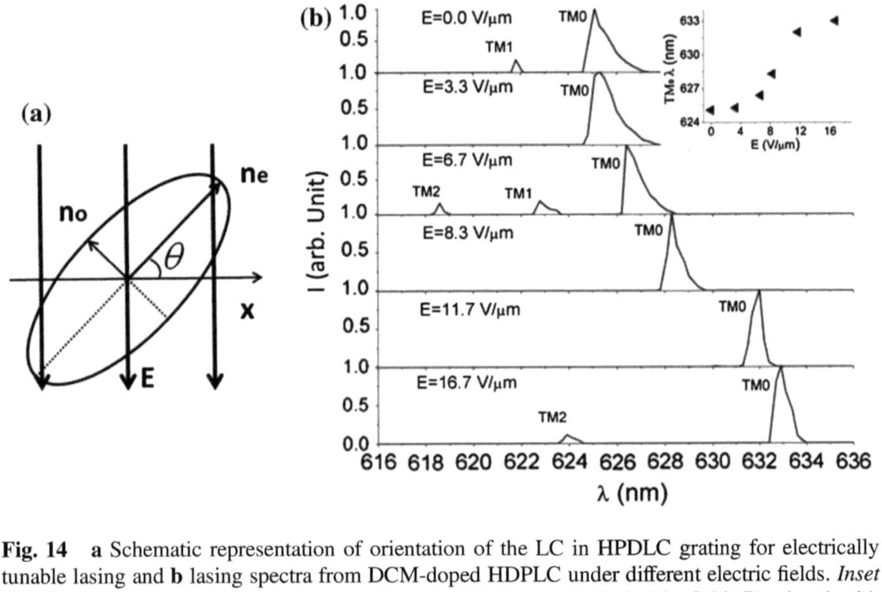

Fig. 14 **a** Schematic representation of orientation of the LC in HPDLC grating for electrically tunable lasing and **b** lasing spectra from DCM-doped HDPLC under different electric fields. *Inset* variations in measured TM0 lasing wavelength with the applied electric field. Reprinted with permission from [104], Copyright 2013, The Japan Society of Applied Physics

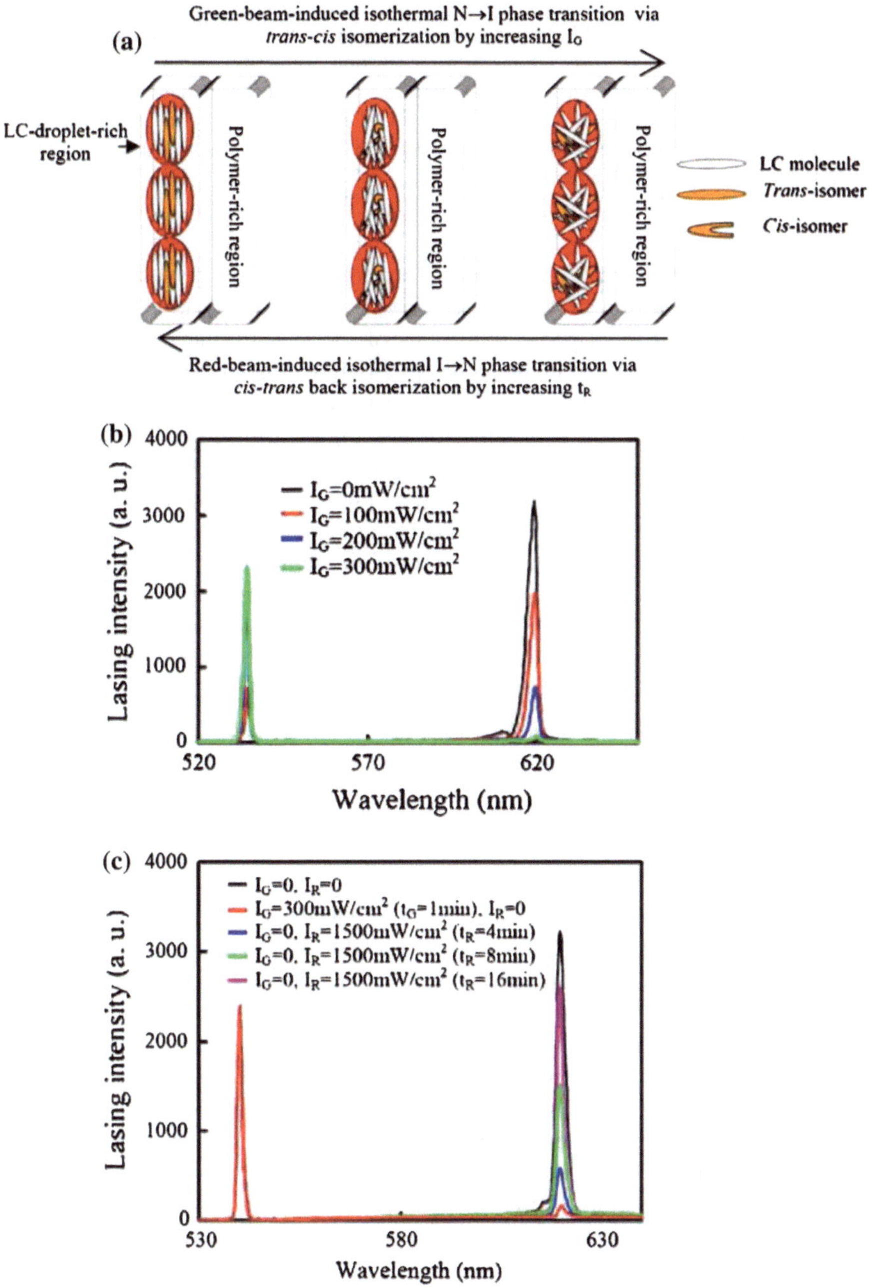

be reduced and increased by increasing the irradiation intensity of one CW circularly-polarized (CP) green beam and the irradiation time of one CW CP

◀ **Fig. 15** **a** Diagram of azo-dye doped HPDLC grating model of green-beam-induced isothermal N → I phase transition and red-beam-induced isothermal I → N phase transition, **b** variations in measured lasing intensity for $\mathbf{E}_{IP}$ with irradiation intensity of CW green beam $(I_G = 0 - 300\,mW/cm^2)$ at a fixed irradiation time $t_G = 1$ min and **c** variation of lasing intensity for $\mathbf{E}_{IP}$ with increasing time (t_R) of irradiation by CW red beam (at a fixed irradiation energy $I_R = 1500\,mW/cm^2$). Incident pumped intensity is set to 25 μJ/pulse. Reprinted with permission from [122], Copyright 2010, OSA.)

redbeam, respectively [122]. The green beam can induce nematic → isotropic (N → I) phase transition of the LCs by *trans* → *cis* isomerization while the red beam can induced I → N phase transitions of the LCs by the *cis* → *trans* back isomerization of the azo-dye in the LC-droplet-rich regions in the HPDLC, which is shown in the grating model in Fig. 15a. The wavelength of pump laser was 532 nm and output lasing peak was at around 620 nm. Figure 15b shows the variations in measured lasing intensity ($\mathbf{E}_{IP}$) with irradiation intensity of CW green beam (I_G) increasing at a fixed irradiation time $(t_G = 1\,min)$. As I_G increased from 0 to 300 mW/cm^2, the concentration of the *cis* D2-dye increased and caused isothermal N → I phase transition. Such a phase transition significantly declined the index modulation, resulting in the significant decay of the lasing emission. Figure 15c shows the intensity of the lasing output ($\mathbf{E}_{IP}$) can be increased with increasing time (t_R) of the irradiation by the CW red beam at a fixed irradiation intensity $(I_R = 1500\,mW/cm^2)$, in which the mechanism of photo control of the lasing emission is inverse process that was *cis-trans* isomerization.

5 Conclusions

Organic DFB lasers based on HPDLCs are introduced and reviewed in this chapter. The materials, fabrication setups and morphologies, which will influence the HPDLC properties and laser properties greatly, are summarized. For organic DFB laser applications, HPDLCs as laser cavities and external light feedbacks and their optical performance are discussed. In addition, as LCs are easy to response to external fields, thermo-, electro- and photo-tuning of HPDLC DFB laser behaviors are presented. From above sections, the progresses and effort of HPDLCs for organic DFB laser are stressed. Because of optical pumping to produce lasers and various methods to alter laser behaviors by opto-optical, electro-optical and thermo-optical modulations, HPDLCs will show more capabilities in multidisciplinary applications with the developments of new materials, optical designs and device configurations.

References

1. Yang, D.-K., Shin-Tson, W.: Fundamentals of Liquid Crystal Devices, 2nd edn. Wiley, London (2014)
2. de Gennes, P.G., Prost, J.: The Physics of Liquid Crystals, 2nd edn. Oxford University Press, New York (1993)
3. Yeh, P., Gu, C.: Optics of Liquid Crystal Displays. Wiley, Hoboken (2009)
4. Singh, S., Dunmur, D.A.: Liquid Crystals Fundamentals. World Scientific, Singapore (2002)
5. Yang, D.-K.: Reflective cholesteric liquid crystal displays. In: Bhowmik, A.K., Li, Z., Bos, P.J. (eds.) Mobile Displays: Technology and Applications. Wiley, London (2008)
6. Yokoyama, H.: Chapter 6. In: Collings, P.J., Patel, J.S. (eds.) Handbook of Liquid Crystal Research. Oxford University Press, New York (1997)
7. Ma, J., Xuan, L.: Towards nanoscale molecular switch-based liquid crystal displays. Displays **34**, 293–300 (2013)
8. Ma, J., Li, Y., White, T., Urbas, A., Li, Q.: Light-driven nanoscale chiral molecular switch: full range color phototuning. Chem. Commun. **46**, 3463–3465 (2010)
9. Chien, L.-C., Lu, S.-Y.: Tunable cholesteric color. In: SPIE Newsroom (2009). doi:10.1117/1112.1200901.1201482
10. Wu, S.-T., Yang, D.-K.: Reflective Liquid Crystal Displays. Wiley, New York (2001)
11. Yang, D.-K., Huang, X.-Y., Zhu, Y.-M.: Bistable cholesteric reflective displays: materials and drive schemes. Annu. Rev. Mater. Sci. **27**, 117–146 (1997)
12. Ge, Z., Wu, S.-T.: Transflective Liquid Crystal Displays. Wiley, London (2010)
13. Atkuri, H.M., Leong, E.S.P., Hwang, J., Palermo, G., Si, G., Wong, J.-M., Chien, L.-C., Ma, J., Zhou, K., Liu, Y.J., Luciano, D.S.: Developing novel liquid crystal technologies for display and photonic applications. Displays **36**, 21–29 (2015)
14. Yang, D.-K., Chien, L.-C., Doane, J.W.: Cholesteric liquid crystal/polymer dispersion for haze free light shutters. Appl. Phys. Lett. **60**, 3102–3104 (1992)
15. Ren, H., Wu, S.T.: Reflective reversed-mode polymer stabilized cholesteric texture light switches. J. Appl. Phys. **92**, 797–800 (2002)
16. Ma, J., Shi, L., Yang, D.-K.: Bistable polymer stabilized cholesteric texture light shutter. Appl. Phys. Express **3**, 021702 (2010)
17. Huang, C.-Y., Chih, Y., Ke, S.: Effect of chiral dopant and monomer concentrations on the electro-optical response of a polymer stabilized cholesteric texture cell. Appl. Phys. B **86**, 123–127 (2007)
18. Ma, J., Zheng, Z., Liu, Y., Xuan, L.: Electro-optical properties of polymer stabilized cholesteric liquid crystal film. Chin. Phys. B **20**, 024212 (2011)
19. Doane, J.W., Vaz, N.A., Wu, B.G., Zumer, S.: Field controlled light scattering from nematic microdroplets. Appl. Phys. Lett. **48**, 269–271 (1986)
20. Drzaic, P.S.: Polymer dispersed nematic liquid crystal for large area displays and light valves. J. Appl. Phys. **60**, 2142–2148 (1986)
21. Sutherland, R.L., Natarajan, L.V., Tondiglia, V.P.: Bragg gratings in an acrylate polymer consisting of periodic polymer-dispersed liquid-crystal planes. Chem. Mater. **5**, 1533–1538 (1993)
22. Sutherland, R.L., Tondiglia, V.P., Natarajan, L.V., Bunning, T.J., Adams, W.W.: Electrically switchable volume gratings in polymer-dispersed liquid crystals. Appl. Phys. Lett. **64**, 1074–1076 (1994)
23. Domash, L., Chen, Y.-M., Gozewski, C., Haugsjaa, P., Oren, M.: Electronically switchable Bragg gratings for large scale NXN fiber optic crossconnects. In: Proceedings of SPIE, vol. 3010, pp. 214–228 (1997)
24. Sutherland, R.L., Natarajan, L.V., Tondiglia, V.P., Siwecki, S.A., Chandra, S., Bunning, T. J.: Switchable holograms for displays and telecommunications. In: Proceedings of SPIE, vol. 4463, pp. 1–10 (2001)

25. Liu, Y.J., Sun, X.W., Liu, J.H., Dai, H.T., Xu, K.S.: A polarization insensitive 2X2 optical switch fabricated by liquid crystal–polymer composite. Appl. Phys. Lett. **86**, 041115 (2005)
26. Li, M.S., Wu, S.-T., Fuh, A.Y.-G.: Superprism phenomenon based on holographic polymer dispersed liquid crystal films. Appl. Phys. Lett. **88**, 091109 (2006)
27. Kakiuchida, H., Tazawa, M., Yoshimura, K., Ogiwara, A.: Optical diffractometry of anisotropic holographic structure composed of liquid crystal and polymer phases with extended Bragg modes. Thin Solid Films **571**, 431–436 (2013)
28. Yuan, H., Colegrove, J., Hu, G., Fiske, T., Lewis, A., Gunther, J., Silverstein, L., Bowley, C., Grawford, G., Chien, L., Kelly, J.: HPDLC color reflective displays. In: Proceedings of SPIE, vol. 3690, pp. 196–206 (1999)
29. Park, M.S., Kim, E.H., Kim, B.K.: Applications of holographic PDLC for full color display. J. Polym. Eng. **28**, 169–178 (2008)
30. Domash, L., Grawford, G., Ashmead, A., Smith, R., Popovich, M., Storey, J.: Holographic PDLC for photonic applications. In: Proceedings of SPIE, vol. 4107, pp. 46–58 (2000)
31. Escuti, M.J., Kossyrev, P., Crawford, G.P., Fiske, T.G., Colegrove, J., Silverstein, L.D.: Expanded viewing-angle reflection from diffuse holographic-polymer dispersed liquid crystal films. Appl. Phys. Lett. **77**, 4262–4264 (2000)
32. Date, M., Takeuchi, Y., Kato, K.: A memory-type holographic polymer dispersed liquid crystal (HPDLC) reflective display device. J. Phys. D Appl. Phys. **31**, 2225 (1998)
33. Wang, K., Zheng, J., Gao, H., Lu, F., Sun, L., Yin, S., Zhuang, S.: Tri-color composite volume H-PDLC grating and its application to 3D color autostereoscopic display. Opt. Express **23**, 31436–31445 (2015)
34. Sio, L.D., Tedesco, A., Tabirian, N., Umeton, C.: Opticall controlled holographic beam splitter. Appl. Phys. Lett. **97**, 183507 (2010)
35. Crawford, G.P.: Electrically switchable Bragg gratings. Opt. Photonics News **14**, 54–59 (2003)
36. Sio, L.D., Tabiryan, N., Caputo, R., Veltri, A., Umeton, C.: POLICRYPS structures as switchable optical phase modulators. Opt. Express **16**, 7619–7624 (2008)
37. Sio, L.D., Veltria, A., Caputo, R., Luca, A.D., Strangi, G., Bartolino, R., Umeton, C.P.: POLICRYPS composite structures: realization, characterization and exploitation for electro-optical and all-optical applications. Liq. Cryst. Rev. **1**, 2–19 (2013)
38. Bunning, T.J., Natarajan, L.V., Tondiglia, V.P., Sutherland, R.L.: Holographic polymer-dispersed liquid crystals (H-PDLCs). Annu. Rev. Mater. Sci. **30**, 83–115 (2000)
39. Ma, J., Huang, W., Xuan, L., Yokoyama, H.: Holographic polymer dispersed liquid crystals: from materials and morphologies to applications. In: Jain, V., Kokil, A. (eds.) Optical Properties of Functional Polymers and Nano Engineering Applications. CRC, Boca Raton, FL (2014)
40. Zheng, Z., Song, J., Liu, Y., Guo, F., Ma, J., Xuan, L.: Single-step exposure for two-dimensional electrically-tuneable diffraction grating based on polymer dispersed liquid crystal. Liq. Cryst. **35**, 489–499 (2008)
41. Zheng, Z., Ma, J., Li, W., Song, J., Liu, Y., Xuan, L.: Improvements in morphological and electro-optical properties of polymer-dispersed liquid crystal grating using a highly fluorine-substituted acrylate monomer. Liq. Cryst. **35**, 885–893 (2008)
42. Wenger, B., Tetreault, N., Welland, M.E., Friend, R.H.: Mechanically tunable conjugated polymer distributed feedback lasers. Appl. Phys. Lett. **97**, 193303 (2010)
43. Vieu, C., Carcenac, F., Pepin, A., Chen, Y., Mejias, M., Lebib, A., Manin-Ferlazzo, L., Couraud, L., Launois, H.: Electron beam lithography: resolution limits and applications. Appl. Surf. Sci. **164**, 111–117 (2000)
44. Chou, S.Y., Krauss, P.R., Renstrom, P.J.: Imprint lithography with 25-nanometer resolution. Science **272**, 85–87 (1996)
45. Vita, F., Lucchetta, D.E., Castagna, R., Criante, L., Simoni, F.: Large-area photonic structures in freestanding films. Appl. Phys. Lett. **91**, 103114 (2007)

46. Kim, E., Woo, J., Kim, B.: Diffraction grating in a holographic polymer dispersed liquid crystal based on polyurethane acrylate. Liq. Cryst. **34**, 79–85 (2007)
47. Zheng, Z., Yao, L., Zhang, R., Zou, Z., Liu, Y., Liu, Y., Xuan, L.: Thermo-stability of acrylate based holographic polymer dispersed liquid crystal gratings. J. Phys. D Appl. Phys. **42**, 115504 (2009)
48. Huang, W., Deng, S., Li, W., Peng, Z., Liu, Y., Hu, L., Xuan, L.: A polarization-independent and low scattering transmission grating for a distributed feedback cavity based on holographic polymer dispersed liquid crystal. J. Opt. **13**, 085501 (2011)
49. Vardanyan, K.K., Qi, J., Eakin, J.N., Sarkar, M.D., Crawford, G.P.: Polymer scaffolding model for holographic polymer-dispersed liquid crystals. Appl. Phys. Lett. **81**, 4736–4738 (2002)
50. Sarkar, M.D., Qi, J., Crawford, G.P.: Influence of partial matrix fluorination on morphology and performance of HPDLC transmission gratings. Polymer **43**, 7335–7344 (2002)
51. Sarkar, M.D., Gill, N.L., Whitehead, J.B., Crawford, G.P.: Effect of monomer functionality on the morphology and performance of the holographic transmission gratings recorded on polymer dispersed liquid crystals. Macromolecules **36**, 630–638 (2003)
52. Kim, B.K., Jang, M.W., Park, H.C., Jeong, H.M., Kim, E.Y.: Effect of graphene doping of holographic polymer-dispersed liquid crystals. J. Polym. Sci. Part A Polym. Chem. **50**, 1418–1423 (2012)
53. Liu, Y., Dai, H., Leong, E., Teng, J., Sun, X.: Electrically switchable two-dimensional photonic crystals made of polymer-dispersed liquid crystals based on the Talbot self-imaging effect. Appl. Phys. B **104**, 659–663 (2011)
54. Park, M.S., Cho, Y.H., Kim, B.K., Jang, J.S.: Fabrication of reflective holographic gratings with polyurethane acrylate (PUA). Curr. Appl. Phys. **2**, 249–252 (2002)
55. Park, M.S., Kim, B.K., Kim, J.C.: Reflective mode of HPDLC with various structures of polyurethane acrylates. Polymer **44**, 1595–1602 (2003)
56. Park, M.S., Kim, B.K.: Transmission holographic gratings produced using networked polyurethane acrylates with various functionalities. Nanotechnology **17**, 2012–2017 (2006)
57. Sutherland, R.L.: Polarization and switching properties of holographic polymer-dispersed liquid-crystal gratings. I. Theoretical model. J. Opt. Soc. Am. B Opt. Phys. **19**, 2995–3003 (2002)
58. Sutherland, R.L., Natarajan, L.V., Tondiglia, V.P., Chandra, S., Shepherd, C.K., Brandelik, D.M., Siwecki, S.A., Bunning, T.J.: Polarization and switching properties of holographic polymer-dispersed liquid-crystal gratings. II. Experimental investigations. J. Opt. Soc. Am. B Opt. Phys. **19**, 3004–3012 (2002)
59. Bowley, C.C., Crawford, G.P., Yuan, H.: Reflection from dual-domains in a holographically-formed polymer-dispersed liquid crystal material. Appl. Phys. Lett. **74**, 3096–3098 (1999)
60. Birnkrant, M.J.: Combining holographic patterning and block copolymer self-assembly to fabricate hierarchical volume gratings. In: Material Science and Engineering, Thesis of Drexel University, Philadelphia (2009)
61. Natarajan, L.V., Shepherd, C.K., Brandelik, D.M., Sutherland, R.L., Chandra, S., Tondiglia, V.P., Timlin, D., Bunning, T.J.: Switchable holographic polymer-dispersed liquid crystal reflection gratings based on thiol-ene photopolymerization. Chem. Mater. **15**, 2477–2484 (2003)
62. Caputo, R., De Sio, L., Veltri, A., Umeton, C., Sukhov, A.V.: Development of a new kind of switchable holographic grating made of liquid-crystal films separated by slices of polymeric material. Opt. Lett. **29**, 1261–1263 (2004)
63. Drevensek-Olenik, I., Jazbinsek, M., Sousa, M.E., Fontecchio, A.K., Crawford, G.P., Eopie, M.: Structural transitions in holographic polymer-dispersed liquid crystals. Phys. Rev. E **69**, 051703 (2004)
64. Natarajan, L.V., Brown, D.P., Wofford, J.M., Tondiglia, V.P., Sutherland, R.L., Lloyd, P., Jakubiak, R., Vaia, R., Bunning, T.J.: Visible light initiated thiol-ene based reflection H-PDLCs. In: Proceedings of SPIE, p. 59360F (2005)

65. Wofford, J.M., Natarajan, L.V., Tondiglia, V.P., Sutherland, R.L., Lloyd, P.F., Siwecki, S. A., Bunning, T.J.: Holographic polymer dispersed liquid crystal (HPDLC) transmission gratings formed by visible light initiated thiol-ene photopolymerization. In: Proceedings of SPIE, p. 63320Q (2006)
66. White, T.J., Liechty, W.B., Natarajan, L.V., Tondiglia, V.P., Bunning, T.J., Guymon, C.A.: The influence of N-vinyl-2-pyrrolidinone in polymerization of holographic polymer dispersed liquid crystals (HPDLCs). Polymer **47**, 2289–2298 (2006)
67. Jakubiak, R., Brown, D.P., Natarajan, L.V., Tondiglia, V., Lloyd, P., Sutherland, R.L., Bunning, T.J., Vaia, R.A.: Influence of morphology on the lasing behavior of pyrromethene 597 in a holographic polymer dispersed liquid crystal reflection grating. In: Proceedings of SPIE, p. 63220A (2006)
68. Jakubiak, R., Bunning, T.J., Vaia, R.A., Natarajan, L.V., Tondiglia, V.P.: Electrically switchable, one-dimensional polymeric resonators from holographic photopolymerization: a new approch for active photonic bandgap materials. Adv. Mater. **15**, 241–244 (2003)
69. Natarajan, L.V., Brown, D.P., Wofford, J.M., Tondiglia, V.P., Sutherland, R.L., Loyd, P.F., Bunning, T.J.: Holographic polymer dispersed liquid crystal reflection gratings formed by visible ligh initiated thiol-ene polymerization. Polymer **47**, 4411–4420 (2006)
70. Pu, H., Yin, D., Gao, B., Gao, H., Dai, H., Liu, J.: Dynamic characterizations of high diffraction efficiency in volume Bragg grating formed by holographic photopolymerization. J. Appl. Phys. **106**, 083111 (2009)
71. Huang, W., Liu, Y., Diao, Z., Yang, C., Yao, L., Ma, J., Xuan, L.: Theory and characteristics of holographic polymer dispersed liquid crystal transmission grating with scaffolding morphology. Appl. Opt. **51**, 4013–4020 (2012)
72. Peng, H., Chen, G., Ni, M., Yan, Y., Zhuang, J., Roy, V., Li, R.K., Xie, X.: Classical photopolymerization kinetics, exceptional gelation, and improved diffraction efficiency and driving voltage in scaffolding morphological H-PDLCs afforded using a photoinitibitor. Polym. Chem. **6**, 8259–8269 (2015)
73. Caputo, R., Luca, A.D., Sio, L.D., Pezzi, L., Strangi, G., Umeton, C., Veltri, A., Asquini, R., d'Alessandro, A., Donisi, D., Beccherelli, R., Sukhov, A.V., Tabiryan, N.V.: POLICRYPS: a liquid crystal composed nano/microstructure with a wide range of optical and electro-optical applications. J. Opt. **11**, 024017 (2009)
74. Caputo, R., Veltri, A., Umeton, C., Sukhov, A.V.: Kogelnik-like model for the diffraction efficiency of POLICRYPS gratings. J. Opt. Soc. Am. B Opt. Phys. **22**, 735–742 (2005)
75. Caputo, R., Sio, L.D., Veltri, A., Umeton, C., Sukhov, A.V.: Realization of POLICRYPS gratings: optical and electro-optical properties. Mol. Cryst. Liq. Cryst. **441**, 111–129 (2005)
76. Caputo, R., Trebisacce, I., Sio, L.D., Umeton, C.P.: Phase Modulator behavior of a wedge-shaped POLICRYPS diffraction grating. Mol. Cryst. and Liq. Cryst. **549**, 29–36 (2011)
77. Strangi, G., Barna, V., Caputo, R., De Luca, A., Versace, C., Scaramuzza, N., Umeton, C., Bartolino, R., Price, G.N.: Color-tunable organic microcavity laser array using distributed feedback. Phys. Rev. Lett. **94**, 063903 (2005)
78. Kogelnik, H., Shank, C.V.: Stimulated emission in a periodic structure. Appl. Phys. Lett. **18**, 152–154 (1971)
79. Siegman, A.E.: Lasers. University Science Press, Mill Valley (1986)
80. Kogelnik, H., Shank, C.V.: Coupled-wave theory of distributed feedback lasers. J. Appl. Phys. **43**, 2327–2335 (1972)
81. Sakhno, O., Stumpe, J., Smirnova, T.: Distributed feedback dye laser holographically induced in improved organic–inorganic photocurable nanocomposites. Appl. Phys. B Lasers Opt. **103**, 907–916 (2011)
82. Tsutsumi, N., Nishida, H.: Tunable distributed feedback lasing with low threshold and high slope efficiency from electroluminescent conjugated polymer waveguide. Opt. Commun. **284**, 3365–3368 (2011)

83. Karnutsch, C., Gyrtner, C., Haug, V., Lemmer, U., Farrell, T., Nehls, B., Scherf, U., Wang, J., Weimann, T., Heliotis, G.: Low threshold blue conjugated polymer lasers with first-and second-order distributed feedback. Appl. Phys. Lett. **89**, 201108 (2006)
84. Karnutsch, C., Pflumm, C., Heliotis, G., Bradley, D., Wang, J., Weimann, T., Haug, V., Gärtner, C., Lemmer, U.: Improved organic semiconductor lasers based on a mixed-order distributed feedback resonator design. Appl. Phys. Lett. **90**, 131104 (2007)
85. Jakubiak, R., Natarajan, L.V., Tondiglia, V., He, G.S., Prasad, P.N., Bunning, T.J., Vaia, R. A.: Electrically switchable lasing from pyrromethene 597 embedded holographic-polymer dispersed liquid crystals. Appl. Phys. Lett. **85**, 6095–6097 (2004)
86. Liu, Y.J., Sun, X.W., Elim, H.I., Ji, W.: Effect of liquid crystal concentration on the lasing properties of dye-doped holographic polymer-dispersed liquid crystal transmission gratings. Appl. Phys. Lett. **90**, 011109 (2007)
87. Lucchetta, D.E., Criante, L., Francescangeli, O., Simoni, F.: Light amplification by dye-doped holographic polymer dispersed liquid crystals. Appl. Phys. Lett. **84**, 4893–4895 (2004)
88. Criante, L., Lucchetta, D.E., Vita, F., Castagna, R., Simoni, F.: Distributed feedback all-organic microlaser based on holographic polymer dispersed liquid crystals. Appl. Phys. Lett. **94**, 111114 (2009)
89. Okamoto, K.: Fundamentals of Optical Waveguides, 2nd edn. Academic Press, Burlington (2010)
90. Hisao, V.K.S., Lu, C., He, G.S., Pan, M., Cartwright, A.N., Prasad, P.N., Jakubiak, R., Vaia, R.A., Bunning, T.J.: High contrast switching of distributed-feedback lasing in dye-doped H-PDLC transmission grating structures. Opt. Express **13**, 3787–3794 (2005)
91. Jakubiak, R., Tondiglia, V.P., Natarajan, L.V., Sutherland, R.L., Lloyd, P., Bunning, T.J., Vaia, R.A.: Dynamic lasing from all-organic two-dimensional photonic crystals. Adv. Mater. **17**, 2807–2811 (2005)
92. Jakubiak, R., Tondiglia, V.P., Natarajan, L.V., Sutherland, R.L., Lloyd, P., Bunning, T.J.: Stimulated emission from pyrromethene 597 in holographic polymer dispersed liquid crystal structures. In: Proceedings of SPIE, pp. 202–207 (2005)
93. Jakubiak, R., Tondiglia, V.P., Natarajan, L.V., Lloyd, P.F., Sutherland, R.L., Vaia, R.A., Bunning, T.J.: Lasing of pyrromethene 597 in 2-D holographic polymer dispersed liquid crystals: influence of columnar conformation. In: Proceedings of SPIE, p. 72320K (2009)
94. Luo, D., Sun, X.W., Dai, H.T., Liu, Y.J., Yang, H.Z., Ji, W.: Two-directional lasing form a dye-doped two-dimensional hexagonal photonic crystal made of holographic polymer-dispersed liquid crystals. Appl. Phys. Lett. **95**, 151115 (2009)
95. Coles, H., Morris, S.: Liquid-crystal lasers. Nat. Photonics **4**, 676–685 (2010)
96. Sakoda, K.: Enhanced light amplification due to group-velocity anomaly peculiar to two-and three-dimensional photonic crystals. Opt. Express **4**, 167–176 (1999)
97. Susa, N.: Threshold gain and gain-enhancement due to distributed-feedback in two-dimensional photonic-crystal lasers. J. Appl. Phys. **89**, 815–823 (2001)
98. Turnbull, G.A., Andrew, P., Barnes, W.L., Samuel, I.D.W.: Operating characteristics of a semiconducting polymer laser pumped by a microchip laser. Appl. Phys. Lett. **82**, 313–315 (2003)
99. Diao, Z., Deng, S., Huang, W., Xuan, L., Hu, L., Liu, Y., Ma, J.: Organic dual-wavelength distributed feedback laser empowered by dye-doped holography. J. Mater. Chem. **22**, 23331–23334 (2012)
100. Diao, Z., Huang, W., Peng, Z., Mu, Q., Liu, Y., Ma, J., Xuan, L.: Anisotropic waveguide theory for electrically tunable distributed feedback laser from dye-doped holographic polymer dispersed liquid crystal. Liq. Cryst. **41**, 239–246 (2014)
101. Huang, W., Diao, Z., Liu, Y., Peng, Z., Yang, C., Ma, J., Xuan, L.: Distributed feedback polymer laser with an external feedback structure fabricated by holographic polymerization technique. Org. Electron. **13**, 2307–2311 (2012)

102. Diao, Z., Xuan, L., Liu, L., Xia, M., Hu, L., Liu, Y., Ma, J.: A dual-wavelength surface-emitting distributed feedback laser from a holographic grating with an organic semiconducting gain and a doped dye. J. Mater. Chem. C **2**, 6177–6182 (2014)
103. Liu, L., Xuan, L., Zhang, G., Liu, M., Hu, L., Liu, Y., Ma, J.: Enhancement of pump efficiency for an organic distributed feedback laser based on a holographic polymer dispersed liquid crystal as an external light feedback layer. J. Mater. Chem. C **3**, 5566–5572 (2015)
104. Huang, W., Diao, Z., Yao, L., Cao, Z., Liu, Y., Ma, J., Xuan, L.: Electrically tunable distributed feedback laser emission from scaffolding morphologic holographic polymer dispersed liquid crystal grating. Appl. Phys. Express **6**, 022702 (2013)
105. Diao, Z., Kong, L., Xuan, L., Ma, J.: Electrical control of the distributed feedback organic semiconductor laser based on holographic polymer dispersed liquid crystal grating. Org. Electron. **27**, 101–106 (2015)
106. Liu, L., Huang, W., Diao, Z., Peng, Z., Mu, Q.Q., Liu, Y., Yang, C., Hu, L., Xuan, L.: Low threshold of distributed feedback lasers based on scaffolding morphologic holographic polymer dispersed liquid crystal gratings: reduced losses through Forster transfer. Liq. Cryst. **41**, 145–152 (2014)
107. Takahashi, H., Naito, H.: Amplified spontaneous emission from fluorene-based copolymer wave guides. Thin Solid Films **477**, 53–56 (2005)
108. Kretsch, K.P., Belton, C., Lipson, S., Blau, W.J., Henari, F.Z., Rost, H., Pfeiffer, S., Teuschel, A., Tillmann, H., Horhold, H.-H.: Amplified spontaneous emission and optical gain spectra from stilbenoid and phenylene vinylene derivative model compounds. J. Appl. Phys. **86**, 6155–6159 (1999)
109. Tsutsumi, N., Kawahira, T., Sakai, W.: Amplified spontaneous emission and distributed feedback lasing from a conjugated compound in various polymer matrices. Appl. Phys. Lett. **83**, 2533–2535 (2003)
110. Forster, T.: In: 10th Spiers Memorial Lecture: Transfer Mechanisms of Electronic Excitation, Discussions of the Faraday Society, vol. 27, pp. 7–17 (1959)
111. May, B., Poteau, X., Yuan, D., Brown, R.G.: A study of a highly efficient resonance energy transfer between 7-N, N-diethylamino-4-methylcoumarin and 9-butyl-4-butylamino-1, 8-naphthalimide. Dyes Pigm. **42**, 79–84 (1999)
112. Liu, Y.J., Sun, X.W., Shum, P., Li, H.P., Mi, J., Ji, W.: Low-threshold and narrow-linewidth lasing from dye-doped holographic polymer-dispersed liquid crystal transmission gratings. Appl. Phys. Lett. **88**, 061107 (2006)
113. Huang, W., Liu, Y., Hu, L., Mu, Q., Peng, Z., Yang, C., Xuan, L.: Second-order distributed feedback polymer laser based on holographic polymer dispersed liquid crystal grating. Org. Electron. **14**, 2299–2305 (2013)
114. Liu, M., Liu, Y., Zhang, G., Liu, L., Diao, Z., Yang, C., Peng, Z., Yao, L., Ma, J., Xuan, L.: Improving conversion efficiency of organic distributed feedback laser by varying solvents of laser gain layer. Liq. Cryst. (2015). doi:10.1080/02678292.02672015.01116628
115. Telle, H.H., Urena, A.G., Donovan, R.J.: Laser Chemistry: Spectroscopy, Dynamics and Applications. Wiley, London (2007)
116. Duarte, F.J.: Tunable Laser Applications, 2nd edn. CRC Press, New York (2009)
117. Bruneau, D., Cazeneuve, H., Loth, C., Pelon, J.: Double-pulse dual-wavelength alexandrite laser for atmospheric water vapor measurement. Appl. Opt. **30**, 3930–3937 (1991)
118. He, X., Fang, X., Liao, C., Wang, D., Sun, J.: A tunable and switchable single-longitudinal-mode dual-wavelength fiber laser with a simple linear cavity. Opt. Express **17**, 21773–21781 (2009)
119. Luo, D., Sun, X.W., Dai, H.T., Demir, H.V., Yang, H.Z., Ji, W.: Temperature effect on the lasing from a dye-doped two-dimensional hexagonal photonic crystal made of holographic polymer-dispersed liquid crystals. J. Appl. Phys. **108**, 013106 (2010)

120. Urbas, A., Tondiglia, V., Natarajan, L., Sutherland, R., Yu, H., Li, J.-H., Bunning, T.: Optically switchable liquid crystal photonic structures. J. Am. Chem. Soc. **126**, 13580–13581 (2004)
121. Liu, Y.J., Zheng, Y.B., Shi, J., Huang, H., Walker, T.R., Huang, T.J.: Optically switchable gratings based on azo-dye-doped, polymer-dispersed liquid crystals. Opt. Lett. **34**, 2351–2353 (2009)
122. Tong, H.-P., Li, Y.-R., Lee, C.-R.: All-optically controllable distributed feedback laser in a dye-doped holographic polymerdispersed liquid crystal grating with a photoisomerizable dye. Opt. Express **18**, 2613–2620 (2010)

MIX
Papier aus verantwortungsvollen Quellen
Paper from responsible sources
FSC® C105338

If you have any concerns about our products,
you can contact us on
ProductSafety@springernature.com

In case Publisher is established outside the EU,
the EU authorized representative is:
Springer Nature Customer Service Center GmbH
Europaplatz 3, 69115 Heidelberg, Germany

Printed by Libri Plureos GmbH
in Hamburg, Germany